System Building with APL + Win

RSP SERIES IN INDUSTRIAL CONTROL, COMPUTERS AND COMMUNICATIONS

Series Editor: **Professor Derek R. Wilson**

Concurrent Engineering – The Agenda for Success
Edited by **Sa'ad Medhat**

Analysis and Testing of Distributed Software Applications
Henryk Krawczyk and **Bogdan Wiszniewski**

Interface Technology: The Leading Edge
Edited by **Janet M. Noyes** and **Malcolm Cook**

CANopen Implementation: Applications to Industrial Networks
Mohammad Farsi and **Manuel Bernado Martins Barbosa**

J: The Natural Language for Analytic Computing
Norman Thomson

Digital Signal Processing: A MATLAB-Based Tutorial Approach
John Leis

Mathematical Computing in J: Introduction, Volume 1
Howard A. Peelle

System Building with APL + Win
Ajay Askoolum

Multimedia Engineering: A Practical Guide for Internet Implementation
A.C.M. Fong and S.C. Hui, with contributions from G. Hong and B. Fong

System Building with APL + Win

Ajay Askoolum
Claybrook Computing Limited, UK

John Wiley & Sons, Ltd

Research Studies Press Limited

Other Wiley Editorial Offices

John Wiley & Sons Inc., 111 River Street, Hoboken, NJ 07030, USA

Jossey-Bass, 989 Market Street, San Francisco, CA 94103-1741, USA

Wiley-VCH Verlag GmbH, Boschstr. 12, D-69469 Weinheim, Germany

John Wiley & Sons Australia Ltd, 42 McDougall Street, Milton, Queensland 4064, Australia

John Wiley & Sons (Asia) Pte Ltd, 2 Clementi Loop #02-01, Jin Xing Distripark, Singapore 129809

John Wiley & Sons Canada Ltd, 22 Worcester Road, Etobicoke, Ontario, Canada M9W 1L1

Wiley also publishes its books in a variety of electronic formats. Some content that appears in print may not be available in electronic books.

British Library Cataloguing in Publication Data

A catalogue record for this book is available from the British Library

ISBN-13   978-0-470-03020-2 (PB)
ISBN-10   0-470-03020-8 (PB)

Typeset in 10/12pt Times New Roman by Laserwords Private Limited, Chennai, India
Printed and bound in Great Britain by Antony Rowe Ltd, Chippenham, Wiltshire
This book is printed on acid-free paper responsibly manufactured from sustainable forestry in which at least two trees are planted for each one used for paper production.

TABLE OF CONTENTS

Editorial Foreword.. xvii
Preface.. xviii
Acknowledgements... xx
Chapter 1 - System Building Overview 1
1.1 Why APL? 1
    1.1.1 APL platforms 2
    1.1.2 APL isolation 3
    1.1.3 APL generations 3
1.2 Which APL? 4
    1.2.1 APL: a renewed promise... 5
    1.2.2 The nature of APL GUI 5
    1.2.3 The user interface... 6
    1.2.4 Design architecture 7
    1.2.5 Component-based software.. 7
    1.2.6 Multi-Language programming 8
1.3 The n-tier model... 8
    1.3.1 Presentation services tier 8
    1.3.2 Business services tier 9
    1.3.3 Data services tier.. 9
    1.3.4 Tier demarcation... 9
1.4 Prevailing design architecture 10
    1.4.1 Client/Server computing 10
    1.4.2 The multi-user client/server 10
    1.4.3 Object Oriented Programming(OOP) 11
    1.4.4 Collaborative processing... 13
1.5 APL interface to components.. 14
    1.5.1 Static (or early) binding 14
    1.5.2 Dynamic (or late) binding... 14
    1.5.3 Object attributes 14
    1.5.4 Object prefixes.. 14
1.6 Structured Query Language (SQL) 16
    1.6.1 Relational Database Management System.. 16
    1.6.2 SQL origin 17
    1.6.3 SQL language 17
1.7 The Windows Registry... 18
1.8 Regional settings.. 19
    1.8.1 The GetLocaleInfo API call.. 20
    1.8.2 API usage... 20
1.9 Software development... 22
    1.9.1 Absence of design certification... 22
    1.9.2 Quality: the crucial measure 22
    1.9.3 Version control 22
    1.9.4 Customisation 23
1.10 APL and Windows API.. 23
    1.10.1 Verifying API definitions 24

1.10.2 Processing API memory pointers...24
1.10.3 Processing API call back 24
1.10.4 Processing API errors 25
1.11 The future challenge...25
Chapter 2 -   Advanced APL Techniques...27
2.1 Removing legacy code clutter...28
　　2.1.1 Boolean scenario (BS) table..28
　　2.1.2 A simple task: copy a file 28
　　2.1.3 An API solution 29
　　2.1.4 An APL solution 29
　　2.1.5 The general BS case..31
　　2.1.6 Auto generating the BS table 32
　　2.1.7 Where to use BS tables? 32
2.2 Bit-wise Boolean techniques...34
　　2.2.1 BitEQV 34
　　2.2.2 BitIMP 35
　　2.2.3 BitAND..35
　　2.2.4 BitXOR 37
　　2.2.5 BitOR...37
　　2.2.6 BitNAND..37
　　2.2.7 BitNOR...37
2.3 Managing workspace variables..37
　　2.3.1 APL is not a relational data source 38
　　2.3.2 Fabricating a record set object 39
2.4 Generating test data...47
　　2.4.1 Dates 47
　　2.4.2 Integers 48
　　2.4.3 Floating point...48
　　2.4.4 Codes 49
　　2.4.5 String 49
　　2.4.6 Binary string 49
　　2.4.7 Numeric vector 50
　　2.4.8 Computations on generated data...50
2.5 APL+Win as an ActiveX Server..52
　　2.5.1 Dynamic properties and methods...52
　　2.5.2 Which class of APL+Win? 53
2.6 Debugging applications..55
　　2.6.1 The CSBlocks function 57
2.7 Functions with methods 60
Chapter 3 -   Application Interface..61
3.1 Managing the hidden interface...61
　　3.1.1 Forcing a session to terminate..62
3.2 The user interface..63
　　3.2.1 Purpose of the user interface..64
　　3.2.2 Hierarchical and sequential..64
　　3.2.3 Invasive interaction..65
3.3 The user interface is the application..65

3.3.1 Where is the interface? 66
3.3.2 Manage user expectations 66
3.3.3 The user interface as a tier 66
3.4 APL+Win design safeguards 67
3.5 Context sensitive help 67
3.5.1 Enabling a help facility in APL+Win 68
3.5.2 Tooltips and prompts 68
3.5.3 What's this help? 69
3.5.4 Help: WINHELP 70
3.5.5 Help: HTML 71
3.6 Help format as a user option 71
3.7 Application messages 72
3.7.1 The language and location of messages 72
3.7.2 Communicating runtime messages 73
3.8 User-defined properties of the system object 78
3.9 The scope of user documentation 78
3.9.1 Types of documentation 79
3.10 Designing menus 80
3.10.1 Auto-generation of the user interface 81
3.10.2 Validating an interface tree 82
3.10.3 Creating menus 83
3.11 Designing forms 86
3.11.1 Enumerating an existing interface tree 86
3.11.2 Resizing forms 92
3.12 Access control 93
3.12.1 File-based applications 94
3.12.2 Database applications 96
3.13 Empower the user 96
3.13.1 Prevent rather than trap errors 96
3.13.2 Validate user entries 96
3.13.3 Intrusive application messages 96
3.13.4 Work with platform features 96
3.13.5 Tools | Options 96
3.13.6 Navigation 96
3.13.7 System legacy 97
3.13.8 Functionality alone is not enough 97
3.14 Sales considerations 97
3.15 Application exit 98
Chapter 4 -   Working with Windows **99**
4.1 The APL legacy 99
4.1.1 Reinventing the wheel 100
4.2 Windows resources 101
4.3 API calls 101
4.3.1 Replacing APL code by API calls 102
4.3.2 API calls to simplify code 102
4.3.3 Formatting date and time 102
4.3.4 Using GetDateFormat and GetTimeFormat APIs 107

4.3.5 APL+Win GetDateFormat 108
4.3.6 Fail safe date format translation? 110
4.4 The Windows Script Host (WSH) 111
4.4.1 Managing the Windows Registry 112
4.4.2 Writing using WScript Shell 112
4.4.3 Reading using WScript Shell 113
4.4.4 Deleting using WScript Shell 113
4.4.5 Writing using Registry API 114
4.4.6 Special folders 116
4.4.7 Environment variables 117
4.4.8 Setting/Reading environment variables 118
4.4.9 Deleting an environment variable 118
4.5 Creating a shortcut 119
4.5.1 Starting another application 120
4.6 Intelligent file operations with API calls 120
4.6.1 DeleteFile API 120
4.6.2 PathFileExists API 121
4.6.3 _lOpen API 121
4.6.4 _lClose API 121
4.6.5 Fail safe deletion status 122
4.7 Universal Naming Convention (UNC) 124
4.7.1 APL and UNC names 124
4.7.2 API calls for handling UNC 124
4.7.3 UNC dynamic mapping 124
4.7.4 Library/UNC mapping 126
4.8 Application configuration 127
4.9 Using INI files with APL 128
4.9.1 Location of the INI file 129
4.9.2 Name of section 129
4.9.3 Name of key 129
4.9.4 Limitations of INI files 129
4.9.5 Implementing the control mechanism 129
4.9.6 INI file: writing a key 130
4.9.7 INI file: reading a key 130
4.9.8 INI file: writing a section 131
4.9.9 INI file : emumerating the names of all sections 131
4.9.10 INI file: reading a section 131
4.9.11 INI file: enumerate names of all keys in a section 132
4.9.12 INI file: enumerate values of all keys in a section 132
4.9.13 Two more APIs 132
4.10 XML files for application configuration 132
4.10.1 The XMLINI function 133
4.10.2 Loading/Creating an XML file 133
4.10.3 XML file: writing a key 135
4.10.4 XML file: reading a key 137
4.10.5 XML file: deleting a key 137
4.10.6 XML file: writing a section 137
4.10.7 XML file: reading a section 138

4.10.8 XML file: commenting a section 138
4.10.9 XML file: deleting a section 139
4.10.10 XML file: enumerate names of all sections 139
4.10.11 XML file: enumerate names of all keys in a section 140
4.10.12 XML file: enumerate values of all keys in a section 140
4.10.13 Saving an XML file 141
4.11 INI/XML comparative advantage 141
4.11.1 Converting INI to XML 141
4.12 The filing system 142
4.12.1 Identifying the filing system 142
4.12.2 The File System Object 143
4.13 Platform enhancements 144
Chapter 5 - The Component Object Model 145
5.1 Objects are global 145
5.2 APL+Win COM event handling 145
5.2.1 COM event arguments 147
5.2.2 Is RPC Server available? 150
5.3 The promise of COM development 151
5.4 Types of COM components 152
5.4.1 Application components 152
5.4.2 GUI components 153
5.4.3 'Silent' or 'slave' components 153
5.4.4 Custom component 153
5.4.5 Component visibility 153
5.5 Maintaining objects 154
5.6 APL+Win and ActiveX components 154
5.6.1 Platform components 156
5.6.2 Opaque APL syntax 157
5.6.3 Anatomy of the APL+Win syntax 157
5.6.4 Hierarchy of objects 157
5.6.5 Data incompatibilities 158
5.6.6 APL index origin 158
5.7 APL+Win post version 4.0 ActiveX syntax 159
5.7.1 Objects hierarchy using redirection 159
5.7.2 Redirection clutter 159
5.7.3 Dynamic syntax 160
5.7.4 Hierarchical syntax 161
5.7.5 Redirection or enumeration? 164
5.8 ActiveX typed parameters 169
5.8.1 Boolean parameters 169
5.8.2 Positional parameters 169
5.8.3 Passing object pointers 170
5.8.4 Typed data parameters 170
5.8.5 Passing selective named parameters 171
5.8.6 Rogue objects 172
5.9 Development environment features 173
5.9.1 User-defined properties 173

5.10 Using ActiveX asynchronously... 174
    5.10.1 Custom properties 175
5.11 Debugging components.. 176
    5.11.1 VB for Application code 176
    5.11.2 VB code 176

Chapter 6 -   Mixed Language Programming.. 179
6.1 Application extension trade-offs 179
6.2 VB ActiveX DLLs 181
6.3 A sample ActiveX DLL project 181
    6.3.1 The DLLFunctionsModule module 181
    6.3.2 The DLLFunctions class.. 182
6.4 Using VBDLLINAPL.DLL 182
    6.4.1 Syntax issues.. 183
    6.4.2 Querying the events in VBDLLINAPL 183
    6.4.3 Querying the properties in VBDLLINAPL...................................... 184
    6.4.4 Querying the methods in VBDLLINAPL 185
    6.4.5 XCurrencySymbol 186
    6.4.6 XDateCompare 186
    6.4.7 XDateScalar 187
    6.4.8 XDateValid 188
    6.4.9 XDaysOfWeek.. 189
    6.4.10 XFindReplace 190
    6.4.11 XGetAgeAttQ 190
    6.4.12 XGetDatePart... 191
    6.4.13 XGetDateTime... 191
    6.4.14 XGetLocal.. 192
    6.4.15 XGetTimeStamp 193
    6.4.16 XMonthsOfYear 194
    6.4.17 XNumberValid 195
    6.4.18 XSpellDate... 195
    6.4.19 XStringCase 196
    6.4.20 Getting help on syntax 197
6.5 Processing APL+Win arrays.. 197
6.6 Deploying ActiveX DLLs... 198
    6.6.1 Name and location 199
    6.6.2 General availability... 199
    6.6.3 Updrading/Replacing ActiveX DLLs 200
    6.6.4 ActiveX DLL coding for APL 201
6.7 Building a DLL for APL using C# Express 2005 202
    6.7.1 Using the C# DLL.. 204
    6.7.2 Deploying C# DLLs... 205

Chapter 7 -   Application Extension using Scripting.................................... 207
7.1 The APL/VBScript affinity 207
    7.1.1 The VBScript built-in functions.. 208
    7.1.2 Adding the Script Control.. 208
    7.1.3 An algebraic expression evaluator 209
7.2 Error trapping... 211

7.3 Exploring the Script Control 211
    7.3.1 The Eval method 213
7.4 Extending the Script Control 215
    7.4.1 What is in SampleCode? 217
7.5 Multi-language programming 222
    7.5.1 Sharing an APL+Win object 222
    7.5.2 Creating own instance of APL+Win object 223
    7.5.3 Processing simple numeric arrays 224
    7.5.4 Passing arguments to built-in functions 228
7.6 Sharing with the APL Grid object 232
7.7 Concurrent sharing with the Script Control 233
7.8 APL+Win and HTML 234
    7.8.1 Creating/Displaying HTML file from APL+Win 236
    7.8.2 Taking control of HTML content 238
    7.8.3 APL+Win and XML 239

Chapter 8 -   Windows Script Components **241**
8.1 Building a Script Component using JavaScript 242
    8.1.1 Coding the methods 246
8.2 Building a Script Component using VBScript 249
8.3 About the VBS file 252
8.4 Runtime errors in script components 253
8.5 Which Scripting language? 253
    8.5.1 A wise choice? 254
8.6 Multi-language Script component 254
8.7 What is in MULTILANGUAGE.WSC? 256
    8.7.1 Is a property read only or read/write? 256
    8.7.2 Firing an event associated with a property 256
    8.7.3 Firing an event associated with a method 257
8.8 Finally, just because it is possible… 258
    8.8.1 Testing with APL+Win as the server application 258
    8.8.2 Testing with VBScript as the client application 259
    8.8.3 JavaScript as the client application 260
    8.8.4 Comparing VBScript and JavaScript 260
8.9 The way forward with script components 261

Chapter 9 -   Working with Excel **263**
9.1 Application or automation server 263
    9.1.1 Using the automation flag 265
9.2 The basic structure of Excel 267
    9.2.1 Orphan Excel sessions 267
    9.2.2 Excel Data representation 268
9.3 APL arrays and Excel ranges 269
    9.3.1 Writing to multiple sheets 270
    9.3.2 Excel Worksheet functions 273
    9.3.3 Excel user-defined functions 274
    9.3.4 Excel dialogues 276
    9.3.5 Excel charts 276
    9.3.6 Excel ad hoc 277

9.4 Object syntax 281
    9.4.1 The FindWindow API... 281
    9.4.2 The GetWindowText API call 283
    9.4.3 Does the ActiveX server still exist? 283
9.5 Excel using APL+Win to retrieve APL data... 283
    9.5.1 Usability models 283
9.6 The Excel Add-In.. 286
    9.6.1 Add-In visiblility 287
    9.6.2 APL server workbook code 288
    9.6.3 APL server module code... 289
    9.6.4 CreateVariable 293
    9.6.5 APLServer class code 293
    9.6.6 The APL server toolbar.. 297
    9.6.7 The initial ActiveX server workspace... 297
9.7 The EWA model in action 298
9.8 Transferring APL+Win data to Excel 302
9.9 Automation issues.. 304
    9.9.1 APL+Win issues 304
9.10 Why use Excel with APL? 304
Chapter 10 - Working with Word 307
10.1 The Word difference 307
10.2 Word templates 308
    10.2.1 Global, user, and workgroup templates 310
10.3 Starting Word.. 311
10.4 Word as a report generation component 312
    10.4.1 Tables.. 313
    10.4.2 Building an APL+Win array.. 313
    10.4.3 APL+Win array to Word table... 317
    10.4.4 Active document random access... 322
    10.4.5 Formula... 325
    10.4.6 DDE automation 325
    10.4.7 INCLUDEPICTURE 327
    10.4.8 INCLUDETEXT... 329
    10.4.9 Form fields.. 331
10.5 Populating form fields... 336
    10.5.1 Error trapping.. 338
10.6 Word vs. Excel for APL+Win automation... 339
10.7 Automation 339
    10.7.1 APL+Win automation issues 339
Chapter 11 - Working with Access 341
11.1 The Access pathways.. 341
    11.1.1 Access smoke and mirrors 342
11.2 The Access object 345
    11.2.1 Dynamic query definition 345
    11.2.2 Queries based on user-defined functions 346
    11.2.3 Deleting Access objects 349
11.3 JET Engine types 350

11.4 Access—below the surface 350
    11.4.1 MDB files 351
    11.4.2 ADP files 352
    11.4.3 The MDB/ADP file menu 353
    11.4.4 Linking tables 354
11.5 Working with many data sources 355
    11.5.1 Troubleshooting databases with linked tables....................... 355
11.6 Troubleshooting data projects 357
    11.6.1 Using an existing MDF file... 358
    11.6.2 Using a new MDF file 359
    11.6.3 Data project: using the ODBC driver................................. 359
    11.6.4 Data project: using the JET provider 359
11.7 The Jet compromise 360
11.8 Unified approach with ΛDO and SQL... 360
    11.8.1 ADOX:ADO Extension for data definition and security 361
11.9 Access SQL... 361
    11.9.1 Access SQL with user-defined function 362
11.10 Database filing 363
    11.10.1 Using an Access database 363
    11.10.2 Storing variables and functions....................................... 364
    11.10.3 Storing Files.. 366
    11.10.4 Deploying database filing 368
11.11 Automation issues.. 368
Chapter 12 - Working with ActiveX Data Object (ADO)........................... 369
12.1 Translating code examples into APL+Win 369
    12.1.1 Simulating On Error Resume Next 369
    12.1.2 Simulating On Error Goto... 370
12.2 The connection object... 371
    12.2.1 Creating an active connection object 371
    12.2.2 Database connection using a connection string—syntax I............ 371
    12.2.3 Database connection using properties—syntax II.................... 373
12.3 The record object 375
    12.3.1 Record object using redirection 375
    12.3.2 Record object without connection object............................. 376
    12.3.3 Creating a record object.. 376
    12.3.4 Cloning a record object... 379
    12.3.5 Tables in a data source 381
    12.3.6 Working with record objects... 383
    12.3.7 Navigating record set objects... 392
    12.3.8 Working with complete record objects 394
12.4 The data source catalogue 404
12.5 Learning ADO 405
Chapter 13 - Data Source Connection Strategies 407
13.1 The application handle 409
13.2 The DSN overhead... 409
    13.2.1 Acquiring a default DSN 410
    13.2.2 Creating a data source.. 410

13.3 Automating user/system DSN creation 411
    13.3.1 With an API call 411
    13.3.2 With the 'Prompt' property of a connection object......................... 414
13.4 The ODBC Data Source Administrator 421
    13.4.1 Enumerating installed drivers 422
13.5 System DSN connection 423
    13.5.1 Creating a SQL Server system DSN 423
    13.5.2 Creating an Oracle system DSN 425
    13.5.3 Configuring a system DSN 425
    13.5.4 Removing a system DSN 426
13.6 User DSN Connection... 426
13.7 DSNManager syntax summary 427
13.8 File DSN Connection 428
    13.8.1 Using a file DSN... 429
    13.8.2 File DSN portability.. 430
13.9 UDL connection... 430
13.10 DSN-less connection.. 434
13.11 Server data sources 434
13.12 Access data sources.. 434
13.13 Excel data sources.. 435
13.14 Text data sources.. 435
    13.14.1 The SCHEMA.INI file... 436
    13.14.2 Creating SCHEMA.INI automatically 437
    13.14.3 Refining content of SCHEMA.INI file 441
13.15 Data source issues 442
13.16 Inward APL+Win issues 442
    13.16.1 Data types 442
    13.16.2 The atomic vector 443
13.17 Outward APL issues.. 444
    13.17.1 CSV files issue.. 444
13.18 The way forward with the data tier 444
Chapter 14 - Structured Query Language... **447**
14.1 SQL statements 447
14.2 SQL prime culprits... 448
    14.2.1 Handling NULL values.. 448
    14.2.2 SQL convention for column names 449
    14.2.3 SQL comments 449
14.3 APL and SQL... 450
    14.3.1 Coping with SQL variations 451
    14.3.2 Using DMBS properties.. 451
    14.3.3 ANSI SQL 454
    14.3.4 Date/Time handling in data sources... 455
14.4 Learning SQL.. 456
    14.4.1 The SQL Data Query Language 457
    14.4.2 The SQL Data Manipulation Language 473
    14.4.3 The SQL Data Definition Language 475
    14.4.4 The SQL Data Control Language 476

14.4.5 The way forward with SQL 476
14.4.6 Debugging SQL 477
14.4.7 Optimising SQL 477
14.4.8 SQL dialect specialisation.. 477

Chapter 15 - Application Evolution.. 479
15.1 Application deployment... 480
15.2 The next release 480
    15.2.1 The schedule of work... 480
    15.2.2 Fault management... 480
    15.2.3 Wish management... 481
    15.2.4 If it works, improve it! 481
    15.2.5 Efficiency management 481
    15.2.6 Small vs large scale improvements................................... 481
15.3 Application workspace... 482
15.4 APL libraries vs UNC names.. 482
15.5 Readability 483
    15.5.1 Style.. 483
15.6 Global variables 484
    15.6.1 Initial values... 484
    15.6.2 Constants.. 485
15.7 Using API calls 485
    15.7.1 Adding missing API calls 485
15.8 Version control... 485
15.9 Change management... 486
15.10 Legacy management 486
    15.10.1 Workspace organisation.. 486
    15.10.2 Modernisation 487
15.11 Indentation 488
    15.11.1 Limitations of indentation... 488
15.12 Documentation... 488
    15.12.1 Context sensitive help and user manuals 489
    15.12.2 Auditing changes 489
15.13 Testing 490
    15.13.1 Functionality changes 491
    15.13.2 Automatic migration 491
15.14 Release 491
    15.14.1 What's new? 491
    15.14.2 Incremental upgrade 491
    15.14.3 Replacement upgrade... 492
15.15 Application listings 492
    15.15.1 Producing the listing 492
15.16 Epilogue 495
Bibliography 497
Index.. 499

EDITORIAL FOREWORD

The challenge facing an Author in today's rapidly changing technical environment is to satisfy the dual needs of readers by writing a book that is both intellectually rigorous and yet application orientated, such that the reader is able to develop their implementation skills, conjointly with their conceptual knowledge.

Ajay Askoolum has satisfied both objectives in this book, *System Building with APL + Win*. The conceptual development of the subject is well-founded in the fact that APL was conceived as a mathematical notation, not as a programming language. In that context APL is a "thinking tool", which enables the system developer to concentrate on the problem to be solved, rather than the machine execution details.

Ajay has demonstrated throughout the book that APL provides a robust development environment, which enables the reader to "LEARN BY DOING". Each step in the process of system design is carefully explained, and demonstrated so that the reader can acquire solution skills.

There is a well-known phrase in systems development that, "one person's system is another's component". APL provides a practical environment that maximizes the opportunities for imaginative and flexible solutions. The facility to assemble a set of current components to deliver a system solution is a key feature of *System Building with APL + Win*.

Ajay notes, quite early in the text that,

"… users rarely know what they want until they know what they have".

The implication of this phrase is that the initial task facing the system developer is often indeterminate and therefore the system builder has to conceive a solution that is organic in nature. It follows therefore that the 'flexibility' of APL+Win is a key feature for professionals seeking to adopt a modern system development environment. The inherent interactive structure of APL maximizes the opportunity to incorporate and apply modern developments, such as multi language programming, shared data sources etc. through the world's most ubiquitous WINDOWS environment.

Enjoy the book; it has been written from the standpoint of a professional system builder with the needs of today's self-learners in mind and its study will be rewarding. You will acquire modern "JOB SKILLS" as well as developing your intellectual insights through *System Building with APL + Win*.

It is a pleasure to thank Ajay, on your behalf, for his comprehensive presentation of the techniques of system building and for the care he has shown in developing practical demonstrations of the techniques throughout the book.

Derek R. Wilson

PREFACE

"In this, the twenty-first century, APL is a valid proposition for a software development tool, not a plea". Discuss!

If the APL community debated this topic, it would certainly be making the case for APL as a development tool. The decisive question is: Use APL exclusively or in conjunction with other tools? Whatever the merits, this would be a misplaced debate: the objective is to vindicate APL as a contemporary development tool, with or without other tools. The graphical user interface and a grid object are not native APL features and yet are used routinely with APL. This yields two substantive dividends: it reduces the span of the development cycle and presents a familiar and acceptable interface to users. The integration of other resources with APL solutions, such as automation servers and databases, increases these benefits dramatically. I am convinced that APL has a natural role in contemporary software development as the agent that binds available solutions and trends.

APL vendors have strived to make APL a contemporary team player: this has been possible because K. E. Iverson's APL is a robust tool of thought, as a concept that transcends computer architecture rather than a computer language. The future depends on the momentum of APL deployment in system building in collaboration with other tools and not as an all-purpose purist solution looking for problems to solve.

Fortunately, I do not have to answer the question 'Why another book on APL?' for there has been a distinct lack of printed material on APL. Like any books that are a collation of knowledge, inevitably with inherent limitations, for the benefit of contemporaries and especially welcome for the next generation this book is my attempt to fulfil this objective with an APL perspective.

For me, writing this book has been a journey of self-discovery. Among the APL communities, there is a tendency to look at software development with an unshakable conviction that APL is the only tool of choice. Today, I think this is tunnel vision, albeit APL was without any doubt the tool of choice at the beginning of its history; APL is just one tool among a rich set of tools.

Paradoxically, APL is without doubt a more difficult language than most of its competitors for programmers with experience of other languages. Yet, based on APL experience, it is possible to deploy other languages more efficiently. With control structures, COM, .NET compliance, WEB services etc APL has made valiant strides forward in survival stakes. Contemporary APL is unique, more so than APL was at the beginning of its history. Its primitive symbols are communication language independent, it's handling of arrays and nested arrays is still unique, it is still unambiguous, and above all, it can now talk to other resources on its platform. Remarkably, core APL is consistent across versions from all APL vendors, in spite of atomic vector differences: unfortunately, extended APL and the bridge to the graphical user interface are not.

With APL, there is no debate about whether the expression *Sex = 'M'* is an assignment or a comparison. With Visual Basic, say, *'M' = Sex* is unequivocally a comparison but *Sex = 'M'* may be either a comparison or assignment. It depends! Similarly, with an expression such as *Price = TotalCost/Quantity*, albeit (correctly) written `Price←TotalCost÷Quantity` in APL, it matters little whether a scalar item is being calculated or an array of items.

There is no doubt that APL vendors have contrived to fit into today's mainstream computing platform seamlessly and successfully. The risk is that the (veteran) community of APL developers do not seize the promised opportunities. Raw application development is without doubt easier with the availability of Windows resources in the workspace of a platform compliant APL.

Should APL developers adopt the credentials of the Windows platform—become compliant Windows citizens—and deliver the promise of rapid application development, APL delivers a renewed promise: it is the tool of choice for contemporary rapid software development, especially so APL + Win with its ability to harness the resources of the Windows platform.

Without any apology, this book is about APL on the Windows platform and how it can be a collaborative tool. My hope is that it somewhat flattens the APL learning curve and that it creates future opportunities for improvements.

Ajay Askoolum

Surrey, England

ACKNOWLEDGEMENTS

I am grateful to Research Studies Press (RSP) for accepting this APL project and especially to Giorgio Martinelli of RSP for his guidance and patience. I also acknowledge the valuable assistance of Derek R. Wilson, an RSP series editor, for sharing his insight during the process of writing this book and indeed for suggesting the title of this book.

I acknowledge the timely assistance of the team at John Wiley in ensuring the fruition of the project: Peter Mitchell, Wendy Hunter, and Debbie Cox.

I am also indebted to my children Meera V, Karuna P, Varuna U, and Akash B for their sustained encouragement in writing this book, especially to Karuna for reading the manuscript and Akash for the initial cover design. Foremost, I dedicate this book to Sobha, my wife.

Chapter 1

System Building Overview

The Microsoft Windows™ platform and other Microsoft tools, such as ActiveX Data Object (**ADO**) for database connectivity, represent the singular de facto contemporary standard for personal and collaborative computing using stand-alone and networked personal computers. The ubiquitous Microsoft Office suite, comprising of **Excel**, **Word**, **Access**, and **Outlook,** reinforces the critical mass of the platform, creating both a model for Windows applications and an influential base of users. The community of Windows developers benefit from the published material, support sites on the Internet, and training resources for tools available on this platform. Vendor-sponsored and independent newsgroups on the Internet that provide free programmer-to-programmer support on techniques and specialist topics. Unfortunately, APL does not have comparable dedicated resources on the Internet.

In order to exploit the Windows environment, the APL developer needs to subscribe to the prevailing (and emerging) standards albeit APL is older than Windows itself. K E Iverson conceived **APL** in his book *A Programming Language*, published in 1962. APL is an original language, conceived as a mathematical notation and without reference to machine or processor architecture and their implied limitations. IBM produced the first implementation of the language at the end of 1966. During the intervening years, APL has been the subject of two mythical and opposing theories, namely,

• It is only a matter of time before APL finds wider recognition of its potential in the realm of software development.

• It is only a matter of time before APL disappears from the realm of software development.

The credentials of the language are such that it eludes both prophecies, with great dexterity. It continues to evolve, on all current computing platforms, in spite of a persistent lack of recognition in the industry, testified by the migration of APL applications to other languages. Ironically, perhaps, the APL mindset is at the heart of this dichotomy. The issue is not whether APL is suitable for system building but the ease with which developers can disregard contemporary industry standards completely. The key to future survival is compliance; users must demonstrate APL as the compliant industry standard development tool that it is. The key to this accomplishment is collaboration and the ability to integrate Windows resources into APL development; the barrier to this goal is the almost complete lack of worked APL examples of techniques. Inevitably, the barrier to understanding available documentation is the ability to understand the jargon of other languages.

1.1 Why APL?

In the real world, user requirements continue to evolve during or after development; they are never frozen. APL is an interactive language suited to the delivery of dynamic requirements and it *is easy* to learn; the barrier to learning is the nature of the language's specialist keyboard. The ability to incorporate dynamic changes ensures that the software is still required by the time it is finished.

The domain of software development is such that what is required is rarely clear at the outset. It is not sufficient to code according to a written and agreed paper specification when the real specification itself changes in the light of evolving requirements. Users rarely know what they want until they know what they have. Even the most intuitive, efficient, and innovative, even beautiful, design is devoid of any intrinsic value unless it satisfies users' needs and conforms to their expectations. APL allows the rapid production and modification of prototypes, almost interactively, and as a result, the cost of APL development tends to be lower. In other words, APL allows the exploration of user requirements with real code and without commitment to a rigid specification.

Yes, APL is easy to learn. The perception that APL is difficult emanates from traditional data processing departments whereas users in business departments, without exposure to formal design standards, are quite adept in deploying the language for ad hoc applications. APL is different for traditional developers because everything about APL is different—keyboard, component files, workspace, arrays, etc. This is precisely why APL just exists; it has neither failed nor succeeded in finding an outright niche in the domain of software development.

APL developers need to learn to comply with contemporary standards; that is, embrace the mainstream approach to software development. Mainstream developers *will* perceive APL differently when it subscribes to industry standard ideologies and delivers solutions involving industry standard technologies. Vendors must address the needs of developer who are completely new to the language. A 'New to APL', or similar, volume can put new developers at ease. Such documentation can do the following:

• Demonstrates the traditional 'Hello World' functionality, that is, compares the APL way with those of other popular tools such as that of **Visual Basic** (**VB**). A cutting-edge innovation would be the provision of a fully-fledged application that demonstrates all the capabilities of APL.

• Provides an APL language dictionary that explains its daily vocabulary—rank, shape, dyadic, monadic, niladic, stack, index origin, namespace, etc.

• Explains the APL keyboard and its layout using interactive keyboard drills in immediate execution mode.

New developers, especially ones with some experience of programming, will have advanced and immediate needs—how to implement a Function and Subroutine in APL, data types, scope of variables, etc.—and will find it easier to grasp the keyboard if they continue to make progress following a painless start. A dedicated APL keyboard is a valid option for building confidence with the APL keyboard but it can be intrusive in the long term.

1.1.1 APL platforms

The period that has seen the rapid emergence of powerful technologies on the Windows platform has also witnessed a gradual decline of APL, leaving the die-hard proponents of the language to keep it alive. The trend may have reversed. Having reinvented itself, APL is a serious contender for application development on the Windows platform. Currently, there are at least six competing commercial APL vendors with offerings for Windows, MAC OS, LINUX, UNIX, and mainframes:

• IBM developed APL2 (www.ibm.com/software/awdtools/apl) for the mainframe and ported it to run virtually unchanged on a Windows PC. This interpreter is also available for AIX and Sun Solaris.

• APLX (www.microapl.co.uk/apl) offer a cross platform APL interpreter with native support for SQL and charts: in theory, its applications transfer seamlessly across Windows, MAC OS, LINUX, and AIX. APLX64 is a soon to be available 64 bit interpreter.

• Dyalog APL (www.dyalog.com) offers an interpreter for UNIX, Windows, and the .NET platform.

• Soliton Associates (www.soliton.com) offer SHARP/APL for the mainframe, LINUX, and UNIX platforms.

• APL2000 (www.apl2000.com) acquired APL*PLUS III from Manugistics and renamed it APL+Win. They continue to develop it for the Windows and other platforms.

IBM has placed TryAPL2 in the public domain: TryAPL2 is a cut-down second-generation APL. Its IDIOMS workspace contains powerful second-generation APL idioms. The original APL*PLUS/PC is also in the public domain as APL/SE. Several international APL conferences take place annually, keeping APL in the frame. There is now a renewed opportunity for APL system development to fall into line with industry standards simply because APL itself is now in line with industry standards.

1.1.2 APL isolation

APL+Win has its roots in APL*PLUS/PC which was first available in 1982 for personal computers—this product was ported from a mainframe version developed during the 1970's. APL development has evolved a tradition for self-containment: at the start, there was neither any application to collaborate with nor any system wide resources such as Windows Application Programming Interfaces (**API**); Windows APIs are a set of pre-programmed functions available on the Windows platform. Contemporary APL is fully empowered to exploit platform resources, right out of the box.

The first generation APL developer had a pathway to any Disk Operating System (**DOS**) function either via dedicated APL system functions or via the command processor gateway (⎕cmd). This enabled control of aspects of hardware such as the screen, keyboard, printer, and file input output. APL component files provided the facility for random access and user privileges. The one thing that APL shared with co-existing applications and the host operating system was text files.

APL still supports some of the legacy features that allow application development in complete isolation. Typically, APL uses text files and component files for data components: applications tend to be self-contained in workspaces. As well as having a strange vocabulary reliant on symbols, the language has rich pathways—several ways of solving any given problem—which has contributed to its reputation for being difficult.

1.1.3 APL generations

Figure 1-1 APL evolution illustrates the composite development of APL. The first generation APL was character based; the second generation provided extensions to the language for dealing with nested arrays, akin to collections in **VB** parlance, as well as limited Graphical User Interface (**GUI**) capability. The third and current generation provides control structures, full support for Windows GUI, component interfacing capability, and very close compatibility with IBM's APL2. Core, or first generation, APL is consistent across APL interpreters from different vendors. Vendors have implemented extensions to the language in their own unique way, including the way the standard Windows GUI is available.

The integration of COM support within the APL interpreter has established APL as a team player on the Windows platform: with this advancement, APL can use the same resources as mainstream tools such as **VB**; APL can act as the glue that holds together an application build by the assembly of custom or off-the-shelf components.

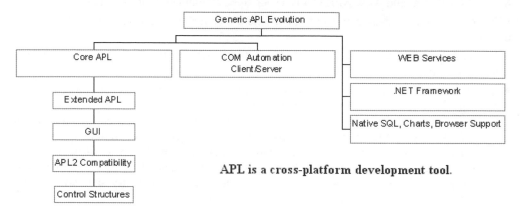

Figure 1-1 APL evolution

1.2 Which APL?

All the interpreters collaborative on their platform and are in conformity with contemporary industry standards. However, the continued enhancement of commercial APL interpreters is on a definite divergent path. Each interpreter has particular strengths. APL2 is the definitive APL language standard and has the advantage of running both on personal computers and mainframes. APLX is a cross platform interpreter that offers native support for SQL and charts; also, there is the prospect of migration to a 64 bit interpreter. APL/W offers multi-threading and integrated .NET Framework support. APL+Win is not only a COM client but also a COM server; its close integration with the Windows operating system resources makes it a very strong player on that platform.

APL+Win pioneered control structures—in the APL language, a feature that competing interpreters adopted very quickly. Control structures have not only made software development more intuitive and transparent but also made code more readable. An example, shown in **Table 1-1 Control structures,** illustrates the difference is coding style using APL with and without control structures.

```
     ▽ With                          ▽ Without option
{1[1]    :select case option      [1]    → (option=1 2 3)/L1,L2,L3
★ [2]         :case 1             [2]    → 0
  [3]              ⍝ Option 1     [3]    L1:
★ [4]         :case 2             [4]         ⍝ Option 1
  [5]              ⍝ Option 2     [5]         → 0
★ [6]         :case 3             [6]    L2:
  [7]              ⍝ Option 3     [7]         ⍝ Option 2
+ [8]         :else               [8]         → 0
  [9]              ⍝ default      [9]    L3:
1}[10]   :endselect              [10]         ⍝ Option 3
     ▽                                 ▽
```

Table 1-1 Control structures

Arguably, the 'With' version is cleaner, easier to debug, and maintain than the 'Without' version—legacy code will tend to be similar to the latter version. The 'Without' version creates three symbols, one for each of its labels, and requires code to end each sub process.

1.2.1 APL: a renewed promise

In stark contrast with other languages, APL is a mathematical notation. The fact that this notation transpired to be computer executable as well is proof of its robustness. The distinguishing feature of APL is that its design concentrates not on machine requirements but on problem solving. APL's efficiency in solving problems unwittingly coerced its deployment in the niche of personal computing—applications developed for own use.

As a Windows development tool, APL is capable of changing and revitalising the APL culture inertia from a personal computing tool to an application development tool. It has the inherent power of a robust notation, deploys a standard Windows graphical interface, and has the ability to reuse industry standard components written in other languages. The interfacing capability of APL enables access to industry-wide technologies like databases, the .NET framework, etc.

APL does have a legacy but not one that ought to handicap its deployment. The natural development of the language in terms of the provision of nested arrays and control structures and the adoption of the Windows GUI has rendered legacy code obsolete; their replacement is straightforward and necessary:

• A single API call can replace several lines of APL code and represent a more efficient implementation of a given functionality.

• APL is COM server capable; that is, it can talk to itself as well as other COM compliant applications and, more to the point, other applications can talk to it.

• With the ability to access industry standard databases, APL applications are no longer workspace bound and the size of applications is limited not by the size of the workspace but by available system resources.

• The 'look and feel' of an APL application is similar, if not identical, to other applications and elicits comparable levels of user acceptance.

• APL is like any other development tool. The implementation of the unified APL keyboard and Windows true type fonts ensure that APL no longer has peculiar set-up requirements.

The APL symbols present a problem for newcomers to the language, especially on a keyboard that does not show them. Over time, this becomes irrelevant for most APL programmers who learn to locate the symbols intuitively. During the process of learning a printed copy of the APL keyboard in conjunction with drill exercises, aimed at using the APL symbols, represents a fail safe way of learning the location of all the symbols. Ideally, vendors should supply a keyboard with engraved APL characters with the interpreter. The APL keyboard with caps lock set is no longer viable. A generic unified APL keyboard from an independent supplier is not viable as each APL has its own keyboard layout.

1.2.2 The nature of APL GUI

APL deploys the standard Windows GUI and takes advantage of native nested and simple arrays, and vector facilities to specify the properties of controls. APL+Win and APLX share a common syntax and make all standard controls available via one system call (⎕wi) providing access to all properties, methods, and events. The five lines of code in the function executed—sequentially—in immediate mode produce the form shown in **Figure 1-2 APL+Win Hello World**.

```
      ∇ HelloWorld;⎕wself
[1]     ⍝ System Building With APL+Win
[2]     ⎕wself←'HelloWorld' ⎕wi 'Create' 'Form'
[3]     ⎕wi 'Set' ('size' 5 20) ('border' 16) ('caption' 'APL+Win')
[4]     0 0ρ'.lbl' ⎕wi 'New' 'Label' ('caption' 'Hello World!') ('style' 1)
      ∇
```

Figure 1-2 APL+Win Hello World

Unlike VB, APL stores the definition of an instance of control objects, forms, menus, etc., not in text files but as executable code that can create the form and its children dynamically:

APL/W uses a different pathway for GUI deployment based on □□□ and it implements the VB-like hierarchical syntax.

System building is not only steeped in jargon but has doggedly laboured under the misapprehension that it is an end in itself; its proponents have promoted it as a mystical art and thrived in large dedicated departments with huge budgets. Users have resolved, slowly but surely, to question the ethics of system building—they have come to demand solutions on a timely and cost effective basis. The software industry has reacted by proposing newer standards for software development that aim to provide solutions that deliver cost effective and timely competitive edge.

In practice, this means that system building is not about computerising a discrete business process—APL has thrived on providing stand-alone mosaic solutions—but about deriving intelligence from business data in whatever forms they exists. This implies that applications need to be able to talk to one another, share resources, and not reinvent solutions to do so. The contemporary approach to system building is generic programming, promoting maximum code reuse. Code reuse has created a new breed of software design: the assembly of pre-written off the shelf software known as components with a standard industry wide user interface.

Historically, APL is at a considerable disadvantage. In the past, APL solutions have tended to be specific or bespoke; that is, APL has tended to deliver precise requirements rather than generic solutions. Moreover, APL solutions have tended to be isolated and encapsulated in the surreal reality of its character set, its workspace, component files, and a user interface that is home grown. The evolution of APL itself has made a lot of older legacy code unsustainable.

APL is fully capable of exploiting the new economics and ethics of software development. This includes producing applications with the industry standard 'look and feel' and subscribing to the contemporary standards for design and code reuse. Modern system development is not about 'growing' code; it is rather like cookery—a process of assembling the correct ingredients (components) and in the correct proportions to deliver a menu of dishes (solutions) that are both complementary and complimentary. These solutions, like the dishes, need to fulfil user-expectations; quite often, these expectations, like 'taste', are volatile and immeasurable.

In this, the age of information, systems provide information, even intelligence, and not simply data. For example, a call centre user who works to a service level agreement that dictates that a query is resolved within three days expects the system to alert him/her of the age of queries rather than simply their log date. Modern solutions need to be agile and adaptable to future expectations.

1.2.3 The user interface

The most visible characteristic of a Windows application is the user interface. The standardisation of the user interface yields two tangible benefits; users are familiar with the interface and providers of development tools, including new programming languages, have clear guidelines to the interface standards. In practice, providers incorporate sophistications that distinguish their product and the trend is to emulate the features of competitors.

A standard GUI and calls to Windows resources via Windows API calls ensure a common 'look and feel' across applications. This ensures that users are able to use systems with a clear expectation of how things should work.

The GUI also includes interaction with hardware components especially the hard disk. The standard API calls manage such things as long names for files and folders and mapped network drives. A new trend is to be able to use Universal Naming Convention (**UNC**) for referring to network drives rather than mapped letters. APL built-in file handling facilities work equally well with mapped drives or UNC names.

1.2.4 Design architecture

Software development according to the prevailing design architecture adopted by the predominant development tools on the Windows platform simplifies the process of developing an application. This includes project control over the software life cycle, system design and testing, version control and application partitioning.

Irrespective of the scale of a project or the language that is used, software development has three drivers: cost, quality, and timeliness. Software is a major capital resource for its sponsor; its cost relates directly to its extendibility, reusability, and adaptability. In other words, designing software for change—driven by technological change and evolving and sometimes unpredictable business needs—reduces cost over the life cycle of the software. The quality of software is measurable by the level of user acceptance. In practice, this means how closely the software functions in line with user expectations as well as how accurately it functions. In other words, software should converge, in terms of 'look and feel' and behaviour, to a predominant platform standard.

APL has thrived as the tool of choice for personal computing tool: users have programmed or automated their own processes for their own specialist needs. Such developments take place outside of computing departments and are oblivious of design principles. This tradition has created a design culture that is resistant to change; this is primarily responsible for the replacement of APL systems. Inability to adapt to change, arising from technological advances or evolving business requirements, reduces the life span of software because the embedded value of software that fails to adapt devalues rapidly. This is especially true of traditional APL systems, which have tended to be one-tier designs. The separation of application logic and data ensures that the application logic is more readily adaptable to changing needs and that the data, in a database, is adaptable to the users business needs in the long term.

Quite simply, if APL can use an independent data source, its relevance as the tool of choice for a software project increases dramatically. The most powerful impediment to APL's credentials as a rapid application development tool is the lack of independence from data sources. APL must lose its 'special' needs: 'special' needs invariably equates to increased costs.

1.2.5 Component-based software

Component-based software uses generic and reusable components which are capable of integration in multiple contexts and which have a common interface for use by multiple systems. Components, servers, expose their functionality, in a standard and consistent way all across clients, to the system that invokes or loads them. The use of components, which have been debugged elsewhere, not only promotes rapid development but also minimizes the cost of development.

The widespread adoption of COM technology has given rise to a proliferation of robust off-the-shelf components and cross component development is now commonplace.

1.2.6 Multi-Language programming

Multi-Language programming is an emerging standard that is gathering momentum. The concept of multi-language programming is simply being able to support several programming languages within any given application development.

A restricted interpretation of multi-language programming is the ability of one language to call code written in another language—this requires a two-way transfer of data; invariably the data types embedded in different languages impede this process. APL can call ActiveX Dynamic Link Libraries (**DLL**) compiled with **VB**, scripting languages such as VBScript and JavaScript, .NET languages and others. Components offer very tangible benefits besides being readily available. For instance, APL can transfer the handling of dates to the scripting languages, which already have the functionality that might be required. However, none of these languages can handle arrays as well as APL.

Most new programming languages are the perfect language at conception—the one language that would meet all programming needs. The fact that new languages continue to appear proves that there is no *perfect* language. In reality, some languages are better—faster or easier to code and maintain—at some tasks than others. No language is universally superior. In practice, it is the proficiency of the developer and not the design of a language or any prescription regarding its usage that produces the critical difference.

Invariably newer languages struggle to establish recognition because existing languages have a valuable legacy of experience and applications. Until multi-language programming becomes truly possible, the emphasis is on being able to use the facilities of other languages, such as existing libraries and scripting languages, within a given environment. Notably, most programming languages already use Structure Query Language (**SQL**) for managing databases, albeit SQL is not a programming language.

1.3 The n-tier model

Figure 1-3 Three-tier design shows a visual representation of the tiers.

The business tier acts as a bridge between the presentation and data tiers.

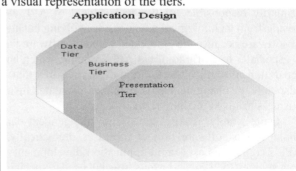

Figure 1-3 Three-tier design

The separation of the data from the application intelligence formalises the concept of tiered systems. The *n* of *n*-tier represents any number higher than two. Typically, systems are three tiered, consisting of a presentation tier, a business tier, and a data tier.

1.3.1 Presentation services tier

This is the user interface, visible part of an application. Users perceive this as the application itself. All data entry or presentation forms, menus, and messages, etc., are contained in this tier. This module will also include simple data validation—ensuring the completion of mandatory fields, the validity of dates, etc.

1.3.2 Business services tier

This tier contains the logic or code for business tasks and rules. It arbitrates the presentation of data from the data tier within the user interface and the updating of data arising in the presentation tier back into the data tier. A business task is any discrete activity such as calculating value added tax, adding a new customer, booking a seat, reporting overdue accounts, etc. Each discrete task is accessible from a menu option in the presentation tier.

This tier also includes business rules. A rule defines the logical sequence for carrying out discrete tasks. For instance, an order for a customer relies on that customer being present in the order dispatch system. Discrete business processes often have a critical path that specifies a predefined order for the completion of tasks; this is the principle that underlies the concept of business workflows.

Business rules may also relate entirely to data. It may be necessary to implement hierarchical privileges for viewing or amending data—for instance, it is legitimate for some personnel to access salary data whereas others, possibly in the same department, may only view such data and others may amend them. The selection of the appropriate data may be an SQL statement or a system of password protection. Another example of a rule relating to data may be the cross validation of data to establish whether individual items of data taken collectively make sense. For example, individual items of data such as the date an individual joins a pension scheme and that individual's date of birth may be correct in that both items are valid dates. The business rule may be that the date at which the individual joins the pension scheme must be at least eighteen years after his/her date of birth.

1.3.3 Data services tier

The data tier manages the persistent data underlying the application and enforces its integrity. A modular design requires well-defined interfaces among the tiers. The advantages are that modules are independent—can be maintained, upgraded, or replaced independently –and each module can work with multiple complementary modules. For instance, the business services layer might work with alternative data services modules such as Oracle, or **Access**, or SQL Server, or other interface compliant database management systems. Likewise, any given data services module is usable by alternative business services layers; a data tier designed for the needs of administration is equally appropriate for marketing.

1.3.4 Tier demarcation

In practice, tier demarcation is blurred and in use, a tiered application is largely indistinguishable from one that is monolithic. However, this is not a valid reason for resisting a tiered design or redesign; this denies clients/users any flexibility deploying the application. A guiding principle for application partitioning into tiers is that each should be independent; that is, capable of being replaced with little or no impact on the others.

The tiered application design is a logical grouping of the individual components of that application—whilst the tier to which a particular function should belong may be obvious, there are no rigid rules for this grouping, and the grouping may be constrained by other factors. The tier to which a particular function should belong may not be clear and therefore its classification may be at the discretion of the designer. For example, controlling access to an application may be in any of the three tiers. Likewise, a complex data query could be in the business tier or in the data tier—arguably, it may be more efficient to keep the query in the data tier as a stored procedure if the data source supported this facility.

A key advantage of partitioning an application into tiers is that each tier can be developed, enhanced, and deployed independently.

1.4 Prevailing design architecture

Advances in technology as well as cumulative experience drive the industry's changing endorsement of the preferred blue print for system building or application partitioning. The compelling lessons learnt are that a modular architecture and code reuse makes for rapid and robust application development enabling the industry to meet rapidly evolving business critical needs.

1.4.1 Client/Server computing

Before the advent of personal computers and local area networks, terminal based applications with their entire data and application intelligence based on a host system; the mainframe was the norm.

Client/Server computing is an arrangement whereby servers host the data and clients, personal computers, host the application intelligence supplanted this norm. The advent of powerful tools such as Open Data Base Connectivity (**ODBC**) and Structured Query Language (**SQL**), that are integral to popular application building tools such as VB, precipitated this transition.

APLX has integral ODBC or SQL capability. However, APL is ODBC and SQL enabled by its ability to use ADO, the current Windows standard for data access. ADO is a component that exposes the data access interface incorporated in relational data sources. Most databases incorporate the ODBC interface. The newer interface is Object Linking and Embedding for databases (**OLE DB**). ADO connects to the data sources either thorough ODBC drivers or OLE DB providers. During the transition period, proprietary database standards such as Microsoft SQL Server and Oracle support both ODBC and OLE DB.

1.4.2 The multi-user client/server

The usual concept is that of a client, i.e. a personal computer that runs an application. Resources existing outside of the client, centralized on the network, serve the application. Therefore, applications use distributed intelligence. The server provides multi-user functionality. It is worth noting a subtlety. The running application, the client, may be using components existing on the personal computer itself. Such components are local and dedicated servers.

1.4.2.1 Fat/Thin client

As a rule of thumb, a thin client is one that relies on the host for most of the processing; a fat client is one that does most of the processing itself. In absolute terms real world applications are neither thin nor fat; they lie somewhere in the thin/fat domain. Traditional APL applications, being tier-less, are fat clients; they manage their own data locally (internally within the workspace or custom component files) and all the processing is done on the client, the computer executing the application. APL systems, notwithstanding the burden of legacy applications, may become closer to the thin client camp on adoption of a modular or tiered design principle.

In the real world, it is extremely difficult to classify an application as thin or fat—for practical purposes, this classification is meaningless. Usually, it is easy to refute any extrapolated conclusions regarding the speed, ease of use, etc of thin or fat client applications.

1.4.2.2 Database servers

APL via its ability to use ADO can use database servers. A database server is *software* that manages one or more databases and has the ability to service concurrent users transparently.

Therefore, APL applications using database servers are two-tier applications comprising of data services and the business logic encapsulated in the programme.

In fact, APL relies on core Windows facilities for its GUI capabilities; APL is capable of producing three-tier applications by the separation of presentation tier from the business logic—the data is already in the data tier.

1.4.2.3 Application servers

Application servers reside in the middle tier—the business services layer. An application server is *software* that acts as a facilitator between the client services and data services layers. The closer the application server is to the data services layer, the more it is optimising the speed and agility of the system and is indistinguishable from a database server; on the other hand, the closer it is to the client services layer, the more it is optimising the efficiency of the business services layer. Code reuse is paramount, above all, in the business services layer.

1.4.2.4 Network servers

A network server is a powerful dedicated computer running a network operating system that allows connected desktop computers to access services and resources—these include automatic regular back up of files, file sharing, messaging, modems, etc.

1.4.3 Object Oriented Programming(OOP)

OOP must be the most slippery concept of contemporary computing, not least because of the abstract jargon. The explanation of the concept usually relies on examples or given by implication. The essence of the concept is code reuse; this in turn implies generic code. Code reuse is desirable because it allows code to evolve and become more robust and promotes rapid application assembly; these characteristics empower the profession to meet the critical needs of business, its financier.

Perceive an object as a black box, at runtime, containing both code (methods or operations) and data (properties or attributes). Although an application can use the functionality that is publicly exposed, the actual code is never exposed. Such an object is useful because it can receive and send messages—it can receive instructions to modify its data and yield its data.

Thus, an object has a state, its properties, or characteristics at any particular stage of existence, and behaviour, the methods, or operations it is capable of carrying out. The object can expose different sets of properties and methods at different stages of its existence. An object begins to exist as soon as an application creates it.

An object is an instance of a class, literally a new copy: the client application can use and modify the instance independently at runtime. A class is software that is packaged and implemented in a predefined manner. An object exists at runtime; a class is a software template, exists both during, and post development.

1.4.3.1 Encapsulation

An object—that is, an instance of a class—is a programme or software; it contains a lot of code that enables its properties, methods, and events to become visible. Literally, encapsulation means to enclose in a capsule. The internal or auxiliary code, or its implementation is not exposed; it is *private* and evades manipulation, or examination, by direct reference. The object has a published interface that exposes only the properties, methods, and events that its developers deem to be *public*.

Its public interface provides the only means of access to an object's properties and methods. Once published, the interface becomes static—publicly, it always works as they it did initially. However, the internal or private code is subject to constantly enhancements to improve performance and errors are fixed but the public interface continues to work in the same manner. This characteristic, formally described as *encapsulation,* allows the object to maintain its visible integrity.

Thus, an object has a set of elements—properties, methods, and events—that that are *private* or *public*; the code underlying private elements are subject to ongoing refinement and change. The *public* elements, irrespective of underlying code, continue to work consistently across releases. This permits new releases of the underlying components without also requiring new releases of the host applications. In addition to private and public elements, an object may also have *protected* elements; these are an integral part of the public interface—akin to read only elements. An object hides its code, it is a black box, and only the public elements can be called arbitrarily from outside.

Runtime changes to an object's state are persistent only as long as the application using it is active. The class of which the object is an instance remains unaltered: this also promotes reusability. Several active applications (consumers or containers of classes) may have separate instances of the class, as objects, and these will have different state and behaviour.

1.4.3.2 Inheritance

This refers to an object defining its public interface generically and exposing it in a hierarchical fashion; each lower level of the hierarchy is able to expose additional properties and methods without affecting the level above. In other words, the nodes in the hierarchy share or inherit the public interface of the nodes above. This saves repeated coding of common properties and methods: sharing promotes consistency. A subtle way to grasp the concept of inheritance is see it as a mechanism that does three things:

• It avoids having to reinvent the wheel—copying existing code and altering it slightly to accomplish a slight change in its functionality. Although expedient, this is a highly inefficient practice. Such copies then exist in their own right and require parallel changes when the original code requires changes.

• It allows the inherited or derived class to be customised, changed, or extended in parts without affecting the parent class but remains linked to the parent.

• It allows the maintenance of the parent class without requiring changes in the derived classes. The derived classes, instances of the parent class, automatically *inherit* the parent's enhancements.

Inheritance promotes code reuse and encourages generic programming.

1.4.3.3 Polymorphism

This refers back to the concept of the object as a black box; it is accessible only by the interface it exposes to its host. Polymorphism literally means taking or occurring in many forms. In the context of software, it means two things:

• The object, that is, the instance of a class, exists in isolation and independently of other objects, even other instances of the same class (objects) that exist within the same application.

• The object is aware and may behave differently in different runtime situations—identified by arguments passed.

The application does not need to be aware of the particular implementation of the public interface. For example, the ADO component does not expose its implementation of ODBC and OLE DB connections. As far as the application is concerned, a connection is made but the component is doing different things internally and behaves differently depending on the state of the connection.

1.4.3.4 APL and OOP

The OOP concept is tortuous and requires a great deal of tenacity before vague comprehension. Fortunately, for APL development, usable components already embody the OOP technology. A component is the assembly of a set of classes—in a predefined way— each with their own properties, methods, and events.

APL is foremost a consumer of OOP objects and, being a procedural language adapted for an event driven environment, it is not an OOP programming language per se. Nonetheless, the principles of encapsulation, inheritance, and polymorphism are usable in a very limited way.

1.4.4 Collaborative processing

The principle involved is the ability to plug in pre-written third party code modules that can interact with the native tool. Such modules can provide any kind of related functionality, with or without a user interface—a word processor, calendar, spell-checker, the APL Grid object, etc. The technical concept is that of the Microsoft Component Object Model (**COM**), which includes evolving standards including COM+, DCOM (distributed COM), and OLE2 (Object Linking and Embedding—a technology for inter-application information access).. A variety of tools, including **VB**, C++, C# etc. can create custom COM objects. However, in deployment COM objects are both language and tool independent: that is, the native language of the server and clients is irrelevant. The applications in Microsoft Office are COM compliant: APL can use **Excel**, although the native language of these applications is different.

COM objects are very diverse and do not readily fall into precise categories. Except for components written specifically for APL, like the Grid object for APL+Win, components will usually be unable to handle arrays, certainly not as well as APL.

1.4.4.1 In-process and out-of-process servers

An important difference is whether the COM object runs in process   where the application and component run as a single process or out-of-process—where the applications and component run as separate processes. An in-process object cannot exist without the host that instantiates it and must exist on the same machine as the host. An out-of-process object can exist independently of the host that instantiates it and can reside on any computer in the network. In general, a COM object implemented as an executable file is an out-of-process server; if implemented as an **OLE** custom control (**OCX**) or dynamic link library (**DLL**), it is an in-process server. That is, it is an out-of-process if it has an extension EXE and is an in-process server if it has an extension OCX or DLL.

In the Windows **Task Manager**, an out-of-process server does not appear within the **Application** tab; it appears on the **Processes** tab. An in-process server appears in both.

1.4.4.1.1 Automation client and automation server

Excel is an example of an out-of-process object or component; in-process components can act as automation servers or automation clients. An instance of **Excel** created by APL sees **Excel** as the automation server; APL itself is the automation client. An instance of APL+Win created by **Excel** reverses the automation roles.

Although of academic interest, the technical intricacies of COM objects are not of any great consequence to APL: it is not possible to write COM objects with APL. However, its ability to use COM objects literally means that components of an APL application already exist even before the developer has sat at the keyboard!

1.5 APL interface to components

In a development environment such as VB, deploy components in one of two ways—early or late binding. The term 'binding' refers to the connection between the application and a component. A component would have several properties, methods, and events. Public functions, which return an explicit result, and subroutines, which do not return a result, become methods of the component; in this context, the terms component, ActiveX object, and automation server are interchangeable. Unlike VB, which can links a specific version of a component using early binding, APL only uses late binding; early binding enables Intellisense, a feature of the **VB** development environment that facilitates code construction.

1.5.1 Static (or early) binding

With static or 'early' binding, the application links a component at compile time. This implies that the connection becomes static. At runtime, an EXE or executable file, produced from a project that refers to a specific component, identified by its internal version number or class id, will seek that same specific component: at runtime, an earlier or later version of that same component is not recognised. This applies to compiled languages and not to interpreted languages such as APL

Early binding confers several advantages, including compile time syntax checking and better performance than 'late' binding. Code is more readable because the application refers to constants within the components by name; with VB, the Intellisense feature extends to the component too.

1.5.2 Dynamic (or late) binding

With dynamic or 'late' binding, the application connects the component at runtime at the point of first use of the component. The application will connect to any version of the automation server or ActiveX object.. 'Late' binding is slower than 'early' binding.

The concept of 'early' or 'late' binding is incongruous in the context of APL, as it does not produce compiled applications. An advantage of 'early' binding in that Intellisense is available: although APL/W has Intellisense too, with APL, all the lists seen via Intellisense can be enumerated.

1.5.3 Object attributes

An object is an instance of a class and inherits its *properties, methods,* and *events*. A property is a characteristic or attribute of the class. It may be static—that is, read only, like colour or version number—or dynamic—that is read and write, like size or content. A method is a means of coercing the class to do something; a method may or may not require parameters and may or may not return a result. An event is an occurrence that the server signals to its client; an event may be a deliberate action, like a key press, or an incidental action like a timeout.

In simplistic and purely APL terms, workspace available, size, index origin, user-defined variables, etc., are properties; available workspace and its size are a read only property whereas the others are read-and-write properties. The APL primitives or user-defined functions are methods; they all do things—create or modify properties or raise events. Runtime errors such as division by zero or workspace full are events. Events may be handled or trapped in order to take remedial action, like `⎕elx←'ErrHandler'`—here, the APL+Win error handler, `⎕elx`, will execute the user-defined function *ErrHandler* when an error occurs.

1.5.4 Object prefixes

APL+Win and APLX, can enumerate the exposed methods, properties, events, and predefined constants of a bound ActiveX component; these may be case sensitive:

```
'SCRIPT' ⎕wi 'Create' 'MSScriptControl.ScriptControl'
'SCRIPT' ⎕wi 'methods'    ⍝ Enumerate methods
'SCRIPT' ⎕wi 'properties' ⍝ Enumerate properties
'SCRIPT' ⎕wi 'events'     ⍝ Enumerate events
```

APL/W has a different syntax for the same purpose.

```
'SCRIPT' ⎕wc 'OLEClient' 'MSScriptControl.ScriptControl'
SCRIPT.MethodList  ⍝ or 'SCRIPT' ⎕WG 'MethodList'
SCRIPT.PropList    ⍝ or 'SCRIPT' ⎕WG 'PropList'
SCRIPT.EventList   ⍝ or 'SCRIPT' ⎕WG 'EventList'
```

In order to remove any possibility of confusion with native (those belonging to the APL+Win or APLX GUI) properties, methods, and events; the properties of the component are prefixed with lowercase 'x', the methods with uppercase 'X' and for APL+Win, the events need to be prefixed with 'onX'. The 'x' and 'X' prefixes may be omitted in circumstances when doing so does not create a conflict with a like named native APL+Win or APLX GUI method, property, or event. Strictly speaking, the prefixes are not necessary unless the object and APL+Win or APLX have like named objects. However, the prefix enables easy identification of an object's public interface and it is good practice to use it because:

• It removes ambiguity and makes the context of properties, methods, and events clear.

• It is easier to distinguish a constant from properties and methods; events are distinguishable by the presence of the 'on' prefix.

APL/W exposes the public interface of the object without any prefixes.

1.5.4.1 Scope

The prefix makes it is visually apparent that the property, method, or event is not native to APL+Win or APLX; it applies to the first level of the ⎕wi argument only. The prefix is *not* required in the second and subsequent levels for two reasons. The scope is clear from the first level of the object name and the prefix is not required within a hierarchical syntax:

```
⎕wself←'⍙XL' ⎕wi 'Create' 'Excel.Application'
⎕wi 'xWorkBooks.Add' ⍝ Note, the method Add does not a prefix
```

The prefix removes ambiguity from APL+Win or APLX code. For example, the **Excel** object has two 'version' properties, the one with the prefix x is native and the other added by APL:

```
ρ¨ ⎕wi ¨'version' 'xVersion'
 0  1
```

In this instance, 'version' and 'xVersion' return different attributes. Always use the prefix with the names of properties, methods, and events. Use the strings *exactly* as enumerated—the exception is with APL+Win where the prefix 'on' needs to be added to events—even though the names of properties, methods, and events may *not* be case sensitive. This enhances code readability and removes ambiguity. The APL+Win editor keeps everything as typed; that is it does not change the indentation or case of text.

1.5.4.2 Help

APL+Win has two special prefixes that invoke levels of help on ActiveX properties, methods, and events from within the workspace. The short help prefix (?) returns the syntax of the call into the workspace whereas the use of the double prefix (??) displays the help file (*.HLP or *.CHM) associated with the component:

```
⎕wself←'xl' ⎕wi 'Create' 'Excel.Application'
⎕wi '?XRun' ⍝ Displays short help; note  that the X prefix is used.
⎕wi '??XRun' ⍝ Displays the topic in the help file
```

⎕wi '?xWorkBooks'	xWorkbooks property: Value@Object_Workbooks ← ⎕WI 'xWorkbooks'

(⎕wi 'xWorkbooks') ⎕wi '?XAdd'	XAdd method: Result@Object_Workbook←⎕WI'XAdd' [Template]
⎕wi '?XRecordMacro'	XRecordMacro method: ⎕WI 'XRecordMacro' [BasicCode [XlmCode]]
⎕wi '?onXSheetActivate'	onXSheetActivate event: ⎕WEVENT ↔ 'XSheetActivate' ⎕WARG ↔ Sh@Object

The short help prefix confirms the classification, indicates the arguments required, if any, together with their type; names enclosed within square brackets indicate optional parameters.

1.5.4.3 Predefined constants

Most components contain a number of predefined constants, which are usually included within code by name as this makes code more readable. However, the named constants are available only with early binding. That is, they are unavailable within APL as it always uses late binding.

With APL+Win, the prefix = forces the evaluation of a constant:	With APLX, the same functionality is present with a different syntax:
(⎕wi '=xlR1C1') ⁻4150	(''xl' Œwi 'ValueOf' 'xlR1C1') ⁻4150

If a constant cannot be resolved, it may be necessary to specify the constants by value:
```
⎕wi 'XApplication.ReferenceStyle' ⁻4150
```
However, this makes code less transparent and therefore harder to maintain. Use the name of the constant for enhanced readability, where available:
```
⎕wi 'XApplication.ReferenceStyle' (⎕wi '=xlR1C1')
```

1.6 Structured Query Language (SQL)

SQL is the standard *language* for retrieving data from relational databases. SQL is a language for talking to databases; it is case insensitive and does not have a user interface. It is neither a propriety software product nor a development environment. The American National Standards Institution (**ANSI**) and International Standards Organisation (**ISO**) manage it. Most databases incorporate an SQL interface based on this standard.

1.6.1 Relational Database Management System

Dr E F Codd of IBM introduced the concept of Relational Database Management System (**RDBMS**) in 1970 in his paper 'A Relational Model of Data for Large Shared Data Banks.' A database is a data store comprising of a collection of two-dimensional named tables, called *relations*; the tables comprise of zero or more rows or records, called *tuples*, and one or more named columns or fields, called *attributes*. The intersection of a row and column is a cell that contains an atomic value, without formatting. This value can be unassigned or missing—a missing value is not equal to any value including a missing value, not even one in the same column. Rows are identifiable by data content, whereas columns are identifiable by name. It is possible to link tables that contain column names in common by data content.

Given this description, it is easy to conclude that a table is synonymous with a spreadsheet. This is incorrect, for the following reasons:

• A cell in a spreadsheet does not necessarily contain an atomic value; it may contain not only formatting but could also be a formula based on other row/column intersections.

• Table rows are independent of each other. Spreadsheet columns do not intrinsically have a name but table columns always do. Spreadsheets may shift cells in any direction on being deleted thereby destroying the independence of rows.

• Logic and not ordinal position or sequence of rows quantifies operations on tables– the reverse applies to spreadsheets.

A given value in a column, comprising a primary key, logically identifies table rows. The atomic value in a primary key would be unique. For example, an employer may have National Insurance Number as a primary key in its staff database; an insurance company may have policy number, etc.

Tables contain simple data types such as character, date, currency, etc., and do not contain pointers, arrays, vectors, or nested values. This is as much an issue for APL, which deals with vectors and arrays naturally, as it is for other languages that deal with pointers and collections.

An RDBMS contains metadata—data about itself—internally. Modification of metadata does not require the reconstruction of the database. That is, the database is dynamically configurable. An application using a relational database simply needs to know its location, name, password, and the interface to use—ODBC drivers or OLE DB providers. Everything else about the database internally held and queried as required upon a successful connection.

Vendors supply custom tools for the management of their specific databases; however, anything that can be done with a database can be achieved with SQL.

1.6.2 SQL origin

SQL, pronounced *ess-cue-ell*, started life as Structured English Query Language, abbreviated to **SEQUEL** in 1974, and pronounced *SEE-qwel*—the adjective 'structured' applied to the noun 'English'. The abbreviation was unusable, because of existing copyright, and was codenamed SQL and is now generally seen as an acronym for Structured Query Language and interchangeably pronounced in either way. The ISO and ANSI standard was drafted in 1986, revised in 1989, 1992, 1996 and 2003.

Database providers have intentionally designed their interface to be unique and therefore distinguishable from the competition; this implies the use of non-standard words as SQL commands and by the provision of customized supplementary commands. This complicates the programmer's life somewhat in that alternative SQL constructions are often necessary to achieve the same result from alternative databases although each may contain the same data. Only the core ANSI subset of SQL is portable across databases.

1.6.3 SQL language

SQL is a complete database management language for relational databases. It is English-like and non-procedural. In other words, with SQL, the desired result is described rather than coding logical steps needed to produce it. Some vendors have added procedural extensions to their proprietary SQL interface. The drivers implement the SQL interface and provide a means of connecting to the database, provided by the vendors.

The received wisdom is that SQL is easy to learn and that proficiency in the construction of queries progresses naturally during software development using relational databases. Nonetheless, SQL may still be an area of special interest concentrating on specific vendors' SQL implementations, on the efficiency of SQL constructions, and on SQL differences where applications support databases from multiple vendors. Specialists, or Database administrators (**DBA**), also use custom vendor tools to co-ordinate structural modifications, manage the security aspects, and implement centralised SQL held within the database known as stored procedures. SQL alone cannot produce applications. While an SQL statement has a precise form, different ways exist for its construction.

1.6.3.1 Interactive SQL

Products such as Microsoft Query Analyser and Oracle SQL*PLUS provide an environment in which SQL statements can be freely entered and executed. The results appear on screen by default. This allows ad hoc databases queries. Programmers often use this environment to construct and test SQL statements before embedding them within application code.

1.6.3.2 Embedded SQL

An SQL statement, usually one that has been debugged in the interactive environment, may be hard coded in the source code of a development language, and will be executed unchanged whenever the application is run. Embedded SQL statements are ideal in circumstances where there is advance knowledge of the desired results. Any change to the embedded SQL will require the re-assembly of the application itself.

1.6.3.3 Dynamic SQL

An application may generate and execute SQL statements internally—this approach is useful whenever it is not possible to specify the desired queries in advance. For example, the desired queries may be dependent on runtime user choices or responses.

1.6.3.4 SQL Data Query Language

The Data Query Language (**DQL**) component of SQL deals with the retrieval of data from a database—it is based on a single command SELECT. This is the most widely-used component of SQL. Although there is a single command, it is probably the most complex of all SQL commands.

1.6.3.5 SQL Data Manipulation Language

The Data Manipulation Language (**DML**) component of SQL deals with the modification of existing data in a database. The primary commands include INSERT, UPDATE, and DELETE.

1.6.3.6 SQL Data Definition Language

The Data Definition Language (**DDL**) component of SQL deals with the modification of the structure of existing databases and the creation of new databases. The primary commands include CREATE TABLE, CREATE DATABASE, ALTER TABLE, DROP TABLE, CREATE INDEX, and DROP INDEX.

1.6.3.7 SQL Data Control Language

The Data Control Language (**DCL**) deals with database access permissions and used by database administrators who have unrestricted access. The commands in this component of SQL are ALTER PASSWORD, GRANT, REVOKE, and CREATE SYNONYM.

There are subtle differences in the syntax of the DQL, DML, and DDL commands depending on the driver; some drivers may not support one or more of these commands.

1.6.3.8 Stored Procedures

Stored Procedures are one or more SQL statements stored by name within the database itself. An application will execute a stored procedure by calling it. A stored procedure can call another stored procedure and can accept runtime parameters. Stored procedures are procedural.

Stored procedures are complex SQL queries stored within the database itself; they are compiled/tokenised upon creation, and therefore execute faster than individual SQL statements. The semantics of stored procedures are vendor specific.

1.7 The Windows Registry

The Windows Registry is a central repository of information, organised as a hierarchical database, and replaces a myriad of INI files, and the AUTOEXEC.BAT and CONFIG.SYS files. System Registry or just Registry is synonymous with Windows Registry. The **Registry** holds information about installed hardware and software, file types, and users. In essence, it is the heart of a Windows computer: a computer will not function with a corrupt or missing **Registry**.

Typically, the Registry is tens of thousands of kilobytes large; in order to back up the registry, clich **Start | Run**, type REGEDIT /E C:\REGBCK.REG and click **OK**. In the event of a disaster, the file can be imported to restore in an attempt to restore it.

The **Registry** contains root keys—or hives—that contain other keys that hold values and other sub-keys and their corresponding values. The root keys are listed in **Table 1-2 The Registry.** The root keys are read-only, because:

● The operating system uses the registry continually.

● The **Registry** is always held on the local hard disk, although key values may make reference to any network hardware and software that the local machine accesses.

● Windows APIs and other software provide read-write access to the **Registry.**

● Programmatic access to some keys is restricted to selected user profiles.

● Applications usually add their own key either to the HKCU or HKLM root keys. At runtime, applications update the sub-keys within their own particular key and update them as necessary at runtime to provide continuity between sessions.

● The value of some keys and sub-keys may be read-only; for example, the class and programme id of ActiveX controls are read-only.

Constant	Short Name	Decimal Value	Notes
HKEY_CLASSES_ROOT	HKCR	2147483648	
HKEY_CURRENT_USER	HKCU	2147483649	
HKEY_LOCAL_MACHINE	IIKLM	2147483650	
HKEY_USERS		2147483651	
HKEY_DYN_DATA		2147483652	Win 9x Only
HKEY_CURRENT_CONFIG		2147483653	
HKEY_PERFORMANCE_DATA		2147483654	NT Only

Table 1-2 The Registry

The **Registry** is also accessible manually via **Start | Run + REGEDIT**; there is no menu option to launch it.

String		Description
REG_SZ	1	String
REG_EXPAND_SZ	2	Expandable string
REG_BINARY	3	VBArray of integers
REG_DWORD	4	Integer
REG_MULTI_SZ	7	VBArray of strings

Table 1-3 Registry data types

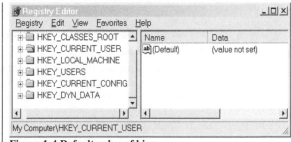

Figure 1-4 Default value of hive

Table 1-3 Registry data types lists the common data types: APL will normally use types 1,2, and 4.

1.8 Regional settings

The use of regional settings is critical in applications destined for use in several countries and it saves the overhead of having to manage hard coded application settings such as date formats, currency symbols, etc. An APL application which is consistent with regional settings is able to share data with other applications running on the same computer—assuming that the other applications are also using the regional settings. An APL application that uses regional settings has the following advantages:

• It appears 'intelligent' to the user by presenting dates and currency amounts in a familiar format.

• It has a standard reference for managing dates, within the workspace, without recourse to an arbitrary format. Of course, data acquisition involving dates may still require a date format if the external source does not hold the dates in the same regional format.

• It has a standard reference for formatting numeric and currency amounts with the appropriate currency symbol and thousands separators.

• System messages, error messages, and text such as names of months or days appear in the local language.

There is a scenario where the use of regional settings may be troublesome, namely, when a user is producing information for consumption in another country. In this case, the developer and tester must set the regional settings to match those of the target country during development and testing.

1.8.1 The GetLocaleInfo API call

This API sets or returns system-wide or user regional settings programmatically. In order to retrieve or set the relevant setting for the local system or local user, specify LOCALE_SYSTEM_DEFAULT or LOCALE_USER_DEFAULT for the *Locale* argument. A single operation usually reads most settings such as date format; others like numeric formats with multiple settings require multiple read operations for each constituent part of the setting.

1.8.2 API usage

The universal benefit of API calls is that they behave consistently across any environment that deploys them; the difference is only in the deployment. For example, the short data format in the three APL environments is retrieved as follows:

```
APLX        ↑↑/GetLocaleInfo 2048 32 '' 256
            dd MMMM yyyy
APL/W       ↑↑/GetLocaleInfo 2048 32 '' 256
            dd MMMM yyyy
APL+Win     ↑↑/⎕wcall 'GetLocaleInfo' 2048 32 (256⍴⎕tcnul) 256
            dd MMMM yyyy
```

The following function uses the local system default:

```
         ∇ Z←L GetLocaleInformation R
  [1]     ⍝ System Building with APL+Win
{1[2]     :if 0=⎕nc 'L'
  [3]         L←'LOCALE_USER_DEFAULT'
1}[4]     :endif
{1[5]     :select L
★ [6]         :case 'LOCALE_SYSTEM_DEFAULT'
  [7]             Z←↑↑/⎕wcall 'GetLocaleInfo' 1024 R (256⍴⎕tcnul) 256
★ [8]         :case'LOCALE_USER_DEFAULT'
  [9]             Z←↑↑/⎕wcall 'GetLocaleInfo' 2048 R (256⍴⎕tcnul) 256
1}[10]    :endselect
         ∇
```

Table 1-4 Locale setting enumerates all the Windows regional settings found with **Start Settings | Control Panel | Regional and Language Options**.

Every aspect of local configuration, for the user or the system, can be queried or set via API calls. The following syntax returns collections such as the names of months of the year and days of the week:

```
⊃GetLocaleInformation ¨(68+0,+\11/1) (49+0,+\6/1)
Jan Feb Mar Apr May Jun Jul Aug Sep Oct Nov Dec
Mon Tue Wed Thu Fri Sat Sun
```

Constant	Value		Setting Returned
	Hex	Decimal	
LOCALE_SYSTEM_DEFAULT	800	2048	
LOCALE_USER_DEFAULT	400	1024	
LOCALE_ILANGUAGE	1	1	Language Id
LOCALE_SLANGUAGE	2	2	Localized Name Of Language
LOCALE_SABBREVLANGNAME	3	3	Abbreviated Language Name
LOCALE_SNATIVELANGNAME	4	4	Native Name Of Language
LOCALE_ICOUNTRY	5	5	Country Code
LOCALE_SCOUNTRY	6	6	Localized Name Of Country
LOCALE_SABBREVCTRYNAME	7	7	Abbreviated Country Name
LOCALE_SNATIVECTRYNAME	8	8	Native Name Of Country
LOCALE_IDEFAULTLANGUAGE	9	9	Default Language Id
LOCALE_IDEFAULTCOUNTRY	A	10	Default Country Code
LOCALE_IDEFAULTCODEPAGE	B	11	Default Code Page
LOCALE_SLIST	C	12	List Item Separator
LOCALE_IMEASURE	D	13	Metric (0); US (1)
LOCALE_SDECIMAL/LOCALE_STHOUSAND	E/F	14/15	Decimal/Thousand Separator
LOCALE_SGROUPING	10	16	Digit Grouping
LOCALE_IDIGITS	11	17	Number Of Fractional Digits
LOCALE_ILZERO	12	18	Leading Zeros For Decimal
LOCALE_SNATIVEDIGITS	13	19	Native ASCII 0-9
LOCALE_SCURRENCY	14	20	Local Monetary Symbol
LOCALE_SINTLSYMBOL	15	21	Int'l Monetary Symbol
LOCALE_SMONDECIMALSEP	16	22	Monetary Decimal Separator
LOCALE_SMONTHOUSANDSEP	17	23	Monetary Thousand Separator
LOCALE_SMONGROUPING	18	24	Monetary Grouping
LOCALE_ICURRDIGITS	19	25	Local Monetary Digits
LOCALE_IINTLCURRDIGITS	1A	26	Int'l Monetary Digits
LOCALE_ICURRENCY	1B	27	Positive Currency Mode
LOCALE_INEGCURR	1C	28	Negative Currency Mode
LOCALE_SDATE/LOCALE_STIME	1D/1E	29/30	Date/Time Separator
LOCALE_SSHORTDATE	1F	31	Short Date Format String
LOCALE_SLONGDATE	20	32	Long Date Format String
LOCALE_IDATE/LOCALE_ILDATE	21/22	33/34	Short/Long Date Format Order
LOCALE_ITIME	23	35	Time Format Specifier
LOCALE_ICENTURY	24	36	Century Format Specifier
LOCALE_ITLZERO	25	37	Leading Zeros In Time Field
LOCALE_IDAYLZERO	26	38	Leading Zeros In Day Field
LOCALE_IMONLZERO	27	39	Leading Zeros In Month Field
LOCALE_S1159/LOCALE_S2359	28/29	40/41	AM/PM Designator
LOCALE_SDAYNAME1-7	2A-30	42-48	Long Name for days

Constant	Value		Setting Returned
	Hex	Decimal	
LOCALE_SABBREVDAYNAME1-7	31-37	49-55	Abbreviated Name for days
LOCALE_SMONTHNAME1-12	38-43	56-67	Long Name-months
LOCALE_SABBREVMONTHNAME1	44-4F	68-79	Abbreviated Name for months
LOCALE_SENGLANGUAGE	1001	4097	English Name Of Language
LOCALE_SENGCOUNTRY	1002	4098	English Name Of Country
LOCALE_STIMEFORMAT	1003	4099	Time Format String

Table 1-4 Locale setting

1.9 Software development

Software development is an evolving process that adopts evolving standards. It is certain that contemporary standards will undergo some form of metamorphosis over time or will be replaced by some other completely radical methodology just as they have replaced earlier standards. All development standards, current and future, have dual objectives:

● Minimize the cost of delivery by managing the complexity of developing and maintaining software; cost is a measure of the resources required to produce and maintain the software and the time span required.

● Maximize the functionality of the software: this involves both the usability of the software in its current context and the potential for its reuse in future contexts.

The difference between competing standards is the relative emphasis placed on these two objectives and the means proposed to achieve them. The contemporary hallmarks are designing for change and rapid development; these criteria dictate code reuse and the use of off-the-shelf components.

1.9.1 Absence of design certification

The industry consensus is that languages such as C++ and Java are OOP development tools. An OOP development tool does not necessarily produce an OOP application—that is, this is not a self-enforcing paradigm. There is no formal and independent design certification and there are no tools for verifying whether an application is OOP compliant, component based, or has a tiered design.

It does not matter how cost efficient and robust any new ideology is it can apply only to future development and not to software that already exists. The ideology must either work with this legacy or co-exist with the ideologies that underpin this legacy. In rare circumstances, it may be viable to re-engineer an application using the new technology.

1.9.2 Quality: the crucial measure

Development standards matter because software development is a cost to business—a means to an end and not an end in itself. Theoretically, it is demonstrable that conformance to OOP and component based development is less costly than the use of any ad hoc methodology over the software life cycle.

Software development is not an exact science if, indeed, it is a science at all. The control, cost and quality of software are far more critical and overriding considerations than any past, present or future ideology of software construction.

1.9.3 Version control

Any application will undergo changes during its life cycle, to fix bugs, to modify existing functionality, or provide new ones, or because of legislation changes. APL development are usually undertaken by small teams, quite often teams of one, reliant on personal initiative to track changes using memory, periodic back up and in-line or off-line documentation. In-line documentation comprises of comments included in code; off-line documentation exists in separate documents independently of code.

Applications changes invariably introduce new bugs and may prove to be unreliable. In such circumstances, it may be necessary to revert to an earlier version of the application and start again. Modern development relies on independent version control software such as Visual Source Safe (**VSS**) to track changes. Unlike VB, which is configurable to enable use of VSS from within its own environment, APL does not have this automatic functionality at present, it is necessary to implement a version control system, manually.

1.9.4 Customisation

Most applications, eventually, provide users with a means of customising their functionality. Essentially, some form of customisation is necessary to extend the application to include bespoke modules. APL, being an interpreted language, ought to be able to implement this quite easily. In practice, this is quite onerous because of the APL keyboard and its right to left execution. APL can deploy the Microsoft Script Control, which has much more familiar execution rules and is free of the APL keyboard shackle.

1.10 APL and Windows API

Windows APIs represent an essential resource for APL: it is debugged, tested, and makes APL applications behave like other applications within the wider domain of user experience. APLX and APL/W define—equivalent to Declare in VB—Windows API calls using the system function ⎕na; however, there are subtle differences in the syntax of the interpreters. In contrast, APL+Win uses its INI files, in the [Call] section, to define API calls. APLWADF.INI is a vendor-provided file that contains a number of standard API calls. Insert any missing, or user-added, API calls into the [Call] section of the APLW.INI file: declarations in this file override entries in the APLWADF.INI file. APLX and APL/W API definitions are workspace bound: definitions are locked monadic functions in the workspace. APL+Win API definitions are available in any workspace via the ⎕wcall system function.

For APL, refer to the VB, rather than C, style API declarations as they are more intuitive from the point of view of translation into APL. An example of such a declaration is:

```
Declare Function GetLocaleInfo Lib "kernel32" Alias "GetLocaleInfoA" (ByVal Locale As Long,
    ByVal LCType As Long, ByVal lpLCData As String, ByVal cchData As Long) As Long
```

The equivalent definitions for APLX and APL/W are as follows:

```
APLX     'GetLocaleInfo' ⍺na 'U Kernel32|GetLocaleInfoA U U >CT[256] U'
APL/W    'GetLocaleInfo'⎕NA'U Kernel32|GetLocaleInfoA U U =OT[] U'
```

For APL+Win, the declaration is:

```
    ⎕wcall 'W_Ini' '[Call]GetLocaleInfo=L(L,L,>C,L) ALIAS GetLocaleInfoA
LIB Kernel32'
```

Alternatively, edit the APLW.INI file and add the following line to the [Call] section:

```
    GetLocaleInfo=L(L,L,>C,L) ALIAS GetLocaleInfoA LIB Kernel32'
```

In all three styles of definition, the API parameters are positional and omit the name of the parameters. Unlike APLX and APL/W, APL+Win can specify the internal name of the parameter also, as follows:

```
    GetLocaleInfo=L(L Locale,L LCType,>C lpLCData,L cchData) ALIAS
GetLocaleInfoA LIB Kernel32'
```

The following function simplifies the APL+Win syntax for fixing API definitions:

```
     ∇ FixAPI R
[1]    ⍝ System Building with APL+Win
[2]    R←⎕wcall 'W_Ini' ('[Call]',R)
     ∇
```

Either insert missing API calls into APLW.INI or use this function as follows:

```
     FixAPI 'PathFileExists=L(*C pszPath) ALIAS PathFileExistsA LIB
shlwapi.dll'
```

Note that entries in the user file, APLW.INI, override corresponding entries in the vendor file, APLWADF.INI: as the vendor entries have been tested, avoid overriding them.

1.10.1 Verifying API definitions

With APLX and APL/W, an existing definition is like any workspace function; it exists if ⎕nl 3 includes it. However, there is the possibility that an application function has the same name as an API definition: name collisions may cause unexpected errors at runtime. This task is more complicated with APL+Win. The following function verifies whether an API is defined:

```
     ∇ Z←L APIExists R
 [1]    ⍝ System Building with APL+Win
 [2]    Z←×⍴⎕wcall 'W_Ini' ('[Call]',R) ⍝ Startup INI File
     ∇
```

In practice, an application will not deploy this function in a runtime environment: it should ensure that up to date INI files from the development environment are distributed.

There is no way of ascertaining whether an API definition in an APL environment is correct except by testing; in practice, several attempts may be necessary.

1.10.2 Processing API memory pointers

In general, API calls represent a single line of APL code and return a result that APL can manipulate directly. However, some API calls require a cover function. For example, the PathFindFileExtension API call returns a memory pointer for its results; APL needs to read the memory location to find the result:

```
     ∇ Z←APIPointer
[1]    ⍝ System Building with APL+Win
[2]    ⍝ Requires: ⎕wcall 'W_Ini' '[Call]PathFindFileExtension=D(*C
          pszPath) ALIAS PathFindFileExtensionA LIB shlwapi.dll'
[3]    ⍝ lstrcpy is defined in APLWADF.INI
[4]    Z←(⎕io+1)⊃⎕wcall 'lstrcpy' (256⍴⎕tcnul) (⎕wcall 'PathFindExtension'
          'c:\X87.PP\Pen.9\Calc.xls')
[5]    Z←Z~⎕tcnul
     ∇
```

Line [4] makes two API calls to return the result:

```
APIPointer
.xls
```

1.10.3 Processing API call back

In the next example, the EnumWindows API call needs a function in the workspace to process its results: think of it as the API call raising an event that the calling environment must handle.

EnumWindows=U(P lpEnumFunc,L lParam) LIB USER32.DLL

The cover function is:

```
     ∇ Z←APICallBack;ptr
[1]    ⍝ System Building with APL+Win
[2]    Z←'' ⍝ Callback function appends to return variable
```

```
[3]    → (0=ptr←⎕wcall 'W_CreateFilter' ('EnumWindows'
          'Z←Z,⊂EnumWindowsCallback'))/0 ⍝ fails
[4]    0 0ρ⎕wcall 'EnumWindows' ptr 0    ⍝ make the call
[5]    0 0ρ⎕wcall 'W_DestroyFilter' ptr ⍝ free the ptr
       ∇
```

It requires the following call back function to enumerate active Windows captions:

```
       ∇ Z←EnumWindowsCallback
[1]    ⍝ System Building with APL+Win
[2]    Z←↑↑/⎕wcall 'GetWindowText' (⊖ρ⎕warg) (256ρ⎕tcnul) 256
       ∇
```

1.10.4 Processing API errors

In the event of an API call failing, the GetLastError API returns the reason for failure and the FormatMessage API translates this code into a more meaningful message. In the next example, the CopyFile API call will fail and the reason for failure is established:

```
0 0ρ⎕wcall 'CopyFile' 'c:\DoesNotExist.APL' 'c:\WillFail.APL' 0
⎕wcall 'FormatMessage' 4096 0 (⎕wcall 'GetLastError') 0 (256ρ⎕tcnul) 256 0
  44 The system cannot find the file specified.
```

API call failure is not catastrophic for the calling environment if it processes the return code appropriately; moreover, APL error handlers are unaware of API call failures.

1.11 The future challenge

Currently, the Windows platform is the most widely used: every aspect of its workings is documented on the Internet, albeit APL specific documentation is very rare. The flagship product of all the APL vendors is for the Windows platform.

The software development landscape is characterised by accepted and evolving technology standards camouflaged in a specialist jargon. The primary driver of software development is cost, measured in terms of both capital expenditure and time. For APL to deliver solutions within the boundaries of cost, it needs to demonstrate its ability to harness the available system building components of the Windows platform.

APL programmers need to break free of the APL cocoon and embrace industry standards. The issue is quite simple: evolve or die. This implies the adoption of industry jargon, databases, components, XML, coding and documentation standards, and application architecture, etc., as well as confining specialist features such as component files to APL history in order to enhance collaboration with other platform tools.

The challenge is to overcome the problem presented by the almost complete lack of any APL documentation of Windows resources. This will ensure that APL is as agile a tool as its competitors. Successful modern software delivers three vital features: it is designed for change, it is easy to use and polite in spite of its underlying complexity, and conforms to user's expectations based on experience of contemporary software.

The current generation of APL providers have reinvented APL and offer access to cutting-edge and contemporary industry technology within their respective interpreters. However, it is in the deployment of such features that lies the future of APL: this is the most pressing challenge that APL developers face today. Adherence to the rules of the platform is not an option but a requirement for legitimacy as an eligible development tool.

Chapter 2

Advanced APL Techniques

APL is a robust development environment. It provides adequate interpreter level resources—capable of exploiting its host operating system as well as other co-existing development tools. APL, like the other competing tools, provides a rich set of pathways for solving problems in any layer of an application, interface, presentation, or data. Herein lay a serious common weakness: the ability to express the solution of a given problem in several ways makes any given solution harder to read as the algorithm is not visually recognisable.

Unlike languages such as **VB**, which have seen incremental enhancements, APL has introduced radical enhancements in the guise of nested arrays, control structures, and COM support. Hence, the burden of legacy code bears more heavily on APL developers than other developers. This is especially so, since the lack of third party APL code examples and reference material has meant that any external reference to coding standards and application design is virtually non-existent. APL programmers have grown accustomed to solving problems in a random pragmatic fashion.

Yet, any theoretical proposition for a prescriptive standard for APL application design is bound to be unworkable unless it is enforceable by the language itself or its editor. APL does not enforce any standards. Nonetheless, control structures do promote a simpler and more readable programme structure and must be mandatory—for both new development and the incorporation of legacy code therein—in the interest of readability.

On a platform where applications are expected to and do share information, APL does have a critical weakness—the lack of a date data type. This problem is aggravated within the APL environment by the fact that different data sources have their own mechanisms for handling dates. In general, other applications store dates as a number of days from a reference point—later dates are positive and earlier dates are negative. However, the reference date varies from application to application. A viable solution is to organise the interchange of dates between APL and other applications in the format specified in the Windows regional settings. However, unlike other applications, which represent dates differently for internal storage and presentation and are capable of treating the two versions as synonyms, APL does not have a built-in internal or string representation for dates. Usually dates are a vector of three numbers or a string in an arbitrary sequence of day, month, and year. The APL+Win Grid does have a date data type and the facility for using the regional setting for date formats—this partially solves the problem for acquiring dates via the native application interface.

The APL design environment is different from that of other development tools. For example, APL stores forms either as a function or as data whereas **VB** requires a form module, a file, for each form. APL has some facilities for optimising the speed of functions comprising an application: for APL+Win refer to the ⎕mf function.

2.1 Removing legacy code clutter

Before the introduction of control structures, algorithms that had different paths relied on branching to labels or absolute line numbers. With control structures, the need for branching no longer exists. However, a series of conditional statements based on the *:if* clause seems just as difficult to read as code with branching. The cleanest code that manages a variable path or flow is based on the *:select* clause.

2.1.1 Boolean scenario (BS) table

Application programming is about managing pathways mapped by data at runtime. A Boolean Scenario (**BS**) table is developed in this book as a means of managing complex programmatic pathways arising in situations where two or more conditions. This technique generates a return code that a calling function can use to control program flow. Consider a simple context where just two conditions are involved. A test of each condition yields the scenarios in **Table 2-1 Boolean scenario table**.

Condition 1	Condition 2	Scenario/Decimal Value
FALSE	FALSE	0
FALSE	TRUE	1
TRUE	FALSE	2
TRUE	TRUE	3

Table 2-1 Boolean scenario table

APL represents **FALSE** and **TRUE** as 0 and 1, respectively. Each row, excluding the final column, represents a binary vector, shown as a decimal number in the final column.

With APL, the conversion of a binary vector to a number is quite straightforward:

Thus, the final row in **Table 2-1 Boolean scenario table** yields:

```
2⊥Binary Vector

2⊥1 1
3
```

2.1.2 A simple task: copy a file

Consider the task of copying a file to another name. This involves three decisions:

- Does the source file exist?

- Does the target file exist?

- Overwrite target file, if it exists?

An algorithm based on a series of *:if ... :else ... :endif* constructions with embedded *and* (^), *or* (∨) and *not* (~) results in a messy and invariably bug-ridden solution especially as the number of conditions increase. The Boolean scenario **Table 2-2 Copy a file** maps all possible states without ambiguity.

Source File Exists	Target File Exists	Overwrite Flag	Decimal Value/ Scenario
0	0	0	0
0	0	1	1
0	1	0	2
0	1	1	3
1	0	0	4
1	0	1	5
1	1	0	6
1	1	1	7

Table 2-2 Copy a file

For the example to hand, scenarios 4, 5, and 7 are valid. Moreover, other considerations yield a further scenario that might prevent the operation from completing—in this case, even though the scenario is 4, 5, or 7, the operation might still fail, because:

- The source file might be in use.
- The target file might be in use.
- The target file name might be invalid.

2.1.3 An API solution

The Windows CopyFile API call is:

```
    CopyFile=B(*C lpExistingFileName, *C lpNewFileName, B bFailIfExists) ALIAS CopyFileA
LIB KERNEL32.DLL
```

It provides a tidy means of copying a file to another name. The arguments are: *lpExistingFileName* is the fully qualified name of the source file, *lpNewFileName* is the fully qualified name of the target/new file and *bFailIfExists* is a Boolean argument—if **FALSE**, *lpNewFileName* is overwritten if it exists already. If the API call succeeds, it returns a non-zero value. If it fails, the return value is zero and another API call GetLastError returns the error code that another API call, FormatMessage, can interpret.

```
GetLastError=D() LIB KERNEL32.DLL
```

```
FormatMessage=D(D dwFlags, *C lpSource, D dwMessageId, D dwLanguageId, >C lpBuffer,
    D nSize, P Arguments) ALIAS FormatMessageA LIB KERNEL32.DLL
```

A simple APL function wraps the whole process including error reporting; it returns 1 if the operation succeeds and 0 if the operation fails. The function is:

```
        ∇ Z←L CopyFile R
   [1]      ⍝ System Building with APL+Win
   [2]      ⍝ R is SourceFile TargetFile
{1[3]      :if 0=⎕nc 'L'
   [4]          L←1 ⍝ Default, fail if TargetFileExists
1}[5]      :endif
   [6]      Z←×⎕wcall (⊂'CopyFile'),R,⊂L
{1[7]      :if 0=Z
   [8]          0 0⍴Msg ↑↑/⎕wcall 'FormatMessage' 4096 0 (⎕wcall
             'GetLastError') 0 (256⍴⎕tcnul) 256 0
1}[9]      :endif
        ∇
```

This function caters for all contingent scenarios; where necessary, it reports an error message indicating the reason why the operation failed.

2.1.4 An APL solution

A solution coded in APL needs to map the results of the three conditions, shown in **Table 2-2 Copy a file**. The syntax of the function is:

```
        [FailIfExist] CopyFileBS SourceFile,TargetFile
        ∇ Z←L CopyFileBS R
   [1]      ⍝ System Building with APL+Win
{1[2]      :if 0=⎕nc 'L'
   [3]          L←1 ⍝ Default, FailIfExist flag
1}[4]      :endif
```

The left-hand argument is optional; when unspecified, it defaults to 1—the target file should not be overwritten if it exists already:

```
{1[5]      :if 2≠≡R
   [6]          R←⊂R
1}[7]      :endif
   [8]      R←2↑R,⊂'' ⍝ Ensure that R has 2 elements, source and target files
```

The right-hand argument must contain two elements, each being a fully qualified file name: source and target file names:

```
[9]    Z←□wcall ¨(⊂⊂'PathFileExists'),¨⊂¨R ⍝ Do files exist?
[10]   Z←Z,L ⍝ Initial scenario
```

Verify the existence of each file and concatenate the FailIfExist flag. This produces one of the rows in **Table 2-2 Copy a file**. The remaining code in this function interrogates other possibilities:

```
     [11]    ⍝ If source file exists, can it be read?
     [12]    L←□wcall '_1Open' (□io⊃R) 0 ⍝ Open Source in read only mode
{1[13]    :if 0=L
     [14]        Z[□io]←¯1
+ [15]    :else
     [16]        0 0ρ□wcall '_1Close' L
1}[17]    :endif
```

It is necessary to establish whether the source file can be read; that is, that the source file is not exclusively tied by another application:

```
{1[18]    :if 0≠Z[□io+1] ⍝ If target file exists ...
     [19]        L←□wcall '_1Open' ((□io+1)⊃R) 0 ⍝ ... Open it in exclusive
         mode
{2[20]        :if 0=L
     [21]            Z[□io+1]←¯1
+ [22]        :else
     [23]            0 0ρ□wcall '_1Close' L
2}[24]        :endif
1}[25]    :endif
```

Equally, if the target file exists, it is necessary to establish that it is usable in exclusive mode. Exclusive mode is necessary if the file is to be overwritten; that is, whether the file can be deleted and recreated:

```
{1[26]    :if 1=Z[□io+2] ⍝ If FailIfExist flag compatible
+ [27]        :andif 0≠Z[□io+1]
     [28]            Z[□io+1]←¯1
1}[29]    :endif
```

If the target is usable in exclusive mode, verify whether the *FailIfExist* flag is compatible:

```
{1[30]    :if ¯1∊Z
     [31]        Z←8
+ [32]    :else
     [33]        Z←2⊥Z
1}[34]    :endif
       ∇
```

The Boolean scenario code is returned; the return code is 8 when one or more of the following are true:

• Read access to the source file is denied.

• Target file exists but exclusive access is denied.

• Target file exists but the FailIfExist flag is set to 0.

2.1.4.1 Preferred solution

That the APL solution is more verbose and less efficient than the API solution is self-evident. Moreover, the APL solution will fail for two more reasons, namely:

• The target file name is unspecified.

• The target file name is the same as the source file name.

Adding validation for these conditions just makes the APL solution worse from the point of view of efficiency. Also, note that there is no appropriate error message for return code 8. These are persuasive reasons for replacing raw APL solutions by API solutions.

2.1.4.2 Handling the return code

One method of handling the return code is to iterate its every possible value, generate a feedback message, and organise further conditional processing. In practice, this verbose method is likely to be impractical when the number of conditions exceeds, say, 3.

In some situations, especially ones involving many conditions, it may not be practical to enumerate every possible outcome. In the example described in **Table 2-2 Copy a file**, return codes 0,1,2,3, and 6 are similar in that they signal failure and 4,5, and 7 are similar in that they signal success and 8 implies failure although the implicit code is one of 4,5 or 7. Consider an alternative CallingFunction:

```
       ∇ CallingFunction2
  [1]    ⍝ System Building With APL+Win
{1[2]    :select OverwriteFlag CopyFileBS Sourcefile TargetFile
= [3]        :caselist 0,1,2,3,6 ⍝ ... Failure
= [4]        :caselist 4,5 7     ⍝ ... Success
= [5]        :case else          ⍝ ... Unknown Error
1}[6]    :endselect
       ∇
```

An even more concise method of handling the return code is to trap only the codes indicating failure. The modified solution is:

```
{1[2]    :if (2⊥OverwriteFlag CopyFileBS Sourcefile TargetFile)∊4 5 7
  [3]    ⍝ ... Success
+ [4]    :else
  [5]    ⍝ ... Unknown Error
1}[6]    :endif
```

The trade-off among the three handling methods is the extent to which an application provides user-friendly feedback. The verbose method provides precise feedback and is appropriate where few conditions are involved; the Boolean method does not indicate the reason for failure and is appropriate where many conditions are involved.

However, the BS method of enumerating every possible return code is all comprehensive and least likely to introduce logic flaws and lends itself to the sound documentation of processes.

2.1.5 The general BS case

The number of return codes will be 2^n+1, where n is the number of conditions. If n is a sufficiently high number, it may not be practical to enumerate every return code. The number of return codes for situations involving 1, 2, 3, 4, 5, 10 conditions is:

```
    1+2⊥¨(1 2 3 4 5 10)⍴¨1
3 5 9 17 33 1025
```

How can a single condition that can be **TRUE** or **FALSE** yield three scenarios? Consider a field 'Sex': it can hold any value including 'M' or 'F':

Sex←'F'	Sex←'M'	Sex←0
2⊥'MF'=↑Sex	2⊥'MF'=↑Sex	2⊥'MF'=↑Sex
1	2	0

The third scenario captures the possibility that the field contains an invalid or missing value—this is a tangible possibility when the value of the field comes from a data source and not from direct input that is subject to validation. The construction of a BS table whenever conditional processing is involved forces sound application documentation and makes the process logic plain. The technique is particularly appropriate to APL as it provides inherent support for Boolean to decimal conversion.

2.1.6 Auto generating the BS table

Given a finite number of conditions, this APL function can generate the BS table:

```
    ∇ Z←AutoBSTable R                                 AutoBSTable 3
[1]   ⍝ ⍉(R/2)⊤0,+\(2⊥R/1)/1 ⍝ Index Origin independent    0 0 0 0
[2]   ⍝ ⍉(R/2)⊤0,⍳R          ⍝ Index Origin is 1           0 0 1 1
[3]   ⍝ ⍉(R/2)⊤¯1+R          ⍝ Index Origin is 0           0 1 0 2
[4]   Z←0,+\(2⊥R/1)/1                                       0 1 1 3
[5]   Z←(⍉(R/2)⊤Z),Z                                        1 0 0 4
    ∇                                                       1 0 1 5
                                                            1 1 0 6
```

The right-hand argument specifies the number of conditions, *n*. For $\quad$ 1 1 1 7
example, the results for *n* being 3 are shown on the right:

This enumerates all eight possible return codes. The calling function must handle the ninth code (8), usually within the *:else* block of a *:select ... :case* construction.

2.1.7 Where to use BS tables?

The short answer is wherever the context involves one or more conditions and each may be visualised as a **TRUE** or **FALSE** state. This promotes robust and exhaustive coding and testing. The enumeration of the 2^n scenarios allows the analysis of the valid and invalid combinations—the business logic will separate them—and the creation of appropriate test data. The $(2^n + 1)$ scenario is a catchall invalid scenario. This approach to testing is more appropriate to unit testing; that is, for testing isolated parts of an application.

2.1.7.1 Raw data validation

Consider an item of data entered via the application interface or retrieved from a database. The data may be empty or null, may be numeric, date, or character, and may be within range or not. For instance, if the data item contains a date of birth, it is valid if it is not empty, is a date, and is earlier than today's date. The three conditions (Not Null, Is Date, IS no later than today's date) yield eight possible return codes of which only a return code of 7 is valid. Analysis of this type makes coding quicker and clearer.

2.1.7.2 Validation for business rules

This involves multiple data items each of which is valid in isolation but is invalid when taken together. Consider three data items: *Frequency*, which is valid if monthly, quarterly, bi-annually, or yearly, and *Amount*, which is valid if numeric, positive, and specified to two decimal places, and *Scheme*, which is one of *Staff* or *Executive*.

Each data item may be valid in that it satisfies raw validation but taken together one or another may be invalid—the Amount must be exactly divisible by the Frequency and a frequency of monthly is valid only if Scheme is *Staff*.

The validation for reasonability results in three conditions, namely: *Frequency* is *monthly*, *Amount* is exactly divisible by *Frequency,* and *Scheme* is *Staff*. Of the nine return codes suggested by the BS table based on the three conditions, only return codes 2, 6, and 7 are valid.

2.1.7.3 Conditional processes

Application programming is all about valid pathways though code based on context. The BS table facilitates the delineation of valid pathways, minimises the potential for logic flaws or bugs and enhances the readability of code. Consider a business example where several scenarios are possible, illustrated in **Figure 2-1 Decisions**.

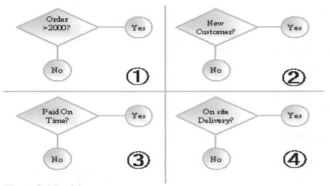

Figure 2-1 Decisions

The rules involve <u>four</u> conditions, each shown individually in **Figure 2-1 Decisions**.

The developer must link the outcome of each condition to the outcomes of the other three; not only is this exercise liable to be painful and bug-ridden but also time-consuming.

An invoice is to be calculated using the following business rules:

• Orders below £2000 accrue a surcharge of 2% of the shortfall but only for new customers requiring on-site delivery.

• Orders above £2000 accrue a discount of 2% or 5% for all customers respectively. If existing customers expect on-site delivery, their discount goes down to 1%: the discount reduces to 0.5% if an existing customer did not settle the previous invoice on time.

• Delivery is a flat charge of £200, applicable to existing customers only.

There are 17 pathways for the calculation of the invoice, $2+2\perp4/1$. It is highly unlikely that a flowchart or raw APL code based on conditional logic will produce an error free and efficient solution. An efficient solution is one that is easy to construct and to maintain. In contrast, a solution based on BS is quite straightforward. Consider this solution:

```
       ∇ Z←L Invoice R
  [1]       ⍝ System Building with APL+Win
  [2]       ⍝ L is a scalar representing the raw amount payable - quantity ×
             price
  [3]       ⍝ R is a Boolean vector of 4 elements
  [4]       ⍝ (Order >£2000),NewCustomer,PaidOnTime,Collect
  [5]       ⍝ Z returns the adjusted amount payable.
{1[6]     :select case 2⊥R
★ [7]         :case    0 ⍝  0 0 0 0
  [8]             ⍝ Will not arise in practice
★ [9]         :case    1 ⍝  0 0 0 1
  [10]            ⍝ No adjustments
★ [11]        :case    2 ⍝  0 0 1 0
  [12]            ⍝ + £200
★ [13]        :case    3 ⍝  0 0 1 1
  [14]            ⍝ No Adjustment
★ [15]        :case    4 ⍝  0 1 0 0
  [16]            ⍝ Surcharge of 2% of shortfall
★ [17]        :case    5 ⍝  0 1 0 1
  [18]            ⍝ No Adjustment
★ [19]        :case    6 ⍝  0 1 1 0
  [20]            ⍝ Surcharge of 2% of shortfall
★ [21]        :case    7 ⍝  0 1 1 1
  [22]            ⍝ No Adjustment
★ [23]        :case    8 ⍝  1 0 0 0
  [24]            ⍝ Less 0.5% discount + £200
```

```
★ [25]        :case   9 ⍝  1 0 0 1
  [26]              ⍝ Less 0.5% discount
★ [27]        :case  10 ⍝  1 0 1 0
  [28]              ⍝ Less 0.5% discount + £200
★ [29]        :case  11 ⍝  1 0 1 1
  [30]              ⍝ Less 1% discount
★ [31]        :case  12 ⍝  1 1 0 0
  [32]              ⍝ Less 5% discount
★ [33]        :case  13 ⍝  1 1 0 1
  [34]              ⍝ Less 5% discount
★ [35]        :case  14 ⍝  1 1 1 0
  [36]              ⍝ Less 5% discount
★ [37]        :case  15 ⍝  1 1 1 1
  [38]              ⍝ Less 5% discount
★ [39]        :case else
  [40]                 ⍝ Return code 16, Will not arise in practice
1}[41]  :endselect
      ∇
```

Programme flow design based on BS table offers two further advantages.

● First, changing or debugging any particular scenario does not affect the logic in the remaining scenarios.

● Second, generic processes become obvious; in order to avoid repeating the same line(s) of code, assemble such processes into separate functions. In this example, "Less n% discount" is a prime candidate for a separate function that takes the amount of the invoice as the left-hand argument and the percentage discount as the right-hand argument and returns the adjusted invoice amount.

2.2 Bit-wise Boolean techniques

Unlike the **VB** operators AND, OR, XOR, etc., APL logical operators fail when applied to non-Boolean data and it does not offer alternative corresponding functions for such data. The **VB** operators represent each argument as a binary number, apply the standard relational operators on each corresponding bit, and convert the resultant Boolean vector as a number in base 10. The **VB** functions tolerate non-integer arguments but rounds them to the nearest whole number internally.

2.2.1 BitEQV

The EQV operator establishes the logical equivalence of two expressions. That is, on an element-by-element basis, the result is **TRUE** only when both the arguments' bits are equal. The result is obscure when either or both arguments are not binary. In fact, the computation is quite simple.

```
    ∇ Z←L BitEQV R                        10 BitEQV ⁻323.78
[1]   ⍝ System Building with APL+Win      329
[2]   Z←323 ⎕dr (11 ⎕dr ⌊L+0.5)=11 ⎕dr ⌊R+0.5
    ∇
```

First, represent each argument as a binary number:
```
    (11 ⎕dr ⌊10+0.5),[0.5] 11 ⎕dr ⌊⁻323.78+0.5
0 0 0 0 1 0 1 0 0 0 0 0 0 0 0 0 0 0 0 0 0 0 0 0 0 0 0 0 0 0 0 0
1 0 1 1 1 1 0 0 1 1 1 1 1 1 1 0 1 1 1 1 1 1 1 1 1 1 1 1 1 1 1 1
```

Second, compare each corresponding bit, returning 1 or 0 depending on whether they are equal or not:
```
    =≠(11 ⎕dr ⌊10+0.5),[0.5] 11 ⎕dr ⌊⁻323.78+0.5
0 1 0 0 1 0 0 1 0 0 0 0 0 0 0 1 0 0 0 0 0 0 0 0 0 0 0 0 0 0 0 0
```

Third, represent the result in base 10:
```
    323 ⎕dr =≠(11 ⎕dr ⌊10+0.5),[0.5] 11 ⎕dr ⌊¯323.78+0.5
329
```
If both arguments are Boolean, EQV returns 1, when both are equal.

2.2.2 BitIMP

The **VB** IMP operator establishes the logical implication of two expressions; it copes with one or both expressions being null. APL does not have a null data type: null indicates a missing value or unassigned variable. The APL character null (") or numeric null (⍬) is not equivalent to a null data type. The function is defined below:

```
      ∇ Z←L BitIMP R                           10 BitIMP ¯323.78
[1]   ⍝ System Building with APL+Win          ¯3
[2]      Z←323 ⎕dr (11  ⎕dr ⌊L+0.5)≤11 ⎕dr ⌊R+0.5
      ∇
```

If both arguments are binary, IMP returns 1 when both are equal, or only the right-hand argument is **TRUE**, see **Table 2-3 Boolean table for VB IMP**.

Condition1	Condition2	Result
0	0	1
0	1	1
1	0	0
1	1	1

Table 2-3 Boolean table for VB IMP

This is a BS table: the two conditions yield four possible outcomes 0, 1, 2, and 3. The programme flow can be organised as follows:

```
      ∇ L FlowIMP R                       ∇ L FlowBS R
[1]      ⍝ System Building with        [1]    ⍝ System Building with
         APL+Win                              APL+Win
{1[2]    :if L BitIMP R                {1[2]    :select 2⊥L,R
[3]          ⍝ true                    ★ [3]        :case 0
+ [4]    :else                         ★ [4]        :case 1
[5]          ⍝ false                   ★ [5]        :case 2
1}[6]    :endif                        [6]          ⍝ false
      ∇                                ★ [7]        :case 3
                                       1}[8]    :endselect
                                          ∇
```

The function FlowBS is more verbose but more intuitive than FlowIMP: a more concise form is shown as FlowBS2.

FlowBS enables tighter control: it enumerates all possible scenarios. In the event of a change in functionality, in the future, isolated changes are possible for each scenario.

```
      ∇ L FlowBS2 R
[1]      ⍝ System Building with
         APL+Win
{1[2]    :select 2⊥L,R
★ [3]        :caselist 0,1,3
[4]          ⍝ true
+ [5]    :else
[6]          ⍝ false
1}[7]    :endselect
      ∇
```

2.2.3 BitAND

Internally, the **VB** AND operator compares the binary representation of its left- and right-hand arguments, returning **TRUE** when each corresponding bit is **TRUE** and returns the result in base 10. The function is defines as follows:

```
      ∇ Z←L BitAND R                           10 BitAND ¯323.78
[1]   ⍝ System Building with APL+Win          8
```

```
[2]    Z←323 ⎕dr (11 ⎕dr ⌊L+0.5)∧11 ⎕dr ⌊R+0.5
     ∇
```

The potential application of this function is unclear: where both arguments are Boolean, it simply evaluates the APL ∧ operator. However, it can provide the answer to this question: Is one argument, representing 2 to the power *n* where *n* is a positive integer greater than or equal to 0, within the other, comprising of the sum of several such numbers?

```
1209 BitAND 64       │ 64 BitAND 1209       │ 32 BitAND 1209
0                    │ 0                    │ 32
1209 BitAND 32       │ 32 BitAND 32         │ ⁻32 BitAND 32
32                   │ 32                   │ 32
```

The result is zero when the condition is **FALSE** and non-zero (the minimum of the two arguments) when **TRUE**. As demonstrated in this example, 32 or 2*5 is included whereas 64 or 2*6 is not.

2.2.3.1 Enumerating logical drives

An interesting use of BitAND is in the enumeration of logical drives; there can be 26 drives, denoted by the letters A-Z. The API call GetLogicalDrives is required:

```
GetLogicalDrives=D() LIB KERNEL32.DLL
```

A simple expression enumerates available logical drives:

```
A←'ABCDEFGHIJKLMNOPQRSTUVWXYZ'
B←⎕wcall 'GetLogicalDrives'
(∈×(⊂B) BitAND ¨2*⁻1++\26/1)/A
ACDEF
```

A more conventional result is:

```
A←'ABCDEFGHIJKLMNOPQRSTUVWXYZ'
((∈×(⊂B)BitAND ¨2*⁻1++\26/1)/A),¨':'
A: C: D: E: F:
```

2.2.3.2 Powers of 2

A given number can be broken down in elements representing 2^n using this function:

```
     ∇ Z←PowersOf2 R
[1]    ⍝ System Building with APL+Win
[2]    Z←⌈2⍟⌈R
[3]    Z←(×R)×⌽(,((Z+1)ρ2)⊤⍒⌈R)/⌽2*0,+\Z/1
     ∇
```

The function can decipher the powers of 2 that represent the sum returned by the API call—drive A is 2^0, B is 2^1 and so on to Z, which is 2^{25}. The result will vary depending on the number of drives on the computer; for this computer, the result is:

```
     ((2*⁻1++\26/1)∈PowersOf2 ⎕wcall 'GetLogicalDrives')/A
ACDEF
```

This function is especially useful for deciphering arguments of some Windows API calls and the value of the properties of some APL GUI objects, such as the style property of a form. Note the expression in line [2] – it calculates the number of digits in the integer portion of a number. Consider the GUI *style* property, which is the sum of one or more values from **Table 2-4 Style values**.

Value	Description
0	Default form.
1	Signal Wait event before Showing form.
2	Make form independent of any other form that is Waiting.
4	Hide form before Waiting.
8	Do not disable for Wait on a secondary form
16	Always on top.
32	Force icon in taskbar even if form is a tool palette (border 512).

Table 2-4 Style values

```
  ∇ MyForm;⎕wself
[1]   ⍝ System Building with APL+Win
[2]   ⎕wself←'MyForm' ⎕wi 'Create' 'Form' 'Close'
[3]   ⎕wi 'style' (2+16+32)
  ∇
```

The code enumerates the values of the style property; however, at runtime, the style property is the sum of the individual values. For example:

The function PowersOf2 will return the individual values:

```
      ⎕wi 'style'
50

      PowersOf2 ⎕wi 'style'
2 16 32
```

2.2.4 BitXOR

The **VB** XOR operator compares the binary representation of its left- and right- hand arguments, returning **TRUE** when each of the corresponding bits is unequal and returns the result in base 10. The function is defined as:

```
  ∇ Z←L BitXOR R                        10 BitXOR 8
[1]   ⍝ System Building with APL+Win     2
[2]   Z←323 ⎕dr (11 ⎕dr ⌊L+0.5)≠11 ⎕dr ⌊R+0.5
  ∇
```

2.2.5 BitOR

The **VB** OR operator compares the binary representation of its left- and right-hand arguments, returning **TRUE** when either of the corresponding bits is **TRUE** and returns the result in base 10. The function is:

```
  ∇ Z←L BitOR R                         10 BitOR 8
[1]   ∩ System Building with APL+Win     10
[2]   Z←323 ⎕dr (11 ⎕dr ⌊L+0.5)∨11 ⎕dr ⌊R+0.5
  ∇
```

2.2.6 BitNAND

This function applies the APL ⍲ operator to the binary representation of its left- and right-hand arguments and returns the result in base 10; it is defined as:

```
  ∇ Z←L BitNAND R                       VB  does   not  have  an
[1]   ⍝ System Building with APL+Win     equivalent NAND operator.
[2]   Z←323 ⎕dr (11 ⎕dr ⌊L+0.5)⍲11 ⎕dr ⌊R+0.5
  ∇
```

2.2.7 BitNOR

This function applies the APL ⍱ operator to the Boolean representation of its left- and right-hand arguments and returns the result in base 10; it is defined as:

```
  ∇ Z←L BitNOR R                        VB  does   not  have  an
[1]   ⍝ System Building with APL+Win     equivalent NOR operator.
[2]   Z←323 ⎕dr (11 ⎕dr ⌊L+0.5)⍱11 ⎕dr ⌊R+0.5
  ∇
```

2.3 Managing workspace variables

APL has rich facilities for managing variables; that is, variables in the workspace. These variables may be global variables that have been acquired independently or results created by the application logic. Workspace variables are saved automatically when the workspace itself is saved, may be copied across workspaces, and may be saved in component files individually or as collections or packages: the properties of the variables—type, shape, and rank—are retained. However, there are some weaknesses in this regime, not least because the application logic and the data reside in the same layer. These are:

• It is not apparent where a global variable comes from. This complicates debugging, especially when a variable is missing or when it is present but has incompatible properties.

• It is difficult to share APL variables with another application as the workspace and component files are unique to APL and not understood by other application.

• It is hard work presenting APL variables; complex GUI interfaces are required to show the variables within the APL environment and complex documents are required for hardcopy.

• It is inefficient to hold all data and results variables within the workspace: they take space and there is always a risk of inadvertent contamination during execution.

2.3.1 APL is not a relational data source

The problems highlighted may be largely addressed by holding the workspace data as a record set. ADO can fabricate a record object without recourse to a data source: an APL workspace or component file is not a data source in the conventional sense. Consider the sample data shown in **Table 2-5 Sample data for fabricating record set**, created within an APL application.

Name	Sex	Dobirth	Salary
Pailles, M	Male	12/03/1986	24000
Laventure, K	Female	16/05/1982	22765
Moka, V	Female	03/09/1972	38400
Flacq, A	Male	07/12/1984	19000
Sobha, F	Female	29/01/1961	22680

Table 2-5 Sample data for fabricating record set

The following functionality is required:

• Allow other applications such as **Excel** or **Word** to share each set of data—all males or all females. A typical APL solution is to write the data to two comma delimited text files, which can be manipulated in **Excel** or **Word**

• Process each row or record individually—all males or all females at a time. This implies selection by sex and a loop. An APL solution must consider all rows but whether a row is processed or not depends on the test based on the variable 'Sex'.

• For presentational purposes, substitute the variable names by headings that are more descriptive; the variables 'Name', 'Sex', and 'Salary' are meaningful but 'Dobirth' is not. An APL solution will painstakingly code the translations of the variable names.

• If required, delete any row identified by name permanently. Usually, this requires a unique identifier or key field; for this example, all the names are unique and serve as a unique identifier. An APL solution will resort to selecting the required rows and reassign the results to the variables.

APL has just two data types: numeric and character. For this example, 'Name', 'Sex', and 'Dobirth' are character and 'Salary' is numeric.

Table 2-6 ADO Data types shows the corresponding ADO data types

Constant	Decimal Value
adChar	129
adVarchar	200
adDouble	5

Table 2-6 ADO Data types

Both the 'adChar' and 'adVarChar' types require the number of characters to be specified: 'adChar' uses memory for the number specified but 'adVarChar' expands up to the number of characters specified and uses memory for the actual number of characters used. **ADO** derives the width of the numeric data type, 'adDouble', internally.

2.3.2 Fabricating a record set object

Given that all the variables, in **Table 2-5 Sample data for fabricating record set**, exist in the workspace, the function FabricateRS creates an ADO record object:

```
        ∇ FabricateRS;⎕wself;i
  [1]    ⍝ System Building with APL+Win
  [2]    ⎕wself←'ADOR' ⎕wi 'Create' 'ADODB.Recordset'
  [3]    ⎕wi 'xFields.Append' 'Name' (⎕wi '=adVarChar') 200 100
  [4]    ⎕wi 'xFields.Append' 'Sex' (⎕wi '=adChar') 6
  [5]    ⎕wi 'xFields.Append' 'Dobirth' (⎕wi '=adChar') 10 ⍝ UK Short date
         format
  [6]    ⎕wi 'xFields.Append' 'Salary' 5
  [7]    ⎕wi 'XOpen'
{1[8]    :for i :in ⍳1↑⍴Name
  [9]        ⎕wi 'XAddNew'
  [10]       ⎕wi 'xFields.Item(0).Value' ('DeleteUB' Strip Name[i;])
  [11]       ⎕wi 'xFields.Item(1).Value' ('DeleteUB' Strip Sex[i;])
  [12]       ⎕wi 'xFields.Item(2).Value' (Dobirth[i;])
  [13]       ⎕wi 'xFields.Item(3).Value' (⍕Salary[i;])
  [14]       ⎕wi 'XUpdateBatch'
1}[15]   :endfor
{1[16]   :if 0≠⎕wcall 'PathFileExists' 'c:\persistrs.xml'
  [17]       0 0⍴⎕wcall 'DeleteFile' 'c:\persistrs.xml'
1}[18]   :endif
  [19]   ⎕wi 'XSave' 'C:\PersistRS.XML' 1 ⍝ 1=adPersistXML
        ∇
```

Some parts of this function need clarification:

● Line [3]—up to two hundred characters are reserved for 'Name' but only as many as required, up to 200, will be used.

● Lines [4] and [5]—the fields 'Sex' and 'Dobirth' have precisely 6 and 10 characters reserved, respectively.

● Line [6]—numeric fields, like 'Salary', do not need their width to be explicitly specified; it is derived internally.

● Lines [10] to [13] *all* values are strings or character before assignment—this does not destroy their inherent type. Moreover, the ADO ActiveX object always uses indices based on index origin 0. It is worthwhile removing leading, trailing, and duplicate blanks from the variables known to be of type character in order to minimise the size of the record object.

● Line [16] to [19] writes the record object, in XML format, to disk—the target file should not already exist. On completion, the record object exists in Windows memory as object ADOR *and* in the file. The XML file is the record object 'persisted'.

This is an alternative version of the same function—the three initial columns show information returned by the ⎕mf function:

```
75  0 1        ∇ FabricateRS2;⎕wself;i
 0  0 0   [1]    ⍝ System Building with APL+Win
 5  5 1   [2]    ⎕wself←'ADOR' ⎕wi 'Create' 'ADODB.Recordset'
 5  5 1   [3]    ⎕wi 'xFields.Append' 'Name' (⎕wi '=adVarChar') 100
 0  0 1   [4]    ⎕wi 'xFields.Append' 'Sex' (⎕wi '=adChar') 6
 0  0 1   [5]    ⎕wi 'xFields.Append' '[Date Of Birth]' 129 10 ⍝ UK Short
          date format
 5  5 1   [6]    ⎕wi 'xFields.Append' 'Salary' 5
 0  0 1   [7]    ⎕wi 'XOpen'
```

```
 0   0 1  {1[8]    :for i :in ιl↑ρName
 0   0 5   [9]            □wi 'XAddNew'
20 15 5   [10]           □wi 'xFields("Name").Value' ('DeleteUB' Strip
          Name[i;])
10 10 5   [11]           □wi 'xFields("Sex").Value' ('DeleteUB' Strip Sex[i;])
15 15 5   [12]           □wi 'xFields("[Date Of Birth]").Value' (Dobirth[i;])
10 10 5   [13]           □wi 'xFields("Salary").Value' (⍎Salary[i;])
 0   0 5   [14]           □wi 'XUpdateBatch'
 0   0 5 1}[15]   :endfor
 0   0 1  {1[16]   :if 0≠□wcall 'PathFileExists' 'c:\persistrs.xml'
 0   0 1   [17]        0 0ρ□wcall 'DeleteFile' 'c:\persistrs.xml'
 0   0 0 1}[18]   :endif
 5   5 1   [19]   □wi 'XSave' 'C:\PersistRS.XML' 1 ⍝ 1=adPersistXML
                ▽
```

The latter version uses a descriptive name for 'Dobirth' and uses the names of the fields, rather than their relative indices, to add their values—thus, making it possible to add fields in any arbitrary order. Enclose names that are valid variable names in square brackets.

The above example uses APL arrays to create the record object. It is likely that user interaction creates each row of the set of arrays individually. The code in line [9] to [14] can create the record object likewise—one record at a time. However, adding one field at a time, Line [10] to [14] adds an unnecessary overhead and sacrifices performance, especially when many fields are involved. The following version adds all fields in one operation:

```
30   0 1          ▽ FabricateRS3;□wself;i;fld;rec
 0   0 0   [1]    ⍝ System Building with APL+Win
10 10 1   [2]    □wself←'ADOR' □wi 'Create' 'ADODB.Recordset'
 0   0 1   [3]    □wi 'xFields.Append' 'Name' (□wi '=adVarChar') 100
 5   5 1   [4]    □wi 'xFields.Append' 'Sex' (□wi '=adChar') 6
 0   0 1   [5]    □wi 'xFields.Append' '[Date Of Birth]' 129 10 ⍝ UK Short
          date format
 0   0 1   [6]    □wi 'xFields.Append' 'Salary' 5
 0   0 1   [7]    □wi 'XOpen'
 5   5 1   [8]    fld←'#' □wi 'VT' (0,⁻1↓+\(□wi 'xFields.Count')/1) 8204
 0   0 1 {1[9]    :for i :in ιl↑ρName
 5   0 5   [10]       rec←('DeleteUB' Strip Name[i;]) ('DeleteUB' Strip
          Sex[i;]) (Dobirth[i;]) (⍎Salary[i;])
 0   0 5   [11]       □wi 'XAddNew' fld ('#' □wi 'VT' rec 8204)
 0   0 5   [12]       □wi 'XUpdateBatch'
 0   0 5 1}[13]   :endfor
 0   0 1  {1[14]   :if 0≠□wcall 'PathFileExists' 'c:\persistrs.xml'
 5   5 1   [15]        0 0ρ□wcall 'DeleteFile' 'c:\persistrs.xml'
 0   0 0 1}[16]   :endif
 0   0 1   [17]   □wi 'XSave' 'C:\PersistRS.XML' 1 ⍝ 1=adPersistXML
                ▽
```

This version takes 0.4 times the duration of the FabricateRS2 version to add the same 5 records. Line [8] creates a vector of the ordinal position of each field—in index origin zero—and converts it to an array. Line [10] converts all values into character and assembles them into an APL nested vector. Line [11] converts this nested vector into an array and adds all the values in one operation. The variable *fld* must reflect the ordinal position of each field. The variable *rec* must reflect the collation order of values, even when the values are collated in the order in which fields are appended. A field has a null value as soon as it is appended; it retains this value until it is reassigned. APL sees a null value as 0ρ' ', or 0/' 0', or 0/0.

The ⎕mf system function provided the timings of each function, using the following cover function to obtain the listing with timing information.

```
      ∇ Z←Benchmark
[1]    ⍝ System Building with APL+Win
[2]    ⎕mf 1 ⍝ Use milliseconds
[3]    1 ⎕mf 'FabricateRS3' ⍝ Switch monitor on
[4]    FabricateRS3
[5]    Z←('List' CSBlocks 'FabricateRS3'),⎕tcnl
[6]    Z←' ',⊃(Z≠⎕tcnl)⊂Z
[7]    Z←((⍕⎕mf 'FabricateRS3'),[⎕io]' '),Z
[8]    0 ⎕mf 'FabricateRS3' ⍝ ⍝ Switch monitor off
[9]    Z←(⊂'DeleteTB') Strip ¨⊂[⎕io+1]Z
[10]   Z←∊Z,¨⎕tcnl
      ∇
```

The APL+Win documentation provides details of the ⎕mf function. Advances in the speed of processors mean that a slow application can run in reasonable time on a faster machine. Nonetheless, it is worthwhile reviewing the structure of core and frequently used functions with a view to improving performance:

● On the Windows platform, all concurrent applications share available system resources. That is, the APL application shares the processor, available memory, etc., with the operating system and other applications. In its turn, that is when it is at the head of the processing queue, an APL application runs fastest when it can finish its logical tasks without involving runtime information being swapped in and out of the virtual disk.

● An application that runs concurrently over a network, or one that runs with a permanent connection to an underlying data source, often lock such resources exclusively, on a first-come first-serve basis, and needs to run fast in order to avoid the impact of queues.

2.3.2.1 Selection and field types

The record object in memory, ⎕wself←'ADOR', behaves just as a record set based on a data source. Thus, particular records are conditionally selectable and the inherent types of fields are preserved; numeric operations are possible on numeric fields. For example:

```
⎕wi 'xFilter' "Name='Sobha, F'"        ⎕wi 'xRecordCount'
                                        1
1.05×⎕wi "xFields('Salary').Value"
23814
```

The 'Filter' property of the record object enables the selection of particular records conditionally—it is necessary to verify that the condition yields one or more records before interrogating the columns in them. Predictably, the 'Salary' field is numeric, confirmed by the numeric operation—even though a character value was assigned. The 'Filter' property hides the records that do not meet the condition specified. The complete record set can be re-exposed as follows:

```
⎕wi 'xFilter' 0                         ⎕wi 'xRecordCount'
                                        5
```

The next example illustrates the selection of records where salary is less than 23,000:

```
⎕wi 'xFilter' "Salary < 23000"          ⎕wi 'XGetRows'
                                        Laventure, K Female 16/05/1982 22765
```

```
                                  │ Flacq, A      Male    07/12/1984 19000
                                  │ Sobha, F      Female  29/01/1961 22680
  ⎕wi 'XMoveFirst'                │ ρ¨⎕wi 'XGetRows'
                                  │   12   6   10
                                  │    8   6   10
                                  │    8   6   10
  The shape of the array Name, used │ ρName
to create the record object was:  │ 5 12
```

However, the shape of the column 'Name' in the record object is variable and fixed for column 'Sex', see the function FabricateRS, although both variables had unnecessary blanks removed. The differences arise because of the ADO type—Name is type 200 (character, width variable, up to 100 characters), whereas Sex is type 129 (character, fixed length 6 characters).

The method 'XMoveFirst' moves the pointer to the first record in the object; the 'XMoveNext' method exposes each record in turn. A typical APL function for processing all records sequentially might look like:

```
       ∇ RecordLoop R;⎕wself
  [1]     ⍝ System Building with APL+Win
  [2]     ⎕wself←R ⍝ R= record set object
  [3]     ⎕wi 'XMoveFirst'
{1[4]     :while ~⎕wi 'xEOF' ⍝ While there are records present
  [5]         ⍝ .... application processing
  [6]         ⎕wi 'XMoveNext' ⍝ point to next record
1}[7]     :endwhile
       ∇
```

The method 'XGetRows' has the following syntax:

```
    array = recordset.GetRows(Rows, Start, Fields)
```

By default, the 'Rows' parameter has value ¯1, which signifies all records from the current record. The 'Start' parameter has a default value of 0 (current record) but can be 1 (first record) or 2 (last record). The 'Fields' parameter is a scalar ordinal number or field name or an array of the ordinal positions or names of the fields to retrieve. For APL applications, the most useful syntax for this method is:

```
  [5]        Array←'ADOR' ⎕wi 'XGetRows' 1
```

This retrieves the current record into a nested variable 'Array'—this is a collection of the field values. The record object contains the names of the field names. Therefore, the function 'RecordLoop' can create APL variables from the elements of a collection, as follows:

```
       ∇ RecordLoop2 R;⎕wself
  [1]     ⍝ System Building with APL+Win
  [2]     ⎕wself←R ⍝ R= record set object
  [3]     ⎕wi 'XMoveFirst'
  [4]     R←((⊂⎕wself) ⎕wi ¨(⊂⊂'xFields.Item'),¨¯1++\(⎕wi
            'xFields.Count')/1) ⎕wi ¨⊂'xName'
  [5]     R←(⊂'ws'),¨'R~¨' '
{1[6]     :while ~⎕wi 'xEOF' ⍝ While there are records present
  [7]         'Assign' Collection R (,⎕wi 'XGetRows' 1 )
  [8]         ⍝ call application to process current record
  [9]         ÷0
  [10]        ⎕wi 'XMoveNext' ⍝ point to next record
1}[11]    :endwhile
       ∇
```

Lines [4] and [5] extract the names of the fields in the record object; the prefix 'ws' is added to all variables names, from which all spaces have been removed in an attempt to ensure that the resulting strings are valid APL variable names. Line [9] contains a deliberate error to stop execution, so that the values can be examined:

```
RecordLoop2 'ADOR'                    │           wsName
DOMAIN ERROR                          │        Pailles, M
RecordLoop2[9]      ÷0                 │           0.5+wsSalary
          ∧                           │        24000.5
```

Line [7] makes the contents of the current record available to APL in named workspace variables—the type of the fields is preserved. This illustrates how APL can process records from a record object, usually created from an external data source.

2.3.2.2 Sharing the data

The XML file contains the contents of the record object created in the workspace using APL workspace variables. An XML file is a text file; however, its content is 'readable' when viewed in Internet Explorer. An XML document contains a tree structure written in an arbitrary format. The file created by ADO has two sections:

First, the *<s:Schema id="RowSchema">* section describes the variables and their respective data types. Each node is indicated with a – (expanded view) or + (collapsed view). In **Figure 2-2 XML view of an ADO record object** *'Dobirth'* is a collapsed node; a click on the + will expand the node. Note that 'Name', declared as *'adVarchar'*, is not a fixed length field whereas 'Sex', declared as *'adChar'*, is. *'Salary'* is a numeric field of type *'float'*.

Figure 2-2 XML view of an ADO record object shows an example of a record set persisted in XML format.

```
- <xml xmlns:s="uuid:BDC6E3F0-6DA3-11d1-A2A3-00AA00C14882" xmlns:dt="uuid:C2F41010-65B3-11d1-
    00AA00C14882" xmlns:rs="urn:schemas-microsoft-com:rowset" xmlns:z="#RowsetSchema">
  - <s:Schema id="RowsetSchema">
    - <s:ElementType name="row" content="eltOnly" rs:updatable="true">
      - <s:AttributeType name="Name" rs:number="1" rs:write="true">
          <s:datatype dt:type="string" rs:dbtype="str" dt:maxLength="100" rs:precision="0"
            rs:maybenull="false" />
        </s:AttributeType>
      - <s:AttributeType name="Sex" rs:number="2" rs:write="true">
          <s:datatype dt:type="string" rs:dbtype="str" dt:maxLength="6" rs:precision="0" rs:fixedlength="true"
            rs:maybenull="false" />
        </s:AttributeType>
      - <s:AttributeType name="Dobirth" rs:number="3" rs:write="true">
          <s:datatype dt:type="string" rs:dbtype="str" dt:maxLength="10" rs:precision="0"
            rs:fixedlength="true" rs:maybenull="false" />
        </s:AttributeType>
      - <s:AttributeType name="Salary" rs:number="4" rs:write="true">
          <s:datatype dt:type="float" dt:maxLength="8" rs:precision="0" rs:fixedlength="true"
            rs:maybenull="false" />
        </s:AttributeType>
        <s:extends type="rs:rowbase" />
      </s:ElementType>
    </s:Schema>
  - <rs:data>
      <z:row Name="Pailles, M" Sex="Male" Dobirth="12/03/1986" Salary="24000" />
      <z:row Name="Laventure, K" Sex="Female" Dobirth="16/05/1982" Salary="22765" />
```

Figure 2-2 XML view of an ADO record object

Second, the *<rs:data>* section holds the column values.

Regretably, this XML schema is not the same as the one used by the APL Grid object. APL and other applications may access the contents of this file using the ADO record object. APL can read any persisted XML file—created by any application, including APL—with the following code:

```
      ∇ XML;⎕wself
[1]      ⍝ System Building with APL+Win
[2]      ⎕wself←'ADOR' ⎕wi 'Create' 'ADODB.Recordset'
[3]      ⎕wi 'XOpen' "c:\PersistRS.xml" 'Provider=MSPERSIST'
      ∇
```

The object ADOR contains all the records—verified by the record count and by displaying the contents of the record object. The XGetString method returns dates as strings:

```
⎕wi 'xRecordCount'          ⎕wi 'XGetString'
5                           Pailles, M     Male    12/03/1986   24000
                            Laventure, K   Female  16/05/1982   22765
                            Moka, V        Female  03/09/1972   38400
                            Flacq, A       Male    07/12/1984   19000
                            Sobha, F       Female  29/01/1961   22680
```

The contents of the record object—whether in memory or populated from the XML file or other data sources—can be sent directly to **Excel**, as shown in **Figure 2-3 Workspace to Excel**.

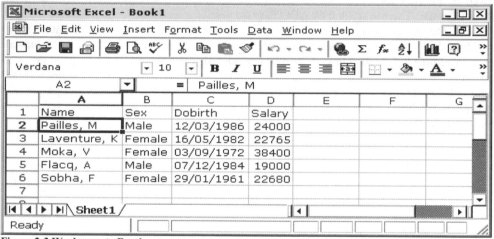

Figure 2-3 Workspace to Excel

The following line of code sends the fabricated record set to **Excel**:

```
      ∇ SendToExcel R;⎕wself
[1]      ⍝ System Building with APL+Win
[2]      ⍝ R=Existing record object
[3]      ⎕wself←'xl' ⎕wi 'Create' 'Excel.Application'
[4]      PageCount←⎕wi 'xSheetsInNewWorkBook' ⍝ User's default
[5]      ⎕wi 'SheetsInNewWorkBook' 1
[6]      ⎕wi 'XWorkBooks.Add'
[7]      ⎕wi 'xSheetsInNewWorkBook' PageCount ⍝ Restore user's default
[8]      Fields←((⊂R) ⎕wi ¨(⊂⊂'xFields.Item'),¯¯1++\(R ⎕wi
            'xFields.Count')/1) ⎕wi¨⊂'xName'
[9]      (⎕wi 'Sheets.Item' ('#' ⎕wi 'VT' 1 8204)) ⎕wi 'xSelect'
[10]     Range←∊'Range("' '")',¨(⎕wi 'Application.ConvertFormula'
            ("R1C1:R1C",⍕⍴Fields) (⎕wi '=xlR1C1') 1) ''
[11]     (⎕wi Range) ⎕wi 'Select'
[12]     ⎕wi 'Selection.FormulaR1C1' Fields
[13]     ⎕wi 'Sheets(1).Select'
[14]     (⎕wi 'Sheets.Item' i) ⎕wi 'Range("A2").CopyFromRecordSet' (R ⎕wi
            'obj') ((¯1+⎕wi 'Rows.Count')⌊R ⎕wi 'xRecordCount') ((⎕wi
            'Columns.Count')⌊θρρFields)
```

```
[15]   (⎕wi 'Sheets.Item' i) ⎕wi 'Columns.autofit'
[16]   (⎕wi 'Sheets.Item' 1) ⎕wi 'Range("A2").Select'
[17]   ⎕wi 'visible' 1
     ∇
```

2.3.2.3 Using XML

The XML format is an ideal method for data acquisition and transfer from APL. The XML file format is text; that means that it is editable (corruptible!) using a text editor and easily transmitted via electronic mail. XML data sources created from fabricated record objects provide APL with a powerful relational data management capability using workspace variables. However, this is not an alternative to using a database for holding the data component of an APL application. The size of an XML data source that APL can handle depends on Windows resources—not the size of the workspace. It is helpful to remove superfluous spaces from character variables using the 'Strip' function. The code is:

```
        ∇ Z←L Strip R
   [1]    ⍝ System Building with APL+Win
   [2]    ⍝ Return R stripped of duplicate scalar (⎕io+1)⊃L, preserving rank
          of R
   [3]    ⍝ L[⎕io+1] defaults to ' '(R=character) or 0(R=numeric)}
   [4]    (Z R)←R ⊖
{1[5]    :if 1≥⍴⍴Z
   [6]        R←↑⍴⍵,[⊖]Z
+ [7]    :else
   [8]        R←↑⍴Z
1}[9]    :endif
{1[10]   :if 1=≡L
   [11]       L←L ( 1↑(1,0=1↑0⍴Z)/' ' 0)
1}[12]   :endif
   [13]   L[⎕io+1]←⊂↑(⎕io+1)⊃L ⍝ force to single character|scalar
{1[14]   :select ⎕io⊃L
★ [15]       :case 'DeleteLB' ⍝ Leading
   [16]           R←(∧\Z=(⎕io+1)⊃L)↓[⎕io]2×⍳R
★ [17]       :case 'DeleteTB' ⍝ Trailing
   [18]           R←(⌽∧\⌽Z=(⎕io+1)⊃L)↓[⎕io]2×⍳R
★ [19]       :case 'DeleteIB' ⍝ Imbedded
   [20]           R←(Z=(⎕io+1)⊃L)↓[⎕io]2×⍳R
★ [21]       :case 'DeleteDB' ⍝ Duplicate
   [22]           R←((2/(⎕io+1)⊃L)∊Z)↓[⎕io]2×⍳R
★ [23]       :case 'DeleteUB' ⍝ Unnecessary i.e. LB, TB, DB
   [24]           R←((∧\Z=(⎕io+1)⊃L)∨(2/(⎕io+1)⊃L)∊Z)↓[⎕io]2×⍳R
★ [25]       :case 'DeleteEB' ⍝ Extraneous i.e. LB, TB
   [26]           R←(∧\Z=(⎕io+1)⊃L)↓[⎕io]2×⍳R
   [27]           R←R+R≠(⌽∧\⌽Z=(⎕io+1)⊃L)↓[⎕io]2×⍳↑⍴R
1}[28]   :endselect
   [29]   (,Z)←(,Z)[⍋,R]
{1[30]   :if 0=1↑0⍴(⎕io+1)⊃L
   [31]       Z←Z×Z≠(⎕io+1)⊃L
1}[32]   :endif
{1[33]   :if 0∊⍴Z
+ [34]   :orif 0=⍴(,Z)~(⎕io+1)⊃L
   [35]       Z←((⍴Z)×(-⍴⍴Z)↑((¯1+⍴⍴R)⍴1),0)⍴Z
+ [36]   :else
   [37]       R←1-L/,(Z=(⎕io+1)⊃L)⊥1
```

```
{2[38]      :if R≠0
  [39]              Z←R↓[(ρρZ)-~⎕io]Z
2}[40]      :endif
1}[41]  :endif
        ∇
```

This function preserves the rank of its right-hand argument; it must be a simple vector or array and either numeric or character. The left-hand argument may be either a keyword—see lines [15], [17], [19], [21], [23] and [25]—or a keyword followed by the character or scalar whose duplicates is to be removed. Some examples of its syntax are:

```
      'DeleteUB' Strip ' This sentence     needs to be tidied up.'
This sentence needs to be tidied up.
      'DeleteUB' 'e' Strip ' This sentence     needs to be tidied up.'
 This sentence     neds to be tidied up.
      'DeleteUB' Strip 1 0 0 1 1 2 3 4 0 3 4 0 0 0 1
1 0 1 1 2 3 4 0 3 4 0 1
      'DeleteUB' 1 Strip 1 0 0 1 1 2 3 4 0 3 4 0 0 0 1
0 0 0 2 3 4 0 3 4 0 0 0 0 0 0
```

2.3.2.3.1 XML with APL nested arrays

So far, fabricated record objects and the XML files use only simple APL variables. In practice, this is highly restrictive, as APL applications generally have nested variables. Purely for use within the APL environment, XML files can contain nested APL variables.

The 'XGetRows' method of the ADOR record object produces a nested array:

```
      ⎕←NestedVariable←'ADOR' ⎕wi 'XGetRows'          ρNestedVariable
Pailles, M    Male    12/03/1986 24000       5 4
Laventure, K Female 16/05/1982 22765              ≡NestedVariable
Moka, V        Female 03/09/1972 38400       2
Flacq, A      Male    07/12/1984 19000
Sobha, F      Female 29/01/1961 22680
```

The function NestXML can write the nested variable to an XML file:

```
      ∇ NestXML;⎕wself
 [1]   ⍝ System Building with APL+Win
 [2]   ⎕wself←'ADOR' ⎕wi 'Create' 'ADODB.Recordset'
 [3]   ⎕wi 'xFields.Append' 'Nest' (⎕wi '=adVarChar') 64000
 [4]   ⎕wi 'XOpen'
 [5]   ⎕wi 'XAddNew'
 [6]   ⎕wi 'xFields.Item(0).Value' (⍕⎕avι'wrapl' ⎕dr NestedVariable)
 [7]   ⎕wi 'XUpdate'
{1[8]   :if 0≠⎕wcall 'PathFileExists' 'c:\nestedrs.xml'
 [9]       0 0ρ⎕wcall 'DeleteFile' 'c:\nestedrs.xml'
1}[10]  :endif
 [11]  ⎕wi 'XSave' 'C:\nestedrs.XML' (⎕wi '=adPersistXML')
        ∇
```

Line [6] of this function is critical; it transforms the nested variable into a character vector, thus:

```
      'wrapl' ⎕dr 'NestedVariable'
§§§§(∊Žρ§⊃⊃NestedVariable⎕⍫(
```

This contains control characters that will corrupt the XML file—the XML format permits such characters only in CDATA sections. In order to circumvent this problem, compute the cardinal sequence of each character, producing a numeric vector. Since an XML file can contain character data only, format the numeric vector as a string and write to the XML file. This transformation has serious weaknesses: the size of the string vector must be known in advance, in this case, it is pre-set to 64,000 characters, and it is dependent on the index origin. This jeopardises the usefulness of the technique.

Reverse the process on acquiring the XML data source in order to reconstitute the original nested variable. The following function recreates a record object from the XML file:

```
     ∇ XML2;⎕wself
[1]   ⍝ System Building with APL+Win
[2]   ⎕wself←'ADOR' ⎕wi 'Create' 'ADODB.Recordset'
[3]   ⎕wi 'XOpen' "c:\nestedRS.xml" 'Provider=MSPERSIST'
     ∇

     'unwrapl' ⎕dr ⎕av[⎕fi 'ADOR' ⎕wi 'xFields(0).Value']
 Pailles, M    Male    12/03/1986 24000
 Laventure, K Female 16/05/1982 22765
 Moka, V       Female 03/09/1972 38400
 Flacq, A      Male    07/12/1984 19000
 Sobha, F      Female 29/01/1961 22680

     NestedVariable≡'unwrapl' ⎕dr ⎕av[⎕fi 'ADOR' ⎕wi 'xFields(0).Value']
1
```

The variable is recreated successfully—this requires that index origin remains consistent. At best, this technique provides an interim means of coping with nested variables; use this technique with care, as it can be tenuous. It is preferable to use data types that are directly transferable between APL and the data source.

2.4 Generating test data

The developer should test the application code. This is the process of 'unit testing', which involves the creation of sample data against which new code or changes to existing code is tested in order to weed out logic flaws and runtime errors. A good strategy is to generate the sample data with some degree of randomness. However, the process of generating character and date data items is usually quite contrived. The following function can generate pseudo random data of the types usually encountered in APL applications.

2.4.1 Dates

Most applications handle date data of one sort or another: typically, such items are dates of birth, delivery dates, maturity dates, etc. The variety of date formats and the vagaries of the number of days in the month make it quite difficult to generate dates. The APL+Win grid object makes it quite easy to generate dates, within the range 1900-9999, in the short date format set in regional settings:

```
       ∇ Z←L GenerateData R;⎕wself
  [1]   ⍝ System Building with APL+Win
  [2]   Z←θ
{1[3]   :select L
★ [4]       :case 'Dates'
  [5]           (1↓R)←1900⌈(1↓R)⌊9999
  [6]           (1↓R)←(1↓R)[⍒(1↓R)]
  [7]           ⎕wself←''APLGrid' ⎕wi 'Create' 'APL.Grid'
  [8]           ⎕wi 'Set' ('Rows' 1) ('Cols' (θρR))
  [9]           ⎕wi 'xText' (1) (⍳2) (1 2ρ(⍕¨'01/01/' '31/12/',¨1↓R)~¨' ')
  [10]          (1↓R)←,⎕wi 'xDate' (1) (⍳2)
{2[11]          :while (ρZ)<1 0 0/R
  [12]              Z←Z,((1 0 0/R)⌊0 0 1/R)?0 0 1/R
2}[13]          :endwhile
  [14]          Z←(1 0 0/R)↑Z
{2[15]          :while 1∈L←Z<0 1 0/R
```

```
  [16]                    (L/Z)←(+/L)?0⊥R
2}[17]                  :endwhile
  [18]                  ⎕wi 'xDate' (1) (⍳⊖⍴R) Z
  [19]                  Z←⊃,⎕wi 'xText' (1) (⍳⊖⍴R)
  [20]                  ''APLGrid' ⎕wi 'Delete'
```

The APL+Win Grid object holds dates as a scalar, 1 being 31/12/1899. This call generates 10 random, and valid, dates between 01/01/1965 and 31/12/2004:

```
     DateOfBirth←'Dates' GenerateData 10 1965 2004
```

Line [10] populates the grid object with the two dates representing the specified range. Line [11] reads the scalar values of the same dates. This provides the range within which the requisite number of dates must reside—if necessary, with replication. Line [19] writes these scalars to the grid object and line [20] reads them back as dates. The result is a two-dimensional character matrix of dates in short date format as set in regional settings. The grid object automatically initialises with the correct locale setting.

2.4.2 Integers

The APL roll primitive function (?) generates positive integers up to and including a maximum number quite easily. Generates 10 integers within the range 20-10000 and return the result as a two-dimensional numeric array, thus:

```
     Salary←'Integers' GenerateData 10 20 10000
```

This function provides the additional facility of restricting the integers to a lower limit. Both limits must be positive and non-zero—the roll primitive does not work with numbers less than 1. Should negative numbers be required, simple code can transform some rows into negative numbers:

```
        R←2?1↑⍴Salary ◇ Salary[R;]←¯1×Salary[R;]
★ [21]           :case 'Integers'
  [22]              (1↓R)←(1↓R)[⍋(1↓R)]
{2[23]             :while (⍴Z)<1 0 0/R
  [24]                 Z←Z,((1 0 0/R)⌊0 0 1/R)?0 0 1/R
2}[25]             :endwhile
  [26]              Z←(1 0 0/R)↑Z
{2[27]             :while 1∊L←Z<0 1 0/R
  [28]                 (L/Z)←(+/L)?0⊥R
2}[29]             :endwhile
  [30]              Z←,[⊖]Z
```

2.4.3 Floating point

The observations with respect to integers apply. Consider this example:

```
     ContRate←'Floating' GenerateData 10 2 5
```

- An array of 10 floating-point numbers in the range 2 to 5 is calculated.

- The range is restricted to positive numbers only.

- Numbers are restricted to two places of decimals.

Should negative numbers be required, apply a technique similar to the one illustrated with integers. The code is:

```
★ [31]           :case 'Floating'
  [32]              (1↓R)←(1↓R)[⍋(1↓R)]
{2[33]             :while (⍴Z)<1 0 0/R
  [34]                 Z←Z,0.01×((1 0 0/R)⌊0 0 1/R)?100×0 0 1/R
2}[35]             :endwhile
  [36]              Z←(1 0 0/R)↑Z
{2[37]             :while 1∊L←Z<0 1 0/R
```

```
   [38]                (L/Z)←0.01×(+/L)?100×0⊥R
2}[39]                 :endwhile
   [40]                Z←,[θ]Z
```

2.4.4 Codes

This type of data would normally be associated with a combo field in the application. It holds one value from a pre-determined list of values. Consider this example:

```
        Status←'Codes' GenerateData 10 'Act' 'Def' 'Pen'
```

The right-hand argument, after the first element, specifies the list of values to select from; this list must contain at least two values. These values would appear in the combo box that elicits the value for the variable 'Status'. The code is:

```
★ [41]              :case 'Codes'
  [42]              Z←(|ρ1↑R)|10⊤(θρR)?⎕rl
{2[43]              :while 1∈L←0=Z
  [44]                  ((<\L)/Z)←?ρ1↓R
2}[45]              :endwhile
  [46]              Z←(1↓R)[Z]
{2[47]              :if 2==Z
  [48]                  Z←⊃Z
+ [49]              :else
  [50]                  Z←,[θ]Z
2}[51]              :endif
```

2.4.5 String

This type of data comprises of an arbitrary string, like name. Consider this example:

```
        Name←'String' GenerateData 10 'Ajay' 'Askoolum'
```

The right-hand argument, after the first element, specifies two strings. Strings are generated from the two strings in the pattern x, y where the length of x (treat it like a surname, with the initial letter in capital) is between the length of the two arguments and y (treat it like initials, all capitals) is between 1 to 3 characters.

The strings are unique (almost) to enable concise reference. The code is:

```
★ [52]              :case 'String'
  [53]              L←⎕wcall 'Charlower' (∈1↓R)
  [54]              (1↓R)←∈ρ¨1↓R
  [55]              (1↓R)←(1↓R)[▲(1↓R)]
  [56]              Z←,⊂L
{2[57]              :while (ρZ)<1 0 0/R
{3[58]                  :if (0 1 0/R)≤ρL←(?0 0 1/R)?0 0 1/R
  [59]                      L←L-~⎕io
  [60]                      Z←Z,⊂(⎕io⊃Z)[L],', ',⎕wcall 'CharUpper'
       ((⎕io⊃Z)[(3⌊1?0 1 0/R)?0 0 1/R])
3}[61]                  :endif
2}[62]              :endwhile
  [63]              Z←⊃1↓Z
  [64]              Z[;⎕io]←⎕wcall 'CharUpper' (Z[;⎕io])
```

2.4.6 Binary string

This type of data is like the codes type but the value can be one of only two options: like gender, which can be male or female. Consider this example:

```
        Sex←'BinaryString' GenerateData 10 'M' 'F'
```

The right-hand argument, after the first element, specifies the two options. The application interface may present the specification of this type of data as a combo box, a frame with two check boxes, or a frame with two radio buttons. The code is:

```
★ [65]        :case 'BinaryString'
  [66]            Z←,[⍬](1↓R)[⎕io+2|(⍵⍴R)?⎕rl]
```

2.4.7 Numeric vector

This type of data is numeric. Consider this example:

```
      BonusRate←'NumericVector' GenerateData 10 3 4 ¯7 0 9
```

● The right-hand argument, after the first element, specifies the maximum value of each element.

● The result comprises of an array of vectors of the same length.

● The result is adjusted to reflect the sign and magnitude of the vector specified as the right-hand argument. The code is:

```
★ [67]        :case 'NumericVector'
  [68]            Z←⊂Z
{2[69]            :while 0≠⍵⍴R
  [70]                Z←Z,⊂(?1⌈|1↓R)××1↓R
  [71]                (⍵⍴R)←(⍵⍴R)-1
2}[72]            :endwhile
  [73]            Z←⊃1↓Z
1}[74]    :endselect
      ∇
```

2.4.8 Computations on generated data

Having generated data randomly, calculations on the data may be necessary. For instance, a variable containing state pension age may be necessary: this is *DateOfBirth* plus 65 and 60 years for males and females respectively. Whilst numeric computations are straightforward with APL, date arithmetic requires custom functions. The Microsoft Script control has a ready-made library of date arithmetic functions. Why not simply use this library?

```
      ∇ Z←L Script R;i;⎕wself
  [1]    ⍝ System Building with APL+Win
  [2]    ⎕wself←'∆SC' ⎕wi 'Create' 'MSScriptControl.ScriptControl'
  [3]    ⎕wi 'xLanguage' ''
  [4]    ⎕wi 'xLanguage' 'VBScript'
  [5]    i←'FormatDateTime(DateAdd("yyyy",★,"&"))'
  [6]    Z←⊂''
{1[7]    :repeat
  [8]        Z←Z,⊂⎕wi 'XEval'  (⍕∊((~i∊'★&')⊂i),¨(L[⎕io;]) (R[⎕io;]) '')
  [9]        L←1 0↓L
1}[10]   :until 0∊⍴R←1 0↓R
  [11]   Z←⊃1↓Z
      ∇
      SPA←(60+5×Sex='M') Script DateOfBirth
```

Unfortunately, VBScript does not process arrays except via a loop. Two VBScript functions are used, FormatDateTime—returns a scalar date in regional date format—and DateAdd—adds an arbitrary number of days, months, or years to a date valid in the regional date format—to return the result as a scalar.

A fabricated record object may hold the sample data. In order to accomplish this, two APL problems must be resolved, namely:

● The record set object does not recognise the APL high minus; the solution is to translate it into a normal minus sign.

• In addition, the record object does not have a data type corresponding to the APL numeric vector; the solution is to hold numeric vectors as character.

On reading the data back from the record object, each solution requires reversal in order to restore the original data. The following function creates the record object:

```
      ∇ SampleRS;⎕wself;i
 [1]   ⍝ System Building with APL+Win
 [2]   ⎕wself←'ADOR' ⎕wi 'Create' 'ADODB.Recordset'
 [3]   ⎕wi 'xFields.Append' 'Name' 129 50
 [4]   ⎕wi 'xFields.Append' 'Sex' 129 1
 [5]   ⎕wi 'xFields.Append' 'DateOfBirth' 129 10 ⍝ UK Short date format
 [6]   ⎕wi 'xFields.Append' 'Salary' 129 10
 [7]   ⎕wi 'xFields.Append' 'Status' 129 15
 [8]   ⎕wi 'xFields.Append' 'ContRate' 5
 [9]   ⎕wi 'xFields.Append' 'BonusRate' 129 30
 [10]  ⎕wi 'xFields.Append' 'SPA' 129 10 ⍝ UK Short data format
 [11]  ⎕wi 'XOpen'
{1[12]  :for i :in ⍳1↑⍴Name
 [13]      ⎕wi 'XAddNew'
 [14]      ⎕wi 'xFields("Name").Value' ('DeleteUB' Strip Name[i;])
 [15]      ⎕wi 'xFields("Sex").Value' ('DeleteUB' Strip Sex[i;])
 [16]      ⎕wi 'xFields("DateOfBirth").Value' (DateOfBirth[i;])
 [17]      ⎕wi 'xFields("Salary").Value' (ConformNumeric Salary[i;])
 [18]      ⎕wi 'xFields("Status").Value' (⍕Status[i;])
 [19]      ⎕wi 'xFields("ContRate").Value' (ConformNumeric ContRate[i;])
 [20]      ⎕wi 'xFields("BonusRate").Value' (ConformNumeric
          BonusRate[i;])
 [21]      ⎕wi 'xFields("SPA").Value' (SPA[i;])
 [22]      ⎕wi 'XUpdateBatch'
1}[23]  :endfor
{1[24]  :if 0≠⎕wcall 'PathFileExists' 'c:\Sample.xml'
 [25]      0 0⍴⎕wcall 'DeleteFile' 'c:\Sample.xml'
1}[26]  :endif
 [27]  ⎕wi 'XSave' 'C:\Sample.XML' 1 ⍝ 1=adPersistXML
      ∇
```

Usually, APL handles the conversion of negative numbers, indicated by the minus sign, in record objects transparently. Other data sources may show such quantities enclosed within round brackets: for these, explicit conversion is required at source. This function delivers the minus sign conversion:

```
      ∇ Z←ConformNumeric R
 [1]   ⍝ System Building with APL+Win
 [2]   Z←⍕R
 [3]   (('‾'=Z)/Z)←'-'
      ∇
```

Swapping the minus signs on either side of the assignment in line [3] reverses this process. **Figure 2-4 Generated data** shows the generated data.

	Name	Sex	DateOfBirth	Salary	Status	ContRate	BonusRate	SPA
1	Aaykaj, AYK	F	05/12/1969	5785	Pen	2.55	3 3 -5 0 9	05/12/2029
2	Aaas, OAY	M	30/11/1972	-5276	Act	4.98	3 2 -4 0 1	30/11/2037
3	Saao, AAJ	F	29/11/1973	7101	Def	3.66	2 3 -7 0 7	29/11/2033
4	Joky, AAK	F	29/03/1982	8948	Pen	4.15	3 3 -6 0 7	29/03/2042
5	Jaysaa, SAA	F	03/11/1966	7834	Act	2.86	1 4 -4 0 1	03/11/2026
6	Ayksaajo, AOS	M	27/09/1978	5403	Act	4.07	3 2 -5 0 2	27/09/2043
7	Akjoasy, AJ	M	21/07/1965	7363	Act	3.54	3 4 -7 0 2	21/07/2030
8	Jaky, Y	M	29/02/1980	-6171	Pen	4.32	2 4 -2 0 3	28/02/2045
9	Oyaj, K	M	31/01/1977	9188	Pen	4.67	2 4 -4 0 8	31/01/2042
10	Yaaasko, SJ	F	07/09/1981	7556	Def	3.25	2 1 -7 0 6	07/09/2041

Figure 2-4 Generated data

Note that the date computations yield valid dates: refer to row 8 in **Figure 2-4 Generated data**.

2.5 APL+Win as an ActiveX Server

APL+Win can act as an ActiveX Server to both APL+Win and other COM compliant applications such as **VB**, **Excel**, and **Word**, etc. The APL+Win manuals provide details of **Registry** entries required to create the APL+Win ActiveX Server.

2.5.1 Dynamic properties and methods

There is a critical difference in the behaviour of the APL+Win server when serving an APL+Win or other client—the APL character set is useable by an APL+Win client and not useable by other clients. This is because APL does not automatically translate its character set from ANSI to ⎕AV, or, vice-versa, when in automation server mode. Consequently, the APL+Win server is unable to expose its dynamic properties and methods to non-APL clients automatically. Dynamic properties and methods are the variables and functions in the active workspace loaded by the server.

2.5.1.1 Enumerating with APL+Win client

Consider the following instance of the server with an APL+Win client that loads the supplied WINDOWS.W3 workspace:

```
     ∇ APLServer;⎕wself
[1]    ⍝ System Building with APL+Win
[2]    ⎕wself←'aplsvr' ⎕wi 'Create' 'APLW.WSEngine'
[3]    ⎕wi 'XSysCommand' 'load D:\PROGRAM FILES\APLWIN\TOOLS\WINDOWS'
     ∇
```

The dynamic properties are:

```
'aplsvr' ⎕wi 'XExec' '⎕nl 2'
```

The dynamic methods are:

```
'aplsvr' ⎕wi 'XExec' '⎕nl 3'
```

Short help—in this case, the function header—on the dynamic methods, or functions, is retrievable if the function is unlocked:

```
'aplsvr' ⎕wi 'XExec' '(⎕cr "I2C")[⎕io;]'
c ← I2C n;⎕io
```

Although this is cryptic, the documentation of the workspace should provide details of the syntax or the actual code will provide the necessary clues, thus:

```
⊃'aplsvr' ⎕wi 'XExec' '⎕vr "I2C"'
    ∇ c ← I2C n;⎕io
[1]   A∇char←I2C int -- Convert integers to character
[2]   AA c ← 82 ⎕dr n A Fails on booleans, etc.
[3]   ⎕IO←0 ◇ c←⎕AV[,⍉⍟256 256 256 256⊤,n]
    ∇
```

2.5.1.2 Enumerating with non-APL client

For instance, **Excel** can create an instance of the APL+Win server and load the supplied WINDOWS.W3 workspace using VBA:

```
Sub APLServer()
    Set aplsvr = CreateObject("APLW.WSEngine")
    aplsvr.syscommand ("load d:\program files\aplwin\tools\windows")
End Sub
```

A function, whose name comprises of the letters a-z, A-Z, and digits 0-9 only, is necessary in the active workspace. This function will return all the variables or properties:

```
    ∇ Z←DynProperties
[1]   A System Building with APL+Win
[2]   Z←⎕nl 2
    ∇
```

Excel can acquire the list of properties in a variant array—not a collection— in option base 0, *irrespective* of the option base setting via the DynProperties function:

```
wsprop=aplsvr.Exec("DynProperties")
?wsprop(0)
ClipFmt'DOC
```

However, the character set problem does not go away: ClipFmt'DOC is in fact ClipFmt△DOC. A similar technique will return the list of dynamic methods. Thus, these conclusions ensue:

• The potential of the EXEC method of the APL+Win server is severely restricted with a non-APL+Win client, as the client cannot pass APL characters to the server.

• Cover functions are required to wrap expressions that a non-APL client needs to evaluate in order to remove the need for entering APL primitive functions.

• A workspace intended for use by a non-APL+Win client must not use the APL character set in the names and content of variables and in the names and results of functions.

2.5.2 Which class of APL+Win?

The GetModuleFileName API call identifies the name of the EXE file—this will return the full path of the executable file that is running. This API identifies the executable file of the environment from which it is called:

```
GetModuleFileName=I(H hModule, >C lpFilename, D nSize) ALIAS GetModuleFileNameA LIB
    KERNEL32.DLL
```

The APL+Win syntax is:

```
    ↑↑/⎕wcall 'GetModuleFileName' 0 (256ρ⎕tcnul) 256
D:\PROGRAM FILES\APLWIN\APLW.EXE
```

However, this will only indicate whether the current session is a development or runtime session. An existing APL+Win session may be one of four types—development system, development system as an Active X server, runtime system and runtime system as an ActiveX server.

The recommendation is that the development system be used during development, with the session visible—optionally, it can be hidden—as an aid to debugging, and the runtime system be used during testing and for deployment. In either mode, the runtime system stays hidden. An APL function may return its identity to its client:

```
       ∇ Z←WhichAPL
  [1]    ⍝ System Building with APL+Win
  [2]    ⍝ Which version of APL is running?
  [3]    Z←⎕sysid,' Version ',((' '= ⎕sysver)⍳1)↑⎕sysver
{1[4]    :select ⎕sys[⎕io+20]
★ [5]        :case 0
  [6]            Z←Z,'Development System '
★ [7]        :case 1
  [8]            Z←Z,'Runtime System '
1}[9]    :endselect
{1[10]   :if 0≠'#' ⎕wi 'server'
  [11]       Z←Z,"As ActiveX Server"
1}[12]   :endif
       ∇
```

The system variable ⎕sys holds either a 0 or 1 at the twenty-first position indicating a development or runtime version, respectively. The value of the 'server' property of the system class ('#' ⎕wi 'server') is non-zero for an ActiveX server session.

In the automated mode of this model, there would be two sessions running—the **Excel** APL+Win ActiveX server and a client APL+Win session automating the **Excel** processes. It would be necessary to be able to establish whether the required session is still running or has stalled and requires termination or re-starting.

2.5.2.1 The Command Line

The target attribute of the menu item or shortcut that starts APL+Win specifies the location of the EXE as the first element followed by several other positional configuration parameters, such as initial workspace size, user number etc. A useful technique is to pass an additional custom parameter that the initial workspace can use for any purpose such overriding built-in application parameters by those passed-in. Obviously, the APL keyboard is unavailable when editing the target attribute: a practical approach is as follows:

"CmdLine;Startup=0;Rate=10.5;id='sbwa';ll=;abc=0"

The whole string is enclosed in double quotes, a semicolon separates lines, the first word is always CmdLine, single quotes enclose literals, and the 'equal to' sign indicates assignment. The following function reads the command line, assigns it to a user defined property of the system object, namely, ∆CmdLine and initialises all the variables as global.

```
       ∇ Z←GetCommandLine;CmdLine;⎕elx
  [1]    ⍝ System Building with APL+Win
  [2]    ⎕elx←'→ ⎕lc+1'
  [3]    Z←∈1↓⎕wcall 'lstrcpy' (1024⍴⎕tcnul) (⎕wcall 'GetCommandLine')
  [4]    '#' ⎕wi '∆CmdLine' (Z←Z~⎕tcnul)
{1[5]    :if 1∈'CmdLine'≤Z
  [6]        Z←1↓(∨\'"CmdLine'≤Z)/Z
  [7]        Z←(~Z∈'"')/Z
  [8]        ((Z='=')/Z)←'←'
  [9]        ⍎⎕def ⊃(';'≠Z)⊂Z
1}[10]   :endif
  [11]   Z←'#' ⎕wi '∆CmdLine'
       ∇
```

Line [3] uses a new API call:

GetCommandLine=P() ALIAS GetCommandLineA LIB KERNEL32.DLL

This function transforms the custom parameter string into a local function and executes it: as the localised error handler in line [2] directs execution to the next line, its scope extends to the local CmdLine function. Therefore, errors in the custom parameter string do not halt processing: in the example, the deliberate error in the penultimate assignment does not prevent the evaluation of the final expression. The function should execute at start-up in order to initialise the variables in the custom parameter; it returns the complete command line for further processing:

```
        GetCommandLine
"C:\Program Files\aplwin\aplw.exe"
"CmdLine;Startup=0;Rate=10.5;id='sbwa';ll=;abc=0"
```

Note that a session that starts by file association, that is, double clicking on a workspace name in the filing system, does not pass the parameters specified in the target attribute of the shortcut or menu item for APL+Win.

2.6 Debugging applications

The APL+Win function editor has a configuration option for automatic indentation of functions. Consider the extreme example shown in **Table 2-7 Indentation**:

`⎕vr 'SampleFunction'`	`'List' CSBlocks ⎕def 'Indent' CSBlocks 'SampleFunction'`
`    ∇ Z←L SampleFunction R`	`    ∇ Z←L SampleFunction R`
`[1]   ⍝ System Building with` `        APL+Win`	`  [1]   ⍝ System Building` `            with APL+Win`
`[2]   Z←''`	`  [2]   Z←''`
`[3]   :while 0≠⍴R`	`{1[3]   :while 0≠⍴R`
`[4]   :select L`	`{2[4]       :select L`
`[5]   :case 'HomeDelivery'`	`★ [5]           :case 'HomeDelivery'`
`[6]   :if R=0`	`{3[6]               :if R=0`
`[7]   Z←Z,L,' Nil'`	`  [7]                   Z←Z,L,' Nil'`
`[8]   :else`	`+ [8]               :else`
`[9]   Z←Z,L,(⍕R),'%'`	`  [9]                   Z←Z,L,(⍕R),'%'`
`[10]  :endif`	`3}[10]              :endif`
`[11]  :case 'Collection'`	`★ [11]          :case 'Collection'`
`[12]  :if R=0`	`{3[12]              :if R=0`
`[13]  Z←Z,L,' Nil'`	`  [13]                  Z←Z,L,' Nil'`
`[14]  :else ⍝ Bonus`	`+ [14]              :else ⍝ Bonus`
`[15]  :if R>8`	`{4[15]                  :if R>8`
`[16]  R←R+0.5`	`  [16]                      R←R+0.5`
`[17]  :endif`	`4}[17]                  :endif`
`[18]  Z←Z,L,(⍕R),'%'`	`  [18]                  Z←Z,L,(⍕R),'%'`
`[19]  :endif`	`3}[19]              :endif`
`[20]  :else`	`+ [20]          :else`
`[21]  0 0⍴Msg 'Error'`	`  [21]              0 0⍴Msg 'Error'`
`[22]  Z←¯1 ⍝ Abort`	`  [22]              Z←¯1 ⍝ Abort`
`[23]  :leave`	`> [23]              :leave`
`[24]  :endselect`	`2}[24]      :endselect`
`[25]  R←1↓R`	`  [25]          R←1↓R`
`[26]  :endwhile`	`1}[26]  :endwhile`
`[27]  :if Z≡¯1`	`{1[27]  :if Z≡¯1`
`[28]  ⎕sa←'OFF'`	`  [28]      ⎕sa←'OFF'`
`[29]  :endif`	`1}[29]  :endif`
`    ∇`	`    ∇`

Table 2-7 Indentation

The usefulness of the APL+Win auto-indentation facility is deficient:

- It simply positions the starting point of a subsequent line at the same column as the start of the previous line.

- It does not indent control structures automatically to show the level of nesting.

- It also lacks the facility for collapsing a function simply to control structure level, thereby allowing the structure of the function to be scrutinised.

Although these facilities might first appear to offer cosmetic benefits only, the lack of them immeasurably complicates the processes of debugging process and documentation especially for long functions. In the listing on in **Table 2-7 Indentation**, the left shows *SampleFunction* as it is, coded by a sloppy developer; APL+Win does not provide any facility from cleaning indentation within the editor. It is surprising how quickly this style of programming—one without proper indentation or one with deliberate or unconscious regard for indentation—becomes the default with functions that span more than one editor window.

The right-hand side in **Table 2-7 Indentation** shows the same functions:

'Strip' CSBlocks 'SampleFunction'	'StripX' CSBlocks 'SampleFunction'
∇ Z←L SampleFunction R	∇ Z←L SampleFunction R
[1] ⍝ System Building with APL+Win	[1] ⍝ System Building with APL+Win
{1[3] :while 0≠⍴R	{1[3] :while 0≠⍴R
{2[4] :select L	{2[4] :select L
⋆ [5] :case 'HomeDelivery'	⋆ [5] :case 'HomeDelivery'
{3[6] :if R=0	{3[6] :if R=0
+ [8] :else	+ [8] :else
3}[10] :endif	3}[10] :endif
⋆ [11] :case 'Collection'	⋆ [11] :case 'Collection'
{3[12] :if R=0	{3[12] :if R=0
+ [14] :else ⍝ Bonus	+ [14] :else ⍝ Bonus
{4[15] :if R>8	{4[15] :if R>8
4}[17] :endif	4}[17] :endif
3}[19] :endif	3}[19] :endif
+ [20] :else	+ [20] :else
> [23] :leave	[22] Z←⁻1 ⍝ Abort
2}[24] :endselect	> [23] :leave
1}[26] :endwhile	2}[24] :endselect
{1[27] :if Z≡⁻1	1}[26] :endwhile
1}[29] :endif	{1[27] :if Z≡⁻1
∇	1}[29] :endif
	∇

Table 2-8 Collapse function

Note the two vital differences:

- First, the start and end of each level of nesting is shown within each block: refer to the markings before the line numbers. Opening and closing curly brackets respectively indicate the start and end of a control structure block. The number following an opening curly bracket indicates the start of a depth of nesting and a number preceding a closing curly bracket indicates the end of that level of nesting. A * indicates a switch within a *:select* structure, a + indicates a condition within an *:if* structure and a > indicate a branch within a *:for, :while* or *:repeat* structure. A '?' on the header line indicates that the function has an *outer syntax* error.

- Second, the function has enhanced readability because of correct indentation.

It would be useful to create a representation of a function at control structure level and comments only, shown in **Table 2-8 Collapse function**.

For a new application, a plausible if not common approach is to structure a function using control structures adding a message at each depth—albeit on a temporary basis. Whether a function enables programme flow in accordance with specification is visually verifiable from the messages. The actual code can replace the message lines subsequently. For an existing function, debugging is more complex. In **Table 2-8 Collapse function**,

The version on the left shows the control structure with nesting markers and includes comment lines—after stripping out all APL code.	The version on the right is the same thing and all lines of APL code with a comment; refer to line [22].

Either version makes it easier to verify whether the function corresponds to the specification and to determine the re-structuring of the function in the event of a specification change. A facility for collapsing functions at control structures start/end points from within the APL+Win editor would enhance the debugging of APL code: this is not available.

2.6.1 The CSBlocks function

The function CSBlocks (control structure blocks) provides the facilities discussed; it is indented and listed using CSBlocks itself. CSBlocks will fail with locked functions as their vector representation is empty. It adds nesting information to the vector representation of a function, as shown:

```
        'List' CSBlocks 'CSBlocks'
        ∇ Z←L CSBlocks R
  [1]     ⍝ System Building with APL+Win
{1[2]     :select L
★ [3]         :case 'List' ⍝ Vector representation showing depth of flow
  [4]             Z←'IdCS' CSBlocks R
{2[5]           :if Z≡θ
  [6]               Z←'Function ',R,' is locked or does not exist.'
+ [7]           :else
  [8]               (Z L R)←Z
  [9]
          Z←(Z≠0)\((~L≠1+¯1ΦL)\'{'),¨(L+L≠1+¯1ΦL),¨(L≠1+¯1ΦL)\'}'
  [10]              (R L)←R
  [11]              Z←⍕¨(0≠L)\(∊(((0≠L)/L)ρ¨1),¨0)⊂(∊(((0≠L)/L)ρ¨1),¨0)\Z
  [12]              (Z R)←(Z,¨' '+★>'[⎕io+((⎕io+1)⊃R)]) (⎕io⊃R)
  [13]              Z←Z~¨' '
  [14]              ((∊'0'=1↑¨Z)/Z)←1↓¨(∊'0'=1↑¨Z)/Z
  [15]              Z←(1+ρZ)↑Z
  [16]              Z[⎕io]←c↑R
  [17]              R←⎕vr R~'?'
  [18]              (R Z)←θ (∊(((⌈/∊ρ¨Z)↑¨Z),¨(R≠⎕tcnl)⊂R),¨⎕tcnl)
2}[19]          :endif
```

The syntax of the code is *'List'* CSBlocks *'functionname'*. This returns a vector representation of the function; CSBlocks does not change the definition of the function. The ⎕vr of the function as it stands is used.

The following section reduces the function, as it stands, to control structure level:

```
★ [20]          :caselist 'Strip' 'StripX' ⍝ Extract Rules in function
  [21]              Z←'List' CSBlocks R
```

```
  [22]              R←(⎕cr R),[⎕io]' '
  [23]              R←(+/∧\' '=R)⌽R
{2[24]              :select L
★ [25]                  :case 'Strip' ⍝ Retain 'other' lines beginning with
       comment
  [26]                      R←(1⌽1 1,1↓¯1↓R[;⎕io]∈':⍝')/⍳↑⍴R
★ [27]                  :case 'StripX'⍝ Retain 'other' lines containing a
       comment anywhere
  [28]                      R←(1⌽1 1,1↓¯1↓(R[;⎕io]∈':')∨Rv.∈'⍝')/⍳↑⍴R
2}[29]              :endselect
  [30]              Z←((⎕io++\Z=⎕tcnl)∈R)/Z
```

Two syntaxes are available for this code:

| 'Strip' CSBlocks 'functionname' | Retains control structure and lines beginning with a comment. |
| 'StripX' CSBlocks 'functionname' | Retains control structures and *any* lines having a comment. |

The following call will correctly indent a function:

```
★ [31]              :case 'Indent' ⍝ Add Wave pattern indentation to function
  [32]              Z←'IdCS' CSBlocks R
{2[33]              :if Z≡θ
  [34]                  Z←'Function ',R,' is locked or does not exist.'
+ [35]              :else
  [36]                  (Z L R)←Z
{3[37]                  :if '?'= ↑⎕io⊃⎕io⊃R
  [38]                      Z←'⍝ Outer Syntax Error in function ',1↓ ⎕io⊃⎕io⊃R
+ [39]                  :else
  [40]                      (⎕io⊃⎕io⊃R)←(⎕io⊃⎕io⊃R)~' '
{4[41]                      :if 2∈(⎕io+1)⊃R
  [42]                          Z←'Function ',(⎕io⊃⎕io⊃R),' contains multi
       statement lines.'
+ [43]                      :else
  [44]                          L←(((⎕io+1)⊃R)×2≠(⎕io+1)⊃R)\L+L≠1+¯1⌽L
  [45]                          Z←'GetIndent' CSBlocks L ⍝ At Control
       Structure Level
  [46]                          Z←Z-1=(⎕io+1)⊃⎕io⊃R
  [47]                          Z←'GetIndentX' CSBlocks Z L ((⎕io+1)⊃⎕io⊃R) ⍝
       Within Control Structures
  [48]                          Z←⊃((Z×4)⍴¨⊂' '),¨⊂[⎕io+1](+/∧\' '=⎕cr
       ⎕io⊃⎕io⊃R)⌽⎕cr ⎕io⊃⎕io⊃R
  [49]                          Z←(~Z∧.=' ')≠Z
4}[50]                      :endif
3}[51]                  :endif
2}[52]              :endif
```

The syntax is *'Indent'* CSBlocks *'functionname'*. The function name, specified as the right-hand argument, is *not* changed—the result is a canonical representation (⎕cr) of the function with correct indentation. If the indented version is to replace the existing version, use the following syntax:

```
⎕def 'Indent' CSBlocks 'functionname'
```

All functions in the workspace can be replaced by correctly indented versions as follows:

```
⎕def ¨(⊂'Indent') CSBlocks ¨⊂[⎕io+1]⎕nl 3
```

Usually, ⎕def ⎕cr *'functionname'* has the effect of reducing the size of *functionname*; this is because ⎕def strips out trailing spaces from each line. The overall effect of using ⎕def with *CSBlocks* is unpredictable—*CSBlocks* adds spaces to create indentation and ⎕def removes trailing spaces.

The 'List', 'Indent', 'Strip', or 'StripX' calls to the function CSBlocks use the remaining lines within it. The *'GetIndent'* call calculates the indentation required *at* each control structure level:

```
★ [53]        :case 'GetIndent' A Internal use only
  [54]           (Z R)←R (⌈/R)
{2[55]            :while 0≠R
{3[56]               :while 1∊L←1=+\R=Z
  [57]                 Z[2↑(R=Z)/⍳⍴Z]←-R
  [58]                 L←1↓L/⍳⍴Z
  [59]                 L←(Z[L]=0)/L
  [60]                 Z[L]←1+R
3}[61]               :endwhile
  [62]              R←R-1
2}[63]           :endwhile
  [64]           Z←(|Z)-×|Z
```

The *'GetIndentX'* call calculates the indentation required *within* each control structure; it refines the results of the *'GetIndent'* call:

```
★ [65]        :case 'GetIndentX' A Internal use only
  [66]           (Z L R)←R
  [67]           R←R×R=2
{2[68]            :while 1∊×L
{3[69]               :while 2∊(1=+\L=⌈/L)/R
  [70]                 (1↓((1=+\L=⌈/L)\1=+\×(1=+\L=⌈
/L)/R)/Z)←1+1↓((1=+\L=⌈/L)\1=+\×(1=+\L=⌈/L)/R)/Z
  [71]                 (((1=+\L=⌈/L)\1=+\×(1=+\L=⌈/L)/R)/R)←0
3}[72]               :endwhile
  [73]                 (2↑(L=⌈/L)/L)←0 A Allow multiple structures at same
        level
2}[74]           :endwhile
```

The *'IdCS'* call calculates the depth of nesting, taking into account comment lines, the case of control structure words and multi-statement lines and returns a numeric vector that is then used, respectively, by *'List'* and *'Indent'* for showing the depth of nesting and indentation:

```
★ [75]        :case 'IdCS' A Identify structures in function
  [76]           Z←⎕cr R
{2[77]            :if ×↑⍴Z
  [78]                 Z←' ',Z
  [79]                 L←,(≠\Z∊'''"A')∨~∨\(Z∊':')∧¯1⌽Z∊' ◇'
  [80]                 (L/,Z)←' '
  [81]                 (Z L)←(⊃(⊂[⎕io+1]Z)~'' ') ⊖
  [82]                 Z←(1⌽'
        :',∊2/⊂'acdefghilnoprstuw')['':ACDEFGHILNOPRSTUWacdefghilnoprstuw'
        ⍳Z]
  [83]
        L←+/(':or'∊Z)+(':and'∊Z)+(+':els'∊Z)+(2×':cas'∊Z)+(3×':con'∊Z)+(3
        ×':lea'∊Z)+(3×':got'∊Z)+3×':ret'∊Z
  [84]
        Z←(':for'∊Z)+(':if'∊Z)+(':repeat'∊Z)+(':select'∊Z)+(':while'∊Z)-
        (':end'∊Z)+':until'∊Z
```

```
   [85]                (Z R)←(,Z) ((R L) (+/0≠×Z))
   [86]                Z←(0≠Z)/Z
   [87]                L←+\Z~0
   [88]                ((2/⎕io)⊃R)←(1↑(0≠0⊥L)/'?'),(2/⎕io)⊃R
   [89]                Z←Z L R
 + [90]            :else
   [91]                Z←θ
2}[92]             :endif
 + [93]        :else
   [94]                Z←'Error - Unknown action ',L
1}[95]     :endselect
        ∇
```

Any attempt to call *CSBlocks* with a left-hand argument of *'IdCS'* or *'GetIndentX'* or *'GetIndent'* is liable to cause the function to crash or may return meaningless results. The appropriate in-context usage calls of these methods are in the *'List'*, *'Indent'*, *'Strip'*, and *'StripX'* sections of the code.

2.7 Functions with methods

Function *CSBlocks* is an example of a function with methods. Any call to the function requires a left-hand argument corresponding to a *:case* keyword, loosely regarded as a method of the function, within the function. For utility functions, this approach has several advantages:

• The function does not have any external dependencies. An application can copy or delete it in isolation.

• The function calls itself, as necessary, to use specialist and in-context routines encapsulated within it. This avoids clutter in the workspace.

• A call to a method from another method can be economical with the requirement for local variables—each call allows the argument variables to be reused.

• The *method* (left-hand argument) corresponding to the *:case* keyword can be descriptive and can add to the readability of the code.

• With APL, the right-hand argument may pass multiple arguments to a function. A call not requiring any arguments requires a dummy argument, such as θ, as the right-hand argument is never optional in APL. Should a function not require a right-hand argument *ever*, use the right-hand argument to specify its internal method.

• Most functions in this book are functions with methods. Such functions need not be any more difficult to debug than usual functions—in both cases, debugging takes place within the context of any problems or runtime errors.

Functions with methods need not necessarily be self-contained. If such a function needs to call a general-purpose routine, it is advisable to keep the routine separate from all the functions that might call it. The sole instance of the routine implies that any changes need to be made in one place only and the functions that call it do not replicate the code.

Chapter 3

Application Interface

In the conventional sense, an application interface is simply the visual aspects of that application; users interact with elements of the user interface to coerce the application into carrying out the tasks for which it was designed. In fact, modern applications, including APL applications, have a two-tier interface:

• The hidden interface: the interaction of the primary development tool with the operating system, other ActiveX components, and applications on the same platform.

• The user interface: users interact with the application using its exposed menu, toolbars, forms, etc., and the hardware devices including the keyboard, mouse, printers, etc.

The implementation of both interfaces affects an application directly; the quality of the hidden interface determines its ability to react to ongoing change and that of the user interface determines its endorsement by the user base. Successful software is no longer simply software that works—produces the expected results—but one that works because it is easy, even pleasant, to use and is timely.

3.1 Managing the hidden interface

An APL application is affected directly by the interaction of APL with the operating system, ActiveX components, other platform applications, and other APL modules. Each aspect of interaction uses either platform resources or APL resources. Consider an APL application that uses **Excel** as a COM server. It creates an instance of **Excel**, opens a workbook. An **Excel** session may exist before the APL **Excel** session is created and one may be created afterwards; therefore:

• The ordinal position of the APL **Excel** session may be random among all existing **Excel** sessions, if any exist.

• **Excel**'s caption includes the name of a workbook. If APL uses several workbooks, it may be difficult to know the caption of an APL **Excel** session, especially within an error recovery situation.

• Any attempt to open a workbook that may already be open will cause a runtime error that may cause the APL application to terminate abruptly. Likewise, if APL opens a workbook and terminates abruptly, the session may continue to persist; in such an event, the APL application may not restart unless either the computer is rebooted or the latent **Excel** session is terminated. Either procedure will free the open workbook.

Taking appropriate steps may mitigate such problems, as detailed below:

• First, do not create objects based on the instance of an object. If this is necessary, use the original APL name of the object as a parent and create all other objects as its children. This ensures that all objects are deleted when the parent is deleted.

• Second, use a single instance of any given COM object in APL using a predetermined APL object name; this may provide access to several workbooks. If this object is deleted or reused, the COM server terminates automatically. ⎕wi 'Create' forces its left-hand argument to be reused:

```
⎕wself←'objXL' ⎕wi 'Create' 'Excel.Application'
SavedCaption←⎕wi 'caption'
```

• Third, change **Excel**'s caption to start with 'APL'; this provides an additional means of identifying the instance of the object created by APL. This must be done before any workbook is open as such an event changes the caption:

```
⎕wi 'caption' 'APL+Win'
```

• Fourth, write the handle of the **Excel** session to a file or database as soon as it is created; this allows a future session of APL to query the existence of that session and terminate it, if necessary. This scenario is likely in the event of the current APL terminates abruptly:

```
⎕wcall 'FindWindow' 0 SavedCaption ⍝ APL+Win in this case
2216
```

Some COM objects, such as those in Office 2003, expose an *hwnd* property that may be used instead of the API call. Whether this **Excel** session still exists—even after APL has been restarted—may be queried using the absolute handle number:

```
0≠⎕wcall 'IsWindow' 2216  │ ↑↑/⎕wcall 'GetClassName' 2216 (256⍴⎕tcnul) 256
1                         │ XLMAIN
```

The class name of Excel is XLMAIN regardless of its version. The name of the workbook that is active is dynamically included in the caption of the **Excel** session and may be queried:

```
↑↑/⎕wcall 'GetWindowText' 2216 (256⍴⎕tcnul) 256
APL+Win
```

In this example, the handle used in the FindWindow call is hard coded; in practice, it is either queried dynamically or read from an external source such as a file to which the handle is written as soon as the object is created so that a future APL session can read it for housekeeping purposes.

3.1.1 Forcing a session to terminate

Usually but not always, a server session is automatically terminated when its APL object name is deleted. However, if APL had terminated abruptly and is restarted, the object name may not exist in a new session but the instance of the object may still exist. It is necessary to terminate orphaned instances of objects for two reasons:

• Such sessions use Windows resources unnecessarily: APL cannot grab the existing session and resume. Indeed, it may not be desirable to grab the existing session, as its state is indeterminate.

• Such sessions may be using files on an exclusive basis thereby preventing a new session from using them.

There are two ways to call a Windows API call to terminate a session, if its handle is known, namely:

```
0 0⍴⎕wcall 'SendMessage' 2216 'WM_CLOSE' 0 0
0 0⍴⎕wcall 'SendMessage' 2216 'WM_SYSCOMMAND' 'SC_CLOSE' 0
```

These calls return 1 if successful or 0 otherwise. In these examples, 2216 is the handle of the session; in practice, the actual handle of the instance of the object can be used. These calls fail if the instance of the object is still in use. Consider the following:

```
⎕wi 'XWorkbooks.Add>wb'   │ 0≠⎕wcall 'IsWindow' 2216
⎕wi 'Delete'              │ 1
↑↑/⎕wcall 'GetWindowText' 2216 (256⍴⎕tcnul) 256 ⍝ Session still exists
APL+Win - Book1
⎕wcall 'SendMessage' 2216 'WM_Close' 0 0        ⍝ Fails to terminate
0
```

Although the APL object name, *objXL*, is deleted, the instance of **Excel** continues to persist because the object *wb* created using redirection is using it:

• Had the latter object been a child of the objXL object, the instance of **Excel** would be terminated as soon as objXL is deleted.

• If the object objXL did not exist in the current session but the **Excel** session still persisted, the SendMessage API call would terminate it.

If a session continues to persist, it can be terminated either via the **Task Manager** or by rebooting the computer as a final resort. Click **Processes Tab** on the **Task Manager**, locate and highlight the session and click the **End Process** button.

3.2 The user interface

It is relatively straightforward to establish whether software works but ease of use is a subjective criterion. Yet, the *perceived* usability of software is the single most important criterion that determines the life span of software. As far as users are concerned, the user interface *is* the application.

Any prescription for usability criteria is inappropriate and feeble because it is neither self-enforcing not can it be enforced transparently. Nonetheless, an understanding of the common factors that affect software lead to better design, as discussed below:

• User consultation at the design stage increases the levels of prospective user acceptance of the eventual design. Notoriously, users' anticipated requirements do not always translate into their actual requirements.

• Users are not a homogeneous entity; broadly speaking, there are novice and expert users. Expectations change during the migration from novice to expert status. The rank of novice users is constantly replenished and their expectations change; expectations are driven by their accumulated experience which varies across individuals.

• All contemporary software is bound by the same constraints; the constraints are those of the platform on which they run. APL is a Windows development tool bound by the Windows culture.

• Windows applications are event driven; the users are capable of finding their own pathways through an application to accomplish their tasks. The interface should not allow the users to raise an event for a task unless the task can be accomplished.

• A robust interface prevents rather than complain about users' errors. An application's complaint is usually in the form of an error message. An error message is an intrusion on the users' objectives; it halts the users' progress.

• Contemporary users expect to be able to use the basic functionality of software based on their accumulated experience alone. User manuals and training, although relevant, are detached from the application itself and are not appropriate tools for guiding the users. The extent to which users can use their intuition to navigate through an application depends directly on how successfully the application hides its complexity.

● Users will avoid an application that implies their incompetence and use one that makes them feel competent, stretching it to its limits. Users feel competent when the application leaves them in control: for example, an application that makes changes to a database might leave users wary unless it also provides an undo facility. A critical feature of an application that leaves users feeling competent is one that encourages users to explore the application without any risk of permanent damage to the application set-up, the underlying database, or the computer itself.

● All software not only share the conventions of the platform on which they run but also have an element of house style incorporated into their design. Most vendors specialise in a particular type of application, which, in turn dictates aspects of the interface.

● No software is universally perfect. However, future versions are bound by the legacy of their earlier versions: a bad design afflicts the software throughout its life cycle. Users expect incremental rather than dramatic changes with evolving software. In theory, a tiered design must allow the user interface to be completely replaced without adverse effects; in practice, such a change is costly—in terms of retraining, documentation, and migration—and rarely acceptable to either users or purchasers.

● The millions of copies of Microsoft Office have made the office application interfaces the accepted standard for the design of any application; this includes the layout and size of controls on a form, toolbars, menus, the help facility, etc. The Office suite is a model reference.

3.2.1 Purpose of the user interface

An instinctive definition of a user interface is that it is a means whereby the user communicates with an application; by implication, it exists for the benefit of the user.

A closer examination quickly points to the conclusion that the user interface, in fact, constrains the user. It is a means of forcing the user to comply with the requirements of the application.

3.2.2 Hierarchical and sequential

Consider a simple application—it is desired to calculate the amount £100 at 5⅜% interest after one year. If the interface were a blank sheet of paper, the following solutions would all be legitimate:

```
100 × 5⅜% = 105.375
100 × 5⅜% = 105.38   © Implicitly rounded to 2 decimal places
100 × 1.05375 = 105.375
100 @ 5⅜% = 105.375
100 + 5⅜% = 105.375
100²⁰⁰ + 5⅜% = 105.375²¹⁰·⁷⁵
```

Each of these expressions can be written with or without a currency symbol prefix with the starting or terminal amount or both and it would still be meaningful. Either can be crossed out and replaced by an alternative value. Several other variations of the expression are possible:

● A computer version of the same solution would not allow this freedom of free expression.

● The starting amount will be expected without the currency symbol, and could not be crossed out retaining a visual trace of a correction.

● The interest rate would be expected as 5.375.

● The terminal amount would be inaccessible as it would be an output field.

• Each entry would be specified sequentially. The starting amount would be the first field, rate of interest the second field and there would be a command button to calculate the terminal value.

Arguably, the user interface introduces a set of rules that the user is forced to abide by. It introduces possible courses of actions, like pressing the command button, which would not exist in the '*blank sheet of paper*' interface, and forces the assimilation of peripheral information such as the positioning of fields, their size, and colour.

A complex application will have several data entry forms, each of which may have one or more mandatory fields. Navigation from form to form, and within each form, from field to field would be constrained by the hierarchical presentation of the forms and fields—the user will need to enter information in the manner presented by the interface.

3.2.3 Invasive interaction

Any user interaction that deviates from the interface's expectation is punished by further invasive action by the interface. The user is forced to make decisions in response to default values being presented in fields, error messages and, in the event of complete frustration, to examine help files.

Of course, this does not mean that applications should not have any user interface. Rather, the interface should facilitate the delivery of the functionality encapsulated in the application—as far as possible, on the user's terms. A viable prescription for a good interface is two fold:

• Prevent, rather than trap, user errors by making menus, buttons, and elements of a toolbar context sensitive: disable ones that lead to invalid course of actions. Keeping these elements visible, albeit disabled in context, allows the user to build a mental picture of the dynamic scope of the application.

• Make the user's next step from any context obvious; in many situations, this simply relates to accepting or overriding system defaults and navigation to the next process. Remember that the user is always aware of the preceding process and reversion to that stage is a valid option—that is an Undo facility is always welcome.

A user interface is effective when the user can control or customise it and when it works *with*, rather than *against*, the user.

3.2.3.1 Blocking user interaction

Rather than controlling the user interface by enabling and disabling its elements, it is possible to prevent the user from interacting with an application by using the following API:

```
BlockInput=L(*B fBlock) Alias BlockInput LIB USER32.DLL
```

This API blocks user interaction if set to TRUE and restores interaction when set to **FALSE**. The APL syntax is:

```
⎕wcall 'BlockInput' 0 ∧ ⎕wcall 'BlockInput' 1
```

This API is somewhat dangerous because:

• It blocks all keyboard and mouse events from reaching **all** applications.

• It can be reset only by the application/thread that sets it. Or, the **Task Manager** may be used to terminate the thread that set it.

• The API should be used only with a timer event that reverses its state.

3.3 The user interface is the application

It is wise to be aware that user confidence is brittle and if shattered it will lead to a perception that the application has an alpha quality interface: this decisively spells the rejection of the application. Users perceive and judge the whole of an application simply by its user interface—the rest of the system is out of view. In order to gain user acceptance, APL systems must adopt the *unwritten* conventions of interface design. This is a challenging proposition for any development but particularly so for APL since APL systems pay little regard to this. The legacy approach for APL systems is incomplete or complete lack of specification that, in turn, invariably promotes an ad hoc and inconsistent design.

Consider a simple application that needs to divide one number by another. A typical reaction to this statement will be a function that takes two arguments and returns a result—customarily offered even before the statement is finished.

```
      ∇ Z←L Divide R
[1]    ⍝ System Building with APL+Win
[2]    Z←L÷R
      ∇
```

3.3.1 Where is the interface?
This APL solution does not, at present, have an interface and makes the following pervasive assumptions:

● All knowledge of how this function is to be called is implicit.

● It is by no means obvious what is to be done with the result.

● The user will understand the answer returned, even when the answer is an error, for instance when the right-hand argument R is zero, or nonsensical as in circumstances when both the left- and right-hand arguments are zero. By default, 0÷0 is 1 in APL!

● The user will always attempt to use this function with atomic left and right arguments or with arguments that conform—in APL, an atomic value may be divided by a vector or matrix and vice versa and vectors of the same dimension can be divided one by the other as can matrices of the same dimensions.

Users do what they do; if they use a system to achieve their goal, they react with the system via an interface. A sound interface enables the user to use the application to full advantage.

3.3.2 Manage user expectations
Had users been consulted about the interface, it is reasonable to assume that a majority of them might have expected a calculator interface. The rules and expected behaviour of the system is known from previous experience and users would expect the APL solution to behave in accordance with this experience.

The implications are that the symbol / will have to mean divide rather than ÷, attempts at division by zero will return an error, and that the dividend is out of view as soon as the divisor is specified.

If an algebraic calculator interface were envisaged, dividing 10 by 5 would entail the following operations:

For a Reverse Polish Notation (**RPN**) calculator, the operations will be:

```
10 is entered
/ is pressed
5 is entered
= is pressed, leaving 2 in the display
```

```
10 is entered
<Enter> is pressed
5 is entered
÷ (or) / is pressed, leaving 2 in the
display.
```

The APL solution is simply 10 Divide 5 and 2 is returned, somehow. In itself, the APL solution is much more capable as illustrated in the following examples:

```
10 20 30 Divide 4
2.5 5 7.5
```

```
10 20 30 Divide 2 3 4
5 6.666666667 7.5
```

However, the working of the APL solution is alien to the user who would react by rejecting it. Like calculators, the design of this solution must incorporate the safeguard that only valid numbers can be entered.

3.3.3 The user interface as a tier

Should the design of an application successfully isolate the user interface, this tier may be adapted or replaced with ease on an ongoing basis. This may be necessary for two reasons:

• Typically, users are unable to quantify their requirements as an abstract design specification but are quite adept at modifying a concrete design. Thus, the developers' initial design can be treated as a prototype. In addition, users' preferences change as they become more experienced. The prime directive is ease of use adaptive to users' experience.

• Legislative changes, software and hardware advancements may necessitate changes to the user interface. Legislative changes may, for instance, lead to the regrouping of user input into separate logical and hierarchical forms. Software and hardware advancements lead to the simplification of the user interface and performance improvements.

3.4 APL+Win design safeguards

In many respects, the APL+Win legacy is counter to platform design principles. In keeping with tradition, **File | Load** reads a workspace: almost all other applications use **File | Open**. However, other improvements, such as the disappearance of ⎕poke and ⎕win, support for long file names, and use of paths instead of library numbers have ensured that the shackle of this legacy is all but broken. Above all, APL deploys the standard Windows Graphical User Interface, albeit its deployment is unique in that inherent APL facilities, such as arrays or nested arrays, can be used.

It is imperative that an APL application user interface shares the 'look and feel' of other applications, irrespective of how it is built. There are fundamental dividends to be reaped in achieving this:

• First, it puts users at ease—denies them an opportunistic argument for dismissing APL.

• Second, it ensures a high degree of user acceptance—the higher the levels of acceptance, the more efficiently will the application be used.

• Third, any degree of customisation creates the overhead of bespoke documentation, training, and coding and creates user resistance.

3.5 Context sensitive help

The hallmark of any Windows application is the provision of context sensitive help; that is, help relating to the context specified by the location of the cursor and invoked by pressing the **F1** key. The provision of a help facility is not optional. On the contrary, it provides the following tangible benefits:

• First, it provides a convenient way for the user to learn about the application.

• Second, the help-authoring package will usually allow a printed user manual to be produced from the same base content.

• Third, if the help facility is constructed during the analysis phase of the application, it can potentially influence the design of the application itself—for the better.

As an integral part of system building, the help facility should also be tested for accuracy and relevance in order to ensure that the help facility is appropriate to users' needs. Herein lays the difficulty: users' need varies depending on whether they are novices or experts. At the very least, the help facility should provide context sensitive help. However, a text search facility and multi-level index in the help file enhances the efficacy of online help.

The ISO is preparing a recommendation–ISO/IEC 18019—for the design and preparation of user documentation.

APL provides several means for guiding users. However, the conventional method is to invoke the help facility on the F1 key being pressed. There are several formats for the help file, including WinHelp (*.HLP), Microsoft HTML Help (*.CHM). The WINHELP API call processes *.HLP files and the HTML API call processes *.CHM files. WebHelp is a cross platform help facility, and pure HTML is a World Wide Web consortium (http://www.w3.org/) proposition—these formats work within a browser. The *.HLP and *.CHM formats are routinely accessible from APL.

3.5.1 Enabling a help facility in APL+Win

It is difficult to gauge user preferences regarding the provision of a help facility, its method of access, and level of detail. Not only do these preferences vary from user to user but also for any given user over time. It is necessary to ensure that the help facility remains unobtrusive:

● The industry uses **F1** for invoking help; some applications also use a floating popup menu for the same purpose. With either mechanism, a deliberate action by the user is required before the help facility is invoked.

● It is viable to empower the user to customize the help facility, by allowing some aspects to be switched on or off as required. For example, a 'tooltip' may be quite helpful to a novice user but an expert user will find them unnecessary and may prefer to switch it off.

With APL, it is possible to provide help via tooltips, status prompts, WINHELP and HTMLHELP APIs, and the MessageBox API.

3.5.2 Tooltips and prompts

All controls have a *tooltip* or *prompt* property. The *tooltip* property is an arbitrary character vector that holds any text that is displayed in a popup window when the mouse pointer hovers over that control if the system object, #, sets *tooltipenabled* property to **TRUE**. In other words, it appears and disappears involuntarily. This facility can be used to direct novice users regarding the purpose of a control. However, experienced users may see it as intrusive. The facility can be switched on or off by setting the *tooltipenabled* property to **TRUE** or **FALSE**, respectively.

The prompt property also holds an arbitrary character vector that may be displayed when a control is highlighted or has focus. In order to make the prompts appear, the following code must have been executed:

```
'#' ⎕wi 'onPrompt' '("MyForm.st1" ⎕wi "SetStatus" 1 (⎕warg ⎕wi "prompt"))'
⎕wself←'MyForm.st1' ⎕wi 'New' 'Status'
```

If *tooltipenabled* is **TRUE** and a handler is specified for the system object's *onPrompt* event, a form may show both a tooltip and a prompt at the same time, for the same or different objects. **Figure 3-1 Tooltips and prompts** shows the prompt for the Name field and the tooltip for Date Of Birth. Refer to the Windows Interface help file for synchronising the display of tooltip and prompt.

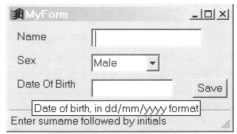

Figure 3-1 Tooltips and prompts

While tooltips are shown for a control when the mouse pointer is over it, irrespective of whether it is enabled or not, a prompt will only be shown for enabled controls. A control that is not enabled will never receive focus and therefore its prompt event will never fire.

3.5.2.1 Tooltips and prompt as a user option

An application can allow the user to switch tooltips on or off and likewise for prompts. The state of tooltips can be a toggle switch:

```
'#' ⎕wi 'tooltipenabled' (~'#' ⎕wi 'tooltipenabled')
```

Prompts can be switched off by turning off the onPrompt event:

```
'#' ⎕wi 'onPrompt' ''
```

In addition, enabling the event can turn it on:

```
'#' ⎕wi 'onPrompt' '("MyForm.st1" ⎕wi "SetStatus" 1 (⎕warg ⎕wi
"prompt"))'
```

The actual line of code is application specific, as it requires the name of a form and a child object, the status control, as arguments.

3.5.3 What's this help?

This facility enables another method of providing help to the user: this is especially appropriate in the event of a validation failure.

The availability of this facility is indicated by the presence of a **?** in the title bar, see **Figure 3-2 What's this help**. If **?** is clicked, the mouse pointer has a question mark and if it is clicked on a control, its onHelp event fires. This can be switched on or off dynamically by setting the border property.

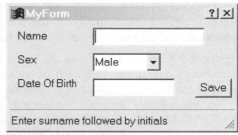

Figure 3-2 What's this help

The current value of the border property can be queried and saved, as follows:

```
'MyForm' ⎕wi 'data' ('MyForm' ⎕wi 'border')
```

Then, the border can be set to enable the "What's this help" facility. This requires the values 16 and 256 to be added to the primary border values 0, 1, or 2. This can be achieved programmatically:

```
    ∇ WhatsThisHelp;R
[1]   ⍝ System Building with APL+Win
[2]   R←PowersOf2 'MyForm' ⎕wi 'data'
[3]   R←+/(R×R≤2),16 256
[4]   'MyForm' ⎕wi 'border' R
    ∇
```

The user has the option of seeing relevant help text by clicking **?** and then clicking the relevant control. The facility is switched off by resetting the border property to its original value:

```
'MyForm' ⎕wi 'border' ('MyForm' ⎕wi 'data')
```

In order to be able to provide specific information in the help text and to make it different from the help text available on pressing **F1**, specify a *context id* in the help property for "What's this help" and in the same in the help context property for **F1** help. When **F1** or the question mark pointer is clicked, the second and third elements of ⎕warg contain the help context and help context ids. In order to determine which type of help is required, query the fifth element of ⎕warg; it is 0 when **F1** is pressed and non-zero otherwise. For each control on a form where help is provided, specify the help and help context properties and the onHelp event handler, as follows:

```
⎕wi 'help' 100        ⎕wi 'help context' 9100    ⎕wi 'onHelp' 'Help ⎕warg'
⍝ What'this help      ⍝ F1 Help
```

The Help function is called with a right-hand argument of ⎕warg whenever **F1** or the question mark pointer is clicked in a control for which the help event has been specified: use different help context ids in the help file for each type of help. The Help function can pick the appropriate context id:

```
       ∇ Help R
  [1]    ⍝ System Building with APL+Win
  [2]    R←⊖ρ(×(⎕io+4)⊃R)↓1↓R
{1[3]    :if 1∊'.CHM'∊'#' ⎕wi '∆HelpFile'
  [4]        ⎕wcall 'HtmlHelp' ('#' ⎕wi 'hwndmain') ('#' ⎕wi '∆HelpFile')
           'HELP_CONTEXT' R
+ [5]    :else
  [6]        ⎕wcall 'WinHelp' ('#' ⎕wi 'hwndmain') ('#' ⎕wi '∆HelpFile')
           'HELP_CONTEXT' R
1}[7]    :endif
       ∇
```

The context ids are unique integers assigned to each topic in the help file; these are either generated automatically by the help compiler or assigned by the help author. The context ids are not visible in the compiled help file.

Figure 3-3 Context IDs shows an example file containing context ids. This must be included in the application documentation/specification.

In this example, the topic Name has a context id of 3000 for help invoked by **F1** and an id of 1001 for help invoked with "What's this help".

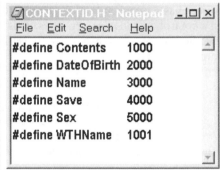

Figure 3-3 Context IDs

3.5.3.1 "What's this help" as a user option
An application can allow the provision of "What's this help" as a user selectable option in order to make it available or unavailable as default. This allows the user to control the visual aspects of the application.

3.5.4 Help: WINHELP
WINHELP relies on a standard Windows API call:

WinHelp=U(hwnd,*C lpHelpFile,U wCommand,D dwData) ALIAS WinHelpA LIB USER32.DLL

The *hwnd* parameter may be 0 or (*frm* ⎕wi 'hwnd') where *frm* is the name of the application form. The *lpHelpFile* parameter is the name of the help file (*.HLP). Unless the help file is located in the Windows HELP folder, this must be the fully qualified name of the help file. The *usCommand* parameter can be one of the values shown in **Table 3-1 WinHelp API constants**.

Constant	Hexadecimal Value	Decimal Value
HELP_CONTEXT	1	1
HELP_QUIT	2	2
HELP_CONTENTS	3	3
HELP_INDEX	3	3
HELP_HELPONHELP	4	4
HELP_SETCONTENTS	5	5
HELP_SETINDEX	5	5

Constant	Hexadecimal Value	Decimal Value
HELP_CONTEXTPOPUP	8	8
HELP_FORCEFILE	9	9
HELP_KEY	101	257
HELP_COMMAND	102	258
HELP_PARTIALKEY	105	261
HELP_MULTIKEY	201	513
HELP_SETWINPOS	203	515

Table 3-1 WinHelp API constants

The *dwData* parameter specifies additional data—this is 0 by default or may be the help context id. The APL+Win syntax for invoking help via WINHELP is:

```
0 0⍴⎕wcall 'WinHelp' 0 'C:\cshelp.hlp' 'HELP_CONTENTS'  ⍬
```

This shows the help file with the default topic visible:

```
0 0⍴⎕wcall 'WinHelp' 0 'C:\CSHELP.HLP' 'HELP_CONTEXT'  3000
```

This shows the topic whose context id is 3000. On closing the application, ensure that its help window is closed also:

```
⎕wcall 'WinHelp' 0 'C:\CSHELP.HLP' 'HELP_QUIT'  ⍬
```

This call can be executed regardless of whether the help window exists. If the help file is missing, WinHelp prompts the user to search for it. If the topic is not found, it returns an error message.

3.5.5 Help: HTML

HTML Help relies on a standard Windows API call:

```
HtmlHelp=H(D hwndCaller, *C pszFile, D uCommand,D dwData ) ALIAS HtmlHelpA LIB
    hhctrl.ocx
```

The *hwndCaller* parameter corresponds to the hwnd parameter for WINIIELP, see above. The *pszFile* parameter specifies the name of the help file. The *uCommand* parameter can be one of values from **Table 3-2 HtmlHelp API constants**.

Constant	Hexadecimal Value	Decimal Value
HH_DISPLAY_TOPIC	0	0
HH_SET_WIN_TYPE	4	4
HH_GET_WIN_TYPE	5	5
HH_GET_WIN_HANDLE	6	6
HH_DISPLAY_TEXT_POPUP	E	14
HH_HELP_CONTEXT	F	15
HH_TP_HELP_CONTEXTMENU	10	16
HH_TP_HELP_WM_HELP	11	17
HH_CLOSE_ALL	12	18

Table 3-2 HtmlHelp API constants

Unless these constants exist in the INI file, specify these parameters as a decimal value.

The *dwData* parameter specifies additional data—this is 0 by default or may be the help context id.

The HtmlHelp API call returns the handle of the window it opens or 0 if it fails. The corresponding call for displaying the default topic is:

```
⎕wcall 'HtmlHelp' 0 'c:\Context Sensitive Help Demo.chm' 0 0 ⍝
HH_DISPLAY_TOPIC
```

The call for displaying a specific topic is:

```
⎕wcall 'HtmlHelp' 0 'c:\Context Sensitive Help Demo.chm' 15 3000 ⍝
HH_HELP_CONTEXT
```

The call for closing the help window is:

```
⎕wcall 'HtmlHelp' 0 'c:\Context Sensitive Help Demo.chm' 18 0 ⍝
HH_CLOSE_ALL
```

Unlike WinHelp, HtmlHelp does not prompt the user to search for the help file if it cannot be found nor does it display an error if the topic is not found.

3.6 Help format as a user option

Some help compilers can produce both HLP and CHM files; thus an application can provide help files in both formats and allow the user to select the desired option. This gives the user a sense of control over the application.

On starting the application or when the user selects the help file format, verify the existence of the file, convert its fully qualified name to upper case, and assign it a user-defined property of the system object, as follows:

```
'#' ⎕wi '∆HelpFile' (⎕wcall 'CharUpper' 'drive:\path\filename.ext')
```

When the user presses **F1** or the "What's this help" cursor, the following lines are executed:

```
{1[3]    :if 1∊'.CHM'∊'#' ⎕wi '∆HelpFile'
  [4]         ⎕wcall 'HtmlHelp' ('#' ⎕wi 'hwndmain') ('#' ⎕wi '∆HelpFile')
              'HELP_CONTEXT' R
+ [5]    :else
  [6]         ⎕wcall 'WinHelp' ('#' ⎕wi 'hwndmain') ('#' ⎕wi '∆HelpFile')
              'HELP_CONTEXT' R
1}[7]    :endif
     ∇
```

If the help file name has extension CHM, the HtmlHelp API is called: otherwise, WINHELP is called—assumes that the properties help and help context are assigned and the onHelp event is assigned. Lines [4] and [6] specify the handle of the system object as the application handle. Since the runtime system does not return this, it is preferable to specify the handle of the main user form instead. If the application has a multiple document interface (**MDI**) form for all its forms, the handle can be established by reference to its name:

```
    myMDIForm ⎕wi 'hwnd'
```

If the application has several forms—that is, it has a single document interface (**SDI**)—ensure that ⎕wself is assigned to the name of each form before it is shown. Then the handle of the form can be established as follows:

```
    ⎕wself ⎕wi 'hwnd'
```

3.7 Application messages

Messages are string literals that an application may communicate to the user, within a given context, and perhaps elicit a response to determine the flow of execution. All messages halt the application. Two issues arise with application messages.

3.7.1 The language and location of messages

One approach to messages is to bury them in the code. This has two adverse effects, namely:

• The language in which the message is written is hard coded; should the application be distributed to another country, another language cannot be accommodated easily.

• The message itself is stored in the code; any changes to error messages require the whole of the application code to be tested and reviewed.

Although all messages are coded in this manner in this book—in order to keep the code cohesive—a better approach is to store messages *outside* the code. Any type of file—INI, XML, native or component—or a database table may be used for this purpose. The application can refer to the file and retrieve the message by name. Using an INI file, messages may be organised as follows:

```
[English]
Err1=File # is currently in use. Please try later.
Err2=No members selected.
```

```
[French]
Err1= Fichier # n'est pas disponible.   Essayer plus tard sil vous plaît.
Err2= Aucuns membres ont été choisis.
```

The structure of the error message file specifies the language and the name of each error message. In this example, English and French are the languages; Err1 and Err2 are error message names. The application can recover the correct message at runtime. The # is a placeholder for text that may be substituted at runtime; see line [4] below:

```
       ∇ Z←L ShowMessage R
  [1]    ⍝ System Building with APL+Win
  [2]    R←↑↑/⎕wcall (⊂'GetPrivateProfileString'),R,'' (256⍴⎕tcnul) 256
         'c:\error.ini'
{1[3]   :if 0≠⎕nc 'L'
  [4]        ((R='#')/R)←⊂⍎L
  [5]        R←∈R
1}[6]   :endif
  [7]   Z←Msg R
       ∇
```

The message shown in **Figure 3-4 English message** is generated by the following code.

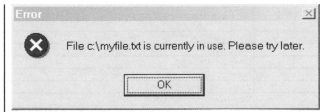

Figure 3-4 English message

```
'c:\myfile.txt' ShowMessage 'English' 'Err1'
```

In practice, the language will be initialised at start-up, as a global variable, say, Language←'French'. However, the error number will be initialised within context, and at runtime, say, Error←'Err2'.

This enables the application to signal messages in any supported language without requiring different versions of the application to be maintained. **Figure 3-5 French message** shows the message in French.

Figure 3-5 French message

```
'c:\myfile.txt' ShowMessage Language Error
```

This approach keeps the code free of literals. The error message file presents all the messages in one location, the literals may be modified readily and another language may be added with ease.

On the negative side, it becomes more likely that error message names within the code are missing from the error message files or that the file contains error message names that are not used in the code. This aspect of maintenance should be undertaken during system testing.

3.7.2 Communicating runtime messages

Windows has a standard API, MessageBox, for applications to provide feedback to users; it returns a value that the application can use to determine the appropriate course of action.

Although this make communication easy, it is wise to remember that messages that halt processing is stopping users from accomplishing their goal: messages need to be purposeful.

The API declaration is:

MessageBox=I(HW hwOwner, *C lpText, *C lpCaption, U uType) ALIAS MessageBoxA LIB
    USER32.DLL

The *hwnd* argument is the handle of the window making the call; this may be 0 or '#' ⎕wi 'hwndmain', the handle of the APL session. The APL+Win handle changes dynamically. The *lpText* argument is the message to be output. A multi-line message must have the lines separated by ⎕tcnl. The *lpCaption* argument specifies the caption of the message box—usually this is the name of the application. The *wType* argument specifies the collection of buttons, the icon, and the particular button that has focus when the message box is displayed. This value is composed from the values specified in **Table 3-3 MessageBox API configuration**.

The *wType* argument comprises of the sum of a *single* value from each of the four groups. This API call returns one of the seven possible values shown in the **Return Code** section.

Group		Hexadecimal Value	Decimal Value
	Buttons		
1	OK only.	0	0
1	OK and Cancel.	1	1
1	Abort, Retry and Ignore.	2	2
1	Yes, No and Cancel.	3	3
1	Yes and No buttons.	4	4
1	Retry and Cancel.	5	5
	Icon		
2	No Icon	0	0
2	Critical Message	10	16
2	Warning Query	20	32
2	Warning Message	30	48
2	Information Message	40	64
	Focus		
3	First button.	0	0
	Second button.	100	256
3	Third button.	200	512
3	Fourth button.	300	768
	Behaviour		
4	Application is suspended until a response to the message box is received.	0	0
4	System is suspended until a response to the message box is received.	1000	4096
4	Adds Help button to the message box.	4000	16384
4	Specifies the message box window as the foreground window	10000	65536
4	Message and caption are right aligned.	80000	524288
4	Message should appear as right-to-left reading on Hebrew and Arabic systems.	100000	1048576
Return Code			
	OK		1
	Cancel		2
	Abort		3
	Retry		4
	Ignore		5
	Yes		6
	No		7

Table 3-3 MessageBox API configuration

The fourth button, Help, cannot be set as the default button and is not available in the Button Focus group. This button creates an event that must be handled appropriately by the calling application. The final value within group four, 1048576, allows messages to appear as right to left, and is not applicable in the UK. The raw APL+Win syntax for calling this API is:

```
       ⎕wcall 'MessageBox' ('#' ⎕wi 'hwndmain') 'This warning may be
ignored.' 'APL+Win' (2+32+512)
```

The message shown in **Figure 3-6 Application message** suggests 'Ignore' as the default course of action.

Figure 3-6 Application message

A call to this API halts the application and awaits user interaction; that is, a user must attend to the executing application.

A means of running applications unattended may add considerable flexibility to an application. Since there is no way to modify this API call to achieve this behaviour, it might be pragmatic to maintain two versions of the application: one that signals all messages and awaits interaction and one that does not. Besides doubling the maintenance overhead, this approach is not viable because it does not resolve a critical problem—that of returning the value of the default button. For the message shown in **Figure 3-6 Application message**, the possible return values are 3 (**Abort**), 4 (**Retry**) and 5 (**Ignore**) and 5 is the default value: the **Ignore** button has got focus.

It would be desirable to have the application return 5 without showing the message; that is, to have the application run unattended. This enables a calculation-intensive application to run overnight, unattended.

3.7.2.1 The APL wrapper function

Although the inability to change the button captions on a message box form is a serious handicap in some circumstances, it is something that all applications have to accept the message box is a hallmark of Windows applications. For APL applications, the critical arguments for this API call are the buttons and the application message itself; the other arguments are static. The following function wraps the API call allowing configuration for unattended runs:

```
       ∇ Z←L Msg R
  [1]    ⍝ System Building with APL+Win
{1[2]    :if 0=⎕nc 'L'
  [3]        L←16
1}[4]    :endif
  [5]    Z←∈0 4096 16384 65536 524288∘.+0 256 512∘.+∈0 16 32 48 64∘.+0 1 2
         3 4 5
  [6]    L←0⊥(L≥Z)/Z
{1[7]    :select '#' ⎕wi '∆AppAuto'
★ [8]        :case 0
  [9]            Z←⎕wcall 'MessageBox' ('#' ⎕wi 'hwndmain') R ('#' ⎕wi
         '∆AppTitle') L
+ [10]       :else
```

```
  [11]              L←L-0⊥(0 4096 16384 65536 524288≤L)/0 4096 16384 65536
        524288
  [12]              Z←0⊥(64 256 512≤L)/0 256 512
  [13]              L←L-Z
  [14]              L←L-0⊥(16 32 48 64≤L)/16 32 48 64
  [15]              L←0⊥(0 1 2 3 4 5≤L)/ι6
  [16]              L←L⊃(1) (1 2) (3 4 5) (6 7 2) (6 7) (4 2)
  [17]              Z←0 256 512ιZL⌈/(ρ,L)↑0 256 512
  [18]              Z←Z⊃,L
{2[19]              :if ('#' ⎕wi 'ΔAppAuto')∈⎕nnums,⎕xnnums
  [20]                 R←(,'G<99/99/9999 99:99:99 {9} >' ⎕fmt
        ((2/100),10000,(3/100),10)⊥⎕ts[(Φι3),3+ι3],Z),R
  [21]                 (1Φ⎕tclf,R,⎕tcnl) ⎕nappend '#' ⎕wi 'ΔAppAuto'
2}[22]              :endif
1}[23]   :endselect
        ∇
```

The optional left-hand argument specifies the buttons and the right-hand argument specifies the error message. The message box caption is read from a user-defined property, '#' ⎕wi ' ΔAppTitle', which defaults to 0 and translates as 'Error'.

3.7.2.2 Using Msg in applications

The *select ... endselect* control structure is ideal for handling calls to the Msg function. An application may call Msg as follows:

```
       ∇ Signal
  [1]    ⍝ System Building with APL+Win
{1[2]    :select (4+32) Msg 'Is this book useful to you?'
★ [3]       :case 6 ⍝ YES clicked
  [4]              ⍝ 'Glad to hear it.'
★ [5]       :case 7 ⍝ NO clicked
  [6]              ⍝ 'Sorry to disappoint you.'
1}[7]    :endselect
        ∇
```

Every call to *Msg* can be trapped in a construction such as *Signal*, making allowance for the potential return values, except where a single button is displayed 'OK' where the call can simply be prefixed by 0 0ρ to absorb the return value from *Msg*. This construction avoids the overhead of assigning the return code.

Automation requires an application wide flag, '#' ⎕wi 'ΔAppAuto', in line [7]; this controls the behaviour of the cover function. If this property is not created, it has a default value of 0 and causes all messages to be output in the normal manner. Usually, an application should provide a menu option that sets the value of this property, as shown in line [12] of the following function:

```
       ∇ Automate;⎕wself
  [1]    ⍝ System Building With APL+Win
  [2]    ⎕wself←'App' ⎕wi 'Create' 'Form' 'Close'
  [3]    ⎕wi 'caption' 'APL+Win Application'
  [4]    ⎕wi 'extent' 3.1875 25.125
  [5]    ⎕wi 'where' 11.75 25
  [6]    ⎕wself←'App.Run' ⎕wi 'New' 'Menu'
  [7]    ⎕wi 'caption' '&Run'
  [8]    ⎕wself←'App.Run.AutoMode' ⎕wi 'New' 'Menu'
  [9]    ⎕wi 'caption' 'Auto&Mode'
  [10]   ⎕wi 'style' 1 ⍝ Toggle (Off|On) menu
```

```
[11]   ⎕wi 'value' 0
[12]   ⎕wi 'onClick' 'SetMode'
[13]   'App' ⎕wi 'Wait'
     ∇
```

Lines [7] – [14] represent the pertinent segment of code illustrated in **Figure 3-7 Auto mode I**.

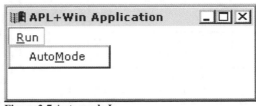
Figure 3-7 Auto mode I

Initially, the user-defined property ⍙AppAuto is unassigned and automation is switched off—this ensures the default behaviour for sending all messages to the user interface.

The presence or absence of a tick mark against the menu option provides visual confirmation of automation status. **Figure 3-8 Auto mode II** shows the menu option switched on.

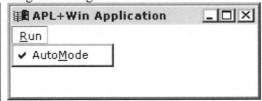
Figure 3-8 Auto mode II

When the menu option is clicked, the event handler function SetMode configures the value of ⍙AppAuto, thus:

```
      ∇ SetMode
  [1]    ⍝ System Building with APL+Win
{1[2]    :select ⎕wi 'value'
★ [3]         :case 0 ⍝ Switch off automation
  [4]             '#' ⎕wi '⍙AppAuto' 0
★ [5]         :case 1 ⍝ Switch on automation
{2[6]            :if ¯1∈⎕nnums,⎕xnnums
  [7]                '#' ⎕wi '⍙AppAuto' ¯1
+ [8]            :else
  [9]                '#' ⎕wi '⍙AppAuto' 1
2}[10]           :endif
1}[11]   :endselect
     ∇
```

If the menu option is switched off, the user-defined property is set to 0 and all messages are sent to the user interface. If the menu option is on and a native file handle ¯1 exists, the property inherits the value of the handle; this allows *Msg* to write all messages to the file. If the file handle does not exist, 1 is assigned to the property and the message is neither displayed nor written to a file. This approach requires a reserved file handle ¯1, or some other fixed handle with a corresponding change in SetMode, when the application initialises. Assign the message box caption within the same block of code:

```
      '#' ⎕wi '⍙AppAuto' 0 ⍝ This is the default
```

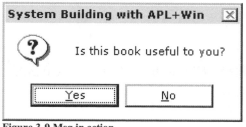

The message appears in the application session. The application halts pending user interaction.

The dialogue shown in **Figure 3-9 Msg in action** highlights the default button, **Yes**. The **Enter** key is equivalent to clicking it.

Figure 3-9 Msg in action

```
'error.log' □ncreate ¯1
'#' □wi 'ΔAppAuto' ¯1 ⍝ Unattended run
Msg 'Is this book useful to you?'
```

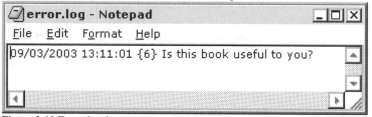

Figure 3-10 Error logging

The Msg function writes the message to a log file, the file tied to ¯1, returns 6, the default button value and the application carries on. The log records the default button's value (6): this is shown in **Figure 3-10 Error logging**. Any non-zero value of ΔAppAuto that is not the handle of a native file causes the value of the default button to be returned: the message is never displayed.

3.8 User-defined properties of the system object

ΔAppTitle and ΔAppAuto are user-defined properties of the APL+Win system object #. These and other universal session parameters may be set in similar fashion at the start of the session. Unlike global variables, which can also hold these properties, user-defined properties are not lost when a new application is loaded: this enables one workspace to pass data to another in the same APL session. Note the following:

• The names of user-defined properties always begin with Δ and are case-sensitive.

• A reference to an undefined user-defined property always returns 0 and not an error. In contrast, a reference to an undefined variable causes a runtime error.

• A user-defined property can be deleted by assigning it a value of zero.

• A system reset ('#' □wi 'Reset') clears user-defined properties also.

3.9 The scope of user documentation

Poor documentation or lack of documentation is costly in several respects:

• It provides competitors with any easy route to gain competitive advantage by providing better documentation.

• It leads to a heavier burden of customer support and may lead to additional staff overheads in manning the help desk. Experienced users are inclined to use applications without preamble and expect to be able to grasp the basic functionality of an application via intuitive navigation.

- It leaves users frustrated and this in turn leads to low levels of user acceptance; if the users are reviewers of a trade magazine, this leads to the product receiving bad press. This may dampen sales and implicitly promote competitors' products.

The net effect is that despite the large capital expenditure in financing the development of the product, its sponsor fails to realise expected returns. Therefore, continued development becomes economically unviable. In such a scenario, the development team becomes the first casualty.

Two versions of documentation are required, an online and a printed version, not least because users prefer one or the other. This may appear to be an unnecessary overhead. However, the product sponsor can use the printed version to good advantage:

- The printed version may use graphics liberally, whereas the potential for doing so may be limited in the online version: the online version must present information one screen at a time.

- Bespoke documentation can be confined to the printed version only, leaving the online help generic.

- The printed version may be sold separately, per user, and generate extra revenue for the sponsor. Also, the printed version may underpin custom training courses provided by the sponsor.

From the point of view of maintenance, it is cost effective to write documentation as a single document and to assemble the different versions from the single source—**Word** is more than capable of managing this. A number of help authoring packages are able to use **Word** documents seamlessly and are able to produce files in both .HLP and .CHM formats, as well as printed manuals.

3.9.1 Types of documentation

The types of documentation depend on the application. A highly specialised niche application may require specialist documentation. In general, a commercial product will have a functional specification, a technical manual, and a user manual. A synthesis of the three documents provides the system specification for developers.

Although it is an ordeal, reiterative revision of the documentation ensures that it is in line with the product, and available on demand. The first version of any documentation is rarely accurate. Up-to-date documentation also ensures that handover of system development and maintenance from one set of developers to another is hassle free. The tool used to create the documentation must provide facilities for version control. This is necessary because the client base may be using different versions of the product and because changes invariably introduce new bugs.

The responsibility for document maintenance is a collaborative effort among the technical authority, users, trainers, and developers. The salutary lesson for developers is that unless a product is documented, they are indefinitely committed to the same project— because it is impossible to hand it over. Initially this leads to career stagnation and ultimately to the product being scrapped: the product is perceived as being too difficult to maintain because it relies on the original developers. Sponsors do not like single points of failure— that is the capital investment in the product being jeopardised by developers leaving.

3.9.1.1 Functional specification

Most organisations initiate a software project by producing a functional specification. Although this is primarily used internally, it can be used for marketing purposes too. A functional specification describes the product in detail.

- It includes details of all input and output as well as any known limitations.
- Explains its interfacing capability with other products, its menus, and dependencies.
- Defines its modularity and tiers.
- Above all, its purpose, the technology that is deployed, and any unique features that gives the product a competitive advantage. For example, if the product has a debugging or trace facility, it would be mentioned.

There is no precise external definition of what a functional specification should contain. However, the contents may be so collated as to enable such a document to fulfil some routines objectives.

- It can provide a template for project management.
- It can be a basis for estimating the duration, and costs, of development and testing.
- It can provide the terms and references for cost benefit analyses of the modules.
- It can provide the framework for planning phases of future enhancements or releases.

If the application is subject to external regulation, the functional specification will also provide an assessment of the product's compliance in every country where it is sold.

This document will also highlight the platform for which the product is intended, the modifications required to the host—such as in the **Registry**, and describe its hardware requirements.

For prospective buyers, this document will also include the vendor's ongoing service level agreements and product support arrangements.

3.9.1.2 Technical manual

A technical manual describes the computations within the product and includes all formulae and details of all critical paths involved. For example, a product that projects asset liability ratios will include the formulae for the models used, the assumptions made, and perhaps a critique of some results, especially intermediate results. This document makes the output of the system verifiable, both internally—during testing—and externally by clients. This document provides a good starting point for future releases as it can highlight the impact of changes and the potential for future enhancements.

3.9.1.3 User manual

A user manual describes how the product is to be used efficiently; that is, how the functionality of the product can be deployed. It purports to transform users into experts in using the product and *not* in what the product does. For example, the user manual of a product that produces pension illustrations will not explain the calculations involved, simply the steps to producing them. The user manual would describe data inputs, validation, error or warning messages, and document the process for setting up calculation bases, printing, etc., and usage shortcuts, where applicable. This document will include a number of images from the user interface; this serves to accelerate the familiarisation process with the application.

3.10 Designing menus

Although this is a precarious and laborious activity, the aesthetic allure of a good design can literally enhance or destroy the appeal of an application.

Consider the REL40.W3 workspace: it has 54 functions to generate the graphical interface, which have 5,810 lines of codes; there are 263 controls in total and no menus. In practice, the volume of code can be generated by]WED, the APL+Win form designer; however this will rarely be perfect and will require editing. The overhead arises in the endeavour to keep the code consistent, especially when controls are added to one or more forms and when controls are relocated from one form to another.

In addition to the volume of code comprising the user interface, there is one further complication: for a distributed application, the users' screen resolution is unknown at design time. The developer has to design for a 'typical' screen resolution (800x600); should the target resolution be different, the look and feel of the application changes too.

3.10.1 Auto-generation of the user interface

The value of every property of any control can be coded by reference except the positional properties such as 'where', which requires determination by visual reference. Specify the user interface as data in, say, an **Excel** spreadsheet? This has several advantages, namely:

• The whole of the user interface is visible, albeit as data, and can be kept consistent much more easily. Any control can be easily relocated—using copy and paste—from one form to another and controls can be added or removed easily.

• Following reiterative changes, the whole of the interface can be regenerated and tested as an isolated exercise.

Storing data definition of user forms and menus presents some restrictions:

• **Excel** cannot store numeric vectors, such as the 'where' or 'size' properties; these have to be stored as character and converted at runtime.

• As **Excel** is columnar, some cells will be superfluous for some classes: for instance, a menu option which is a separator will not have an onClick event; these cells must be left blank and ignored at runtime.

• It is undesirable to specify the ordinal position of each menu and at the same time, it is impossible to determine the correct order programmatically.

• **Excel** does not allow APL characters to be entered from the keyboard: thus, all columns, especially Name and onClick, cannot include APL characters.

A sample sheet for menus is shown in **Figure 3-11 Menu sheet**. The first row contains a subset of the properties and events of the APL+Win menu class, except for the first column where the name of the form to which the menu will be added is shown. If MDI forms are in use, the fully qualified names of child forms must be shown.

The example in **Figure 3-11 Menu sheet** assumes that an SDI form is used.

	C	D	E	F	G	H	I	J	K	
1	Name	caption	style	enabled	visible	separator	shortcut	prompt	value	help
2	File	&File	0	1	1	0		Input/Output	0	
3	File.Open	&Open	0	1	1	0		Open an existi	0	
4	Run.Automo	&Automoc	1	1	1	0		Run Mode	1	
5	File.Save	&Save	0	0	1	0	'S' 4	Open an existi	0	
6	Run	&Run	0	1	1	0		Run Mode	1	
7	File.S1		0	0	1	1			0	
8	File.Exit	&Exit	0	1	1	0		End	0	

APL+Win Menu in Excel

Figure 3-11 Menu sheet

Some precautions are necessary in order to avoid any problems with the acquisition of data from **Excel**, namely:

• The first row shows the names of columns: these names are the property names of the APL+Win menu class and must be spelt consistently. However, the columns may appear in any order.

• Do not leave blank rows in the used **Excel** range: although blank rows are disregarded when read, the JET driver includes them when determining the data type of each column.

• The fully qualified names of menu items are shown in the 'Name' column. Any character cell that is inappropriate or inapplicable—as in the shortcut column—is left blank.

• A numeric column is fully populated, with a default value of 0. If this value is inappropriate, the menu generation function will ignore it. For example, the value column is relevant only when the style value is 1 but it must be fully populated.

• Any cell that contains a string in quotes, like cell I5, is preceded with a single quote, thereby the two single quotes open the string, but only one closes it. This is a quirk in **Excel**, which uses a single quote to signify a literal.

3.10.2 Validating an interface tree

The immediate problem is that the names of controls may not be valid for two reasons:

• The hierarchical names specified may not be consistent; in the example shown in **Figure 3-11 Menu sheet**, the name *Edit.Undo* is invalid because the parent *Edit* does not exist or has not been specified.

• Invalid names may be used within the hierarchy—names must be valid APL names.

This can be resolved by the following function:

```
          ∇ Z←L VTree R
   [1]    ⍝ System Building with APL+Win
   [2]    Z←(⍴R)⍴0
{1[3]     :for L :in R
   [4]        Z←Z+↑¨(⊂L)∊¨R
1}[5]     :endfor
   [6]    L←Z>+/¨'.'=¨R
   [7]    L←L∧∧/¨(⎕nc ¨⊃¨(~'.'=R)⊂¨R)≠4
   [8]    L←L∧(R⍳R)=⍳⍴R
   [9]    Z←Z×L
          ∇
       VTree 'Help' 'Edit.Undo' 'File' 'File.Open' 'File.4Exit' 'Help'
'File.Save'
2 0 1 2 0 0 2
```

In this example, the second and fifth names are invalid (indicated by 0). The second name is invalid because the parent object *Edit* is not specified, the fifth name contains an invalid name—established in line [7]—in the second level, and the sixth name is invalid because it is recurring—established in line [8]. The names of APL objects are case sensitive. This function copes with names specified in any order—child and parent objects may be specified in any order. This illustrates the problem relating to the order in which controls should be created: there are no inherent characteristics that can be used to determine the logical order.

The function returns a non-zero integer indicating the order in which the child control can be created. It is pure coincidence that 'Help' is specified twice and ends up being created after 'File'. However, if 'Help' is specified once only, it has an order of 1—it will be created before 'File', thus:

```
      VTree 'Help' 'Edit.Undo' 'File' 'File.Open' 'File.4Exit' 'File.Save'
1 0 1 2 0 2
```

3.10.3 Creating menus

The CreateMenu function creates the menus specified in the **Excel** worksheet:

```
      ∇ Z←L CreateMenu R;Cnn;Sql;⎕wself
 [1]    ⍝ System Building with APL+Win
```

The right-hand argument is the name of an existing APL+Win form.

```
 [2]    Cnn←'Provider=Microsoft.Jet.OLEDB.4.0;Data
        Source=c:\menu.xls;Extended Properties=Excel 8.0;'
 [3]    Sql←"SELECT ⋆ FROM [APL+Win Menu in Excel$] WHERE FORM = '@' AND
        CLASS = 'Menu'"
 [4]    ((Sql='@')/Sql)←⊂R
 [5]    Sql←⊂Sql
 [6]    ⎕wself←'∆MenuRS' ⎕wi 'Create' 'ADODB.Recordset'
 [7]    ⎕wi 'XOpen' Sql Cnn (⎕wi '=adOpenStatic') (⎕wi
        '=adLockBatchOptimistic') (⎕wi '=adCmdText')
 [8]    Z←''
```

ADO uses the JET provider to read the **Excel** worksheet. In line [2] the location and name of the workbook is specified as *source*; the *Properties* value may need to be reviewed depending on the version of **Excel** used to save the workbook—in this instance, **Excel** 2000 is used. A temporary ADO record object is created in line [6]. The following block of code extracts the menus for a particular form:

```
{1[9]     :if 0≠⎕wi 'xRecordCount'
{2[10]        :if 0≠⍴R ⎕wi 'self'
 [11]            L←,⎕wi 'XGetRows' (⎕wi '=adGetRowsRest') (⎕wi
        '=adBookmarkFirst') 'Name'
 [12]            L←(VTree L) L
{3[13]            :while 0≠⍴((⎕io⊃L)=L/(⎕io⊃L)~0)/(⎕io+1)⊃L
 [14]                ⎕wi 'xFilter'
        ("Name='",(∈(<\(⎕io⊃L)=L/(⎕io⊃L)~0)/(⎕io+1)⊃L),"'")
```

If menus exist for the form, specified as the right-hand argument, and the form itself exists as an object, the menu structure is verified using the VTree function, see line [12].

Menu items specified are processed sequentially in the following block of code:

```
{4[15]                :while ~⎕wi 'xEOF'
{5[16]                    :if 0=⍴(R,⎕wi 'xFields().Value' 'Name') ⎕wi 'self'
 [17]                        R←R,'.',⎕wi 'xFields().Value' 'Name'
 [18]                        0 0⍴R ⎕wi 'New' 'Menu'
{6[19]                        :if 0=(⎕wi 'xFields().Value' 'separator')
 [20]                            R ⎕wi 'caption' (⎕wi 'xFields().Value'
        'caption')
 [21]                            R ⎕wi 'style' (↑⎕wi 'xFields().Value'
        'style')
 [22]                            R ⎕wi 'enabled' (↑⎕wi 'xFields().Value'
        'enabled')
 [23]                            R ⎕wi 'visible' (↑⎕wi 'xFields().Value'
        'visible')
{7[24]                            :if 0≠⍴(⎕wi 'xFields().Value'
        'shortcut')~' '
 [25]                                R ⎕wi 'shortcut' (⍋⎕wi
        'xFields().Value' 'shortcut')
7}[26]                            :endif
```

```
   [27]                          R ⎕wi 'prompt' (⎕wi 'xFields().Value'
      'prompt')
   [28]                          R ⎕wi 'help context' (↑⎕wi
      'xFields().Value' 'help context')
{7[29]                             :if 0≠⍴(⎕wi 'xFields().Value' 'onClick')
   [30]                             R ⎕wi 'onClick' (⎕wi 'xFields().Value'
      'onClick')
7}[31]                             :endif
{7[32]                             :if 1=(↑⎕wi 'xFields().Value' 'style')
   [33]                             R ⎕wi 'value' (↑⎕wi 'xFields().Value'
      'value')
7}[34]                             :endif
+ [35]                          :else
   [36]                             R ⎕wi 'separator' 1
6}[37]                          :endif
5}[38]                       :endif
   [39]                       Z←Z,⊂R
   [40]                       R←(∧\R≠'.')/R
   [41]                       ⎕wi 'XMoveNext'
4}[42]                    :endwhile
   [43]                    ((<\(⎕io⊃L)=⌊/(⎕io⊃L)~0)/⎕io⊃L)←0
   [44]                    ⎕wi 'xFilter' ''
3}[45]                 :endwhile
```

Any invalid menus—that is, child menus whose name is invalid or whose parent does not exist—are discarded. Menus are processed parent down although they may have been specified in any order in the worksheet. If additional columns are used in the worksheet, say a help column, this block of code needs to be amended to use it.

If the form does not exist or no menus for it are found, error messages are given:

```
+ [46]        :else
   [47]           Z←Msg 'Unable to find object ',R
2}[48]        :endif
+ [49]     :else
   [50]        Z←Msg 'Unable to locate menus for form ',R
1}[51]     :endif
   [52]  ⎕wi 'XClose'
   [53]  ⎕wi 'Delete'
        ∇
```

The return value from this function is a nested vector of the menus created or 1: this is the value of the OK button. As the form must exist, it is possible that it contains menu items. If any existing menu item is the same as one in the worksheet, the existing item is retained.

The form and its menus, as shown in **Figure 3-11 Menu sheet**, are added by the following code:

```
'ABC' ⎕wi 'Create' 'Form' 'Hide'
   CreateMenu 'ABC'
```

In order to verify the definition of the form as code, use the]WED editor:

```
   ABC_def←'ABC' ⎕wi 'def'
   ]wed ABC
```

```
   ∇ ABC_Make;x;⎕wself
[1]    ⍝∇ABC_Make -- Created 11/02/03
          at 17:43:22
[2]    'ABC' ⎕wi 'Delete'
[3]
[4]    ⎕wself←'ABC' ⎕wi 'New' 'Form'
          'Close'
[5]    ⎕wi 'caption' 'ABC'
[6]    ⎕wi 'where' 9.6 25.6 19.2 51.2
[7]
[8]    ⎕wself←'ABC.File' ⎕wi 'New'
```

Click **Form | Define ABC_Make** and then **Form | Exit**. This creates the function ABC_Make, see right.

Observe lines [20] and [21] creating a caption and shortcut.

```
         'Menu'
[9]      ⎕wi 'caption' '&File'
[10]     ⎕wi 'help context' 9001
[11]     ⎕wi 'prompt' 'Input/Output
         operations'
[12]
[13]     ⎕wself←'ABC.File.Open' ⎕wi
         'New' 'Menu'
[14]     ⎕wi 'caption' '&Open'
[15]     ⎕wi 'help context' 9002
[16]     ⎕wi 'prompt' 'Open an existing
         file'
[17]     ⎕wi 'onClick' 'Msg ''Clicked
         File.Open'''
[18]
[19]     ⎕wself←'ABC.File.Save' ⎕wi
         'New' 'Menu'
```

The CreateMenu function modifies the definition of the form object. Calling the Wait method displays the form as shown in **Figure 3-12 The ABC form with menus**.

`'ABC' ⎕wi 'Wait'`

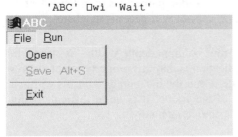

Figure 3-12 The ABC form with menus

```
[20]     ⎕wi 'caption'
         ('&Save',⎕TCHT,'Alt+S')
[21]     ⎕wi 'shortcut' 'S' 4
[22]     ⎕wi 'enabled' 0
[23]     ⎕wi 'help context' 9003
[24]     ⎕wi 'prompt' 'Open an
         existing file'
[25]     ⎕wi 'onClick' 'Msg
         ''Clicked File.Save'''
[26]
```

If the revised definition of the form is to be kept, either a **make** function can be created, see ABC_Make, or the **def** property of the object may be stored. If the form is created with other properties such as the ReSize event handler, both the 'make' function and the 'def' variable will retain it.

The facility for dynamic addition of menus to a form has several advantages:

● The menu structure can be edited externally; that is, outside the APL environment. This provides an ideal opportunity for involving users.

● The structure of the menu for the whole application may be modified in one place, the **Excel** worksheet in this example. This makes it easy to verify for consistency.

● The presentation of the menu structure within the application interface may be altered with minimum fuss.

● There is no need for repetitive code, and there is little, if any, compromise in performance.

```
[27]     ⎕wself←'ABC.File.S1'
         ⎕wi 'New' 'Menu'
[28]     ⎕wi 'caption' 'S1'
[29]     ⎕wi 'separator' 1
[30]
[31]     ⎕wself←'ABC.File.Exit'
         ⎕wi 'New' 'Menu'
[32]     ⎕wi 'caption' '&Exit'
[33]     ⎕wi 'help context'
         9002
[34]     ⎕wi 'prompt' 'End'
[35]     ⎕wi 'onClick' 'Exit'
[36]
[37]     ⎕wself←'ABC.Run' ⎕wi
         'New' 'Menu'
[38]     ⎕wi 'caption' '&Run'
[39]     ⎕wi 'help context'
         9004
[40]     ⎕wi 'prompt' 'Run
         Mode'
[41]
[42]
         ⎕wself←'ABC.Run.Aut
         omode' ⎕wi 'New'
         'Menu'
```

```
[43]   ⎕wi 'caption'
          '&Automode'
[44]   ⎕wi 'help context'
          9004
[45]   ⎕wi 'prompt' 'Run
          Mode'
[46]   ⎕wi 'style' 1
[47]   ⎕wi 'value' 1
  ∇
```

3.11 Designing forms

Code for GUI forms is just as repetitive as code for menus and can be automated using a similar technique. However, the task of generating forms is more complex.

• Each class has its own subset of properties and events; a common set of columns for each property and event must be shown on row 1. This means that the number of cells that may be superfluous for a given class increases. This introduces a bigger scope for errors.

• Controls on a form must have a navigation order, usually from top left to bottom right. It is no longer sufficient to create the first parent and its children and then the next and it is not viable to specify the order property for each control. The 'order' property is interdependent; it can easily be incorrect and a change in any particular control will necessitate changes to all other controls on the form.

• The correct specification of the 'where' property—row, column, depth, width—is critical in ensuring a visually appealing form. A further complication is that this property is not always specified in relation to the top left-hand corner of the form: the location of controls within a frame is calculated from the top left-hand corner of the frame. It is quite difficult if not impossible to calculate this property and it must be specified manually.

3.11.1 Enumerating an existing interface tree

The function VTree validates a tree that does not yet exist. A tree that exists can be enumerated by the following function:

```
      ∇ Z←Tree R
[1]    ⍝ System Building with APL+Win
[2]    Z←⊂R
[3]    R←R ⎕wi 'children'
{1[4]   :while 0≠⍴R
[5]        Z←Z,Tree ⎕io⊃R
[6]        R←1↓R
1}[7]   :endwhile
      ∇
```

The REL40.W3 workspace's *Open* function creates a form fmDemo; this form has 263 controls:

```
⍴Tree 'fmDemo'
263
```

3.11.1.1 The Tree control

The function Tree is easily modified to produce a hierarchical tree suitable as the value of the list property of an APL+Win Tree object, thus:

```
      ∇ Z←L TreeView R
[1]    ⍝ System Building with APL+Win
{1[2]   :if 0=⎕nc 'L'
[3]        L←1
1}[4]   :endif
[5]    Z←⊂L R
[6]    R←R ⎕wi 'children'
```

```
{1[7]    :while 0≠ρR
  [8]       Z←Z,(L+1) TreeView ⎕io⊃R
  [9]       R←1↓R
1}[10]   :endwhile
     ▽
```

This too is a recursive function, based on Tree; the left-hand argument is used internally only and must not be specified.

3.11.1.1.1 Tree control: with GUI objects

The function Show_fmDemo illustrates the use of the TreeView function to display the hierarchy of GUI controls:

```
     ▽ Show_fmDemo;⎕wself
[1]    ⍝ System Buiilding with APL+Win
[2]    ⎕wself←'fma' ⎕wi 'Create' 'Form' 'Close'
[3]    ⎕wi 'caption' (,'fmDemo as a Tree view')
[4]    ⎕wi 'where' 9.375 24.625 17.75 41.625
[5]    ⎕wself←'fma.tr' ⎕wi 'New' 'Tree'
[6]    ⎕wi 'list' (0,⊃TreeView 'fmDemo')
[7]    ⎕wi 'where' 0.5 0 14 39.5
[8]    'fma' ⎕wi 'Wait'
     ▽
```

This produces the basic structure required by the list property of the Tree control.

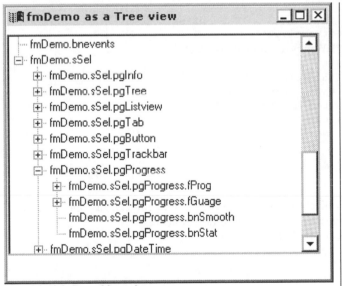

The tree object is shown in **Figure 3-13 fmDemo hierarchy** after expansion.

If the right-hand argument is specified as the system object, the tree of all the GUI objects defined is returned by:

```
TreeView '#'
```

Figure 3-13 fmDemo hierarchy

3.11.1.1.2 Tree control: with APL+Win arrays

Unfortunately, the Tree control only accommodates a single column—the technique for showing an associated column of information based on the node selected in the Tree control is illustrated in the REL40.W3 workspace. Windows Explorer is also an example of showing additional data relating to a selected node.

A tree structure may be built from an APL array and used to populate a tree control. Consider an ADO record set object:

```
     ▽ Z←GetRecordset;ADORS;Cnn;Sql
```

```
[1]    ⍝ System Building with APL+Win
[2]    'ADORS' ⎕wi 'Create' 'ADODB.Recordset'
[3]    Cnn←'Driver={SQL Server};Server=AJAY
         ASKOOLUM;Database=pubs;UID=SA;PWD=;'
[4]    Sql←'Select a.*,b.title from sales a, titles b where
         a.title_id=b.title_id'
[5]    'ADORS' ⎕wi 'XOpen' Sql  Cnn
[6]    Z←'ADORS' ⎕wi 'GetRows'
     ∇
```

This function gets an ADO record set; the first column is used to build a Tree control's list property value. To start, run the GetRecordset function:

```
      0 0⍴GetRecordset
ADORS
```

Now the record set object exists. The first column is used to populate a tree object. Next, locate the record pointer to the top:

```
      'ADORS' ⎕wi 'XMoveFirst'
```

The list property is created by the following function:

```
     ∇ Z←BuildTreeList R
[1]    ⍝ System Building with APL+Win
[2]    Z←((,R)⍳R)=⍳⍴,R
[3]    ((0=Z)/Z)←2
[4]    Z←0,(,[⌽]Z),R
     ∇
```

```
'ADORS' ⎕wi 'MoveFirst'                      | ⍴FirstColumn  | ≡FirstColumn
FirstColumn←'ADORS' ⎕wi 'GetRows(,,0)'       | 21 1          | 2
```

Another function Show_APLWin is defined as follows:

```
     ∇ Show_APLWin R;⎕wself
[1]    ⍝ System Building with APL+Win
[2]    ⎕wself←'fma' ⎕wi 'Create' 'Form' 'Close'
[3]    ⎕wi 'caption' ('A nested matrix as a Tree')
[4]    ⎕wi 'where' 9.375 24.625 17.75 41.625
[5]    ⎕wself←'fma.tr' ⎕wi 'New' 'Tree'
[6]    ⎕wi 'list' R
[7]    ⎕wi 'where' 0.5 0 14 39.5
[8]    'fma' ⎕wi 'Wait'
     ∇
```

The content of the record object can be displayed in a tree control: the objective is to enable the user to construct the desired view of the information:

```
      Show_APLWin BuildTreeList FirstColumn
```

This BuildTreeList function expects a nested array of rank 2 as the right-hand argument; an ADO record set object holds each column as a nested array of rank 2. A simple array is converted to a nested array using the following code:

```
array←c[⎕io+1]array
```

The variable *array* is an APL matrix of rank 2.

Although a Tree hierarchy—such as the one shown in **Figure 3-14 Nested matrix as a tree** does condense data for presentation, it is terse if not supplemented by complementary data in another control that is updated dynamically.

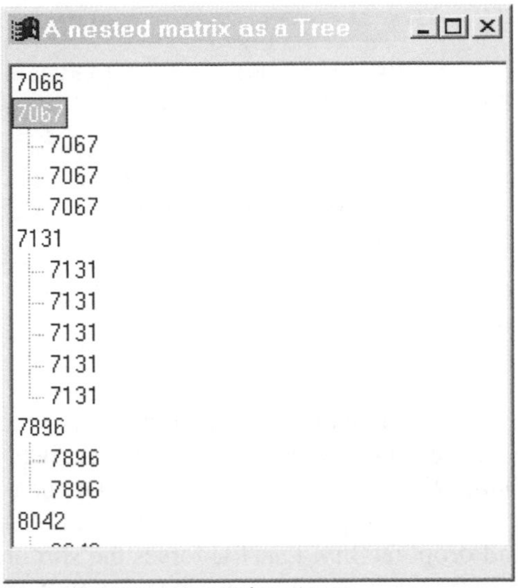

Figure 3-14 Nested matrix as a tree

3.11.1.2 The ListView and APL Grid controls

Like the Tree control, the *ListView* control requires the value of its list property to be constructed—as opposed to it accepting APL simple or nested arrays; however, the *ListView* control cannot show a hierarchical tree, because:

• Each column is a nested character array of rank 2.

• The value of all columns must be specified in one operation; while rows can be deleted or inserted, there is no means of inserting, or deleting rows.

The APL Grid object's column values may be populated using simple or nested arrays.

The developer now has a choice for displaying columnar data, in either a LISTVIEW control or the APL + Win Grid. Although both controls display their contents in a grid like structure, there are differences in the inherent properties of each object: this is summarised in **Table 3-4 ListView Grid feature comparison**. In situations where either control may be used, it is viable to give the user a choice.

Facility	ListView	APL Grid
XML	No	Yes
Populate from ADO Record set	Yes	Yes
Auto determine number of rows	Yes	No
Add/Delete Rows	Yes	Yes
Add/Delete Columns	No	Yes
Reorder columns	Yes/Set property	User code
Print content	No	No
Populate by Column	No	Yes
Populate whole content	Yes	No
Sort Columns by double click	Yes/Set property	User Code
Read only	Default, column one editable	User Code
Only column one editable	Yes/Set Property	User code
Drag & Drop	Yes/Property & Method	No

Table 3-4 ListView Grid feature comparison

3.11.1.2.1 Using ListView with a record set

The following function shows the record set created above in a *ListView* control:

```
  ∇ RSinListView R;⎕wself
[1]    ⍝ System Building with APL+Win
```

```
[2]    ⎕wself←'∆frm' ⎕wi 'Create' 'Form' 'Hide' ('caption' 'ADO Record set
         in ListView')
[3]    0 0⍴'.lst' ⎕wi 'New' 'Listview' ('where' 0 0 20 50)  ('style' 1024)
[4]    '.lst' ⎕wi 'onColClick' '⎕wself ⎕wi ''sortorder''  (Φ⎕wself ⎕wi
         ''sortorder'')'
[5]    '.lst' ⎕wi 'columndisplay' (⊃(⊂¨(⊂R) ⎕wi
         ¨(⊂⊂'xFields().Name'),¨¯1+⍳(R ⎕wi
         'xFields.Count')),¨⊂(⊂0),⊂'left'  )
[6]    R ⎕wi 'xMoveFirst'
[7]    '.lst' ⎕wi 'list' (0,1,⍢¨R ⎕wi 'GetRows')
[8]    '.lst' ⎕WI 'AutoFit' 'header' 'all'
[9]    'Set' ReSize ⎕wself
[10]   ⎕wi 'Wait'
     ∇
```

The orientation of the columns is specified as 'left' irrespective of their data type: in practice, numeric columns should be right justified. See Chapter 12 *Working with ActiveX Data Object (ADO)* for details of ADO data types.

The style property specified in line [3] allows columns to be relocated using the 'drag and drop' facility. Line [4] forces the sort order of the contents to be based on the *initial* first column, *stor_id*, and then on any column that is clicked. Refer to the APL+Win online help for implementing column independent sort order. Line [5] sets all columns as left justified.

In line [7], the complete record set is sent to the ListView control in a single operation. Although both the record set object and the ListView control use memory outside of the workspace, the record set content is briefly acquired into the workspace: this may cause a runtime error. Refer to the APL+Win documentation for writing rows sequentially; that is, one at a time, to the ListView control. The record set is shown in a list view control in **Figure 3-15 Record set in ListView**.

Note the date column: APL does not have a date data type and the dates are coerced into integers representing the number of days from a reference date.

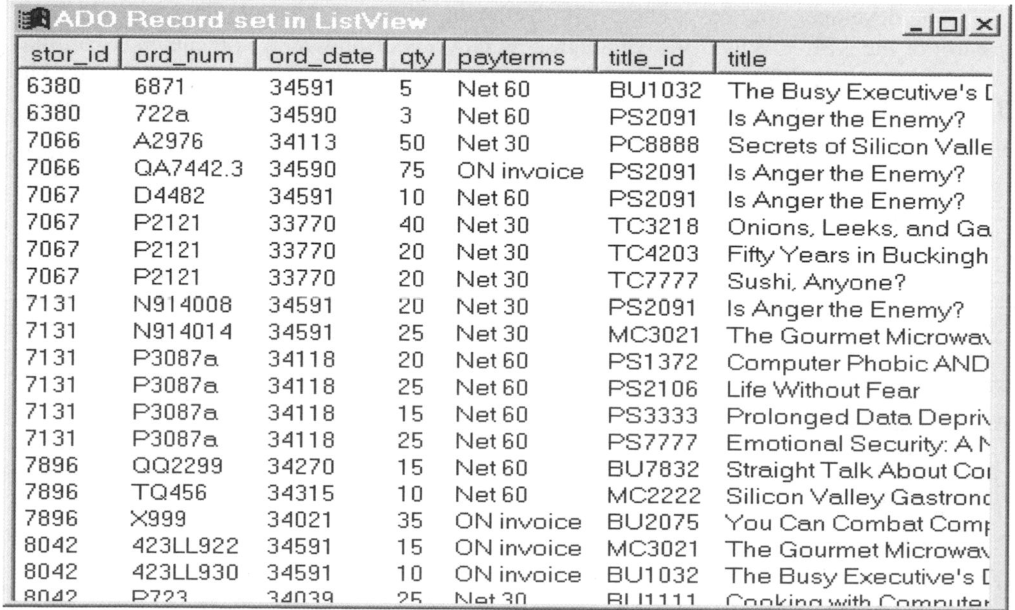

stor_id	ord_num	ord_date	qty	payterms	title_id	title
6380	6871	34591	5	Net 60	BU1032	The Busy Executive's [
6380	722a	34590	3	Net 60	PS2091	Is Anger the Enemy?
7066	A2976	34113	50	Net 30	PC8888	Secrets of Silicon Valle
7066	QA7442.3	34590	75	ON invoice	PS2091	Is Anger the Enemy?
7067	D4482	34591	10	Net 60	PS2091	Is Anger the Enemy?
7067	P2121	33770	40	Net 30	TC3218	Onions, Leeks, and Ga
7067	P2121	33770	20	Net 30	TC4203	Fifty Years in Buckingh
7067	P2121	33770	20	Net 30	TC7777	Sushi, Anyone?
7131	N914008	34591	20	Net 30	PS2091	Is Anger the Enemy?
7131	N914014	34591	25	Net 30	MC3021	The Gourmet Microwav
7131	P3087a	34118	20	Net 60	PS1372	Computer Phobic AND
7131	P3087a	34118	25	Net 60	PS2106	Life Without Fear
7131	P3087a	34118	15	Net 60	PS3333	Prolonged Data Depriv
7131	P3087a	34118	25	Net 60	PS7777	Emotional Security: A N
7896	QQ2299	34270	15	Net 60	BU7832	Straight Talk About Coi
7896	TQ456	34315	10	Net 60	MC2222	Silicon Valley Gastronc
7896	X999	34021	35	ON invoice	BU2075	You Can Combat Comp
8042	423LL922	34591	15	ON invoice	MC3021	The Gourmet Microwav
8042	423LL930	34591	10	ON invoice	BU1032	The Busy Executive's [
8042	P723	34039	25	Net 30	BU1111	Cooking with Computer

Figure 3-15 Record set in ListView

3.11.1.2.2 Using ListView with APL+Win arrays

Consider the following variables for showing in the *ListView* control:

```
      Functions←⎕nl 3
      Size←,[⊖]⎕size Functions
      Lines←⊃1↑¨ρ¨⎕cr ¨⊂[⎕io+1]Functions
   ∇ APLinListView R;⎕wself
[1]   ⍝ System Building with APL+Win
[2]   ⎕wself←'∆frm' ⎕wi 'Create' 'Form' 'Hide' ('caption' 'APL variables
         in ListView')
[3]   '.lst' ⎕wi 'New' 'Listview' ('where' 0 0 20 50)  ('style' 1024)
[4]   '.lst' ⎕wi 'onColClick' '⎕warg ◊ ⎕wself ⎕wi ''sortorder''  (Φ⎕wself
         ⎕wi ''sortorder'')'
[5]   '.lst' ⎕wi 'columndisplay' (⊃(⊂¨ R),¨0,¨⊂¨↑¨(0=¨1↑¨0ρ¨⍒¨R)↓¨⊂'left'
         'right')
[6]   '.lst' ⎕wi 'list' (0,1,⍒¨⍵⊂[(-~⎕io)+ρρR]R←⊃,[⊖]¨⊂[⎕io+1]¨⍒¨R)
[7]   '.lst' ⎕WI 'AutoFit' 'header' 'all'
[8]   'Set' ReSize ⎕wself
[9]   ⎕wi 'Wait'
   ∇
```

APL variables may be shown in the ListView control using APLinListView; the variable names are specified as the right-hand argument.

```
      APLinListView 'Functions' 'Size' 'Lines'
```

The result is shown in **Figure 3-16 APL variables in ListView**. Note that the numeric columns are right justified whereas the character column is left justified. The names of the APL variables are used as column headers.

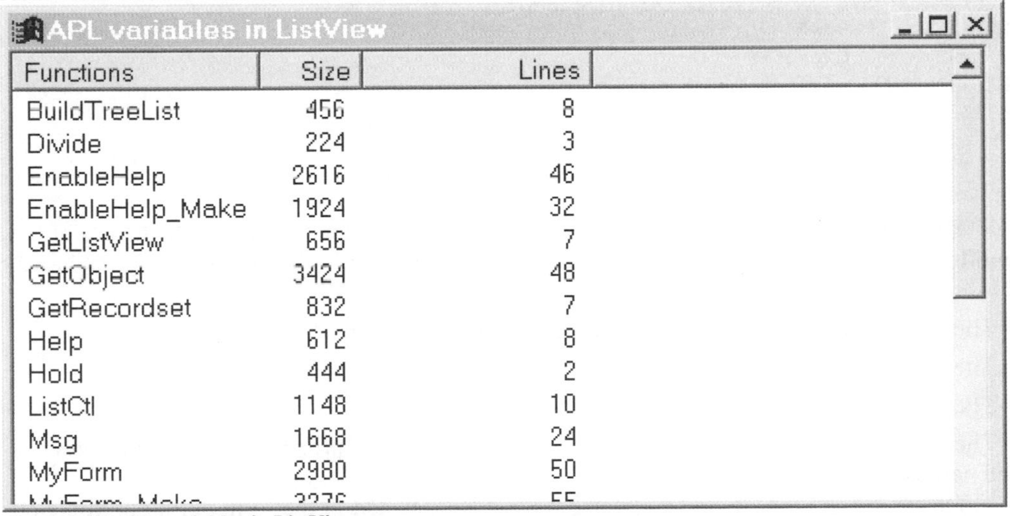

Functions	Size	Lines	
BuildTreeList	456	8	
Divide	224	3	
EnableHelp	2616	46	
EnableHelp_Make	1924	32	
GetListView	656	7	
GetObject	3424	48	
GetRecordset	832	7	
Help	612	8	
Hold	444	2	
ListCtl	1148	10	
Msg	1668	24	
MyForm	2980	50	
MyForm_Make	3376	55	

Figure 3-16 APL variables in ListView

Some useful APL techniques are used in APLinListView, namely:

● The type of a variable is returned by $0=1↑0ρVar$; the result is 1 if numeric or 0 if character. This corresponds to the APL+Win ⎕type function.

● The outer most level of nesting is removed by ⊃Var; this corresponds to the APL+Win specific ⎕mix function.

● The rank of any type of variable is increased by 1 using ,[⊖]Var; for numeric variables, Var∘.×,1 achieves the same purpose.

• A simple matrix of any type can be transformed into a nested vector using ⊂[(-
~⎕io)ρρVar]Var; if the matrix is known to be of rank 2, a simpler expression may be used
⊂[⎕io+1]Var. This also reduces the rank of the argument by 1. This corresponds to the
⎕split function.

The quad functions mentioned are unique to APL+Win: they are a legacy of the attempt
to provide compatibility with APL2 when the primitive functions behaved differently. Now
that APL+Win is consistent with APL2—assuming that evolution level is 2, the default—
there is little point in using these quad functions. The use of core APL primitives keeps code
meaningful to programmers of other vendors' APL.

3.11.2 Resizing forms

A constant source of frustration on the Windows platform is that the form Minimize and
Maximize buttons do not adjust the size of controls proportionately. It is common to assume
a 'universal' screen resolution, usually 800 x 600 pixels, at design time; however, unless
measures are in place for resizing the individual controls, the Minimize and Maximize
buttons destroy the aesthetic appearance of the form—with APL+Win, all the controls retain
their original size and are clustered in the top left-hand corner. The function ReSize forces
individual controls to be resized when a form's size is changed.

In order that ReSize acts on a form, a line is added in the function defining the form,
before the Wait statement—see line [9] of RSinListView. The Resize function is defined
below:

```
       ∇ L ReSize R;⎕wself
   [1]    ⍝ System Building with APL+Win
{1[2]     :if 0≠⍴R ⎕wi 'self'
   [3]        ⎕wself←R
{2[4]        :select L
* [5]            :case 'Enum'
   [6]               R←R ⎕wi 'children'
   [7]               ⎕wi ':data' ((⎕wi ':data'),R)
   [8]               (⊂'Enum') ReSize ¨R
```

This 'Enum' block of code is used internally; the data property is given the names of the
children of the form. Any controls added to the form after it is shown will not be included.
This property is a nested vector with the following elements:

• The starting location of the form; that is, its 'where' property.

• The starting extent of the form; that is, its 'extent' property.

• The names of all the child controls on the form.

• The starting location of each child control that has a 'where' property—in the same order as
the names—on the form.

The 'Set' ReSize ⎕wself statement must be included in the form definition function
just before the Wait method is invoked: this ensures that all the controls on the form have
been created and their properties are available. The block of code below captures the initial
size and relative location of individual controls on the form:

```
* [9]            :caselist 'Set' '.Set'
{3[10]              :select L
* [11]                 :case 'Set'
   [12]                    ⎕wi 'onMouseDouble' '''Restore'' ReSize
        ⎕wself'
3}[13]              :endselect
```

```
  [14]                 ⎕wi 'Set' ('onResize' '''ReSize'' ReSize ⎕wself')
         ('data' (⊂⎕wself))
  [15]                 'Enum' ReSize ⎕wself
  [16]                 L←1↓⎕wi 'data'
  [17]                 L←(~(L ⎕wi ¨⊂'class')∊⊂'Page')/L ⍝ Only children
         resized
  [18]                 L←(~∨≠↑¨(((L ⎕wi ¨⊂'class')∊'Status'
         'CommandBar')/L)∘.≤L)/L ⍝ None resized
  [19]                 L←((⊂⊂'where')∊¨L ⎕wi ¨⊂'properties')/L
  [20]                 L←L (0.005⌈4↑¨L ⎕wi ¨⊂'where')
{3[21]                 :if 0=⎕wi 'opened'
  [22]                     ⎕wi 'Hide'
3}[23]                 :endif
  [24]                 R←⎕wi 'Ref' 'where' 'size'
  [25]                 L[⎕io+1]←L[⎕io+1]÷⊂⊂4⍴(⎕io+1)⊃R
  [26]                 ⎕wi 'data' (R,L)
```

Either the 'Set' or '.Set' keyword may be used to initialise resizing; the former sets the onMouseDouble event to trigger a call to ReSize forcing the form to assume its original position and size, and the latter does not, in case this event is already used:

```
★ [27]                :case 'ReSize'
  [28]                (L R)←(0⊥0,(0 1 2=0⊥⎕warg)/1 0 1) (⎕wi 'data')
{3[29]                :if L
+ [30]                    :andif 3==R
  [31]                        ((⎕io+2)⊃R) ⎕wi
         ¨(⊂⊂'where'),¨0.1×⌊10×((⎕io+3)⊃R)×⊂4⍴⎕wi 'size'
3}[32]                :endif
```

This call recalculates the size of each control, increasing or decreasing the original size as a ratio of the original and current size of the form:

```
★ [33]                :case 'Restore'
  [34]                ⎕wi (⊂'Set'),(⊂¨'where' 'size'),¨2↑⎕wi 'data'
2}[35]        :endselect
1}[36]  :endif
         ∇
```

If enabled, the double-click event will call this block of code—it restores the form and its children to their original aspect ratio.

This function is not perfect and will require maintenance; its appeal is that it requires a single line in a form function rather than the modification of each control's definition. It is not perfect because there are no rules for resizing and mean different things to different users; for example, one school of thought is that some controls such as buttons, radio buttons, etc., should not be resized at all. It will require maintenance because the 'where' property of some controls is read only; that is, cannot be set. At present, the Status and CommandBar controls fall into this category. APL+Win does not have a facility for determining whether the 'where' property of a control is read only; lines [17] – [18] enumerates the class of such controls.

3.12 Access control

Access to an application—that is, permission to use it—requires some form of password protection scheme. A simple implementation can comprise of an edit box with style 128.

A constant source of frustration with multi-user applications is the need to control access to underlying resources, especially when the application relies on files rather than a database for input and output. By default, files cannot be used for multi-user concurrent access on the Windows platform.

3.12.1 File-based applications

For a file-based application, it is advisable to 'lock' the folder where the files are found on an exclusive 'first come first serve' basis in order to eliminate conflicts arising from concurrent access. Unless folders on local drives are shared, 'locking' applies to network drives only. The following function locks a file or folder:

```
        ∇ Z←SecureAccess R
   [1]    ⍝ System Building with APL+Win
{1[2]    :if 0≠⎕wcall 'PathFileExists' R
   [3]       R←R,'\secure.txt'
{2[4]       :if 0=⎕wcall 'PathFileExists' R
   [5]          Z←⁻1+⌊/0,⎕xnnums,⎕nnums
   [6]          R ⎕ncreate Z
   [7]          ⎕nuntie Z
2}[8]       :endif
   [9]       Z←⎕wcall '_lOpen' R 16
{2[10]      :if ⁻1=Z
   [11]         Z←0×Msg 'Folder ',R,' is in use. Please select another.'
+ [12]      :else
{3[13]         :if 0≠'#' ⎕wi '∆secure'
   [14]            0 0⍴wcall '_lClose' ('#' ⎕wi '∆secure')
3}[15]         :endif
   [16]         '#' ⎕wi '∆secure' Z
   [17]         Z←×Z
2}[18]      :endif
+ [19]   :else
   [20]      Z←0×Msg 'Unable to locate folder ',R
1}[21]   :endif
        ∇
```

The right-hand argument of this function is the name of the folder to which access is required—it is specified without a trailing \. Line [2] verifies the existence of the folder; if it is not found, an error message is signalled in line [17]. An alterative to signalling an error, when the target folder does not exist, is to create the folder specified using the MakeSureDirectoryPathExists. Delete lines [19] - [21] and replace line [2] with the following lines:

```
{1[2]    :if 0=⎕wcall 'PathFileExists' R
   [2.1]       0 0⍴⎕wcall 'MakeSureDirectoryPathExists' R
1}[2.2]   :endif
```

Line [4] of the original function verifies the existence of the file SECURE.TXT; if this file is not found, it is created. Thus, the application may be initialised in a new folder. Line [9] opens SECURE.TXT in exclusive mode; if this fails, the error in line [11] is signalled. Otherwise, the handle of the file is written to a user-defined property of the system object, in line [16], after an existing lock, if one exists, is cleared.

In this scenario, it is not possible to advise the user or computer that is using the target folder. If this is desirable, the name of the user or computer needs to be written to an INI file and the error message in line [11] may be modified to include this name.

3.12.1.1 Securing individual files

If access to a folder is based on access to the file SECURE.TXT located within it, all other files in that folder are still accessible by other applications. For instance, **Excel** may still open an XLS file in a 'secured' folder. In order to avoid conflict with other applications, the **Excel** file must first be secured on an exclusive basis using _lOpen; if the file is available, close the file using _lClose and use it immediately. Refer to Chapter 4 *Working with Windows* for further discussions of _lOpen and _lClose. This applies to folders on both local (i.e. unshared) and network (i.e. shared) drives.

```
        ∇ Z←SecureFile R
 [1]    ⍝ System Building with APL+win
{1[2]   :if 0≠⎕wcall 'PathFileExists' R
 [3]        Z←⎕wcall '_lOpen' R 16
{2[4]        :if ¯1≠Z
 [5]            0 0⍴⎕wcall '_lClose' Z
 [6]            Z←1
+ [7]        :else
 [8]            Z←0
2}[9]        :endif
+ [10]  :else
 [11]       Z←0
1}[12]  :endif
        ∇
```

This function returns 1 if the file is available or 0 otherwise. The following function determines availability and then uses a file:

```
        ∇ Z←App
 [1]    ⍝ System Building with APL+Win
{1[2]   :if SecureFile 'c:\schema.ini'
 [3]        Z←¯1+⌊/0.⎕nnums,⎕xnnums
 [4]        'c:\schema.ini' ⎕xntie Z
+ [5]   :else
 [6]        Z←0×Msg 'File ','c:\schema.ini',' is currently in use'
1}[7]   :endif
        ∇
```

Theoretically, this function is not foolproof in that during the moment the file is found to be available and its use, another application might grab the file; this might happen between line [3] and line [4]. In practice, this is a negligible risk.

3.12.1.2 Session inheritance

In order to ensure that an application resumes with the same configuration, it is necessary to record session parameters in a file as configuration changes are made. Usually, an INI or XML file located in the same folder is used for this purpose. What is recorded depends on the application. A typical set of entries is as follows:

```
[Session]
BasisFile=CLIENTX.SF
OutputCounter=3
```

Note that the folder name is not recorded with file names: if the INI or XML file can be accessed, then the folder name is already known. Any number of entries may be held. This mechanism allows the same or next user to resume with exactly the same configuration as in the previous session—a 'polite' application enjoys a higher degree of user acceptance.

3.12.2 Database applications

The management of concurrent access is much simpler with applications that rely of databases, especially server databases. The inherent facilities of the database access protocol may be used to eliminate conflicts. Should such an application also use the filing system, it is necessary to implement 'locking' facilities, as described.

3.13 Empower the user

All applications impoverish users by coercing them along pathways defined within them—this is the nature of computerised solutions and is acceptable. A successful application minimises the degree of coercion and does not dictate users' actions but accommodates them.

3.13.1 Prevent rather than trap errors

A common pitfall is to leave menu and toolbar options that are invalid within context still enabled, and then, to trap their invocation and to rebuke the user upon their invocation. A more preferable approach is to enable only the menu and toolbar options that are valid. This ensures that the user is visibly informed about what courses of actions are permissible without having to embark on them.

3.13.2 Validate user entries

User entries must be validated to eliminate runtime errors, usually at a later stage. One approach is to validate each entry after it is complete—completion is signalled by the act of exiting the field. Another approach is to validate all entries before closing a form. Both approaches are valid; exit validation ensures that each field entry is valid in its own right, whereas form validation ensures that all entries are mutually consistent. Where appropriate and acceptable, fields must have default entries in order to avoid the intensity of data entry.

3.13.3 Intrusive application messages

Any application message stops users' activity and demands user interaction. Such behaviour is welcome in some situations—for example in the event of a course of action that makes permanent changes such as deleting data from a database or deleting a file. On the other hand, a message that warns the user that they have not saved their data is an irritation unless it offers the user the choice of doing so.

3.13.4 Work with platform features

The Windows platform allows users to customise their **Desktop**, colour schemes, sound, and fonts. In addition, there are several Windows features that customise the hardware, such as sound sentry, sticky keys, magnifier, etc. The application should be receptive to these choices; for APL, this means using the system settings throughout. One exception might be to force the font to an APL font where APL symbols are necessary—in this situation it is courteous to provide an on-screen keyboard.

3.13.5 Tools | Options

A good model for giving users control over the configuration of an application is the **Tools | Options** facility in Office products. It is valid for applications to make assumptions about session parameter such as folders, file names, tooltips, system prompts, help style, etc. However, users must be able to override these assumptions in one convenient location—refer to the **Tools | Options** menu in **Word**. Users who are able to configure the application environment in line with their particular preferences use that application more effectively.

3.13.6 Navigation

Windows application menus are hierarchical in nature. The system design must be aware of the effort required to invoke the functionality of an application. Successful applications usually implement shortcuts to commonly-used system functions: usually, toolbar buttons are used. For example, the menu **File | Save** may save the current file—two steps—and the same may be achieved by clicking one button on a tool bar.

A more important consideration is the need to make the users' next step obvious. An experienced user knows the pathways within an application. All users know the reverse pathway—that is, how to get to the previous stage—and successful applications leave this option open. This is akin to the 'Undo' feature. Users master an application much more readily if they have the confidence that any change is not permanently destructive and any change may be re-attempted should they want to.

3.13.7 System legacy

All applications, except their first version, have a legacy in terms of the data they have created. Should the existing data become invalid in a future version, the application must ensure that it migrates the historical data seamlessly. Such structural changes are rare but do arise either as consequence of legislation changes or changes in internal practices or structural changes in system design.

Should an application need to migrate historical data, it should leave the source intact and rewrite the changes to another location, leaving the user to revert to a previous incarnation of the data/application if necessary. It is the application that needs to work with the objectives of the user and not the user who has to comply with the requirements of the application.

3.13.8 Functionality alone is not enough

A user interface that frustrates the user whether this is by design or incidentally guarantees the same outcome: a low level of user acceptance that causes the application to be rated as beta quality. An application that does not have user acceptance is not used efficiently whereas one that does tends to be used to its limits.

Applications need not only hide their complexity but also avoid snaring the user.

3.14 Sales considerations

Given the highly competitive nature of software, it is imperative that the design of an application maximizes the impact of a sales demonstration. To this end, the robustness of the application and its speed, the deployment of the Tree and ListView controls, a resizing and undo facilities, a data tier that supports the prominent industry databases, the ability to run unattended, and documentation are vital building blocks—all these are visible aspects of an application. This is especially useful for calculation intensive applications: unattended mode allows the user to leave the application running overnight.

Like users, decision makers cannot see behind the user interface. It is impossible to gauge the quality of the design, the elegance of the code, or the complexity of the calculations carried out by an application from its interface; therefore, these factors do not influence a purchasing decision. Nonetheless, it is quite easy to ask detailed questions about hardware requirement, concurrent usage, the three-tier design architecture, the configuration and installation of the application, to observe whether the application leaves enabled menus and buttons that are invalid within any given context, and to count the number of steps or clicks required to access the functionality that the application delivers. Prospective clients' assessment of all these metrics and their perception of the vendors' ability and disposable resources to provide training, on-site support when required, and ability to continue to enhance the application and fix bugs directly affect the continued viability of any software development project.

3.15 Application exit

An application exit routine must ensure that all session bound parameters are archived to a file—to enable resumption—and that all system resources are released.

Session bound parameters should be recorded as soon as they change. Although system resources are released automatically when an application terminates, some complementary house-keeping is usually necessary:

- Close any help sessions that may still be open.
- Close all native and component files.
- Clear any file/folder 'lock' using the _lClose API call.
- Where appropriate, terminate any independent applications started by the APL application.
- Delete any temporary files created by the application.
- Delete all GUI elements and use '#' ⎕wi 'ReleaseObjects'.

Chapter 4

Working with Windows

With APL, it is quite easy to start a session and ignore the operating system completely without any compromise of functionality. With first generation APL, every application had its own hallmark and each could do things in its own way. In essence, this was the era of process automation—it mattered little whether an application was easy to use or whether it was easier to use than the next application.

APL tended to provide a system function for virtually every operating system functionality, including hardware management, within the APL environment. The penalty for building such self-contained applications is that the developer has to provide 100% of the code, 100% of the maintenance, and had 100% responsibility for debugging the application.

Windows 3.1 heralded a new era for application development. It empowered users to have expectations about how an application should look and be used.

• It introduced **D**ynamic **D**ata **E**xchange (**DDE**)—a mechanism for applications to exchange data.

• It provided a rich set of **A**pplication **P**rogramming **I**nterface (**API**) calls, which meant that an application developer needed to provide less than 100% of the application code.

• It prescribed a standard interface for user interaction.

• Above all, an application is no longer simply the automation of a process. An application needs to provide incremental business value; that is, it needs to work with information on the same platform in any format and generate business information both for sharing with other applications and in its own right.

Windows itself has continued to evolve against a background of rapid technological advances in CPU speed and architecture as well as hard disk capacities; the concept underlying **DDE** evolved into other technologies, various guises of **O**bject **L**inking and **E**mbedding (**OLE**) and the **C**omponent **O**bject **M**odel (**COM**). APL*PLUS/PC too evolved into APL+Win, a compliant citizen of the Windows standard. Technology will continue to evolve, making the burden of legacy APL less and less viable.

4.1 The APL legacy

Throughout its history, APL has a continued to evolve. Today's APL empowers developers to use any available Windows resource. There remain two impediments to ensuring that APL has a legitimate role as a Windows applications development tool.

The APL protagonist will quite candidly argue that the APL way is better and more concise than any other language. In practice, this means that solutions which run quite fast can be produced much more quickly using APL than another language.

The APL mindset is self-indulgent, with the consequence that APL has shrunk during this period of isolation. In turn, this has implied that it has not been commercially viable to publish reference material on APL: the cascading effect is a slow uptake of the language and a lack of APL skills to maintain and enhance APL applications.

A common manifestation of the APL mindset is a strong reluctance to embrace the new features of APL itself. The rationale is *"if it works, don't fix it"*. As a result, the legacy code handed down via the tortuous route starting with APL*PLUS/PC is never revamped in line with the newer features of APL+Win. This includes a reticence in adopting control structures for new development and incorporating them in existing code. Without the guiding principle of *"innovate or die"*, APL may fail to realise the potential it is empowered to deliver.

4.1.1 Reinventing the wheel

The plea is not simply to discard legacy APL code in favour of versions that newer versions of APL have made possible—this is akin to reinventing the wheel—but to adopt/favour solutions if they are already available on the Windows platform and to improve those implemented in APL code.

Consider this example: an insurance application that needs to allow an optional paragraph on a document at the discretion of the underwriter. It is quite straightforward to provide a text box for collecting the actual text; this text must also be checked for spelling. APL does not have any spell checker! Although it is no doubt possible to write a spell checker in APL, it would be impertinent to contemplate doing so. APL can seamlessly call the **Word** spell checker with just a few lines of code. During this call, **Word** is not visible but the spell checker dialogue is shown within the APL application and allows corrections and substitutions to be made. The original or revised text is returned to APL depending on whether **Cancel** was clicked or revisions were made. The code to invoke the spellchecker is:

```
     ∇ Z←SpellCheck R;⎕wself
[1]    ⍝ System Building with APL+Win
[2]    ⎕wself←'Word' ⎕wi 'Create' 'Word.Application'
[3]    ⎕wi 'xDisplayAlerts' 0
[4]    0 0⍴⎕wi 'xDocuments.Add'
[5]    ⎕wi 'xSelection.text' R
[6]    0 0⍴⎕wi 'xDialogs(828).Show()' ⍝ 828 =
          wdDialogToolsSpellingAndGrammar
[7]    :if 1≠⍴⎕wi 'xSelection.Text'
[8]        Z←⎕wi 'xSelection.Text'
[9]    :else
[10]       Z←R
[11]   :endif
[12]   0 0⍴⎕wi 'xActiveDocument.Close(0)' ⍝ 0 = wdDoNotSaveChanges
[13]   0 0⍴⎕wi 'XQuit'
[14]   ⎕wgive 0
[15]   ⎕wi 'Delete'
     ∇
```

The standard **Word** spellchecker dialogue is shown in **Figure 4-1 Using the Word spell checker.** All the command buttons on the dialogue retain their default functionality; other settings such as auto-correction settings and dictionary are also retained:

```
     SpellCheck 'this is a cample sentence' ⍝ Corrections made
This is a sample sentence.

     SpellCheck 'this is a cample sentence' ⍝ Cancel clicked
this is a cample sentence
```

Although this demonstrates the basic facility, in practice, the application should provide a multi-line text box where the text may be entered and returned.

This is a powerful demonstration of how APL can deploy the power of existing tools. Note the **Options** button that allows further configuration of the task of grammar and spell checking.

Without any doubt, a chest comprising of just one tool can produce robust applications, but such applications wield a powerful negative and unacceptable anachronism in the view of decision makers—the applications are deemed to be non-standard applications.

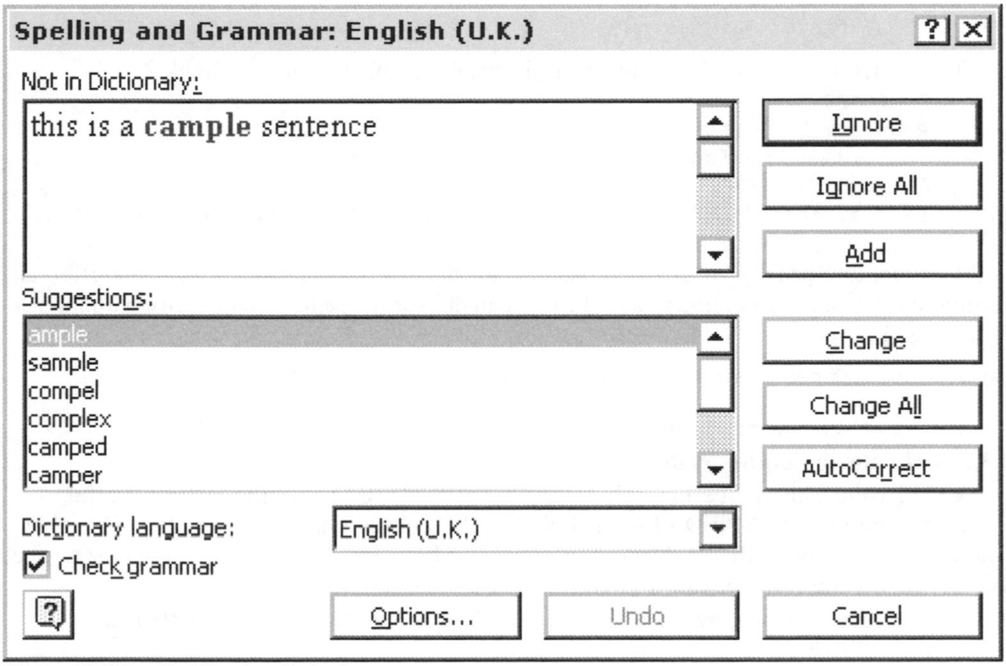

Figure 4-1 Using the Word spell checker

4.2 Windows resources

Three platform resources underpin the behaviour of all contemporary applications: API calls, the filing system, and application configuration via the **Registry**. EXtensible Mark-up Language (**XML**) is a preferred standard for cross-platform data storage and interchange.

4.3 API calls

API calls are a set of library routines within the operating system. An application development environment, such as APL, can call these routines to request the operating system to supply information and to direct the operating system to work with custom application settings. These are the important factors to note:

● The set of routines are commonly available to all development environments, thereby promoting consistency in the way applications are developed with different tools.

● The maintenance of existing and provision of new API calls is external to the development environment. Usually partial upgrades to the operating system or an upgrade to a new version of the operating system takes care of maintenance issues.

● API calls save a great deal of coding effort. The use of an API call instead of APL code to deliver a process can save the costs associated with the development and testing of the APL code as well as the overheads of packaging such code in an application. Moreover, the API calls are more efficient.

4.3.1 Replacing APL code by API calls

An example showing a do-it-yourself solution, typical of legacy APL systems and a corresponding API call can illustrate the contrast in coding style:

```
UPCASE ⊃'1. first' '2. second'           │ 1. FIRST
                                          │ 2. SECOND
UPCASE ¨'1. first' '2. second'           │ 1. FIRST 2. SECOND
```

This function occupies 408 bytes; its right-hand argument an array of rank 2:

```
      ∇ Z←UPCASE R
[1]    ⍝ System Building with APL+Win
[2]    Z←(~(,R)∊'abcdefghijklmnopqrstuvwxyz')/,R←⍪R
[3]    Z←((Z⍳Z)=⍳⍴Z)/Z
[4]    Z←(Z,'ABCDEFGHIJKLMNOPQRSTUVWXYZ')[(Z,'abcdefghijklmnopqrstuvwxyz')⍳R]
      ∇
```

The tangible penalty in this approach is not the size of the function but the effort in coding and testing it. In contrast, an API call can deliver the same functionality without any of the penalties:

```
□wcall 'CharUpper'  (⊃'1. first' '2. second')    │ 1. FIRST
                                                 │ 2. SECOND
□wcall ¨(⊂⊂'CharUpper'),¨⊂¨'1. first'  '2. second' │ 1. FIRST 2. SECOND
```

4.3.2 API calls to simplify code

Some API calls can replace multiple lines of APL code. Consider the task of creating the folder structure C:\MYAPP\DATA\CLIENT1\; APL does not have any facility for checking whether any or all folders in this hierarchy exist. In contrast, the MakeSureDirectoryPathExists API makes the task quite simple:

MakeSureDirectoryPathExists=L(*C lpPath) ALIAS MakeSureDirectoryPathExists LIB
    imagehlp.dll

Irrespective of its depth, a folder can be created by a single API call:

```
      □wcall 'MakeSureDirectoryPathExists' 'C:\MyApp\Data\Client1\'
```

Note that the path is terminated by backslash (\) otherwise the final level is treated as a file name and is discarded.

4.3.3 Formatting date and time

A familiar system requirement is to show a time stamp on reports. Surprisingly, this is a more complex problem than would appear at first. A number of decisions are involved: Use the system or user defaults for date and time format? Show long or short dates? Use the 12 or 24-hour clock? Cater for arbitrary date formats, such as DDD DD MM YYYY, or not? Show leading zeros in months, days, and years or not? The following function presents one solution:

```
       ∇ Z←L Now R
  [1]    ⍝ System Building with APL+Win
{1[2]    :if 0=□nc 'L'
  [3]        L←'LOCAL_USER_DEFAULT'
1}[4]    :endif
  [5]    Z←'LOCAL_USER_DEFAULT' 'LOCAL_SYSTEM_DEFAULT'
  [6]    L←0⊥(1,Z∊⊂L)/2 1/1024 2048
  [7]    R←0⊥(1,'Long' 'Short'∊⊂R)/2 1/32 31
```

```
   [8]    Z←⎕wcall 'CharUpper' (↑↑/⎕wcall 'GetLocaleInfo' L R (256⍴⎕tcnul)
          256)
   [9]    Z←Z,'+',⎕wcall 'CharLower' (↑↑/⎕wcall 'GetLocaleInfo' L 4099
          (256⍴⎕tcnul) 256)
  [10]    (Z R)←'On ' (Z~⎕tcnul)
{1[11]    :if 1∊'tt'∊R
  [12]        Z←(' ',⎕wcall 'CharLower' (↑↑/⎕wcall 'GetLocaleInfo' (⊖⍴L)
          (0⊥(1,120000<100⊥3↑3↓⎕ts)/40 41) (256⍴⎕tcnul) 256)),Z
  [13]        R←R~'t'
1}[14]    :endif
  [15]    R←∊'' ' At ',¨(R≠'+')⊂R
{1[16]    :while 0≠⍴R
{2[17]        :select ↑R
★ [18]            :case 'D'
{3[19]                :select +/∧\'D'=R
★ [20]                    :case 1
  [21]                        Z←Z,⍕1↑2↓⎕ts
★ [22]                    :case 2
  [23]                        Z←Z,,'G<99>' ⎕fmt 1↑2↓⎕ts
★ [24]                    :case 3
  [25]                        Z←Z,∊(¯1⌽↑¨↑/¨⎕wcall¨(⊂'GetLocaleInfo'
          L),¨(49+0,+\6/1),¨⊂(256⍴⎕tcnul) 256)[('WeekDay' FromISODate
          100⊥3↑⎕ts)-~⎕io]
+ [26]                    :else
  [27]                        Z←Z,∊(¯1⌽↑¨↑/¨⎕wcall¨(⊂'GetLocaleInfo'
          L),¨(42+0,+\6/1),¨⊂(256⍴⎕tcnul) 256)[('WeekDay' FromISODate
          100⊥3↑⎕ts)-~⎕io]
3}[28]                :endselect
  [29]                R←(~∧\'D'=R)/R
★ [30]            :case 'M'
{3[31]                :select +/∧\'M'=R
★ [32]                    :case 1
  [33]                        Z←Z,⍕1↑1↓⎕ts
★ [34]                    :case 2
  [35]                        Z←Z,,'G<99>' ⎕fmt 1↑1↓⎕ts
★ [36]                    :case 3
  [37]                        Z←Z,∊(↑¨↑/¨⎕wcall¨(⊂'GetLocaleInfo'
          L),¨(68+0,+\11/1),¨⊂(256⍴⎕tcnul) 256)[(1↑1↓⎕ts)-~⎕io]
+ [38]                    :else
  [39]                        Z←Z,∊(↑¨↑/¨⎕wcall¨(⊂'GetLocaleInfo'
          L),¨(56+0,+\11/1),¨⊂(256⍴⎕tcnul) 256)[(1↑1↓⎕ts)-~⎕io]
3}[40]                :endselect
  [41]                R←(~∧\'M'=R)/R
★ [42]            :case 'Y'
{3[43]                :select +/∧\'Y'=R
★ [44]                    :case 4
  [45]                        Z←Z,,'G<9999>' ⎕fmt 1↑⎕ts
+ [46]                    :else
  [47]                        Z←Z,,'G<99>' ⎕fmt 100|1↑⎕ts
3}[48]                :endselect
  [49]                R←(~∧\'Y'=R)/R
★ [50]            :case 'h'
{3[51]                :select +/∧\'h'=R
★ [52]                    :case 1
  [53]                        Z←Z,⍕1↑3↓⎕ts
```

```
★ [54]                        :case 2
  [55]                            Z←Z,,'G<99>' ⎕fmt 1↑3↓⎕ts
3}[56]                      :endselect
  [57]                      R←(~∧\'h'=R)/R
★ [58]                  :case 'm'
{3[59]                  :select +/∧\'m'=R
★ [60]                      :case 1
  [61]                          Z←Z,⍕1↑4↓⎕ts
★ [62]                      :case 2
  [63]                          Z←Z,,'G<99>' ⎕fmt 1↑4↓⎕ts
3}[64]                      :endselect
  [65]                      R←(~∧\'m'=R)/R
★ [66]                  :case 's'
{3[67]                  :select +/∧\'s'=R
★ [68]                      :case 1
  [69]                          Z←Z,⍕1↑5↓⎕ts
★ [70]                      :case 2
  [71]                          Z←Z,,'G<99>' ⎕fmt 1↑5↓⎕ts
3}[72]                      :endselect
  [73]                      R←(~∧\'s'=R)/R
+ [74]                  :else
  [75]                      (Z R)←(Z,↑R) (1↓R)
2}[76]              :endselect
1}[77]      :endwhile
  [78]    Z←(3×' '=⊖ρZ)⌽Z
  [79]    Z←(~' '⍷Z)/Z
        ∇
```

This function has the following characteristics:

• It circumvents the potential confusion between the M of month, that of minute by converting the date format into uppercase, and the time format into lowercase.

• It allows arbitrary formats and uses the separators specified in regional settings.

• It allows the user to select local system or user defaults, and long or short date formats.

The convention for regional time format is shown in Figure 4-2 Timestamp rules.

The AM/PM indicator may be inaccurate if the format includes 't' and the system does not show a 24-hour clock.

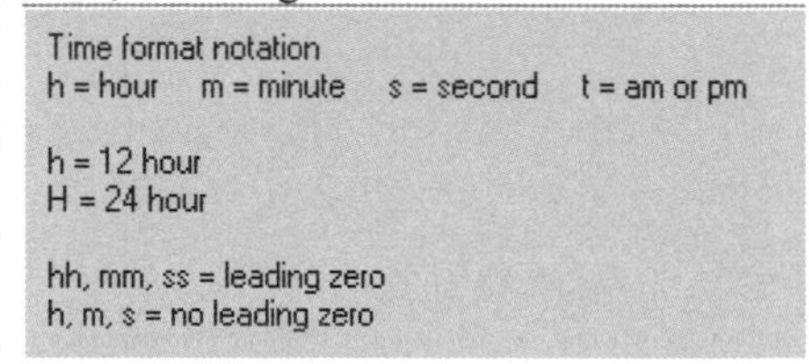

Time format notation
h = hour   m = minute   s = second   t = am or pm

h = 12 hour
H = 24 hour

hh, mm, ss = leading zero
h, m, s = no leading zero

Figure 4-2 Timestamp rules

Date formats shown in regional setting can be case sensitive; three Ds or Ms for day and month indicates a three-letter abbreviation for the day of the week or month of the year. Four or more Ds or Ms forces the full name to be shown. By default, the function returns the long date format and time using the local user defaults as follows:

Syntax	Result
Now ''	On 25 June 2003 At 02:07:50am
'LOCAL_SYSTEM_DEFAULTS' Now 'Short'	On 25/06/2003 At 02:10:08am
Now 'Long' ⍝ Long date format set to d MMMM d yyyy	On 25 June 25 2003 At 02:11:41am
Now 'Short' ⍝ Short date format set to yyyymmdd	On 20030625 At 02:14:07am

Should the date format include three or more D's, APL needs to calculate the day of the week. **VB** calculates the day of the week for a given date: it represents the day of the week sequentially: 1 for Sunday, 2 for Monday, 3 for Tuesday, 4 for Wednesday, 5 for Thursday, 6 for Friday, and 7 for Saturday. APL can adopt the same convention.

The function Now uses ⎕ts. That is, it works with a scalar date. For date manipulation, the ISO date format (yyyymmdd) is ideal for APL in that it can be held as a number and makes date comparisons straightforward. The function FromISODate uses ISO dates, as shown:

```
      ∇ Z←L FromISODate R
[1]    ⍝ System Building with APL+Win
[2]    ⍝ R=Date in ISO format yyyymmdd, numeric & valid
[3]    R←⍦10000 100 100⊤,R
{1[4]   :select L
★ [5]       :caselist 'Long' 'Short'
[6]             L←0⊥(1,'Long' 'Short'∊⊂L)/2 1/32 31
[7]             L←⎕wcall 'CharUpper' (↑↑/⎕wcall 'GetLocaleInfo' 1024 L
      (256⍴⎕tcnul) 256)
[8]         Z←(1 0×⍴R)⍴''
{2[9]       :while  0≠⍴L
{3[10]         :select ↑L
★ [11]            :case 'D'
{4[12]               :select +/∧\'D'=L
★ [13]                  :case 1
[14]                      Z←Z,⍕0 0 1/R
★ [15]                  :case 2
[16]                      Z←Z,'G<99>' ⎕fmt 0 0 1/R
★ [17]                  :case 3
[18]                          Z←Z,⊃(¯1↑↑¨↑/¨⎕wcall¨(⊂'GetLocaleInfo'
      1024),¨(49+0,+\6/1),¨⊂(256⍴⎕tcnul) 256)[,('WeekDay' FromISODate
      100⊥⍦R)-~⎕io]
+ [19]                       :else
[20]                             Z←Z,⊃(¯1↑↑¨↑/¨⎕wcall¨(⊂'GetLocaleInfo'
      1024),¨(42+0,+\6/1),¨⊂(256⍴⎕tcnul) 256)[,('WeekDay' FromISODate
      100⊥⍦R)-~⎕io]
4}[21]                   :endselect
[22]                     L←(~∧\'D'=L)/L
★ [23]             :case 'M'
{4[24]               :select +/∧\'M'=L
★ [25]                  :case 1
[26]                      Z←Z,⍕0 1 0/R
★ [27]                  :case 2
[28]                      Z←Z,'G<99>' ⎕fmt 0 1 0/R
★ [29]                  :case 3
[30]                          Z←Z,⊃(↑¨↑/¨⎕wcall¨(⊂'GetLocaleInfo'
      1024),¨(68+0,+\11/1),¨⊂(256⍴⎕tcnul) 256)[,(0 1 0/R)-~⎕io]
+ [31]                       :else
[32]                             Z←Z,⊃(↑¨↑/¨⎕wcall¨(⊂'GetLocaleInfo'
      1024),¨(56+0,+\11/1),¨⊂(256⍴⎕tcnul) 256)[,(0 1 0/R)-~⎕io]
4}[33]                   :endselect
[34]                     L←(~∧\'M'=L)/L
★ [35]             :case 'Y'
{4[36]               :select +/∧\'Y'=L
★ [37]                  :case 4
```

```
    [38]                              Z←Z,'G<9999>' ⎕fmt 1 0 0/R
 + [39]                          :else
    [40]                              Z←Z,'G<99>' ⎕fmt 100|1 0 0/R
4}[41]                          :endselect
    [42]                     L←(~∧\'Y'=L)/L
 + [43]                     :else
    [44]                     (Z L)←(Z,↑L) (1↓L)
3}[45]                  :endselect
2}[46]              :endwhile
    [47]              Z←⊃∈¨(⊂[⎕io+1]~(' '⍷Z)∨' ',⍷Z)⊂¨⊂[⎕io+1]Z
 ★ [48]         :case 'Day'
    [49]              Z←0 0 1/R
 ★ [50]         :case 'Month'
    [51]              Z←0 1 0/R
 ★ [52]         :case 'Year'
    [53]              Z←1 0 0/R
 ★ [54]         :case 'WeekDay' ⍝ Adapted from the algorithm at
        http://cybertips.topcities.com
    [55]              L←⍒×-/0=⍉4 100 400 4000∘.|1 0 0/R
    [56]              Z←¯1+(1 0 0/R)+-/⌊(1 0 0/R)∘.÷4 100 400 4000
    [57]              Z←(7|Z)-L×2≥0 1 0/R
    [58]              Z←,Z+(1 1 2 3 4 4 4 5 6 6 7 7)[1 10 5 8 2 3 11 6 9 12 4
        7⍳0 1 0/R]
    [59]              Z←(7|Z)⌽¨⊂+\7/1
    [60]              Z←(,0 0 1/R)⊃¨Z,¨(,0 0 1/R)⍴¨⊂+\7/1 ⍝ Sunday=1, Saturday=7
        like VB
1}[61]  :endselect
        ∇
```

This function disregards the time element and generalises the formatting of dates—scalar and arrays—in short or long date formats specified in regional settings. The following examples illustrate the use of FromISODate:

```
⊂[⎕io+1]'Long' FromISODate 3 1⍴18090212 18691002 17910922
 12 February 1809   02 October 1869    22 September 1791
 ⊂[⎕io+1]'Day' FromISODate 3 1⍴18090212 18691002 17910922
 12  2  22
⊂[⎕io+1]'Month' FromISODate 3 1⍴18090212 18691002 17910922
 2  10  9
⊂[⎕io+1]'Year' FromISODate 3 1⍴18090212 18691002 17910922
 1809  1869  1791
'WeekDay' FromISODate 3 1⍴18090212 18691002 17910922
1 7 5
```

With the short date format set to ddd d MMMM yyyy, it will also return the day of the week. An example is shown in **Table 4-1 Sample dates in regional format**.

```
        'Short' FromISODate 6 1⍴18090212 18691002 17910922 20040406 19290115
```

Sun 12 February 1809	Abraham Lincoln's birthday
Sat 2 October 1869	Mahatma Gandhi's birthday
Thu 22 September 1791	Michael Faraday's birthday
Tue 6 April 2004	Start of tax year 2004
Tue 15 January 1929	Martin Luther King's birthday

Table 4-1 Sample dates in regional format

The reverse process that converts dates from any arbitrary format to ISO format is more complex: the existing format can be varied. This function converts scalar dates to ISO format and the result is numeric:

```
        ∇ Z←L ToISODate R
   [1]    ⍝ System Building with APL+Win
   [2]    (L R)←(⎕wcall 'CharUpper' L) (,R)
{1[3]    :if 0∊R∊'0123456789'
   [4]        Z←100⊥⎕fi ∊' ',¨((R∊'0123456789')⊂R)['DMY'⍳'YMD']
+ [5]    :else
   [6]        Z←⍋'DMY'⍳L
   [7]        (L R)←(L[Z]) (R[Z])
   [8]        Z←⍋'YMD'⍳L
   [9]        (L R)←(L[Z]) (R[Z])
   [10]       Z←100⊥⎕fi (∊(1+L≠1⌽L)↑¨⍳1)\R
1}[11]   :endif
         ∇
```

Line [3] determines whether the date format contains delimiters:
```
'dd/mm/yyyy' ToISODate '01/12/2003'
20031201
```
Multiple dates, in the same format, can be converted using the following syntax.
```
(⊂'dd/mm/yyyy') ToISODate ¨'01/12/2003' '12/01/2003'
20031201 20030112
```
Multiple dates, each with their own format, can be converted using the following syntax:
```
'ddmmyyyy' 'yyyyddmm' ToISODate ¨'01012000' '19990105'
20000101 19990501
```

This function will not cope with dates that contain the names of days or months.

4.3.4 Using GetDateFormat and GetTimeFormat APIs

A neater alternative solution is to use Windows APIs albeit with the compromise that APIs can deal with a single timestamp at a time. Two APIs are required:

GetDateFormat=L(L Locale,L dwFlags,*SYSTEMTIME lpDate,*C lpFormat,>C lpDateStr,L cchDate) Alias GetDateFormatA LIB KERNEL32.DLL

GetTimeFormat=L(L Locale,L dwFlags,*SYSTEMTIME lpDate,*C lpFormat,>C lpDateStr,L cchDate) Alias GetTimeFormatA LIB KERNEL32.DLL

Both these API calls rely on the SYSTEMTIME data structure, defined in APLWADF.INI file:

SYSTEMTIME={W wYear, W wMonth, W wDayOfWeek, W wDay, W wHour, W wMinute, W wSecond, W wMilliseconds}

The SYSTEMTIME structure is the same as the ⎕ts structure except that it has one additional element at the third position. This parameter is the day of the week and has a default value of 0—forcing the API to calculate it as necessary.

4.3.4.1 Locale

See **Error! Reference source not found.** System Building Overview for details of *Locale*.

4.3.4.2 dwFlags

If the *lpFormat* argument is specified (i.e. is non-zero), then this argument has one of the values shown in **Table 4-2 Date/Time options**.

GetDateFormat		GetTimeFormat	
1	Use Regional Short Date format	1	Do not use minutes and seconds
2	Use Regional Long Date format	2	Do not use seconds

GetDateFormat		GetTimeFormat	
8	Use year/month format—i.e. do not use day. This *cannot* be added to 1 or 2.	4	Do not use time marker—am or pm
		8	Use 24-hour time format

Table 4-2 Date/Time options

If this argument is non-zero, then lpFormat must be zero.

4.3.4.3 lpDate and lpTime

These values correspond to ⎕ts with an element inserted between the second and third elements:

```
        1 1 0 1 1 1 1 1\⎕ts
2003 7 0 2 22 33 4 160
```

4.3.4.4 lpFormat

The value depends on whether the context requires a date or time format; the range of values is shown in **Table 4-3 Date/Time formats**.

GetDateFormat		GetTimeFormat	
dd or d	Day of month with or without leading zero.	Hh or h	Hours with or without a leading zero, using the 12-hour clock.
ddd or dddd	Abbreviated (3-letter) or full name of the day of week.	HH or H	Hours with or without a leading zero, using the 24-hour clock.
MM or M	Must be uppercase. Month of year with or without leading zero.	Mm or m	Must be lowercase. Minutes with or without a leading zero.
MMM or MMMM	Must be uppercase. Abbreviated (3-letter) or full name of month of year.	Ss or s	Seconds with or without a leading zero.
yy or y or yyyy	Year as two digits (with or without leading zero) or four digits.	Tt or t	Multi or single character time marker: A or P and AM or PM respectively.

Table 4-3 Date/Time formats

The names of the day of week and month of year are taken from the locale setting of the computer. All characters other than those listed in the first column of **Table 4-3 Date/Time formats** are returned in the same position in the formatted date or time string.

4.3.4.5 lpDatestr and lpTimestr

This specifies the name of the buffer into which the formatted date or time string will be written. The buffer must be long enough to accommodate the string.

4.3.4.6 cchDate and cchTime

This specifies the length, that is, the shape (ρ) of the buffer.

4.3.5 APL+Win GetDateFormat

The API call can be implemented as an APL function:

```
       ∇ Z←L GetDateFormat R
  [1]     ⍝ System Building with APL+Win
{1[2]    :select L
★ [3]        :case 'Short'
  [4]            Z←↑↑/⎕wcall 'GetDateFormat' 1024 1 R 0 (256ρ⎕tcnul) 256
★ [5]        :case 'Long'
  [6]            Z←↑↑/⎕wcall 'GetDateFormat' 1024 2 R 0 (256ρ⎕tcnul) 256
★ [7]        :case 'YearMonth'
  [8]            Z←↑↑/⎕wcall 'GetDateFormat' 1024 8 R 0 (256ρ⎕tcnul) 256
+ [9]        :else
  [10]           Z←↑↑/⎕wcall 'GetDateFormat' 1024 0 R L (256ρ⎕tcnul) 256
1}[11]   :endselect
       ∇
```

This function provides the options detailed in **Table 4-2 Date/Time options**. The interesting part of the function is in line [10], which allows user-defined formats. Thus if the YEARMONTH format is desired, it can be accomplished by:

```
     'MM yyyy' GetDateFormat 1 1 0 1 1 1 1 1\⎕ts
07 2003
```

4.3.5.1 APL GetTimeFormat

The APL version of the GetTimeFormat API call is:

```
     ∇ Z←L GetTimeFormat R
  [1]    ⍝ System Building with APL+Win
{1[2]    :select L
★ [3]        :case 'NoMinutesOrSeconds'
  [4]            Z←↑↑/⎕wcall 'GetTimeFormat' 1024 1 R 0 (256ρ⎕tcnul) 256
★ [5]        :case 'NoSeconds'
  [6]            Z←↑↑/⎕wcall 'GetTimeFormat' 1024 2 R 0 (256ρ⎕tcnul) 256
★ [7]        :case '24HourFormat'
  [8]            Z←↑↑/⎕wcall 'GetTimeFormat' 1024 4 R 0 (256ρ⎕tcnul) 256
★ [9]        :case 'NoTimeMarker'
  [10]           Z←↑↑/⎕wcall 'GetTimeFormat' 1024 8 R 0 (256ρ⎕tcnul) 256
★ [11]       :case 'Local'
  [12]           L←↑↑/⎕wcall 'GetLocaleInfo' 1024 4099 (256ρ⎕tcnul) 256
  [13]           Z←↑↑/⎕wcall 'GetTimeFormat' 1024 0 R L (256ρ⎕tcnul) 256
+ [14]       :else
  [15]           Z←↑↑/⎕wcall 'GetTimeFormat' 1024 0 R L (256ρ⎕tcnul) 256
        1}
  [16]   :endselect
     ∇
```

In addition to the options specified in **Table 4-2 Date/Time options**, this function allows the time to be formatted as in regional settings, lines [13] and [14], and in a custom format, line [15].

4.3.5.2 Formatting numeric and currency amounts

The APL ⎕fmt function is extremely powerful in transforming numeric data and, in conjunction with the GetLocaleInfo API, it would be possible to render numeric amounts in the format specified for numbers or currency in regional settings. However, this is hard work! The process is much simpler using two API calls:

GetNumberFormat=L(L Locale,L dwFlags,*C lpValue,*C lpFormat, >C lpNumberStr, L
    cchNumber) Alias GetNumberFormatA LIB KERNEL32.DLL

GetCurrencyFormat=L(L Locale,L dwFlags,*C lpValue,*C lpFormat, >C lpCurrencyStr, L
    cchCurrency) Alias GetCurrencyFormatA LIB KERNEL32.DLL

The lpFormat parameter (it is a structure) enables the regional settings to be overridden. Fortunately, the objective is to comply with regional settings and the lpFormat parameter can be specified as 0. The APL syntax for GetNumberFormat and GetCurrencyFormat is:

```
a←0 (256ρ⎕tcnul) 256 ⍝ Define as constant                │
↑↑/⎕wcall 'GetNumberFormat' 1024 0 '2333.678',a          │ 2,333.68
↑↑/⎕wcall 'GetNumberFormat' 1024 0 '-2333.678',a         │ -2,333.68
↑↑/⎕wcall 'GetCurrencyFormat' 1024 0 '2333.678',a        │ ú2,333.68
↑↑/⎕wcall 'GetCurrencyFormat' 1024 0 '-2333.678',a       │ -ú2,333.68
      ANSI2AV ↑↑/⎕wcall 'GetCurrencyFormat' 1024 0 '-2333.678',a
-£2,333.68
```

Some noticeable irregularities are: The API calls cope with scalar amounts only; it will be desirable to format arrays in APL. The amount must be specified as a string, with negative amounts showing the normal minus sign rather than the APL high minus. For currency formatting, the currency symbol is not visually meaningful for the UK until the ANSI2AV translation is made. A cover function resolves these minor irritations, as shown:

```
       ∇ Z←L RegionalFmt R
  [1]    ⍝ System Building with APL+Win
  [2]    R←⍕R
  [3]    ((('¯'=,R)/,R)←'-' ⍝ Replace high minus
{1[4]    :select L
★ [5]       :case 'Number'
  [6]          Z←↑↑/⎕wcall 'GetNumberFormat' 1024 0 R 0 (256⍴⎕tcnul) 256
★ [7]       :case 'Currency'
  [8]          Z←↑↑/⎕wcall 'GetCurrencyFormat' 1024 0 R 0 (256⍴⎕tcnul)
       256
  [9]             (('ú'=,Z)/,Z)←'£'
1}[10]   :endselect
       ∇
```

Numeric formatting can be done using the following syntax:

`'Number' RegionalFmt ¯2378.17 ⍝ Scalar`	-2,378.17
`(⊂'Number') RegionalFmt¨ ¯2378.17 100.34 ⍝ Vector`	-2,378.17 100.34
`(⊂'Number') RegionalFmt¨⊂[⎕io+1]2 1⍴ ¯2378.17 100.34 ⍝ Array`	-2,378.17 100.34
`'Currency' RegionalFmt ¯2378.17 ⍝ Scalar`	-£2,378.17
`(⊂'Currency') RegionalFmt¨ ¯2378.17 100.34 ⍝ Vector`	-£2,378.17 £100.34
`(⊂'Currency') RegionalFmt¨⊂[⎕io+1]2 1⍴ ¯2378.17 100.34 ⍝ Array`	-£2,378.17 £100.34

4.3.6 Fail safe date format translation?

Consider the following dates in US format, m/d/yyyy: 25/12/2000, 1/1/1999, and 12/8/2003. The first date is valid in the UK but clearly invalid for the US.

Unlike numbers and strings, dates always need to be considered with a given format albeit the format may be implicit; that is, it may be considered to be, say, the short date format specified in regional settings. Note the following:

• An APL solution will be too complex and may not handle all originating formats likely to be encountered within an international application.

• An alternative solution based on the APL+Win grid object is more viable—it has a locale property that is applicable to a range—as this yields an implicit format for dates.

• It is desirable to work with the regional short date format if it includes the century in the specification of the year. This implies that if the format must include the century: it can be reset using the SetLocaleInfo API call.

```
SetLocaleInfo=B(D Locale, D LCType, *C lpLCData) ALIAS SetLocaleInfoA LIB
    KERNEL32.DLL
```

The four-digit integer referred to may be interrogated on the local machine.

If the UK short date format specified in regional settings does not include the century in the year, it may be changed programmatically by reference to the LCID:

```
⎕wcall 'GetSystemDefaultLCID' ⍝ For the UK        | 2057
⎕wcall 'SetLocaleInfo' 2057 31 'dd/mm/yyyy'
```

Therefore, an APL application can, with the user's consent, change it silently. For the given problem, the US format dates may be translated, using the grid object, by the following function:

```
     ∇ Z←L LocaleDate R;⎕wself
[1]    ⍝ System Building with APL+Win
[2]    ⎕wself←'∆APLGrid' ⎕wi 'Create' 'APL.Grid'
[3]    Z←1↑⍴R
[4]    ⎕wi 'Set' ('Rows' Z) ('Cols' 1)
[5]    Z←+\Z/1
[6]    ⎕wi 'Set' ('xText' Z 1 R) ('xLocale' Z 1 (⊖⍴L))
[7]    R←(⎕wi 'xDate' Z 1)
[8]    ⎕wi 'Set' ('xLocale' Z 1 (0⊥L)) ('xDate' Z 1 R)
[9]    Z←⎕wi 'xText' Z 1
[10]   ⎕wi 'Delete'
     ∇
```

The left-hand argument specifies the originating and target locale and the right-hand argument specifies a nested vector of the dates that require translation:

```
1033 2057 LocaleDate 2 1⍴'25/12/2000' '12/8/2003'   │ 25/12/2000
                                                     │ 08/12/2003
```

Note that the first date is retained—as it is valid in the UK format—although it is invalid in US format, without a runtime error. The second date is formatted from the US format to the UK format. Any dates that are invalid in both the originating and target formats will cause a runtime error. Ideally, applications should translate dates to the ISO format—namely, yyyymmdd. The function ToISODate can transform the result of the LocaleDate function into the ISO format.

4.4 The Windows Script Host (WSH)

Subject to licensing requirements, WSH if free from http://msdn.microsoft.com/scripting. The Script Shell may be initialised in APL as follows:

```
APL+Win & APLX     ⎕wself←'∆Wsh' ⎕wi 'Create' 'WScript.Shell'
APL/W              '∆WSH' ⎕wc 'OLEClient' 'WScript.Shell'
```

This component exposes just one property, *CurrentDirectory*, and one method, *Exec*, and no events. Its other properties and methods are not exposed. *CurrentDirectory* is a read/write property; it can either return the current directory or set it:

```
APL+Win & APLX     ⎕wi 'xCurrentDirectory' 'c:\'           ⍝ Change
                   ⎕wi 'xCurrentDirectory'                 ⍝ Query
                   c:\
APL/W              ∆WSH.CurrentDirectory←'c:\'             ⍝ Change
                   ∆WSH.CurrentDirectory                   ⍝ Query
                   c:\
```

The *Exec* method starts an executable. Consider the following example:

```
     ∇ Notepad R;⎕wself
  [1]    ⍝ System Building with APL+Win
{1[2]    :if 0≠⎕wcall 'PathFileExists' R
{2[3]        :if 0≠⍴'∆notepad' ⎕wi 'self'
{3[4]            :if 0 = '∆notepad' ⎕wi 'status'
  [5]                '∆notepad' ⎕wi 'XTerminate'
3}[6]            :endif
2}[7]        :endif
  [8]        ⎕wself←'∆Wsh' ⎕wi 'Create' 'WScript.Shell'
```

```
 [9]       0 0ρ'∆notepad' ⎕wi 'Create' (⎕wi 'XExec' ('notepad.exe','
           ',R))
 [10]      ⍝ ⎕wcall 'SetWindowText' (⎕wcall 'FindWindow' 'Notepad' 0)
           (R,' - APL+Win')
+[11]   :else
 [12]      → 0×Msg 'Unable to find file ' ,R
1}[13]  :endif
         ∇
```

In this case, although NOTEPAD is not a COM object, the Script Shell allows an instance of NOTEPAD to be created as an APL object. Any EXE file may be started in the same manner although it might be necessary to use its fully qualified name: all instances created in this manner expose the same set of properties and methods; no events are exposed.

• The *Status* property is 0 if the EXE is still active and 1 if it has been terminated.

• The *ExitCode* property is 0 if the EXE is still active otherwise, it holds the exist code returned by the EXE on termination.

• The *ProcessId* property returns the process id of the EXE.

The *Terminate* method allows the EXE to be abruptly terminated. The NOTEPAD session can be terminated with the following command—it does not create a run tine error if the user has already terminated the EXE manually:

```
'∆notepad' ⎕wi 'XTerminate'
```

The Script Shell may be used to make the instance of the EXE the topmost window:

```
0 0ρ⎕wi 'AppActivate' ('∆notepad' ⎕wi 'xProcessId')
```

This may be deployed within the APL GUI, either via a button or via a menu option—since the APL GUI does not allow OLE object embedding, instances of external applications may be shown independently only.

4.4.1 Managing the Windows Registry

In essence, the **Registry** is the heart of a Windows computer. It is a hierarchical database containing information relating to all aspects of installed hardware and software, file types, and users' profiles. Applications dynamically create and update keys and sub-keys in order to hold configuration settings and provide continuity between sessions. The **Registry** is more secure than INI files and allows session management at system or user level.

Unlike the registry Win32 API calls, the WScript Shell provides simple to use methods for writing, reading, and deleting registry keys. APL can use either the API calls or the Script Shell for managing Registry entries. **Table 4-4 Managing the Registry** summarises the scope of a demonstration.

Key	Type	Value	Write	Read	Delete
ISBN	REG_SZ	0470030208	APL+Win	APLX	APL/W
PRICE	REG_DWORD	40	APLX	APLW	APL+Win
TITLE	REG_EXPAND_SZ	%TITLE%\Edition 1.0	APLW	APL+Win	APLX

Table 4-4 Managing the Registry

Write the three values to HKEY_LOCAL_MACHINE\SOFTWARE\BOOK using the Script Shell: the TITLE key value is an environment variable, created as follows:

```
⎕wcall 'SetEnvironmentVariable' 'TITLE' 'System Building with APL?'
```

This environment variable is volatile: it exists for the duration of the active session only. Create the APL instance of the Script Shell as follows:

```
APL+Win       ⎕wself←'∆Wsh' ⎕wi 'Create' 'WScript.Shell'
APLX          'WSS' ⍺wi 'Create' 'WScript.Shell'
APL/W         'WSS' ⎕WC 'OLEClient' 'Wscript.Shell'
```

4.4.2 Writing using WScript Shell
The generic syntax is: object.RegWrite(strName, anyValue [,strType]).

● 'strType' is optional; by default, it is REG_SZ.

● 'anyValue' is a value consistent with 'strType'; the WScript Schell enforces consistency if possible.

● 'strName' is a string specifying a key name within a hive; if it terminates with a \, the key name is (Default) otherwise, it is the last level in this string. The following expression writes the default value:

```
⎕wi 'XRegWrite' 'HKLM\SOFTWARE\BOOK\ISBN\' 'X0470030208X' 'REG_SZ'
```

This demonstrates the writing operations from **Table 4-4 Managing the Registry**:

```
⎕wi 'XRegWrite' 'HKLM\SOFTWARE\BOOK\ISBN' '0470030208' 'REG_SZ'
'WSS' ⎕wi 'XRegWrite' 'HKLM\SOFTWARE\BOOK\PRICE' 40 'REG_DWORD'
WSS.RegWrite 'HKLM\SOFTWARE\BOOK\TITLE' '%TITLE%\Edition 1.0' 'REG_EXPAND_SZ'
```

4.4.3 Reading using WScript Shell
The generic syntax is: object.RegRead(strName). The following expression reads the default value:

```
⎕wi 'XRegRead' 'HKLM\SOFTWARE\BOOK\ISBN\'
X0470030208X
```

This demonstrates the reading operations from **Table 4-4 Managing the Registry**:

'WSS' ⎕wi 'XRegRead' 'HKLM\SOFTWARE\BOOK\ISBN'	0470030208
10×WSS.RegRead 'HKLM\SOFTWARE\BOOK\PRICE'	400
⎕wi 'XRegRead' 'HKLM\SOFTWARE\BOOK\TITLE'	%TITLE%Edition 1.0

The numeric key value, PRICE, returns as a number. However, the key value TITLE is not resolved: this requires a further operation:

```
⎕wi 'ExpandEnvironmentStrings' (⎕wi 'XRegRead' 'HKLM\SOFTWARE\BOOK\TITLE')
System Building with APL?\Edition 1.0
```

4.4.4 Deleting using WScript Shell
The generic syntax is: object.RegDelete(strName). Before deleting the entries, it is worth seeing the changes to the registry hive; these are a REG file that looks as follows:

```
Windows Registry Editor Version 5.00

[HKEY_LOCAL_MACHINE\SOFTWARE\BOOK]
"ISBN"="0470030208"
"PRICE"=dword:00000028
"TITLE"=hex(2):25,00,54,00,49,00,54,00,4c,00,45,00,25,00,45,00,64,00,69,00,74,\
   00,69,00,6f,00,6e,00,20,00,31,00,2e,00,30,00,00,00

[HKEY_LOCAL_MACHINE\SOFTWARE\BOOK\ISBN]
@="X0470030208X"
```

An application may distribute the REG file, created from the development computer, as a means of updating other computers running that application; this is an alternative to writing the entries afresh.

This demonstrates the deletion operations from **Table 4-4 Managing the Registry**:

```
⎕wi 'XRegDelete' 'HKLM\SOFTWARE\BOOK\PRICE'
'WSS' ⎕wi 'XRegDelete' 'HKLM\SOFTWARE\BOOK\PRICE'
WSS.RegDelete 'HKLM\SOFTWARE\BOOK\TITLE'
```

At this point, only the entry for the default value for ISBN exists, as well as the hive HLKM\SOFTWARE\BOOK\. The following expressions remove the entries:

```
□wi 'XRegDelete' 'HKLM\SOFTWARE\BOOK\ISBN\'
□wi 'XRegDelete' 'HKLM\SOFTWARE\BOOK\'
```

4.4.5 Writing using Registry API

The API calls for managing the registry are much more complicated. Keys must be open before they can be updates, read, or deleted: thus, each operation must open the key, conclude its operation, and close it. This function encapsulates the write operation:

```
       ∇ Z←L RegWrite R
   [1]    ⍝ System Building With APL+Win
{1[2]    :if 0=□nc 'L'
   [3]       L←'REG_SZ'
1}[4]    :endif
{1[5]    :if (⊂L←□wcall 'CharUpper' L)∊'REG_SZ' 'REG_EXPAND_SZ'
         'REG_BINARY' 'REG_DWORD'
   [6]       (Z R L)←(⊂L),R ⍝ Key Value
   [7]       R[R⍳'\']←□tcnul
{2[8]       :if '\'∊R
   [9]          ((⌽R)[(⌽R)⍳'\'])←□tcnul
2}[10]       :endif
   [11]      R←(3↑((R≠□tcnul)⊂R),⊂''),⊂Z
{2[12]      :if 0=↑Z←□wcall 'RegCreateKeyEx' (□io⊃R) ((□io+1)⊃R)
         ((□io+3)⊃R) 'REG_OPTION_NON_VOLATILE' 'KEY_ALL_ACCESS' 0 θ θ
{3[13]         :select (□io+3)⊃R
=  [14]            :case 'REG_SZ'
   [15]               L←(θρ323 □dr (□io+1)⊃Z) ((□io+2)⊃R) 'REG_SZ' L
         (θρρL←⍕L)
=  [16]            :case 'REG_EXPAND_SZ'
   [17]               L←(θρ323 □dr (□io+1)⊃Z) ((□io+2)⊃R)
         'REG_EXPAND_SZ' L (θρρL←⍕L)
=  [18]            :case 'REG_BINARY'
   [19]               L←((2+⌊256⍟L)/256)⊤L←⌊⍳L ⍝ must be >0
   [20]               L←(θρ323 □dr (□io+1)⊃Z) ((□io+2)⊃R) 'REG_BINARY'
         (L) (θρρL)
=  [21]            :case 'REG_DWORD'
   [22]               L←(θρ323 □dr (□io+1)⊃Z) ((□io+2)⊃R) 'REG_DWORD'
         (⌽(4/256)⊤⌊L) 4
3}[23]         :endselect
   [24]         Z←↑□wcall (⊂'RegSetValueEx'),L
   [25]         0 0ρ□wcall 'RegCloseKey' (□io⊃L)
2}[26]      :endif
+  [27]   :else
   [28]      Z←0×Msg 'Unable to recognise data type ',L
1}[29]   :endif
       ∇
```

The left-hand argument defaults to REG_SZ; it may be one of REG_SZ, REG_EXPAND_SZ, REG_BINARY, or REG_DWORD. The right-hand argument specifies the hive and the value to write. The next function encapsulates the read operation.

```
       ∇ Z←L RegRead R
   [1]    ⍝ System Building With APL+Win
   [2]    R[R⍳'\']←□tcnul
{1[3]    :if '\'∊R
   [4]       ((⌽R)[(⌽R)⍳'\'])←□tcnul
```

```
1}[5]      :endif
  [6]      R←3↑((R≠⎕tcnul)⊂R),⊂''
{1[7]      :if 0=↑Z←⎕wcall (⊂'RegOpenKeyEx'),R[⍳2],0 'KEY_ALL_ACCESS' ⋄
  [8]          Z[⎕io+1]←⊂⊖⍴323 ⎕dr (⎕io+1)⊃Z
  [9]          L←(⎕io+1)⊃Z ⍝ Handle to open key
  [10]         R←(⎕io+2)⊃R
{2[11]         :if 0=↑Z←⎕wcall 'RegQueryValueEx' L R ⋄ 0 ⋄
{3[12]             :select ⊖⍴(⎕io+1)⊃Z
= [13]                 :case 1  ⍝ REG_SZ
  [14]                     Z←¯1↓(⎕io+2)⊃⎕wcall 'RegQueryValueEx' L R ⋄
         (((⎕io+3)⊃Z)⍴⎕tcnul) ((⎕io+3)⊃Z)
= [15]                 :case 2  ⍝ REG_EXPAND_SZ
  [16]                     Z←¯1↓(⎕io+2)⊃⎕wcall 'RegQueryValueEx' L R ⋄
         (((⎕io+3)⊃Z)⍴⎕tcnul) ((⎕io+3)⊃Z)
  [17]                     Z←↑↑/⎕wcall 'ExpandEnvironmentStrings' Z
         (256⍴⎕tcnul) 256
= [18]                 :case 3  ⍝ REG_BINARY
  [19]                     Z←2⊥11 ⎕dr(⎕io+2)⊃⎕wcall 'RegQueryValueEx' L R ⋄
         (((⎕io+3)⊃Z)/⎕tcnul) ((⎕io+3)⊃Z)
= [20]                 :case 4  ⍝ REG_DWORD
  [21]                     Z←↑¯1↓(⎕io+2)⊃⎕wcall 'RegQueryValueEx' L R ⋄ Z
         (323 ⎕dr (⎕io+3)⊃Z)
+ [22]                 :else
  [23]                     Z←⍬'
3}[24]             :endselect
+ [25]         :else
  [26]             Z←⍬
2}[27]         :endif
  [28]         0 0⍴⎕wcall 'RegCloseKey' L
+ [29]     :else
  [30]         Z←⍬
1}[31]     :endif
        ∇
```

This function attempts to convert the values it retrieves to numbers, if possible; see lines [14] to [16]. Finally, this function encapsulates the delete operation:

```
    ∇ Z←L RegDelete R
  [1]   ⍝ System Building With APL+Win
  [2]   R[R⍳'\']←⎕tcnul
  [3]   R←(R≠⎕tcnul)⊂R
{1[4]   :if 0=↑Z←⎕wcall (⊂'RegOpenKeyEx'),R,0 'KEY_ALL_ACCESS' (4⍴⎕tcnul)
  [5]       0 0⍴⎕wcall 'RegDeleteKey' (⊖⍴323 ⎕dr (⎕io+1)⊃Z) ''
+ [6]   :else
  [7]       ((Φ(⎕io+1)⊃R)[(Φ(⎕io+1)⊃R)⍳'\'])←⎕tcnul
  [8]       R←R[⎕io],(⎕tcnul≠(⎕io+1)⊃R)⊂(⎕io+1)⊃R
{2[9]       :if 0=↑Z←⎕wcall (⊂'RegOpenKeyEx'),R[⍳2], 0 'KEY_ALL_ACCESS'
         (4⍴⎕tcnul)
  [10]          0 0⍴⎕wcall 'RegDeleteValue' (⊖⍴323 ⎕dr (⎕io+1)⊃Z)
         ((⎕io+2)⊃R)
2}[11]      :endif
1}[12]  :endif
  [13]  0 0⍴⎕wcall 'RegCloseKey' (⊖⍴323 ⎕dr (⎕io+1)⊃Z)
  [14]  Z←↑Z
      ∇
```

The syntax of the API functions corresponds to those of the WScript Shell methods. All functions return 0, if successful, and require the full name of the root keys because the short name is missing from APLWADF.INI. The writing, reading, and deletion syntax are as follows:

```
RegWrite 'HKEY_LOCAL_MACHINE\SOFTWARE\MYBOOK\TITLE' '%TITLE%\Ver 1.0'
RegRead  'HKEY_LOCAL_MACHINE\SOFTWARE\MYBOOK\TITLE'
%TITLE%\Ver 1.0
```

In order to expand the environmental variable, this API is necessary:

```
ExpandEnvironmentStrings=D(*C lpSrc,>C lpDst, D nSize) ALIAS ExpandEnvir
    onmentStringsA LIB KERNEL32.DLL

A←RegRead  'HKEY_LOCAL_MACHINE\SOFTWARE\MYBOOK\TITLE'
↑↑/⎕wcall 'ExpandEnvironmentStrings' A (256⍴⎕tcnul) 256
System Building with APL?\Ver 1.0
```

Finally, the delete operation is as follows:

```
RegDelete 'HKEY_LOCAL_MACHINE\SOFTWARE\MYBOOK\TITLE'
```

4.4.6 Special folders

The Script Host's *SpecialFolders* method can return the values of special folders on the computer.

```
        ↑'∆Wsh' ⎕wi 'XSpecialFolders("Templates")'
C:\WINNT\Profiles\Administrator\Templates
```

Dedicated API calls exist for some but not all special folders:

```
GetTempPath=D(D,>C) ALIAS GetTempPathA LIB KERNEL32.DLL
        ↑↑/⎕wcall 'GetTempPath' 255 (255⍴⎕tcnul)
C:\WINNT\Profiles\ADMINI~1\LOCALS~1\Temp\
GetSystemDirectory=U(>C,U) ALIAS GetSystemDirectoryA LIB KERNEL32.DLL
        ↑↑/⎕wcall 'GetSystemDirectory' (255⍴⎕tcnul) 255
C:\WINNT\System32
```

The GetSpecialFolders function enumerates all folders or any named folders:

```
      ∇ Z←L GetSpecialFolders R;⎕wself
 [1]    ⍝ System Building with APL+Win
 [2]    ⎕wself←(R,'.Tmp') ⎕wi 'Create' (↑R ⎕wi 'XSpecialFolders')
{1[3]   :if 0=⎕nc 'L'
 [4]        L←'Named'
1}[5]   :endif
```

This function requires a right-hand argument—the name of an existing APL Script Shell object. The left-hand argument is optional; if a left-hand argument is specified, the valid values are *All* or *Named*:

```
{1[6]   :select L
* [7]       :case 'All'
 [8]            Z←⊂''
 [9]            0 0⍴⎕wi 'EnumStart'
{2[10]          :while 0≠⍴L←∊⎕wi 'EnumNext'
 [11]               Z←Z,⊂L
2}[12]          :endwhile
 [13]           Z←1↓Z
 [14]           Z←(((⊂'SpecialFolder'),¨⍕¨+\(⍴Z)/1)~¨' ') Z
 [15]           L←⍬ GetSpecialFolders R
 [16]           ((⎕io⊃Z)[(2⊃Z)⍳2⊃L])←⎕io⊃L
+ [17]      :else
 [18]           Z←'AllUsersDesktop' 'AllUsersStartMenu' 'AllUsersPrograms'
        'AllUsersStartup'
```

```
     [19]            Z←Z,'Desktop' 'Favorites' 'Fonts' 'MyDocuments' 'NetHood'
            'PrintHood'
     [20]            Z←Z,'Programs' 'Recent' 'SendTo' 'StartMenu' 'Startup'
            'Templates'
     [21]            Z←Z (⎕io⊃¨⎕wi ¨(⊂⊂'xItem'),¨Z)
1}[22]    :endselect
{1[23]    :if L≡0
> [24]        :return
+ [25]    :else
  [26]        (R,'.Tmp') ⎕wi 'Delete'
  [27]        '#' ⎕wi 'ReleaseObjects'
{2[28]        :if 0≠⍴Z
  [29]            (⎕io⊃Z)←'∆',¨⎕io⊃Z
2}[30]        :endif
1}[31]    :endif
       ∇
```

The list of special folders may be fixed in the workspace by the following expression:
```
      a←'All' GetSpecialFolders '∆Wsh'
      ⍴⎕io⊃a
18
```

4.4.7 Environment variables

The Script Shell's *Environment* method can return the values of environment variables on the computer by type: the default value for the optional argument of the *Environment* method is *System* for Windows NT and 2000 and *Process* for other versions; it may be one of *System*, *User*, *Process*, or *Volatile*. The collection of names returned depends on the argument. Volatile variables contain only the variables that are session bound; that is, available only to the session that creates them.

The GetEnvironmentVariables function enumerates all folders or any named folders:
```
      ∇ Z←L GetEnvironmentVariables R;⎕wself
  [1]     ⍝ System Building with APL+Win
{1[2]     :if 0=⎕nc 'L'
  [3]         L←'System'
1}[4]     :endif
{1[5]     :select L
* [6]         :case 'Process'
  [7]            ⎕wself←(R,'.Tmp') ⎕wi 'Create' (↑R ⎕wi
            'XEnvironment("Process")')
* [8]         :case 'User'
  [9]            ⎕wself←(R,'.Tmp') ⎕wi 'Create' (↑R ⎕wi
            'XEnvironment("User")')
* [10]        :case 'Volatile'
  [11]           ⎕wself←(R,'.Tmp') ⎕wi 'Create' (↑R ⎕wi
            'XEnvironment("Volatile")')
+ [12]        :else
  [13]           ⎕wself←(R,'.Tmp') ⎕wi 'Create' (↑R ⎕wi
            'XEnvironment("System")')
1}[14]    :endselect
  [15]    Z←⊂''
  [16]    0 0⍴⎕wi 'EnumStart'
{1[17]    :while 0≠⍴L←∈⎕wi 'EnumNext'
  [18]        Z←Z,⊂R  ⎕wi 'XExpandEnvironmentStrings' L
1}[19]    :endwhile
  [20]    Z←1↓Z
```

```
{1[21]    :if  0≠ρZ
  [22]         Z←(Z≠¨'=')⊂¨Z
  [23]         Z←(⎕io⊃¨Z)((⎕io+1)⊃¨Z)
  [24]         (⎕io⊃Z)←'∆',¨⎕io⊃Z
1}[25]    :endif
  [26]    (R,'.Tmp') ⎕wi 'Delete'
  [27]    '#' ⎕wi 'ReleaseObjects'
         ∇
```

The optional left-hand argument, default System, can be Process, User, Volatile or System; the right-hand argument is the name of the APL Script Shell object. This block of code enumerates the available environment variables. A single variable can be returned directly:

```
'∆Wsh' ⎕wi 'XEnvironment.Item("WINDIR")'
%SystemRoot%
```

Variables that are coded or encrypted may be decoded by the Script Shell's *ExpandEnvironmentStrings* method.

```
'∆Wsh' ⎕wi 'XExpandEnvironmentStrings' '%SystemRoot%'
C:\WINNT
```

4.4.8 Setting/Reading environment variables

Environment variables may be interrogated and reassigned arbitrarily: their names have the following characteristics:

- They are not case sensitive.

- They can only hold string values.

- They need *not* be valid variable names.

```
'∆Wsh' ⎕wi 'XEnvironment.Item("apl+win")' 'System Building'
```

This syntax is somewhat restrictive in that the name of the variable is integral to the first element of the right-hand argument of ⎕wi. A general form of the syntax is:

```
'∆Wsh' ⎕wi 'XEnvironment().Item()' Environment Name Value
```

The fully qualified syntax for reading the variable is:

```
'∆Wsh' ⎕wi 'XEnvironment().Item()' Environment Name
```

The variable *apl+win* may be read with this line:

```
'∆Wsh' ⎕wi 'XEnvironment().Item()' 'System' 'apl+win'
System Building
```

Variables may be set or read using either the Script Shell or Windows API:

```
SetEnvironmentVariable=B(*C lpName, *C lpValue) ALIAS SetEnvironmentVariableA LIB
    KERNEL32.DLL

GetEnvironmentVariable=D(*C,>C,D) ALIAS GetEnvironmentVariableA LIB KERNEL32.DLL
```

The APL syntax is:

```
⎕wcall 'SetEnvironmentVariable' 'apl+win' 'New Value'
↑↑/⎕wcall 'GetEnvironmentVariable' 'APL+WIN' (256ρ⎕tcnul) 256
New Value
```

The API calls work with the *Process* environment only. Both the Script Shell and the API calls return a string of length zero when a non-existent variable is queried.

4.4.9 Deleting an environment variable

Unlike the API call, which works with the *Process* environment only, the Script Shell can delete a variable from any environment. The general syntax is:

```
'∆Wsh' ⎕wi 'XEnvironment().Remove' Environment Variable
```

This can be encapsulated in a function:

```
      ∇ L DeleteEnvironmentVariable R;⎕wself
[1]    ⍝ System Building with APL+Win
[2]    ⎕wself←R
[3]    ⎕wi (⊂'XEnvironment().Remove'), L
      ∇
```

The right-hand argument is the name of an existing APL Script Shell object. The left-hand is a nested vector of *Environment* and *Variable*. An example:

```
    'Process' 'apl+win' DeleteEnvironmentVariable '∆Wsh'
```

4.5 Creating a shortcut

A powerful demonstration of a sophisticated application is its ability to create a shortcut on the **Desktop**. Usually, this would be done with the user's consent; that is if the user responds in the affirmative to a prompt regarding the creation of the shortcut on the **Desktop**.

The following function creates a shortcut; its right-hand argument is the name of an EXE and its left-hand argument is the argument that is passed to the EXE on start up:

```
        ∇ Z←L CreateShortcut R;⎕wself;ExePath;FileName;FilePath;Desktop
   [1]    ⍝ System Building with APL+Win
   [2]    ⍝ L=Fully qualified name of file R=Name of Shortcut
   [3]    ExePath←(↑↑/⎕wcall 'FindExecutable' R '' (1024ρ⎕tcnul) )~⎕tcnul
{1[4]    :if 0≠ρExePath
   [5]        ExePath←⎕wcall 'CharUpper' ExePath
{2[6]        :if Z←1∊'EXE'∊ExePath
   [7]            ⎕wself←'Shell' ⎕wi 'Create' "WScript.Shell"
   [8]            Desktop←⎕io⊃⎕wi 'SpecialFolders("Desktop")'
   [9]            R←-('\'≠⌽ExePath)ι0
   [10]           FilePath←(R↓ExePath),'\'
   [11]           FileName←⎕io⊃(~(R↑ExePath)∊'\.')⊂R↑ExePath
   [12]           ⎕wi 'CreateShortcut>∆Lnk'    (Desktop,'\',FileName,'.lnk')
   [13]           ⎕wself←'∆Lnk'
   [14]           ⎕wi 'xArguments' L
   [15]           ⎕wi 'xDescription'  ('Shortcut to ',ExePath)
   [16]           ⎕wi 'xHotKey' ''
   [17]           ⎕wi 'xIconLocation' (ExePath,',2')
   [18]           ⎕wi 'xTargetPath'   ExePath
   [19]           ⎕wi 'xWindowStyle' 3
   [20]           ⎕wi 'xWorkingDirectory' FilePath
   [21]           ⎕wi 'XSave'
   [22]           ⎕wi 'Delete'
   [23]           'Shell' ⎕wi 'Delete'
2}[24]       :endif
+ [25]   :else
   [26]       → 0×Msg 'Unable to find ',R
1}[27]   :endif
        ∇
```

This function uses an API call to recover the fully qualified path and name of the EXE, if possible:

```
FindExecutable=H(*C lpFile,*C lpDirectory,>C lpResult) ALIAS FindExecutableA LIB
     Shell32.DLL
       ↑↑/⎕wcall 'FindExecutable' 'APLW.EXE' '' (1024ρ⎕tcnul)
D:\Program Files\APLWIN\APLW.EXE
```

Thus, if an application workspace is saved in a certain state and will need to be restarted later, a shortcut may be placed on the **Desktop** both as a reminder to the user and as a convenient means of launching it:

```
      'C:\MYAPP\SPECIAL.W3' CreateShortcut 'APLW.EXE'
1
```

4.5.1 Starting another application

The FindExecutable API call is useful for retrieving the fully qualified location of an EXE; however, it does not always work:

```
      ↑↑/⎕wcall 'FindExecutable' 'NOTEPAD.EXE' '' (1024ρ⎕tcnul)
C:\WINDOWS\NOTEPAD.EXE
      ρ(↑↑/⎕wcall 'FindExecutable' 'EXCEL.EXE' '' (1024ρ⎕tcnul))~⎕tcnul
0
```

Although EXCEL.EXE does exist, its location is not found. Should an APL application need to start another application, it will need to locate the EXE prior to calling the WinExec API call.

```
WinExec=U(*C lpCmdLine,U mCmdShow) LIB KERNEL32.DLL
```

For example, it is simpler for APL to call NOTEPAD to print a text file than to do so via an instance of the APL Printer object:

```
⎕wcall 'WinExec' 'C:\WINDOWS\NOTEPAD.EXE /p "c:\aa.vbs"''SW_SHOWMINIMIZED'
```

A return code greater than 31 indicates a successful call. This opens NOTEPAD and prints the file to the default printer. However, the instance of NOTEPAD persists and must be terminated:

```
⎕wcall 'SendMessage' (⎕wcall 'FindWindow' 'Notepad' 0) 'WM_CLOSE' 0 0
```

4.6 Intelligent file operations with API calls

As mentioned already, APL lacks the facility—and, FSO does not provide it—for determining whether a file is available for access. The usual technique for determining whether a file is available for access is to tie or open it with an error trap—on failure, the error trap returns a value that the file is currently unavailable. There is some ambiguity as it is unclear whether the tie or open operation failed because the file is in use or nonexistent.

How are some applications able to produce a message such as the one shown in **Figure 4-3 File locking**? Note that the reason for the file's unavailability is given.

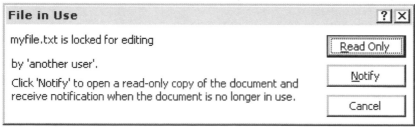

Figure 4-3 File locking

A number of standard Windows API calls provide the building blocks for intelligent file handling.

4.6.1 DeleteFile API

```
DeleteFile=B(*C lpFileName) ALIAS DeleteFileA LIB KERNEL32.DLL
```

This API returns 1 on deletion and 0 on failure to delete – failure to delete may occur if the file does not exist in the first place or if the file is in use, as shown:

```
      ⎕wcall 'DeleteFile' 'c:\aa.sf'
0
```

```
        'c:\myfile' ⎕fcreate 10
        ⎕wcall 'DeleteFile' 'c:\myfile.sf'
0
        ⎕funtie 10
        ⎕wcall 'DeleteFile' 'c:\myfile.sf'
1
```

Although the API call is simpler than having to tie the file and then deleting it, it is not safe unless it returns 1. A result of 0 indicates either that the file did not exist or that it is in use. A related API call is:

MoveFile=D(*C lpExistingFileName,*C lpNewFileName) ALIAS MoveFileA LIB KERNEL32.DLL

4.6.2 PathFileExists API

PathFileExists=L(*C pszPath) ALIAS PathFileExistsA LIB shlwapi.dll

This API will return 1 if a file or path exists—irrespective of whether it is in use—and 0 otherwise:

```
        'c:\myfile' ⎕fcreate 10
        ⎕wcall 'PathFileExists' 'c:\myfile.sf'
1
```

Note that the DeleteFile and PathFileExists API calls take the fully qualified names of files—DeleteFile cannot delete folders but PathFileExists can query the existence of both files and folders. APL extended component file functions do not require the extension under normal circumstances. In addition, APL may refer to a path by a library number—the API calls cannot make any sense of library numbers.

4.6.3 _lOpen API

The API definition is:

lOpen=I(*C lpPathName,I iReadWrite) ALIAS _lOpen LIB KERNEL32.DLL

The iReadWrite access and share mode parameters are mutually exclusive within their own respective sets. The _lOpen API returns a positive integer greater than zero in the event of success—this number represents a file handle, which can be passed to _lClose to close the file. In the event of failure, it returns ⁻1.

Table 4-5 lOpen API constants lists the access and share modes for the iReadWrite parameter, which is the sum of one or more values.

	Decimal Value	Description
Access Mode		
READ	0	Open file with read access only
READ_WRITE	1	Open file with read and write access
WRITE	2	Open file with write access only
PLUS Optional Share mode		
OF_SHARE_COMPAT	0	Open file in compatibility mode, allow other applications on the same machine to open the file
OF_SHARE_EXCLUSIVE	16	Open file in exclusive mode, deny other applications read and write access
OF_SHARE_DENY_WRITE	32	Open file, and deny other programs write access
OF_SHARE_DENY_READ	48	Open file, and deny other programs read access
OF_SHARE_DENY_NONE	64	Open file, and allow other applications read or write access

Table 4-5 lOpen API constants

4.6.4 _lClose API

The API definition is:

lClose=I(I hFile) ALIAS _lClose LIB KERNEL32.DLL

This API returns 0 in the event of success and ¯1 otherwise. The result is ¯1 when a non-existent or invalid handle is passed to _lClose:

```
⎕wcall 'lOpen' 'c:\aa.vbs' 16 │ ⎕wcall 'lClose' 12 │ ⎕wcall 'lClose' (?100)
12                              │ 0                   │ ¯1
```
```
      ↑↑/⎕wcall 'FormatMessage' 4096 0 (⎕wcall 'GetLastError') 0
(200ρ⎕tcnul) 200 0
The handle is invalid.
```

4.6.5 Fail safe deletion status

It may be desirable to delete a file for two reasons – to be able to recreate it afresh or to remove it from the filing system. In other words, it is necessary to know if the file still exists after its deletion; if it still exists, any attempt to recreate it will generate an error. The APL function DeleteFile returns a Boolean result: 1 if the file no longer exists, and 0 if the file still exists; that is, it could not be deleted:

```
     ∇ Z←DeleteFile R
[1]    ⍝ Return 1 if file no longer exists, 0 if it does
[2]    Z←⎕wcall 'PathFileExists' R
[3]    :if Z
[4]        Z←⎕wcall '_lOpen' R 16
[5]        ⎕wgive 0 ⍝ Give windows a chance to open the file
[6]        :if ¯1≠Z
[7]            Z←0 ⍝ File is in use, can't be deleted
[8]        :else
[9]            Z←⎕wcall '_lClose' Z
[10]           ⎕wgive 0 ⍝ Give windows a chance to close the file
[11]           :if ¯1=Z
[12]               ⎕wcall 'DeleteFile' R
[13]               Z←1 ⍝ File has been deleted
[14]           :else
[15]               Z←0 ⍝ File could not be deleted for some other reason
[16]           :endif
[17]       :endif
[18]   :else
[19]       Z←1 ⍝ File did not exist in the first place
[20]   :endif
     ∇
```

This function is simply a wrapper around the API call—refer to the comments in the function listing to see how it adds 'intelligence' to the API call to return an unambiguous result that the calling environment may use.

Note that the function DeleteFile (or the API call) will return 0 if the target file is in use. This applies even when the same environment that is using the file makes the API call.

4.6.5.1 Fail safe file access

In designing a multi-user application, it might seem reasonable to make all file accesses read/write and non-exclusive. Indeed, this will remove most routine access problems if files are tied just before reading/writing and untied after. However, access remains a problem under three circumstances:

● APL cannot access some files on a shared basis; that is, all native file access is exclusive.

● It is necessary to compact component files using ⎕fdup not only to keep such files tidy but also to make access efficient. The system function ⎕fdup requires exclusive access.

• Any file that APL wants to access may be in use by another application – this includes native files as well as custom files such as **Excel** or **Word** files. It is unlikely that other applications will use component files.

Any attempt to use a file that is locked for exclusive use by another user or application will encounter an error. The APL SeekFileAccess function may be used to determine if a file is available with the desired access privileges before any attempt to use it:

```
       ∇ Z←L SeekFileAccess R
   [1]    ⍝ System Building with APL+Win
   [2]    ⍝ Return 1 if successful, otherwise 0
   [3]    ⍝ R is file name. L = Access mode + Share mode (default 17)
{1[4]   :if 0=⎕nc 'L'
   [5]       L←17 ⍝ Exclusive read/write access
1}[6]   :endif
   [7]    (Z L)←(⎕wcall 'PathFileExists' R) (1↑+/L)
{1[8]   :if Z
{2[9]      :if L←∊0 1 2∘.+0 16 32 48 64
  [10]         Z←⎕wcall '_lOpen'  R
  [11]         ⎕wgive 0
{3[12]         :if ¯1≠Z
  [13]            Z←⎕wcall '_lClose' Z
  [14]            ⎕wgive 0
  [15]            Z←1 ⍝ Access mode granted
+ [16]         :else
  [17]            Z←0 ⍝ Access mode denied for some reason
3}[18]         :endif
+ [19]      :else
  [20]
2}[21]      :endif
+ [22]  :else
  [23]      Z←1 ⍝ File does not exist, can be created with desired access
1}[24]  :endif
       ∇
```

The left-hand argument is constructed from the possible values listed in **Table 4-5 lOpen API constants**. The right-hand argument is a fully qualified file name. The function returns 1 if the desired access mode is possible, 0 otherwise. A result of 0 may be returned if the files does not exist, is already in use, or if the file exists and is not in use but access is denied. Network volumes also control file access—files may be non-shareable or a user may simply be denied access.

Depending on the context in which the functions DeleteFile and SeekFileAccess are used, it may be desirable to alert the user to the cause of failure, thus:

```
DeleteFile[15]              Z←SignalAPIError
SeekFileAccess[16]              Z←SignalAPIError
```

The function SignalAPIError is:

```
       ∇ Z←SignalAPIError
  [1]    ⍝ Signal last error in a message box
  [2]    Z←⎕wcall 'GetLastError'
  [3]    :if 0≠Z
  [4]       Z←↑↑/⎕wcall 'FormatMessage' 4096 0 Z 0 (200⍴⎕tcnul) 200 0
  [5]       Z←0×Msg Z
  [6]    :endif
       ∇
```

Note that a GetLastError code of 0 implies that no error has occurred—if SignalAPIError detects this, no message is signalled.

4.7 Universal Naming Convention (UNC)

An early approach with network disks whereby a partition or folder on that disk is mapped to a drive letter is becoming increasingly unworkable for two reasons:

• First, an application can easily become confused when different users map the same partition or folder to a different drive letter. It is practically impossible to enforce a uniform mapping convention.

• Second, there are a finite number of drive letters and it may not be possible to retain permanently mapped drives—application errors occur when an expected drive is unavailable because it is not mapped or not mapped to the expected letter.

4.7.1 APL and UNC names

UNC names have the following pattern, \\SERVERNAME\FOLDER, and are available only on computers connected to a network. Like the FSO, APL+Win's integral file handling functions ⎕xn★, ⎕xf★ handle UNC names without any problems. However, both lack facilities for manipulating fully qualified file names involving UNC.

A regressive approach that APL can adopt is to use library numbers with UNC volumes. The library number schema for file access is in the following format, 'libnum filename' ⎕f. This will avoid the problems associated with limited drive letters if and only if component files with short names only are used – this is an unnecessary restriction. The use of library numbers is a dated approach and is inadvisable, because:

• It introduces the overhead of having to configure the library numbers up front, consistently for all users and makes the APL code unnecessarily complex and ambiguous.

• Native file functions cannot use library numbers – they require fully qualified names. Any reference to a filename alone forces the location of that file to be the current folder, which can be volatile depending on how APL is started.

• Component file functions dealing with long file names cannot distinguish the library number from the name of the file.

• The use of the current folder and its common prefixes is equally likely to cause confusion, not least, because the current folder may change mid-session and without warning.

An application that does not use fully qualified file names is an unreliable or unpredictable application. Although a reference to a file in the current folder does not need to be qualified by the current folder name, using the fully qualified name removes any uncertainty about its location. The APL function ⎕chdir ⍬ returns the current folder.

4.7.2 API calls for handling UNC

The file SHLWAPI.DLL (Shell Light Weight API) is part of the Windows system and has a rich set of calls for handling both UNC names and those based on mapped drive letters. Unfortunately, these calls are missing from the APLWADF.INI file and need to be added to the APLW.INI file if APL were to use them. PathFileExists is one such function.

4.7.3 UNC dynamic mapping

The Windows platform has 26 drive letters, A-Z, of which A-C are usually in use. If UNC names cannot be used and the remaining 23 drive letters are not enough to map UNC names uniquely, UNC names may be mapped dynamically.

4.7.3.1 Are UNC paths available?

It is necessary to establish whether UNC paths are available; these are available when the computer is connected to a local or wide area network. The following API is used:

IsNetworkAlive=L(>L lpdwFlags) LIB SENSAPI.DLL

This API returns 0 if the computer is not logged on, or 1 when connected to a local area network or 2 if connected to a wide area network.

4.7.3.2 Working with UNC paths

The interface that allows users to roam the available paths is **Windows Explorer**; it allows drive mapping and can enumerate available network volumes. An application needs to be able to manage access programmatically. The following function implements one approach:

```
        ∇ Z←L Network R;⎕wself
  [1]     ⍝ System Building with APL+Win
{1[2]     :if 0≠⎕io⊃⎕wcall 'IsNetworkAlive' ⋄
{2[3]         :if 0=ρ'∆WshNet' ⎕wi 'self'
  [4]             ⎕wself←'∆WshNet' ⎕wi 'Create' 'WScript.Network'
+ [5]         :else
  [6]             ⎕wself←'∆WshNet'
2}[7]     :endif
{2[8]     :select L
```

The Windows Script Host provides access to all network drive mapping facilities. Lines [3] – [7] create an instance of the Script Network object if one does not exist already. The next block of code establishes whether a live network connection exists:

```
★ [9]          :case 'IsNetworkAlive'
  [10]              Z←⎕io⊃⎕wcall 'IsNetworkAlive' ⋄
```

Since network drives are accessible only when the computer is logged on to the network, all facilities rely on an active connection. In line [10], the result is 1 when an active connection exists. The next block of code queries all existing mapped drives:

```
★ [11]          :case 'MappedDrives'
  [12]              ⎕wi 'EnumNetworkDrives>.Drv'
  [13]              ⎕wself←'∆WshNet.Drv'
{3[14]              :if 0≠⎕wi 'xCount'
  [15]                  Z←↑¨(⊂⎕wself) ⎕wi ¨(⊂⊂'xItem()'),¨(-⎕io)+⍳⎕wi
       'xCount'
  [16]                  Z←((2|+\(ρZ)/1)/Z) ((~2|+\(ρZ)/1)/Z)
+ [17]              :else
  [18]                  Z←2/⊂0ρMsg 'No mapped drives are available.'
3}[19]              :endif
  [20]              '∆WshNet.Drv' ⎕wi 'Delete'
```

The *MappedDrives* call returns a nested vector of two elements; the first hold the drive letters and the second the UNC path names:

```
        ⊃¨'MappedDrives' Network ⋄
  E:  \\sales\2003\Quarter1\STATEMENTS
  G:  \\sales\2003\Quarter1\Projections
  I:  \\sales\budget\2002
  J:  \\product\pricing\endowment
  K:  \\Conv\mailshot\cpp
  H:  \\product\NHS
  P:  \\product\target\volume
  V:  \\personnel\staff2003\newentrant
★ [21]          :case 'MapDrive'
  [22]              (L R)←R
```

```
[23]              L←1⌽':',↑⎕wcall 'CharUpper' (,∊L)
[24]              0 0⍴'DisconnectDrive' Network L
[25]              0 0⍴⎕wi 'MapNetworkDrive' L R
```

This call maps a drive letter to a UNC path. The syntax is:
```
'MapDrive' Network driveletter uncpath
```
If the drive letter is in use, it is disconnected first; the drive letter must be one that is not in use and must not be one already used by Windows. The next block of code disconnects a mapped drive:
```
★ [26]          :case 'DisconnectDrive'
  [27]              R←1⌽':',↑⎕wcall 'CharUpper' (,∊R)
{3[28]                :if Z←(⊂R)∊⎕io⊃'MappedDrives' Network θ
  [29]                  0 0⍴⎕wi 'RemoveNetworkDrive' R
3}[30]              :endif
```
This call disconnects a mapped drive. The syntax is:
```
'DisconnectDrive' Network driveletter
```
```
★ [31]          :case 'CurrentMap'
  [32]              R←1⌽':',↑⎕wcall 'CharUpper' (,∊R)
{3[33]                :if 1∊L←(⎕io⊃Z←'MappedDrives' Network θ)∊⊂R
  [34]                  Z←L/¨Z
+ [35]              :else
  [36]                  Z←2/⊂0⍴Msg 'Drive ',R,' is not mapped.'
3}[37]              :endif
```
This call returns the UNC path for a mapped drive or an error if the drive is not mapped. The syntax is:
```
'CurrentMap' Network driveletter
```
```
★ [38]          :case 'FreeLetters'
  [39]              Z←↑¨⎕io⊃'MappedDrives' Network θ
  [40]              Z←('CDEFGHIJKLMNOPQRSTUVWXYZ'~↑¨('DrivesEx' FSO
      θ),Z),¨':'
2}[41]      :endselect
+ [42]      :else
  [43]          Z←0×Msg 'This computer is not connected to a network.'
1}[44]      :endif
          ∇
```
The final call returns the drive letters that are available. The File System Object is used to collate the drives used by Windows and the Script Shell object is used to collate the mapped drives.

UNC ensures that the fully qualified name of any given file on one computer is identical to the fully qualified name of the same file on any other computer connected to the same network. This promotes a more robust application.

4.7.4 Library/UNC mapping
Libraries may be defined either using Universal Naming Convention (UNC) or drive letters or both. Consider the following session:
```
⎕libd '0 \\dev\\staff\\'
⎕libd '1 \\dev\\staff\\state\\aa\\'
tie1←'0 afile.text' ⎕ncreate 0 ⍝ Library with native files
tie2←'0 state\aa\bfile.txt' ⎕ncreate 0 ⍝ Path & library
tie3←'1 cfile.txt' ⎕ncreate 0
tie4←'2 dfile.txt' ⎕ncreate 0
⎕nnums
```

```
1 2 3 4
     ⎕nnames ⍝ Refers to UNC names, not libraries
\\dev\\staff\\afile.text ⍝ unc name returned
\\dev\\staff\\state\aa\bfile.txt ⍝ path correctly appended
\\dev\\staff\\state\\aa\\cfile.txt
c:\ajay\dfile.txt ⍝ fixed path returned
```

Library numbers and UNC or fixed paths can be used interchangeably with tied files:

```
'\\cdev\\staff\\afile.text' ⊃nerase tie1
'0 staff\aa\bfile.txt' ⊃nerase tie2 ⍝ library + path
'1 cfile.txt' ⊃nerase tie3 ⍝ library instead of path
'c:\ajay\dfile.txt' ⊃nerase tie4 ⍝ path instead of library
```

This is a session in MicroAPL's APLX version 3.0: the code will **not** work in APL+Win, as it is unable to assign library numbers to UNC paths. As well as being able to generate the next available tic number, APLX's component file functions include the facility for inserting components and dropping components anywhere between the first and last components. Native file tie numbers may be either negative or positive.

APL's file handling functions need to be revamped, to lose replace the function names with their counterpart having the prefix *x*. The APLX model provides a very powerful solution for working with UNC paths.

A promising but somewhat dangerous workaround is to use an API call to set the current directory to a UNC path and for APL to work with files and folders without reference to an explicit path. The following APIs may be used:

SetCurrentDirectory=B(*C lpPathName) ALIAS SetCurrentDirectoryA LIB KERNEL32.DLL

SetCurrentDirectory=B(*C lpPathNAme) ALIAS SetCurrentDirectoryA LIB KERNEL32.DLL

GetWindowsDirectory=U(>C lpBuffer,U nSize) ALIAS GetWindowsDirectoryA LIB
     KERNEL32.DLL

GetFullPathName−D(*C lpFileName, D nBufferLength, >C lpBuffer, >P lpFilePart) ALIAS
     GetFullPathNameA LIB Kernel32

Although the API calls work, APL's functions ⎕chdir, ⎕libd, ⎕lib, etc. do not return the expected results:

```
⎕chdir ''                    ↑↑⌿⎕wcall 'GetCurrentDirectory' 256 (256⍴⎕tcnul)
C:\Program Files\aplwin       C:\Program Files\aplwin
                              ⎕wcall 'SetCurrentDirectory' '\\QUANTUM\SALES'
                              ↑↑/⎕wcall 'GetCurrentDirectory' 256 (256⍴⎕tcnul)
                              \\QUANTUM\SALES
                              ⎕chdir ''
                              DRIVE NOT READY
```

However, the current directory has been changed: any file created without explicit reference to a path is located in the current directory set by the SetCurrentDirectory API.

4.8 Application configuration

In line with industry standard applications, APL applications must also be configurable for user-selected paths, in respect of both the installation of the system and the creation of user files at runtime. At runtime, users may elect to store files for each client or for specific time spans in separate folders and demand the facility for switching between these folders via the application itself.

This raises the question of how references to file names should be coded– these would be hard coded within the application and yet are volatile and unpredictable during runtime. An obvious solution is to use indirect reference to the path. For example,

```
SysFolder←'a:\new folder for files'
```

```
        (SysFolder,'\MyClient.dat') ⎕xfcreate 500
        ⎕xfnames
a:\new folder for files\MyClient.dat
```

Here, a variable SysFolder holds the fully qualified path name and is used indirectly in creating a file MYCLIENT.DAT at that location. The general principle is to initialise all the variables holding path names at the start of the application.

The control mechanism, allowing for concurrent users, may work as follows:

● Find the Windows folder, using GetWindowsDirectory.

● Verify that MyApp.INI exists, using PathFileExists.

● Query the user's name – the person whose profile is currently active, using either GetUserName or GetComputerName.

● Read the name of the set of folders last used by current user, using GetPrivateProfileString.

● Verify that this set is not currently in use – if it is, offer a choice of the existing set names for folders, using GetPrivateProfileSection. The concept is to allow a single user to use a set of folders exclusively.

● Update the INI file, marking the set as being in use, and activate the selected session, using WritePrivateProfileString.

This may seem like a daunting task; in fact, it is quite simple to achieve using an INI file. All the API calls needed are already included in the APLWADF.INI file.

The alternative is to hold this information in the **Registry**. The latter approach adds to the complexity of the task and offers nothing more than better protection of the configuration information.

A workable solution is to hold all the application configuration variables in an INI file in a folder that will always exist and can be readily identified—such as the folder where Windows is installed.

The name of the INI file needs to be determined in advance and hard coded within the APL code. The system wide parameters must be held in isolated section names, which cannot be changed by users.

Note the three sections in the INI file shown in **Table 4-6 The makeup of an INI file**. Key values that are string do not requite quotation marks.

Table 4-6 The makeup of an INI file

4.9 Using INI files with APL

Windows does not specify any constraints for INI files except that section names should be included in square brackets ([]) and that keys, within sections, should be preceded by the equal to (=) sign. An INI file can be altered easily with any text editor, such as NOTEPAD.

Potentially, this may make an application less reliable. Alternatives to storing application settings in an INI file are storing them in an XML file or the **Registry**.

4.9.1 Location of the INI file

The name of an INI file does not have to have the INI extension; in fact, it does not have to have an extension at all. However, the INI extension is recommended as it makes the nature of the file immediately apparent.

The usual technique is to locate the INI file at a location can that be queried dynamically, such as the Windows folder, using a Windows API call. That is, the INI file like the **Registry** resides on the local hard disk. Therefore, each user of the system has a personal copy of the INI file or **Registry** entries.

4.9.2 Name of section

The name of a section does not have to be a word—it can be a phrase, for example, [my section] is perfectly valid. However, it is advisable to name sections using valid APL names such that a section can be read into an APL variable of the same name. The order in which sections appear in an INI file is inconsequential.

4.9.3 Name of key

The name of a key does not have to be a word—it can be a phrase, for example, *my key* = is perfectly valid. However, it is advisable to name keys using valid APL names such that a key can be read into an APL variable of the same name. The order in which keys appear in an INI file is inconsequential.

4.9.4 Limitations of INI files

The main limitation is that since the launch of Windows 95, Microsoft's stance is that application developers should use the **Registry** rather than INI files to store application settings. The rationale is that INI files:

• Are text files and can be easily altered; this has the potential of making applications less reliable.

• Cannot exceed 64K in size.

• The value of individual string keys is restricted to 255 characters only, depending on the version of Windows.

Consequently, Microsoft has marked the INI file API calls as superseded. Nonetheless, it is highly unlikely that the APIs will disappear in the near future. If these limitations are not an issue, INI files can still be used for configuring applications.

4.9.5 Implementing the control mechanism

There are eight INI file API calls. These are encapsulated in an APL function. This function is deliberately structured such that the syntax of calls is identical, wherever possible, to that of the XMLINI function:

```
      ∇ Z←L APIINI R
 [1]    ⍝ System Building With APL+Win
{1[2]    :if 0=⎕nc 'L'
> [3]        :return ⍝ signal error in calling environment
1}[4]    :endif
{1[5]    :select L
= [6]        :case 'Load'
 [7]            R←↑↑/2↑⎕wcall 'GetFullPathName' R 256 (256⍴⎕tcnul) ''
{2[8]            :if ⎕wcall 'PathFileExists' R
 [9]                '#' ⎕wi '∆INIFile' R
+ [10]           :else
```

```
  [11]                        R←'\',(⎕io⊃(R≠'.')⊂R←∈¯1↑(R≠'\')⊂R),'.ini'
  [12]                        Z←↑↑/⎕wcall 'GetEnvironmentVariable' 'APLHome'
          (256ρ⎕tcnul) 256
{3[13]                        :if 0=ρZ
  [14]                            Z←↑↑/⎕wcall 'GetWindowsDirectory' (256ρ⎕tcnul) 256
3}[15]                        :endif
{3[16]                        :if 0≠⎕wcall 'PathFileExists' (Z,R)
  [17]                            '#' ⎕wi '∆INIFile' (Z,R)
+ [18]                        :else
  [19]                            L←¯1+⌊/0,⎕nnums,⎕xnnums
  [20]                            (Z,R) ⎕xncreate L
  [21]                            (';Created',(Now ⊖),⎕tcnl,⎕tclf) ⎕nappend L
  [22]                            ⎕nuntie L
  [23]                            '#' ⎕wi '∆INIFile' (Z,R)
3}[24]                          :endif
2}[25]                    :endif
  [26]                Z←'#' ⎕wi '∆INIFile'
> [27]                :return
1}[28]            :endselect
  [29]  Z←'#' ⎕wi '∆INIFile'
```

The INI file is referenced by the following call:
```
    'Load' APIINI 'sbwa'
C:\WINDOWS\sbwa.ini
```

The *Load* call verifies whether the INI file exists in the current folder, the location specified by the APLHOME environment variable or in the Windows system folder. If it does not, it is created in the APLHOME location, if it exists, or in the Windows system folder.

4.9.6 INI file: writing a key

The general syntax is

```
        'WriteKey' APIINI Section Key Value
{1[30]    :if 0≠⎕wcall 'PathFileExists' Z
{2[31]      :select L
= [32]        :case 'WriteKey'
  [33]            Z←⎕wcall (⊂'WritePrivateProfileString'),(⍕¨R),⊂Z
```

The following API called is used to write the key:
```
WritePrivateProfileString=U(*C,*C,*C,*C) ALIAS WritePrivateProfileStringA LIB
     KERNEL32.DLL
     'WriteKey' APIINI 'Client1' 'Name' 'Writing to INI File'
1
```

If the *Section* or *Key* does not exist, they are automatically created and the *Value* is written. A section may contain several keys.

4.9.7 INI file: reading a key

The general syntax is:
```
      ★ 'GetKey' APIINI Section Key
 = [34]          :case 'GetKey'
   [35]              Z←↑↑/⎕wcall (⊂'GetPrivateProfileString'),R, ''
          (256ρ⎕tcnul) 256 Z
```

The following API is used to read the key:
```
GetPrivateProfileString=I(*C,*C,*C,>C,I,*C) ALIAS GetPrivateProfileStringA Lib
     KERNEL32.DLL
     'GetKey' APIINI 'Client1' 'Name'
```

```
Writing to INI File
```
There is no API call to delete a key in an INI file: at best, a null value may be written to the key.

4.9.8 INI file: writing a section

The general syntax is

```
        'WriteSection' APIINI Section (Key1 Key2) (Value1 Value2)
 =[36]              :case 'WriteSection'
  [37]                   R←R[⎕io],(⁻1↓∊,¨/(⍕¨¨¨1↓R),¨¨¨'=' ⎕tcnul) Z
  [38]                   Z←⎕wcall (⊂'WritePrivateProfileSection'),R
```

The right-hand argument is a three element nested vector comprising of the name of the section, the list of keys' names and a list of their corresponding values.

```
        'WriteSection' APIINI 'NewClient' ('Name' 'ID') ('ABC Plc' 'PRE97')
1
```

Internally, the following API is used to write the values:

```
WritePrivateProfileSection=B(*C lpAppName, *C lpString, *C lpFileName) ALIAS
    WritePrivateProfileSectionA LIB KERNEL32.DLL
```

4.9.9 INI file : emumerating the names of all sections

The general syntax is

```
        'GetSectionsName' APIINI θ
= [39]              :case 'GetSectionsName'
  [40]                   Z←↑↑/⎕wcall 'GetPrivateProfileSectionNames'
          (8192⍴⎕tcnul) 8192 Z
  [41]                   Z←(Z≠⎕tcnul)⊂Z
```

The following API is used to enumerate all the section names:

```
GetPrivateProfileSectionNames=D(>C lpReturnBuffer, D nSize, *C lpFileName) ALIAS
    GetPrivateProfileSectionNamesA LIB KERNEL32.DLL
        'GetSectionsName' APIINI θ
Client1 NewClient
```

With this call, it is especially visible that the syntax of the APIINI function does not include the name of the file—it is held in the user-defined property ΔINIFILE of the system object, see APIINI line [13].

4.9.10 INI file: reading a section

The general syntax is

```
        'GetSection' APIINI Section
= [42]              :caselist 'GetSection' 'GetSectionKeysName'
          'GetSectionKeysValue'
  [43]                   Z←↑↑/⎕wcall 'GetPrivateProfileSection' R (8192⍴⎕tcnul)
          8192 Z
{3[44]               :if  0=⍴Z
  [45]                   Z←'=',⎕tcnul
3}[46]               :endif
  [47]               Z←(Z≠⎕tcnul)⊂Z
  [48]               Z←(Z∨.∊¨¨'=')/Z
  [49]               Z←(Z≠'=')⊂¨Z
  [50]               Z←R (↑¨Z) (1↓¨Z)
{3[51]               :select L
= [52]                   :case 'GetSectionKeysName'
  [53]                       Z←(⎕io+1)⊃Z
= [54]                   :case 'GetSectionKeysValue'
  [55]                       Z←(⎕io+2)⊃Z
3}[56]               :endselect
```

An entire section can be read. Internally, the following API call is used:

GetPrivateProfileSection=D(*C lpAppName, >C lpReturnedString, D nSize, *C lpFileName)
       ALIAS GetPrivateProfileSectionA LIB KERNEL32.DLL

This API call returns the keys and their corresponding values. The function call adds the name of the section:

```
      ρ'GetSection' APIINI 'NewClient'
3
```

A nested vector of three elements is returned, comprising of the name of the section, the list of keys, and the list of corresponding values. Lines [35] and [38] impose an arbitrary limit of 8,192 bytes on the size of the buffer, which may be inadequate for large INI files: increase the buffer size as required.

4.9.11 INI file: enumerate names of all keys in a section

The list of keys' names can be extracted from the same block of code. The syntax is:

```
      'GetSectionKeysName' APIINI 'NewClient'
 Name ID
```

4.9.12 INI file: enumerate values of all keys in a section

The same block of code yields all the values in a section. The syntax is:

```
      'GetSectionKeysValue' APIINI 'NewClient'
  ABC Plc   PRE97
```

4.9.13 Two more APIs

APLINI contains two more API calls; these extract the name of the current user and the name of the computer:

```
★ [52]            = [57]              :caselist 'GetUserName' 'GetComputerName'
  [58]                 R←(32ρ256ρ⎕tcnul) (⎕av[⎕io+,Φ⍨256 256 256 256⊤32])
  [59]                 Z←(⎕io+1)⊃⎕wcall (⊂L),R
+ [60]          :else
  [61]               Z←0×Msg 'Unable to find action ',L
2}[62]        :endselect
+ [63]    :else
  [64]        Z←0×Msg 'Unable to find file ',Z
1}[65]    :endif
       ∇
```

The API calls are:

GetUserName=B(>C lpNameBuffer, >D nBufferSize) ALIAS GetUserNameA LIB Advapi32.dll
GetComputerName=B(>C,>D) ALIAS GetComputerNameA LIB KERNEL32.DLL

```
'GetUserName' APIINI θ          │ 'GetComputerName' APIINI θ
Ajay Askoolum                   │ AJAY ASKOOLUM
```

These APIs transparently identify the current user and the computer; these details may be used to hold personal entries in the INI file.

4.10 XML files for application configuration

EXtensible Mark-up Language (**XML**) may be used to provide similar capabilities as INI files without the size limitations on the size of files or keys. XML is highly complex and can be used to store the data types of keys together with ancillary information such as annotations or comments, time stamps, etc.

The APL+Win grid works better with the Microsoft XML parser, available as a free download, installed. This installs a DLL that APL can use to create, read, and write XML files. XML files are text files too but contain tags; the overall appearance of an XML file is quite complex, making it less likely that it will be casually changed with a text editor. XML is case sensitive.

4.10.1 The XMLINI function

This self-contained function can create replacement INI files in XML format. In addition, it provides facilities for reading, writing, and deleting sections and keys. A crucial difference between INI and XML files is that the content of XML files is case-sensitive and certain characters or combination of characters have intrinsic meaning and cannot be used for user data. A hassle free approach is to use lowercase characters only for the names of sections and keys and to construct these names without embedded spaces.

4.10.2 Loading/Creating an XML file

In order to use XML files for storing dynamic application configurations, Microsoft's MSXML 4.0, must be installed on the computer. The logical location of a configuration is a location specified by the application: this is held in an environment variable named *aplapp*. If the file is not found in the folder specified by *aplapp*, it is sought in the Windows system folder.

The syntax for creating a new file or using an existing one is:

```
'Load' XMLINI 'sbwa'
C:\WINDOWS\sbwa.xml
```

The right-hand argument comprises of just the name of the file. If the file is not found, it is automatically created. As shown in **Figure 4-4 New XML File,** it is not very readable but the parser can make sense of it!

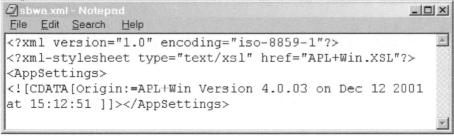

Figure 4-4 New XML File

Note the reference to APL+WIN.XSL. Extensible Stylesheet Language: Transformation (**XSLT ALSO XSL**) is a language for transforming one XML document into another. APL+Win.XLS is: this is an XML style sheet that is expected in the same location as the XML file and is used for formatting the contents in the browser. **Table 4-7 XSL template** lists the XSL file; line numbers have been added as an aid to readability:

```
 1  <?xml version="1.0" encoding="iso-8859-1"?>
 2  <xsl:stylesheet
 3      version="1.0"
 4      xmlns:xsl="http://www.w3.org/1999/XSL/Transform">
 5      <xsl:template match="/">
 6          <html>
 7              <body>
 8                  <table>
 9                          <xsl:apply-templates select="AppSettings/section"/>
10                  </table>
11              </body>
12          </html>
13      </xsl:template>
14      <xsl:template match="AppSettings/section">
15          <tr>
16              <td>
17                  <b>
```

```
18                      [
19                      <xsl:value-of select="@name"/>
20                      ]
21                      <xsl:apply-templates select="key"/>
22                    </b>
23                </td>
24            </tr>
25        </xsl:template>
26        <xsl:template match="key">
27            <tr>
28                <td>
29                        <i>
30                        <xsl:value-of select="@name"/>=<xsl:value-of
select="text()"/>
31                        </i>
32                </td>
33            </tr>
34        </xsl:template>
35 </xsl:stylesheet>
```

Table 4-7 XSL template

The function XMLINI shows the code that initialises the XML parser, creates, and loads a file. The first step is to initialise the XML parser:

```
      ∇ Z←L XMLINI R;⎕wself
  [1]    ⍝ System Building with APL+Win
{1[2]    :if 0=⎕nc 'L'
> [3]       :return ⍝ signal error in calling environment
1}[4]    :endif
{1[5]    :if 0=ρ'∆XML' ⎕wi 'self'
{2[6]       :if 0∊ρ'#' ⎕wi 'XInfo' 'MSXML.FreeThreadedDOMDocument'
  [7]          Z←0×Msg 'MSXML Parser Not Found On This PC'
> [8]          :return
+ [9]       :else
  [10]          Z←'∆XML' ⎕wi 'Create' 'MSXML.FreeThreadedDOMDocument'
2}[11]      :endif
1}[12]   :endif
  [13]   ⎕wself←'∆XML'
{1[14]   :if 'ActiveObject MSXML.FreeThreadedDOMDocument'≡'∆XML' ⎕wi
         'class'
{2[15]      :select L
```

The *Load* call verifies whether the XML file exists in the location specified by the *aplapp* environment variable or in the Windows system folder. If it does not, it is created in the *aplapp* location, if it exists, or in the Windows system folder:

```
* [16]             :caselist 'Load' 'GetSignature'
{3[17]                :select L
* [18]                   :case 'Load'
  [19]                      R←'\',(⎕io⊃(R≠'.')⊂R←∊¯1↑(R≠'\')⊂R),'.xml'
  [20]                      Z←↑↑/⎕wcall 'GetEnvironmentVariable' 'aplapp'
         (256ρ⎕tcnul) 256
{4[21]                     :if 0=ρZ
  [22]                         Z←↑↑/⎕wcall 'GetWindowsDirectory'
         (256ρ⎕tcnul) 256
4}[23]                     :endif
{4[24]                     :if 0=⎕wcall 'PathFileExists' (Z,R)
  [25]                         L←¯1+⌊/0,⎕nnums,⎕xnnums
  [26]                         (Z,R) ⎕xncreate L
```

```
     [27]                        ('<?xml version="1.0" encoding="iso-8859-
          1"?>',⎕tcnl,⎕tclf) ⎕nappend L
     [28]                        ('<?xml-stylesheet type="text/xsl"
          href="APL+Win.XSL"?>',⎕tcnl,⎕tclf) ⎕nappend L
     [29]                        ('<AppSettings>',⎕tcnl,⎕tclf) ⎕nappend L
     [30]                        ('GetSignature' XMLINI 0) ⎕nappend L
     [31]                        ('</AppSettings>',⎕tcnl,⎕tclf) ⎕nappend L
     [32]                        ⎕nuntie L
4}[33]                  :endif
     [34]              '#' ⎕wi '∆XMLFile' (Z,R)
     [35]              Z←'#' ⎕wi '∆XMLFile'
{4[36]                  :if ⎕wi 'Xload' Z
{5[37]                      :if 0=(⎕wi 'xparseError') ⎕wi 'xerrorCode'
     [38]                          R←(⎕wi 'ximplementation') ⎕wi
          'XhasFeature' 'XML' '1.0'
     [39]                          R←R∧'AppSettings'≡(⎕wi
          'xdocumentElement') ⎕wi 'xnodename'
{6[40]                          :if 0=R
     [41]                              Z←0×Msg 'Document ' ,Z, ' is
          invalid'
6}[42]                          :endif
+ [43]                      :else
     [44]                          Z←0×Msg 'Error parsing document ',Z
5}[45]                      :endif
+ [46]                  :else
     [47]                      Z←0×Msg 'Failed to load document ',R
4}[48]                  :endif
{4[49]                  :if 0≡Z
     [50]                      ⎕wi 'Delete'
4}[51]                  :endif
```

The *GetSignature* call is used internally to acquire comment strings for the XML file when it is created and from each section:

```
★ [52]                  :case 'GetSignature' ⍝ Internal use only
{4[53]                  :if R=0
     [54]                          Z←'Origin:=APL+Win',⍒'Version ' 'on ' ''
          '' 'at ' ,¨¯1↓( ' '≠ ⎕sysver)⊂⎕sysver
     [55]                          Z←(⎕wi 'XcreateCDATASection' Z) ⎕wi 'xxml'
+ [56]                  :else
     [57]                          Z←⎕wi 'XcreateComment' ('Created by
          APL+Win',Now ⍬)
4}[58]                  :endif
3}[59]              :endselect
> [60]              :return
2}[61]          :endselect
```

4.10.3 XML file: writing a key

The names of sections and keys are case sensitive: thus, *client1* and *Client1* will be seen as distinct sections and *SCHEMEID* and *SchemeId* will be seen as distinct keys. The syntax for writing a key is:

```
     'WriteKey' XMLINI Section Key Value
```

If the section or key does not exist, it is created automatically and the new value is written to the file:

```
     'WriteKey' XMLINI 'Client1' 'Name' 'XYZ Plc'
```

1

After the key is added, a double click on the file name opens it in the browser and it looks as shown in **Figure 4-5 Formatted XML File.**

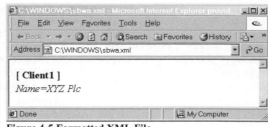

Figure 4-5 Formatted XML File

The XSL file listed in **Table 4-7 XSL template** renders the XML file as a familiar INI file. A return value of 1 indicates that the key has been written. A section may contain several keys:

```
      'WriteKey' XMLINI 'Client1' 'SchemeID' 102.99
1
```

The code for writing a key:

```
{2[62]        :if 0≠⍴⎕wi 'xxml'
{3[63]            :select L
★ [64]               :case 'WriteKey'
  [65]                  R←⍕¨R
{4[66]                  :if 0=⎕wi 'XselectSingleNode'
      ('AppSettings/section[@name="',(⎕io⊃R),'"]')
  [67]                     Z←⎕wi 'XcreateElement' 'section'
{5[68]                     :if 4=⍴R
  [69]                        0 0⍴Z ⎕wi 'appendChild' (⎕wi
      'xcreateComment' ((⎕io+3)⊃R))
+ [70]                     :else
  [71]                        0 0⍴Z ⎕wi 'appendChild' ('GetSignature'
      XMLINI 1)
5}[72]                     :endif
  [73]                     L←⎕wi 'XcreateAttribute' 'name'
  [74]                     0 0⍴L ⎕wi 'value' (⎕io⊃R)
  [75]                     0 0⍴Z ⎕wi 'XAttributes.setNamedItem' L
  [76]                     Z←⎕wi 'xdocumentElement.appendChild' Z
4}[77]                  :endif
{4[78]                  :if 0=⎕wi 'XselectSingleNode' ('GetSectionKey'
      XMLINI R)
  [79]                     Z←⎕wi 'XcreateElement' 'key'
  [80]                     L←⎕wi 'XcreateAttribute' 'name'
  [81]                     0 0⍴L ⎕wi 'value' ((⎕io+1)⊃R)
  [82]                     0 0⍴Z ⎕wi 'XAttributes.setNamedItem' L
  [83]                     0 0⍴Z ⎕wi 'Text' ((⎕io+2)⊃R)
  [84]                     L←⎕wi 'XselectSingleNode'
      ('AppSettings/section[@name="',(⎕io⊃R),'"]')
  [85]                     Z←L ⎕wi 'XappendChild' Z
+ [86]                  :else
  [87]                     Z←⎕wi 'XselectSingleNode' ('GetSectionKey'
      XMLINI R)
  [88]                     Z←Z ⎕wi 'xchildnodes(0).nodevalue' ((⎕io+2)⊃R)
4}[89]                  :endif
  [90]                  Z←'Save' XMLINI ⊖
```

The GetSectionKey call is for internal use only. The code is:

```
★ [91]                  :case 'GetSectionKey' ⍝ Internal use only
```

```
   [92]                         Z←'AppSettings/section[@name="' θ '"]/key[@name="'
            θ '"]'
   [93]                         ((∈0=ρ¨Z)/Z)←2↑R
   [94]                         Z←∈Z
```

4.10.4 XML file: reading a key

The syntax for writing a key is:

```
       'GetKey' XMLINI Section Key
```

The latter entry may be read as follows:

```
       'GetKey' XMLINI 'Client1' 'SchemeID'
102.99
```

The code that retrieves the value of a section/key pair is:

```
★ [95]                 :case 'GetKey'
  [96]                         Z←⎕wi 'XselectSingleNode' ('GetSectionKey' XMLINI
        R)
{4[97]                             :if 0=Z
  [98]                                 Z←θ
+ [99]                             :else
  [100]                                Z←Z ⎕wi 'xchildnodes(0).nodevalue'
4}[101]                            :endif
```

4.10.5 XML file: deleting a key

The syntax for deleting a key is:

```
       'DeleteKey' XMLINI Section Key
```

For example, the code to remove the key *SchemeId* from the section *Client1* is:

```
       'DeleteKey' XMLINI 'Client1' 'SchemeID'
1
```

The code for this call is:

```
★ [102]                :case 'DeleteKey'
  [103]                        L←⎕wi 'XselectSingleNode'
        ('AppSettings/section[@name="',(⎕io⊃R),'"]')
{4[104]                            :if Z←0≠L
  [105]                                Z←⎕wi 'XselectSingleNode' ('GetSectionKey'
        XMLINI R)
{5[106]                                :if Z≠0
  [107]                                    Z←L ⎕wi 'XremoveChild' Z
  [108]                                    Z←'Save' XMLINI θ
5}[109]                                :endif
4}[110]                            :endif
```

If the specified section or the key does not exist or the key could not be removed, the result is 0.

4.10.6 XML file: writing a section

A new section comprising of several keys may be written to the file in a single operation. The general syntax is:

```
       'WriteSection' XMLINI Section (key1 key2) (value1 value2)
```

The right-hand argument comprises of a nested variable with three elements, the names of all keys and the values of all keys:

```
       'WriteSection' XMLINI 'Client2' ('Name' 'ID') ('Staff Scheme'
2003.03)
1
```

• Given the syntax of this call, a section cannot be created without at least one key.

• If the section exists already, it is deleted first—thereby removing all its keys—prior to the new keys being written.

The code for the *WriteSection* call is:

```
★ [111]                :case 'WriteSection'
  [112]                    Z←'DeleteSection' XMLINI ⎕io⊃R
{4[113]                    :while 0≠ρ(⎕io+1)⊃R
{5[114]                        :if 0=Z←'WriteKey' XMLINI R[⎕io],⍪¨∊¨1↑¨1↓R
> [115]                            :leave
5}[116]                        :endif
  [117]                        R[⎕io+⍳2]←1↓¨1↓R
4}[118]                    :endwhile
```

If the operation fails, the return code is 0.

4.10.7 XML file: reading a section

All the keys in a section may be read in a single operation. The syntax is:

```
'GetKey' XMLINI  Section
```

This returns a nested variable with three elements: the name of the section, the names of all keys and the values of all keys:

```
'GetSection' XMLINI 'Client1'
```

The code is:

```
★ [119]                :case 'GetSection'
  [120]                    Z←R ('GetSectionKeysName' XMLINI R)
        ('GetSectionKeysValue' XMLINI R)
```

This is an example of a call within a given function making other calls within the same function. The code for existing calls—*GetSectionKeysName* and *GetSectionKeysValue*—is reused to good effect.

4.10.8 XML file: commenting a section

In order to provide a means of documenting a section, arbitrary text can be added as a comment to any section. The syntax is:

```
'SectionComment' XMLINI Section [Comment]
```

The right-hand argument comprises of the name of the section and an optional comment. If the comment is omitted, it is returned and if it is specified, a return value of 1 indicates that it is updated:

```
'SectionComment' XMLINI 'Client2' 'Added by A Askoolum on 13/09/2002'
'SectionComment' XMLINI 'Client2'
Added by A Askoolum on 13/09/2002
```

The code is:

```
★ [121]                :case 'SectionComment'
{4[122]                    :if 1==≡R
  [123]                        R←,⊂R
4}[124]                    :endif
  [125]                    Z←⎕wi 'XselectSingleNode'
        ('AppSettings/section[@name="',(⎕io⊃R),'"]')
{4[126]                    :if 0=Z
  [127]                        Z←0×Msg 'Unable to find section ',⎕io⊃R
+ [128]                    :else
{5[129]                        :if ⎕wi 'xhasChildNodes'
{6[130]                            :for L :in ¯1↓+\0,(Z ⎕wi
        'xchildNodes.length')/1
{7[131]                                :if 8=((Z ⎕wi 'xChildNodes') ⎕wi
        'xItem' L) ⎕wi 'xnodeType'
{8[132]                                    :if 1=ρR
  [133]                                        Z←((Z ⎕wi 'xChildNodes') ⎕wi
        'xItem' L) ⎕wi 'xtext'
```

```
+ [134]                                    :else
  [135]                                         0 0ρ((Z ⎕wi 'xChildNodes') ⎕wi
        'xItem' L) ⎕wi 'xtext' ((⎕io+1)⊃R)
  [136]                                             Z←'Save' XMLINI θ
8}[137]                                        :endif
> [138]                                        :leave
7}[139]                                      :endif
6}[140]                                    :endfor
+ [141]                                :else
  [142]                                     Z←0×Msg 'Unable to update comment for
        section ',⎕io⊃R
5}[143]                                     :endif
4}[144]                                 :endif
```

4.10.9 XML file: deleting a section

A section, including all its keys and values may be deleted in a single operation:

```
        'DeleteSection' XMLINI Section
```

If the section specified does not exist or it could not be deleted the result is 0, otherwise, it is 1:

```
        'DeleteSection' XMLINI 'Client1'
1
```

The code is:

```
★ [145]                    :case 'DeleteSection'
  [146]                    Z←⎕wi 'XselectSingleNode'
        ('AppSettings/section[@name="',R,'"]')
{4[147]                        :if Z≠0
  [148]                            Z←⎕wi 'xdocumentElement.removeChild' Z
  [149]                            Z←'Save' XMLINI θ
4}[150]                        :endif
```

The return code is 0 if the section specified does not exist or could not be deleted.

4.10.10 XML file: enumerate names of all sections

The names of all existing sections may be enumerated with the following syntax:

```
        'GetSectionsName' XMLINI θ
```

Note that the right-hand argument is null and not the name of the XML file; this functionality is available only after the file has been loaded:

```
        'GetSectionsName' XMLINI θ
 Client1 client1 Client2
```

The result is a nested vector comprising of as many elements as sections in the file; section names are case sensitive. The code is:

```
★ [151]                    :case 'GetSectionsName'
  [152]                    (Z L)← θ (⎕wi 'XselectNodes'
        'AppSettings/section')
{4[153]                        :if 0≠L
{5[154]                            :if 0<L ⎕wi 'xlength'
{6[155]                                :for R :in ¯1↓+\0,(L ⎕wi 'xlength')/1
  [156]                                    Z←Z,⊂(L ⎕wi 'xitem' R) ⎕wi
        'xAttributes.getNamedItem("name").nodevalue'
6}[157]                                :endfor
5}[158]                            :endif
4}[159]                        :endif
```

A null vector is returned if the XML file does not contain any sections.

4.10.11 XML file: enumerate names of all keys in a section

The names of all keys in an existing section may be enumerated with the following syntax:

```
'GetSectionKeysName' XMLINI Section
```

A null value is returned if the section specified does not exist:

```
'GetSectionKeysName' XMLINI 'Client1'
Name name
```

The result is a nested vector of as many elements as there are keys in the specified section; the name of keys is case sensitive. The code is:

```
* [160]                 :case 'GetSectionKeysName'
  [161]                     (Z L)←θ (⎕wi 'XselectSingleNode'
       ('AppSettings/section[@name="',R,'"]'))
{4[162]                       :if Z←0≠L
{5[163]                         :if L ⎕wi 'xhasChildNodes'
  [164]                           L←L ⎕wi 'XselectNodes("key")'
{6[165]                             :for R :in ¯1↓+\0,(L ⎕wi 'xlength')/1
  [166]                               Z←Z,⊂(L ⎕wi 'xitem' R) ⎕wi
       'xAttributes.getNamedItem("name").nodevalue'
6}[167]                             :endfor
5}[168]                         :endif
4}[169]                       :endif
```

If the section specified does not exist or it is empty—does not have any keys—a null value is returned.

4.10.12 XML file: enumerate values of all keys in a section

The values of all keys in an existing section may be enumerated with the following syntax:

```
'GetSectionKeysValue' XMLINI Section
```

The result is a nested vector of as many elements as there are keys in the specified section. The order in which the values are collated corresponds to the order in which the keys exist in the XML file, as shown:

```
'GetSectionKeysValue' XMLINI 'Client1'
XYZ Plc Added In Error
```

The code is:

```
* [170]                 :case 'GetSectionKeysValue'
  [171]                     (Z L)←θ (⎕wi 'XselectSingleNode'
       ('AppSettings/section[@name="',R,'"]'))
{4[172]                       :if 0≠L
{5[173]                         :if L ⎕wi 'xhasChildNodes'
{6[174]                           :for R :in ¯1↓+\0,(L ⎕wi
       'xchildNodes.length')/1
{7[175]                             :if 1=(L ⎕wi 'xchildNodes.item' R) ⎕wi
       'xnodetype'
  [176]                               Z←Z,⊂(L ⎕wi 'xchildNodes.item' R)
       ⎕wi 'xtext'
7}[177]                             :endif
6}[178]                           :endfor
5}[179]                         :endif
4}[180]                       :endif
```

If the specified section does not exist or it is empty—does not have any keys—a null value is returned.

4.10.13 Saving an XML file
The XML file that is loaded resides in memory. All writing operations achieved within the XMLINI function commit changes to the source file immediately. This is achieved by calling:

```
'Save' XMLINI ◊
```

All changes are saves to the same file. The code is:

```
★ [181]              :case 'Save' ⍝ Internal use only
{4[182]                  :if Z←0≠↑⍴'#' ⎕wi '∆XMLFile'
 [183]                      ⎕wi 'Xsave' ('#' ⎕wi '∆XMLFile')
4}[184]                  :endif
+ [185]              :else
 [186]                  Z←0×Msg 'Unable to recognise action ', L, ' within
       XMLINI'
3}[187]          :endselect
+ [188]      :else
 [189]          Z←0×Msg ('#' ⎕wi '∆XMLFile'),' must be re-loaded'
2}[190]      :endif
+ [191] :else
 [192]      Z←0×Msg 'MSXML Object name XML is in use as an instance of
       another object'
1}[193] :endif
        ∇
```

4.11 INI/XML comparative advantage
Both INI files and the implementation of XML files presented above hold character values only. Unlike INI files, which have a limit of 255 characters for the value of any key, XML files can accommodate larger values for keys. However, the XML file is held in memory; therefore, some care is required to ensure that there are no resource problems.

Nonetheless, an XML file is better able to hold APL typed values because it can hold more data per key. A technique similar to the one used in the *RegWrite* and *RegRead* functions may be used. In the example below, index origin of 1 is enforced:

```
⎕io←1 ◊ Value←(2 5⍴⍳10) 'System Building'
'WriteKey' XMLINI 'APL+Win' 'NestedValue' (⎕av⍳'wrapl' ⎕dr Value)
ValueBack←'GetKey' XMLINI 'APL+Win' 'NestedValue'
⎕io←1 ◊ Value≡'unwrapl' ⎕dr ⎕av[⎕fi ValueBack]
1
```

4.11.1 Converting INI to XML
Where common, the syntax of the functions APIINI and XMLINI calls is identical. This makes it quite easy to rewrite INI files in XML format:

```
    ∇ INItoXML;Sections
 [1]   ⍝ System Building with APL+Win
 [2]   ⍝ L = INI file specification
 [3]   ⍝ R = XML file specification
 [4]   0 0⍴⎕wcall 'SetEnvironmentVariable' 'appapl' 'd:\program
       files\aplwin'
 [5]   0 0⍴'Load' APIINI 'aplw.ini'
 [6]   0 0⍴'Load' XMLINI 'aplwini.xml'
 [7]   Sections←'GetSectionsName' APIINI ◊
{1[8]  :while 0≠⍴Sections
 [9]       0 0⍴'WriteSection' XMLINI 'GetSection' APIINI ⎕io⊃Sections
 [10]      Sections←1↓Sections
1}[11] :endwhile
        ∇
```

This rewrites the APL+Win's APLW.INI file to APLINI.XML. The XML file is shown in **Figure 4-6 XML view**

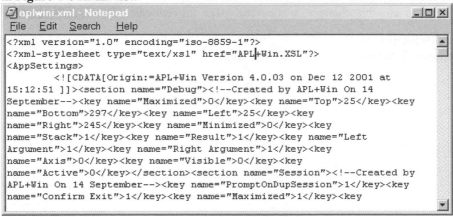

Figure 4-6 XML view

The XML file contains the following additional information:

● Includes a comment stating the origin of the file and a timestamp stating its creation date.

● A comment stating the origin of each section and a partial time stamp is included.

Figure 4-7 XSL view shows it as the style sheet renders it; as the *href* statement, in line 2 of the XML file, simply refers to the XSL file, this file is expected in the same location as the XML file.

If the XSL file is held elsewhere, its fully qualified location must be stated.

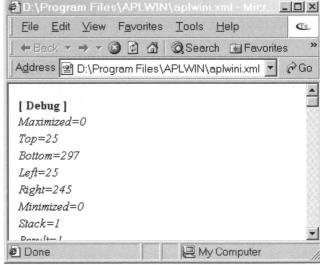

Figure 4-7 XSL view

4.12 The filing system

Windows filing system is increasingly more complex, allowing user selectable options on its structure—FAT16, FAT32, NTFS, CDFS, etc. Local area networks introduce Universal Naming Convention (**UNC**)—a means of referring to files and folders without recourse to a drive letter. The Internet has imposed the concept of Uniform Resource Locator (**URL**)—a 'virtual path' protocol for accessing a network server and its files

4.12.1 Identifying the filing system

The GetVolumeInformation API call returns information about a volume, including its file system. The API, defined in APLWADF.INI, is:

GetVolumeInformation=B(*C lpRootPathName, >C lpVolumeNameBuffer, D
    nVolumeNameSize, >D lpVolumeSerialNumber, >D lpMaximumComponentLength,
    >D lpFileSystemFlags, >C lpFileSystemNameBuffer, D nFileSystemNameSize) ALIAS
    GetVolumeInformationA LIB KERNEL32.DLL

Sample calls and results are shown in **Table 4-8 Volume information**.

Drive letter	Network map
`GetVolumeInformation 'c:\'`	`GetVolumeInformation '\\quantum\sales\'`
` Drive      c:\`	` Drive      \\quantum\sales\`
` Label      DRIVE_C`	` Label      SALES`
` Serial     1032286326`	` Serial     139396866`
` FileSystem FAT32`	` FileSystem NTFS`

Table 4-8 Volume information

The function GetVolumeInformation simplifies the API call: a right-hand argument is expected. The drive name must be specified with a trailing \: it may be either a drive letter or a network map. The function is defined thus:

```
      ∇ Z←GetVolumeInformation R
 [1]    ⍝ System Building with APL+Win
 [2]    Z←⎕wcall 'GetVolumeInformation' R (255⍴⎕tcnul) 255 0  0 0
          (255⍴⎕tcnul) 255
{1[3]   :if ↑Z
 [4]        Z←⊤¨0 1 1 0 0 1/Z
 [5]        Z←((⎕tcnul=Z)⍳¨1)↑¨Z
+ [6]   :else
 [7]        Z←3/⊂' '
1}[8]   :endif
 [9]    Z←⊤⊃¨('Drive' 'Label' 'Serial' 'FileSystem') ((⊂,R),Z)
      ∇
```

4.12.2 The File System Object

The File System Object (**FSO**) is part of SCRRUN.DLL—the Microsoft Scripting Runtime Library. FSO provides access to the filing system—files, folders, and drives—and means of reading and writing sequential text files. FSO works within the limits of the access privileges of the user account that creates the instance of the object.

APL lacks support for the management of the filing system – there are no 'clean' methods for verifying whether a file exists, a drive is ready, the amount of space on a drive, etc. The sparse file management facilities provided by APL may be replaced by the FSO functions except for those relating to the creating, reading, writing, and compressing component files. FSO cannot read component files; in fact, it cannot read any type of binary file. In addition, unlike APL+Win, which has duplicate functions for use with short and long name files, FSO copes with file and folder names transparently.

A number of the methods offered by FSO achieve the same function as a standard Windows API call; where appropriate, the API call will be included when discussing the methods of the FSO object. FSO exposes four groups of properties, methods, and events – those relating to disk drives, the file system, folders, and files.

The APL cover function FSO, discussed below, implements the properties and methods of the FSO object. The FSO object does not expose any events. Some methods return objects that also expose properties and events – FSO is a hierarchy of objects, which sometimes have common properties and methods. In the cover function, pathways to common properties and methods are suppressed where possible in the interest of simplicity and where functionality is compromised.

Especially for APL application, FSO provides:

- Convenient access to the filing system.

- Many facilities that are not available in APL and would require complex code to achieve.

4.13 Platform enhancements

The operating system is under constant development at nil cost to any particular organisation. Therefore, it can be safely assumed that the tools will continue to evolve and may be deployed within the APL environment. This ensures that an APL application behaves like other Windows applications—this is a key consideration of users—and is able to provide comparable facilities.

Like APL, Windows offers pathways; that is, several ways of solving a given problem. For example, the existence of a file may be established with either the PathFileExists API call or the FileExists method of the FileSystemObject. An additional bonus is that Windows liberates APL from having to solve such routine problems in a robust and efficient way; APL development should use platform solutions and remove custom APL solutions wherever possible.

Chapter 5

The Component Object Model

An overriding consideration with legacy APL systems is the maintenance of a library of routines, which could be reused, optimised, and debugged overtime. In the APL world, two such libraries, the FinnAPL and IBM Idioms libraries are widely used; these libraries save a great deal of time and help to standardize code. The same principles underpin the modern concept of components. However, unlike the APL libraries whose code needed to be integrated into the application workspace, like native code, a component is not integral to the host application.

The Component Object Model (**COM**) is a language-independent model that allows software development to be implemented as units called components. Components are self-contained software—collections of objects—that are capable of exposing their functionality, in a standard way, to a host development environment and other compliant components regardless of their native respective languages.

COM defines the standards or rules for the public interface that a component exposes. Software that subscribes to this standard is a COM component, also interchangeably called ActiveX Server or ActiveX Component. A COM component is either an in-process or an out-of-process server.

5.1 Objects are global

All APL+Win's instances of ActiveX components are global and cannot be localised; the localisation of the APL+Win ⎕wself variable simply localises the reference to the instance and not the instance itself. Consider this example:

```
    ∇ Global;⎕wself
[1]   ⍝ System Building with APL+Win
[2]   ⎕wself←'ComDlg' ⎕wi 'Create' 'MSCOMDLG.COMMONDIALOG'
[3]   ⍝ Common dialogues. This is a commercial Microsoft control for
        standard dialogues.
[4]   ⍝ It may be present on a machine but unusable unless licensed.
[5]   ⍝ Try the XShowColor XShowFont XShowHelp XShowOpen XShowPrinter
        XShowSave methods
    ∇
```

In this example, the APL+Win context ⎕wself is localised but the object *ComDlg* is global. If this object existed before Global is run, it is permanently redefined in line [2]. Thus, instances of ActiveX classes cannot be used recursively in APL+Win.

5.2 APL+Win COM event handling

APL provides a **unique** way to handle events raised by out-of-process COM automation servers, like **Excel** or **Word**: an APL function may be run on an event being raised by the automation server. This is unique because it is not possible to replicate this with other environments like VB/VBA; that is, it is not possible to run a VB/VBA subroutine in the client environment.

Consider this APL session log from a new session that does not contain any objects:

```
      ⎕wself←'wd' ⎕wi 'Create' 'Word.Application'
      ⎕wi 'XQuit'
¯2147352572 ¯2147352572 ¯2147352572
      ⎕wi 'caption'
⎕WI ERROR: 800706BA The RPC server is unavailable.
      ⎕wi 'caption'
      ∧
      '#' ⎕wi 'children'
 wd
```

Although the instance of **Word** has been terminated, the APL object still persists but is non-functional. It might be good practice to delete the APL object on the instance being terminated:

```
      ⎕wself←'wd' ⎕wi 'Create' 'Word.Application'
      ⎕wi 'onXQuit' "'wd' ⎕wi 'Delete'"
      ⎕wi 'xApplication.Quit'
¯2147352572 ¯2147352572 ¯2147352572
      ρ'#' ⎕wi 'children'
0
```

To create an event handler for an event of a COM object such as **Excel**, refer to the **Excel** VBA help topic "Using Events with the Application Object". The procedure is defined in the code shown in **Figure 5-1 COM events**. The code is saved within the workbook, or in case of **Word**, within the document and not the template; in other words, its effect is isolated.

In **Excel**, on saving the active workbook, both the App_WorkbookBeforeSave and WorkbookBeforeSave events will fire, the latter will fire first. This may be readily tested by inserting an MsgBox statement in each subroutine. Note the following:

• The App_WorkbookBeforeSave event will not exist without the code in the modules, EventModule and CodeModule,

• The Workbook_BeforeSave event will not exist without the code in the ThisWorkbook module.

• Note that the arguments of each subroutine are different.

With APL, all code, such as those shown in **Figure 5-1 COM events**, is superfluous. This is highly advantageous in that APL does not need to add this code to a new workbook.

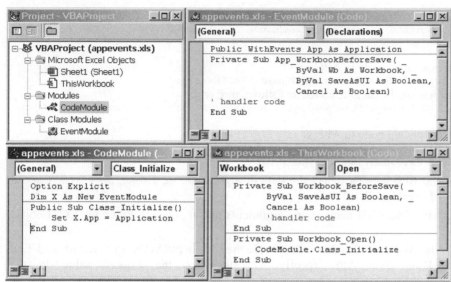

Figure 5-1 COM events

Given that only one event, WorkBookBeforeClose, is listed by ⎕wi 'events', a good question is, which event calls back APL?

```
    ⎕wself←'xl' ⎕wi 'Create' 'Excel.Application'
    0 0⍴⎕wi 'xWorkbooks.Add'
    ⎕wi 'xActiveWorkbook.Name'
Book2
    ⎕wi 'onXWorkbookBeforeSave' '⎕warg'
    ⎕wi 'xActiveWorkbook.Save'
0 1 7146800
```

The first two elements of ⎕warg are Boolean; the third appears to be an object pointer. On the face of it, this is consistent with neither event, until it is realised that ⎕warg collates the arguments in reverse order, compared to **Excel**. It is obvious that it fits the App_WorkbookBeforeSave event, <u>without</u> it being explicitly coded.

It is easily proven that the object pointer in ⎕warg is a workbook object:

```
    ⎕wi 'onXWorkbookBeforeSave' "'xl.wb' ⎕wi 'Create' (¯1↑⎕warg)"
    ⎕wi 'xActiveWorkbook.Save'
    'xl.wb' ⎕wi 'xFullName'
D:\ajay\Excel\Book2.xls
```

5.2.1 COM event arguments

The names of the events can be enumerated with ⎕wi 'events'. In this example, only the custom events are enumerated:

```
    3↑(∈'X'=1↑¨⎕wi 'events')/⎕wi 'events'
XNewWorkbook XSheetActivate XSheetBeforeDoubleClick
```

A prefix 'on' must be added to the name when coding event handlers. Consider some examples:

```
    ⎕wself←'xl' ⎕wi 'Create' 'Excel.Application'
    0 0⍴⎕wi 'xWorkbooks.Add'
    ⎕wi 'onXSheetSelectionChange' '⍳10'
    ⎕wi 'xCells(,).Select' 10 10 ⍝ expect the event to fire
1 2 3 4 5 6 7 8 9 10
```

Consider this example, coded as follows in **Excel:**
```
Private Sub Workbook_SheetSelectionChange(ByVal Sh As Object, ByVal Target As
     Range)
End Sub
```
The corresponding code in APL is:
```
     ⎕wi 'onXSheetSelectionChange' 'APLHandler'
```
Consider a second example, coded as follows in **Excel**:
```
Private Sub Workbook_BeforeSave(ByVal SaveAsUI As Boolean, Cancel As Boolean)
End Sub
```
The corresponding code in APL is
```
     ⎕wi 'onXWorkbookBeforeSave' 'APLHandler'
```
This raises three interesting questions:

• How does the APL handler detect the default arguments, *sh, Target, SaveAsUI, Cancel*, and the like, of event handlers?

• How does APL detect and manage any arguments that are objects within the COM object hierarchy?

• Obviously, the APL event handler will provide the code between the 'Private Sub' and 'End Sub' lines. How does APL specify a value like the *Cancel* value in the Workbook_BeforeSave event? When *Cancel* is set to **FALSE**, the workbook will not be saved.

5.2.1.1 COM default arguments
Every property, method, and event in a COM object is unique: each of these, if used, must be catered for individually within the APL environment. Therefore, a generic and automated or programmatic solution is unviable. A generic manual method is to use the short help query for the property, method, or event.

• For properties, named arguments are reported in the format name@type. For any property that is read only, the line with the APL assignment symbol (←) is relevant. For read/write properties, the name and type of argument is reported on the ⎕wi line. The name 'Value' is a generic name.

• For methods, the ⎕wi line reports the name and type of argument. The type of a variable is implicit when the name@type pattern is not used; an implicit data type is a Variant.

• For events, the ⎕warg line enumerates the arguments passed to the event handler and ⎕wres line enumerates the arguments that the event handler passes back to the COM server.

5.2.1.2 COM default argument types
Where applicable, the type of an argument is specified using the name@type pattern, subject to the following:

• Where the @type information is missing, the type is Variant.

• **Table 5-3 APL+Win external data type** lists the data types supported by APL.

• Names within square brackets are optional.

Wherever the @type indicates *Object*, the argument is an object pointer. The APL+Win system variable ⎕warg holds the object pointers for the arguments. It is necessary to create APL objects for each argument; this is best handled in a function:
```
     ⎕wi 'onXSheetSelectionChange' 'XLEvent ⎕wself'
```
The APL function XLEvent is now set as the handler for the event. It is defined as:
```
     ∇ XLEvent R;obj;typ
[1]    ⍝ System Building with APL+Win
[2]    obj←Φ(ρ⎕warg)↑(⊂R,'.obj'),¨(⍕¨⍳¨+\(ρ⎕warg)/1)~' '
```

```
[3]    Φobj ⎕wi ¨(⊂⊂'Create'),¨,¨⎕warg
     ∇
```

Line [2] overcomes the immediate problem: how many arguments are passed in? The answer is as many as the shape of the ⎕warg variable. This variable holds the pointer for the arguments in <u>reverse</u> order when compared with **Excel**. In order to maintain correspondence with **Excel**, the vector of names is reversed, as seen by:

```
     ⎕wi 'xCells(,).Select' 1 5 ⍝ Trigger event
 xl.obj1 xl.obj2
```

Two new objects, *xl.obj1*, xl.obj2 corresponding to sh, and *Target*, are created. This is confirmed by querying a typical property of each object: *sh* must have a name property and *Target* must have an address property. This is confirmed as follows:

```
     'xl.obj1' ⎕wi 'xName'
Sheet1
     'xl.obj2' ⎕wi 'xAddress'
$E$1
     (⊂'xl.obj2') ⎕wi ¨'xRow' 'xColumn'
1 5
```

This confirms that an APL event handler can recover the arguments passed:

● In practice, the APL event handler function must delete the transient objects it creates.

● The event may pass objects and values: thus, a cover function such as XLEvent is not a suitable approach. A function that handles each event individually is required: this will allow the fixing of values and objects to be hard coded.

5.2.1.3 Passing COM default values

Arguments are passed back to the calling environment by assignment to the ⎕wres variable. If the value being passed back is not an object, a simple assignment is adequate. If an object is passed back, the syntax is:

```
     ⎕wres←'object' ⎕wi 'obj' ⍝ object is the name of the object
```

5.2.1.4 COM Event handling demonstration

As noted, it is impractical to code a generic event handler because:

● ⎕warg contains a variable number of items and of different types.

● Every event, that is handled, requires specific code.

Consider this example: using an **Excel** workbook, it is desirable to prevent changes from being saved unless the user is AJAY ASKOOLUM:

```
     ∇ DemoXLEvent;⎕wself
[1]    ⍝ System Building with APL+Win
[2]    0 0ρ⎕wcall 'DeleteFile' 'c:\demoxl.xls'
[3]    ⎕wself←'xl' ⎕wi 'Create' 'Excel.Application'
[4]    0 0ρ⎕wi 'xWorkbooks.Add'
[5]    ⎕wi 'xActiveSheet.Cells(,).Value' 10 5 'Testing XL Event handling'
[6]    0 0ρ⎕wi 'xActiveWorkbook.SaveAs(Filename)' 'c:\demoxl.xls'
[7]    ⎕wi 'onXWorkbookBeforeSave' 'COMEvent ⎕wself'
[8]    ⎕wi 'xActiveSheet.Cells(,).Value' 10 5 'Changing a cell value'
[9]    ⎕wi 'xActiveWorkbook.Save'
[10]   0 0ρ⎕wi 'xActiveWorkbook.Close(SaveChanges)' 0
[11]   0 0ρ⎕wi 'xWorkbooks.Open' 'c:\demoxl.xls'
[12]   'Cell(10,5).Value=',⎕wi 'xActiveSheet.Cells(,).Value' 10 5 ⍝ Reading
          value
[13]   ⎕wi 'XQuit'
[14]   ⎕wself ⎕wi 'Delete'
     ∇
```

In line [7], the event trap is set: it would call a function COMEvent, defined as follows:

```
      ∇ COMEvent R;user
 [1]     ⍝ System Building with APL+Win
{1[2]    :select ⎕wevent
★ [3]        :case 'XWorkbookBeforeSave'
 [4]            ⍝ Fix arguments
 [5]            0 0⍴(R,'.wb') ⎕wi 'Create' (¯1↑⎕warg)
 [6]            (SaveAsUI Cancel)←⌽2↑⎕warg
 [7]            ⍝ Process event
 [8]            user←(⎕io+1)⊃⎕wcall (⊂'GetComputerName'),(32⍴256⍴⎕tcnul)
     (⎕av[⎕io+,⌽⍒256 256 256 256⊤32])
 [9]            user←⎕io⊃(user≠⎕tcnul)⊂user
{2[10]           :if user≡'AASKOOLUM' ⍝ user is 'AJAY ASKOOLUM'
 [11]                ⎕wres←0 ⍝ Cancel = False
 [12]                'Changes have been saved!'
+ [13]           :else
 [14]                ⎕wres←1 ⍝ Cancel = True
 [15]                'Access denied!'
2}[16]           :endif
 [17]            ⍝ Tidy up
 [18]            (R,'.wb') ⎕wi 'Delete'
 [19]            0 0⍴⎕ex ⊃'SaveAsUI' 'Cancel'
★ [20]       :case 'XSheetSelectionChange'
 [21]            ⍝ Fix arguments
 [22]            0 0⍴((⊂R),¨'.sh' '.Target') ⎕wi¨ (⊂⊂'Create'),¨⌽⎕warg
 [23]            ⍝ Process event
 [24]
 [25]            ⍝ Tidy up
 [26]            ((⊂R),¨'.sh' '.Target') ⎕wi¨⊂⊂'Delete'
+ [27]       :else
1}[28]   :endselect
      ∇
```

The function COMEvent traps two events: the WorkbookBeforeSave event is fully coded. In line [10], the user name is deliberately coded differently from the name returned by lines [8] and [9]; also, the comparison in line [10] does not make any allowance for case sensitivity. The result of running the DemoXLEvent function is:

```
      DemoXLEvent
Access denied!
Cell(10,5).Value=Testing XL Event handling
```

Change COMEvent[10] to match the user name, as follows:

```
{2[10]           :if user≡'AJAY ASKOOLUM'
```

Now, the result of running the DemoXLEvent function is:

```
      DemoXLEvent
Changes have been saved!
Cell(10,5).Value=Changing a cell value
```

5.2.2 Is RPC Server available?

The automation server is referred to as the Remote Procedure Call (**RPC**) server. APL does not provide an explicit method for querying the status of an automation server. A runtime error occurs if the server cannot be located:

```
      ⎕wself←'wd' ⎕wi 'Create' 'Word.Application'
      ⎕wi 'XQuit'
¯2147352572 ¯2147352572 ¯2147352572
```

```
      '#' ⎕wi 'children'  ⍝ wd still exists
 wd
      ⎕wi 'caption'  ⍝ Querying a typical property
⎕WI ERROR: 800706BA The RPC server is unavailable.
```

In this example, the APL instance of **Word** exists even after **Word** has been terminated using the Quit method; any attempt to use the APL object creates a runtime error. This function may be used to establish the status of an automation server:

```
          ∇ Z←IsRPCAvailable R;⎕wself
    [1]    ⍝ System Building with APL+Win
    [2]    Z←0≠⍴R ⎕wi 'self'
{1[3]    :if Z
{2[4]        :select R ⎕wi '∆hwnd'
★ [5]          :case 0
   [6]             ⎕wself←'∆rpc' ⎕wi 'Create' (R ⎕wi 'interface')
   [7]             Z←('x'∊∊1↑¨⎕wi 'properties')∧'X'∊∊1↑¨⎕wi 'methods'
{3[8]             :if 0=Z
   [9]                 R ⎕wi 'Delete'
3}[10]            :endif
   [11]           ⎕wself ⎕wi 'Delete'
+ [12]          :else
   [13]             Z←⎕wcall 'IsWindow' (R ⎕wi '∆hwnd')
2}[14]       :endselect
1}[15]   :endif
          ∇
```

The result is 1 if the RPC Server is available and 0 otherwise. Line [2] verifies whether the object exists. If the object exists and has a non-zero value in the user-defined property ∆hwnd, an API call is used to establish its status in line [13]. Otherwise, if the object exists, it will have at least one property and one method, line [7]. If it is not available, the APL object is deleted, see line [9]. Whether an RPC server is available can be established thus:

```
      IsRPCAvailable 'wd'
0
```

It is more reliable to use the handle of the instance of an automation server, if it can be established. Some servers, or more accurately, some versions of servers have a property that returns their handle. If this property is missing and if the server has a caption or its class is known, its handle can be queried using an API call:

```
      ⎕wself←'wd' ⎕wi 'Create' 'Word.Application'
      ⎕wcall 'FindWindow' 0 'Microsoft Word'  ⍝ Using caption
1308
      ⎕wcall 'FindWindow' 'OpusApp' 0  ⍝ Using class
1308
      ⎕wcall 'FindWindow' 'OpusApp' 'Microsoft Word'  ⍝ using both
1308
      'wd' ⎕wi '∆hwnd' 1308
```

If instances of the server might exist independently of APL, it may be necessary to change the caption, if possible, of the APL instance in order to recover the correct handle.

5.3 The promise of COM development

The promise that Component Based Development (**CBD**) holds out is rapid application assembly—the developer writes the code to use or glue together pre-written and debugged components to implement complex business functionality. In theory, rapid application assembly provides business with cutting-edge advantages:

• Components are available off the shelf—both industry-wide components and those developed in-house with previous projects. With industry standard components, the cost of development and support is external.

• The prototype of an application built using components can be produced much more quickly because the components are pre-written. A prototype provides a tangible and sound basis for defining the scope of a project to all interested parties—the business analysts, users, testers, developers and resource or project planners. An understanding of the scope of a project is vital to the accurate or reliable estimation of the cost and development cycle of a project.

• Applications can be built using multiple languages since the COM standard enables components in diverse languages to work together. It is no secret that some languages are more efficient at some tasks than others. In general, third party COM components do not expose their code. The developer can only work with the interface that the component exposes through its documentation.

• Multi-language development comes into its own when COM components are developed specifically for the host development environment—such as the APL2000's APL Grid component, which is specifically optimised for APL array facilities.

• Components promote code reuse, which in turn offers the dual benefits of consistency and debugged code. The look and feel of a component is consistent across applications—irrespective of whether the applications use the same or different languages. Components rapidly become very robust over time—the industry wide developer community deploys, debugs, and documents them, with fully coded working examples.

• Applications reusing components present a familiar interface to their users—this creates confident users. Users' endorsement of an application is the most telling accolade that an application might get. An application that is liked by its users is a successful application; it becomes the one tool for all jobs.

• Tools such as **VB** can be used to write components from scratch. For APL development, this is an attractive proposition in that a component can be written very quickly and specifically to provide key facilities—such as date handling—that are missing from APL. Custom components re-introduce the overhead of development and testing. Nonetheless, they warrant consideration especially in the realm of APL development, which lacks core facilities such as date handling.

5.4 Types of COM components

For practical purposes, a classification of components is a futile exercise—components are too diverse in nature to be categorised. Nonetheless, they have key characteristics that distinguish them apart.

5.4.1 Application components

Components such as **Excel, Word,** and APL+Win are applications in their own right:

• These components have a user interface that can be exposed or hidden by the host. Components that have a user interface permit ad hoc user interaction independently of the host on being visible. Both **Excel** and **Word** have application level *DisplayAlerts, ScreenUpdating, and UserControl* properties that enable an automation client to control their user interface events; in addition, documents, templates and workbooks have a *saved* property that may be used to disable application dialogues from appearing.

• They can be started and terminated independently or terminated independently if started by a host.

• They are capable of supporting bespoke code, which can be added interactively by the host or via existing custom files containing the code.

5.4.2 GUI components

Components, such as the APL Grid, cannot expose a user interface except via a host user interface. Components without an independent user interface are not applications in their own right, because:

• They have an indirect public interface, one that needs to be embedded in the interface provided by the host to enable user interaction.

• They are terminated via the termination of the host user interface. For example,

```
      ∇ APLGrid
[1]    ⍝ System Building with APL⍳Win
[2]    'APLGrid' ⎕wi 'Create' 'APL.Grid' ('Rows' 10) ('Cols' 5) ('visible'
         1)
      ∇
```

This function will create an instance of the APL Grid object but will not expose it—even when made visible—for user interaction. User interaction requires the parent object to have the *Wait* method. The APL Grid object itself does not support the 'Wait' method; it relies on the parent object for exposing it.

5.4.3 'Silent' or 'slave' components

Silent or slave components, such as a spell checker, do not have a user interface; they expose methods and properties to a host application. These tend to be specialist components dedicated to specific environments.

5.4.4 Custom component

Industry standard components are designed for the development community at large. Custom components are code modules compliant with the COM standard and written in-house for a specific purpose. The cost of ensuring the quality of the code—development standard and debugging—is borne in-house. **VB** provides the means of producing such components.

All components 'serve' their host or client environment. In general, components hide their code, implementation, and techniques from the user—they are black box code containers of binary code. This applies to components that are custom built—such as those built with **VB**—in that, they too are binary.

In a very loose sense, the FinnAPL and IBM Idioms libraries can also be seen as custom components—they are open source components designed specifically for the community of APL developers. The peculiar nature of these particular components is that they use the same language as the host for which they are destined; they do not comply with the COM standard.

5.4.5 Component visibility

By default, instances of components are not visible; they become visible either when the parent object, of which they are children, is made visible, like GUI components, or are made visible by setting their visible property to 1, like application components. 'Silent' components determine their own visibility depending on context. It is more efficient to keep components hidden for two reasons:

• Should an application component be visible, the user may terminate it thereby causing errors in the host application; therefore, either leave the component hidden or verify its continued existence before using any of its functionality. This adds a costly overhead in terms of the application's performance.

● A component needs to be visible to allow user interaction. If this interaction takes place within a dialogue shown by the component, the automation client cannot query or set any properties or call any methods <u>while</u> the dialogue is still active; an error occurs if it does.

Should the component need to be made visible for any purpose, make it visible for the duration required to accomplish that purpose and then hide it. This minimises the possibility of the component being terminated inadvertently. In practice, periodic verification of the components existence is advisable since the user may terminate the component abruptly via the **Task Manager**.

5.5 Maintaining objects

There is no enforceable naming convention for instances of COM components. However, some discipline in the naming convention will simplify application maintenance. Two procedures are recommended:

● First, use the prefix ∆ with the names of all objects that will not or cannot be deployed as part of the APL GUI order to identify instances of such objects. Thus, these objects must be children of the system object, #. The list of objects present may be queried:

```
('∆'=↑¨'#' ⎕wi 'children')/'#' ⎕wi 'children'
```

● Second, use the *ReleaseObjects* method to free resources used by instances of objects created within the workspace. '#' ⎕wi 'ReleaseObjects' does not delete instances of any active objects: the application code should terminate instances of objects that are no longer required, in the appropriate manner. For example, an instance of the Script Control can simply be deleted whereas an instance of **Excel** must call its *Quit* method before being deleted. Failure to terminate objects correctly will leave sessions running, that is using resources, as would failure to call *ReleaseObjects*.

5.6 APL+Win and ActiveX components

APL+Win is an ActiveX Server—capable of serving itself as well as other COM aware environments such as **VB**, **Excel**, etc.

In theory, the native languages used to implement components are irrelevant in that the COM standard relieves the developer from having to understand those languages. However, the developer needs to understand the documentation that describes a component's publicly-known interface—the properties, methods, and events that the component encapsulates.

The APL developer faces several issues with the deployment of components; both of them add a time overhead to coding. As a rule:

● There is the lack of APL documentation. This is a serious problem especially when a component exposes a hierarchy of objects each with their own set of methods, properties, and events. Documentation, especially worked examples, speeds up the learning curve, and leads to the rapid adaptation of solutions.

● Any available documentation needs to be translated into APL: sometimes a translation does not exist.

● APL always uses late binding; this method of creating an instance of an ActiveX control does not expose all the facilities that early binding does. Consider, for example, the ActiveX Data Object control.

Figure 5-2 Early binding collection shows VBA using early binding, the constants collections are available and used by Intellisense within the IDE.

● With late binding, these collections are not available. APL uses late binding.

Thus, the option of querying any particular item in the collection programmatically is unavailable:

```
'ADOConstants' ⎕wi 'Create' 'ADODB.CursorTypeEnum'
⎕WI CREATION ERROR: Class not registered "ADODB.CursorTypeEnum"
```

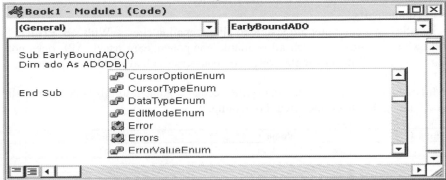

Figure 5-2 Early binding collection

Neither is the option of creating an instance of the ADODB.CursorTypeEnum object available to APL, as it is in VB, see **Figure 5-2 Early binding collection**. The objects shown in the drop-down box are collections of predefined constants. With APL, each constant must be resolved individually.

The constant *adUseClient* is an item in the *CursorTypeEnum* collection. Such collections *cannot* be enumerated in APL.

```
'Cursor' ⎕wi 'Create' 'ADODB.adUseClient'
⎕WI CREATION ERROR: Class not registered "ADODB.adUseClient"
```

A value can be assigned to this variable programmatically, when early binding is used; see **Figure 5-3 Early binding constants**. With APL, the value of such constants must be looked up and hard coded within the application or, preferably, looked up with a special syntax:

```
⎕wself←'ADO' ⎕wi 'Create' 'ADODB.Connection'
⎕wi '=adUseClient'
```

3

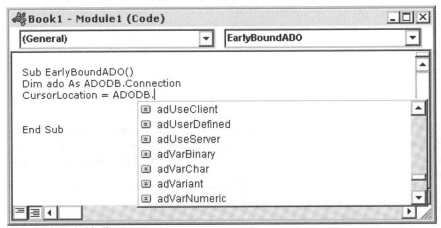

Figure 5-3 Early binding constants

Unlike APL variables, the name of the constants may not be case sensitive. The documentation of a component will usually list all predefined constants; they might also be listed in an INC file, such as ADOVBS.INC intended for VBScript. Using the named constant rather than its value makes the code more readable.

5.6.1 Platform components

Components installed on a computer are visible to all automation clients. With **VB**, **Project | References** shows a dialogue in which all available components are listed. APL does not have such a facility; the following simple expression enumerates all available components:

```
'#' ⎕wi 'XInfo'
```

This produces a three column nested matrix. A second argument to this code returns all components containing the string specified. For example:

```
'#' ⎕wi 'XInfo' 'apl'
```

Column 1	Column 2	Column3
APL.Grid	ActiveControl	APL Grid Control
APL.Grid.1	ActiveControl	APL Grid Control
APL2000.Grid	ActiveControl	APL2000 Grid Control
APL2000.Grid.1	ActiveControl	APL2000 Grid Control
APLW.WSEngine	ActiveObject	WorkspaceEngine Class
APLW.WSEngine.3	ActiveObject	WorkspaceEngine Class

Table 5-1 XInfo columns

The second column in **Table 5-1 XInfo columns** is a classification of the component. Either the first or the third column of `'#' ⎕wi 'XInfo'` can be used to specify the class. Consider the File System Object. The first method is:

```
⎕wself←'∆FSO' ⎕wi 'Create' 'Scripting.FileSystemObject'
⎕wi 'class'
ActiveObject Scripting.FileSystemObject
```

The second method is:

```
⎕wself←'∆FSO' ⎕wi 'Create' 'FileSystem Object'
'∆FSO' ⎕wi 'class'
ActiveObject FileSystem Object
```

Generally speaking, the two instances would be identical in terms of the properties, methods and events that they expose. However, if an application verifies that an APL object is an instance of the expected component, the creation method affects the verification code. For the first method, the code is:

```
'ActiveObject Scripting.FileSystemObject'≡'FSO' ⎕wi 'class'
1
```

For the second method, the code is:

```
'ActiveObject FileSystem Object'≡'FSO' ⎕wi 'class'
1
```

If an APL cover function validates a name against a class, the name must be created in a consistent way in order to avoid recreating an instance of the class.

Predictably, instances of components using the first column are not identical to those created using the third column for all components; **Excel** is an example of such a component. It has two objects as shown in **Table 5-2 XInfo for Excel Application Objects**.

Column 1	Column 2	Column 3
Excel.Application	ActiveObject	Microsoft Excel Application
Excel.Application.9	ActiveObject	Microsoft Excel Application

Table 5-2 XInfo for Excel Application Objects

An instance of **Excel** using the third column exposes fewer methods than one using the first column, as shown below:

```
      ⎕wself←'∆xl1' ⎕wi 'Create' 'Excel.Application'
      ρ⎕wi 'methods'
65

      ⎕wself←'∆xl2' ⎕wi 'Create' 'Microsoft Excel Application'
      ρ'∆xl2' ⎕wi 'methods'
28
```

In order to be consistent with other applications, use the first column; **VB** cannot use the literals in the third column to create instances of classes.

5.6.2 Opaque APL syntax

Public domain and documented examples that can be readily adapted are vital in promoting routine component deployment in application development. The APL developer is a victim in this respect in that APL examples just do not exist and the APL syntax is 'strange'. For instance, documentation aimed at the **VB** developer may state that an instance of **Excel** may be created as follows:

```
      Set objXL = CreateObject("Excel.Application")
```

The corresponding APL syntax is:

```
      'objXL' ⎕wi 'Create' 'Excel.Application'
```

The transition is by no means obvious and deters the APL developer from deploying components. APL2000 started providing examples of APL code using ActiveX components after release 3.0—very little else exists in the public domain.

5.6.3 Anatomy of the APL+Win syntax

The APL interface to components relies primarily on a single system function, ⎕wi, which takes two arguments. The left argument is a string vector representing the name of an object. The left-hand argument is optional when the context of the ⎕wi call is pre-specified: the system function ⎕wself fixes the context; this permits the omission of the left-hand argument. Also, the first element of the right-hand argument for ⎕wi:

- Does <u>not</u> require the full stop prefix.

- Is not specified as an assignment.

- Is always specified as a string.

- And, is prefixed by x or X.

With APL, the right-hand argument is either a string vector or a composite (nested) string vector and numeric scalar/vector/array. Although **TRUE** is -1 in **VB** and **VB** for Application (**VBA**), it is specified as 1 via APL. The key point is that ⎕wi works with strings—besides being enclosed in quotation marks, strings are also case sensitive. Although the names of natively supported APL objects are not case sensitive and must be specified precisely. The names of ActiveX objects' properties, methods, and events tend to be case insensitive.

Code is significantly more readable when specified *exactly* as reported by the object— this includes the prefix. Unlike APL, the **VB** IDE makes this transformation automatically when the object is created using early binding.

5.6.4 Hierarchy of objects

For each node of a hierarchical object, APL requires an instance of the node before exposing its properties, methods, and events. APL creates instances of hierarchical objects using redirection.

Excel is a hierarchical object—the application has workbooks, which have worksheets, which have ranges, which have cells, etc. In order to open a workbook in **Excel**, two objects may be required:

```
    'objXL' ⎕wi 'WorkBooks > .wkb'        ⍝ Create Workbooks object
    'objXL.wkb' ⎕wi 'Open' 'easter.xls' ⍝ Open workbook
```

In contrast, the **VB** documentation would suggest that this might be achieved as follows:	`objXL.WorkBooks.Open "easter.xls"`
APL/W supports the same syntax:	`'objXL' ⎕wc 'OLEClient' 'Excel.Application'` `objXL.Workbooks.Open 'easter.xls'`
APL+Win version 4.0 onwards supports the VB-like syntax	`'objXL' ⎕wi 'XWorkbooks.Open' 'easter.xls`

5.6.5 Data incompatibilities

Data incompatibilities between APL and components arise on two counts–type and structure. APL has just two data types: character and numeric. Components will generally have many data types such as string, date, integer, long, etc. By way of data structures, APL has scalars, vectors, matrices and nested vectors and matrices. Components will generally have vectors and matrices—occasionally, components may have user-defined data types.

The APL+Win ⎕wi function will usually manage the transfer of data between APL and components seamlessly. This makes working with components quite straightforward. An example of this is the Boolean data types— ‾1 and 0 in components but 1 and 0 in APL.

5.6.6 APL index origin

APL allows index origin ⎕io to be localised and within scope, indexing works either in base 0 or 1. Thus, index origin is global to the workspace and may, optionally, be set locally within every function in the workspace. In contrast, **VB** allows Options Base—equivalent of index origin—to be set at module level. An **Excel** workbook has the following types of modules:

• Workbook module: one in every workbook.

• Sheet module: one for every sheet that exists in the workbook.

• Modules independent of the workbook and sheets, added to the workbook using **Insert | Module** within the IDE. As many of these exist as have been added deliberately.

• Class modules, also independent of the other types of modules, added to the workbook using **Insert | Class Module** within the IDE. As many of these exist as have been added deliberately.

The scope of Option Base extends to the module in which the declaration appears; that is, to **all** subroutines and functions that it contains.

The scope of the APL ⎕io **never** extends to the components it is using. Components *always* work with their internal option base; this could be 0 or 1. It is a coincidence when both the APL index origin and the components option base are identical; the index origin may be changed without affecting the component's option base. Unless the APL developer guards against this, APL code working with components may encounter spurious runtime errors:

```
    ⎕io←0 ◇ 'objXL' ⎕wi 'xActiveWorkbook.xWorksheets(1).xName'
Sheet1
    ⎕io←0 ◇ 'objXL' ⎕wi 'xActiveWorkbook.xWorksheets(0).xName'
⎕WI ERROR: exception 8002000B
        ◇ 'objXL' ⎕wi 'xActiveWorkbook.xWorksheets(0).xName'
        ^
```

5.7 APL+Win post version 4.0 ActiveX syntax

APL+Win version 4.0 implements an enhancement to the way it handles ActiveX objects, bringing the APL syntax closer to the **VB** syntax—thereby making it easier to follow documentation written for **VB**. The simplification that version 4.0 has introduced produces simpler code and requires fewer objects.

5.7.1 Objects hierarchy using redirection

Redirection refers to the instance of an object arising within an expression being explicitly created within the scope of the same expression. Until version 4.0, redirection was the only method for accessing components exposing a hierarchy of objects. For example, a worksheet object is inaccessible without a preceding reference to a workbook object.

5.7.1.1 Redirection syntax I

With versions before 4.0, obtaining the name of the first worksheet in the active workbook would have required several steps:

```
'objXL' ⎕wi 'xActiveWorkBook > .wb' ⍝ Creates object objXL.wb
'objXL.wb' ⎕wi 'xWorksheets(1) > .ws' ⍝ Creates objXL.wb.ws
'objXL.wb.ws' ⎕wi 'xName'
```
Sheet1

The 'redirection' happens with the code shown in italics below:

```
'objXL' ⎕wi 'xActiveWorkBook > .wb' ⍝ Creates object
```

The hierarchy of objects created is:

```
'objXL' ⎕wi 'children'
 objXL.wb
'objXL.wb' ⎕wi 'children'
 objXL.wb.ws
```

The advantage in creating a hierarchy of objects is that objects below a node may be removed by deleting that node:

```
'objXL.wb' ⎕wi 'Delete' ⍝ Remove objXL.wb with its children
```

Redirection always creates an object, even if it already exists. This it is equivalent to:

```
'obj' ⎕wi 'Create' '... activexObject ...'
```

In contrast, the alternative expression does not:

```
'obj' ⎕wi 'New' '... activexObject ...'
```

The latter does not overwrite the object if it already exists. The difference between *'Create'* and *'New'* is that *'New'* will fail if the object already exists—safeguards the existing object—whereas 'Create' will not.

5.7.1.2 Redirection syntax II

An alternative syntax for redirection may be used:

```
'objXL.wb' ⎕wi 'Create' ('objXL' ⎕wi 'xActiveWorkBook')
objXL.wb
   ⎕wself←'objXL.wb.ws' ⎕wi 'Create' ('objXL.wb' ⎕wi 'xWorksheets(1)')
   ⎕wi 'xName'
Sheet1
```

Syntax 2 offers two advantages, namely:

• It allows the context for ⎕wi to be set with ⎕wself← in the same expression, thereby simplifying subsequent reference to the object just created.

• It offers a choice as to whether a previous instance of an object is overwritten or not, simply by using *'New'* instead of *'Create'*

5.7.2 Redirection clutter

The redirection technique does require a hierarchy of objects to be created—objects can be created or replaced freely. Consequently, the workspace may end up in a clutter of objects that make development and runtime workspace maintenance rather more complicated that necessary for these reasons:

• Whether an intermediate object, one higher up in the hierarchy, exists or needs to be created is not self-evident.

• If it is known that such an object exists, it is not clear whether its reference is correct—for example, it is not visibly clear which workbook objXL.wb is referring to. Whether an object exists can be established by *'object'* ⎕wi *'self'*—the shape of the result is zero if it does not exist.

• Instances of APL objects encapsulate the properties of the component at the time of creation and do not refresh automatically when changes to the component are made. *See Rogue Objects below.*

• Redirection does not enforce an intended hierarchy.

In the context of the latter issue, consider:

```
      'objXL' ⎕wi 'xActiveWorkBook > objWB' ⍝ Creates object objWB
      'objWB' ⎕wi 'xWorksheets(1) > objWS' ⍝ Creates objWS
      'objWS' ⎕wi 'xName'
Sheet1
```

All the objects created are at the same level—and each will require to be deleted independently–as can be seen:

```
      '#' ⎕wi 'children'
 objXL objWB objWS
```

Objects are *always* global to the session—they cannot be localised within functions in the same way as variables can—and cannot be expunged or erased as variables using ⎕ex or)erase or reassigned. An object must be deleted using the **Delete** method of ⎕wi—for example, *'objXL'* ⎕wi *'Delete'.* An object can be reassigned by the **Create** method of ⎕wi and the new value must always be another object—for example 'objXL' ⎕wi 'Create' *class*.

5.7.3 Dynamic syntax

Dynamic syntax refers to a scope where an object created by an intermediate expression is used by a subsequent expression in the same statement—the objects created while the statement is being executed evaporate—unless explicitly preserved—as soon as execution of the statement is complete.

If the names of all worksheets were to be enumerated, objXL.wb.ws (or objWB and objWS depending on how redirection is used) would need to be created as many times as there were worksheets in the active workbook. The creation and deletion of the transient objects degrade runtime performance. It is advisable, and indeed possible, to avoid creating the transient objects:

```
(('objXL' ⎕wi 'xActiveWorkBook') ⎕wi 'xWorksheets(1)') ⎕wi 'xName'
Sheet1
```

The object objXL exists. The code within each set of round brackets creates an object that survives until the whole expression is evaluated. On completion or failure of the expression, the transient objects no longer exist. This promotes cleaner code and avoids the pitfalls of redirection. The code is cleaner in that the intermediate objects do not persist. This saves having to track objects within a session—this can be pretty well impossible if objects are created within conditional blocks of code. It is easily verifiable that the expression within each set of round bracket creates an object:

```
      ⎕wself←'objXL'
      (⎕wi 'xActiveWorkBook') ⎕wi 'class'
ActiveObject
```

```
      ((⎕wi 'xActiveWorkBook') ⎕wi 'xWorkSheets(1)') ⎕wi 'class'
ActiveObject
      ⍴'objXL' ⎕wi 'children'
0
```

However, the initial object, objXL, is unaffected but the intermediate objects do *not* persist; this is easily verifiable:

```
      '#' ⎕wi 'children'
 objXL
```

This form of dynamic syntax with ActiveX objects relies on forcing the order of evaluation to the innermost brackets first. APL uses dynamic syntax in a different form, not reliant on evaluation order, with native objects. For instance:

```
      'MyForm' ⎕wi 'Create' 'Form' ('visible' 1)
```

In this example, the expression within the brackets is clearly *not* evaluated first—if it had, it would fail. Within the scope of the bracket the object, whose visible property is being set, would not have been created. Clearly, the expression works!

5.7.4 Hierarchical syntax

APL+Win version 4.0 onwards still supports both redirection and dynamic syntax. Dynamic syntax does produce clearer code than redirection but it still requires several levels of nesting to force a particular order of execution. The new hierarchical syntax introduced by version 4.0 makes code less verbose and clearer and, potentially, easier to maintain:

```
      'objXL' ⎕wi 'xActiveWorkbook.Worksheets(1).Name'
Sheet1
```

This hierarchical syntax using '.' (dot) to separate objects known *to exist but not created* applies to ActiveX objects only. Code using the hierarchical syntax is obviously clearer than equivalent code using redirection or dynamic syntax. The most tangible benefit of hierarchical syntax is that it produces code that is very close to **VB** syntax—this makes the translation of **VB** examples into APL a lot more straightforward.

The dot separation is used differently with native objects. If used in the left-hand argument, it specifies a child object *to be* created. If used in the right-hand argument, it specifies a child object that exists; that is, that has been created. In essence, the dot is used as a shortcut reference. As such, it is arguable that it makes code less transparent.

Consider an example showing the dot notation in the left-hand argument of ⎕wi:

```
    ∇ NativeHierarchy
[1]    ⍝ System Building with APL+Win
[2]    ⎕wself←'MyForm' ⎕wi 'Create' 'Form' ('visible' 0)
[3]    ⎕wself←'.Frame' ⎕wi 'New' 'Frame'
[4]    ⎕wself←'.Edit' ⎕wi 'New' 'Edit'
    ∇
```

In the left-hand argument, the dot creates an object *below* the current context, ⎕wself, as is shown by the names of the objects created. ⎕wself is reassigned on each line where an object is created. Only one dot can be used:

```
      'MyForm' ⎕wi 'children'
MyForm.Frame
      'MyForm.Frame' ⎕wi 'children'
MyForm.Frame.Edit
```

The following illustrates the use of dot in the right-hand argument of ⎕wi:

```
      ⎕wi¨ '....class' '....name'
System #
      ⎕wi¨ '...class' '...name'
Form MyForm
      ⎕wi¨ '..class' '..name'
```

```
Frame Frame
     ⎕wi¨ '.class' '.name'
Edit Edit
     'MyForm' ⎕wi '.Frame.Edit.enabled'
1
```

In the right-hand argument, each hanging dot—without a class prefix—refers to one level *above* in the hierarchy of objects that exist. It is all too easy to specify too many dots. The hierarchical syntax with native APL objects is contrived and opaque.

5.7.4.1 Dynamic syntax with Excel

Excel VBA stores code that is independent of the workbook in modules and classes. However, the code is saved with the workbook. A module is added by pressing **Alt + F11** followed by either **Insert | Module** for a module or **Insert | Class Module** for a class.

In this example, an **Excel** object—a Range—is passed as an argument to a VBA function. The following user-defined function is added to a module within an **Excel** workbook:

```
Public Function DynSyntax(ByVal anObject As Variant) As String
    DynSyntax = "Range has " & anObject.Rows.Count & _
                " Rows and " & anObject.Columns.Count & _
                " Columns."
End Function
```

In a worksheet, cell A4 contains the formula *=DynSyntax(A1:A10)* and cell A5 contains *=DynSyntax(A2:A12)*: **Figure 5-4 Formula view I** shows the formulae.

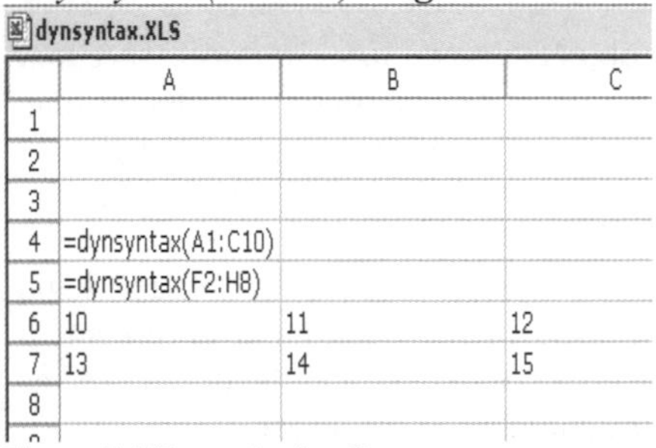

📄 dynsyntax.XLS			
	A	B	C
1			
2			
3			
4	=dynsyntax(A1:C10)		
5	=dynsyntax(F2:H8)		
6	10	11	12
7	13	14	15
8			

Figure 5-4 Formula view I

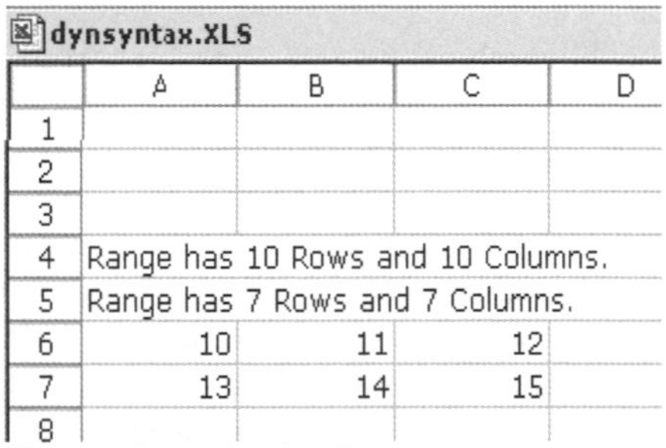

📄 dynsyntax.XLS				
	A	B	C	D
1				
2				
3				
4	Range has 10 Rows and 10 Columns.			
5	Range has 7 Rows and 7 Columns.			
6	10	11	12	
7	13	14	15	
8				

Figure 5-5 Results view I

The formulae evaluate to produce the expected results, shown in **Figure 5-5 Results view I**, confirming that the argument is passed to the function as an object.

Note that the argument to the function DynSyntax: function names are not case-sensitive in **Excel** and are specified *without* quotation marks. The objective is to query the number of rows and columns in a range passed as argument *anObject* to the function *DynSyntax*. Note that an object is not explicitly created within this function—the object is passed in by the formula in the cell—but a hierarchical syntax is used to query the number of rows and columns in the range.

5.7.4.2 Dynamic syntax for Excel objects within APL+Win

In the previous section, dynamic syntax was demonstrated within the *same* environment, that is, within **Excel** itself. The new objective is to demonstrate that it is possible to pass an object as an argument *across* environments—the object is the range A6:C7, containing the matrix 2 3⍴11 12 13 14 15 16.

The host or client environment is **Excel** and the target or server environment is APL with a workspace *dynsyntax.w3* loaded. The function, *DynSyntax*, is in this workspace. The APL function is:

```
    ∇ Z←DynSyntax R
```

```
[1]    ⍝ Return shape of range passed in as object R from Excel
[2]    R←⍉R ⎕wi 'xValue'
[3]    Z←⍕ρ↑¨R
   ∇
```

In **Excel**, a module contains the following code:

```
 1 Global APL As Object
 2 Public Function StartAPLServer() ' Start APL+Win ActiveX Server & load
        workspace
 3     Set APL = CreateObject("APLW.WSEngine")
 4     APL.syscommand ("load " & Application.DefaultFilePath & "\dynsyntax")
 5     StartAPLServer = APL.Visible
 6 End Function
 7 Function RunAPLFn(ByVal APLFn As String, _
 8                   Optional ByVal RightArg As Variant)
 9     RunAPLFn = APL.Call(APLFn, RightArg)
10 End Function
```

The **Excel** function StartAPLServer starts the APL+Win ActiveX Server and loads the initial workspace DYNSYNTAX.W3. The function RunAPLFn calls a function within the ActiveX Server active workspace, passes one argument to it and returns its result. As expected, the sole argument is passed as the right-hand argument.

The formulae are shown in **Figure 5-6 Formula view II**. Note:

• Inserting a user-defined formula in cell A2 starts the APL+Win ActiveX server: the server must be started before a call is made to a function.

• The APL function *DynSyntax* is called in cell A5.

• Function and variable names *are* case-sensitive in APL and must be spelt exactly as they appear in the server workspace.

• In **Excel**, function and variable names may be entered freely as Intellisense automatically writes them correctly.

• The results of these user-defined formulae are shown in **Figure 5-7 Results view II**.

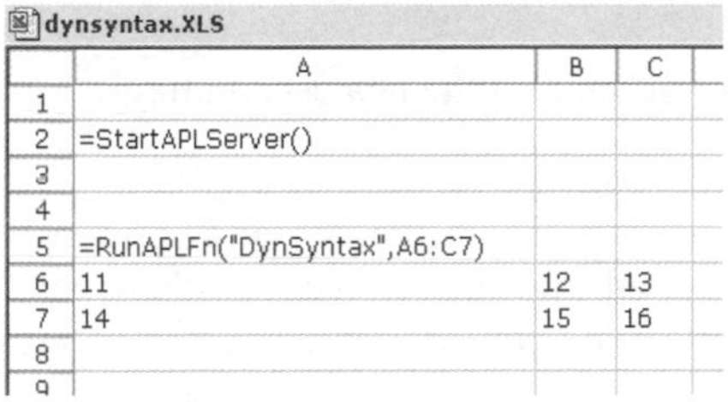

	A	B	C
1			
2	=StartAPLServer()		
3			
4			
5	=RunAPLFn("DynSyntax",A6:C7)		
6	11	12	13
7	14	15	16
8			
9			

	A	B	C	D
1				
2	0			
3				
4				
5	2 3			
6		11	12	13
7		14	15	16
8				
9				

Figure 5-6 Formula view II 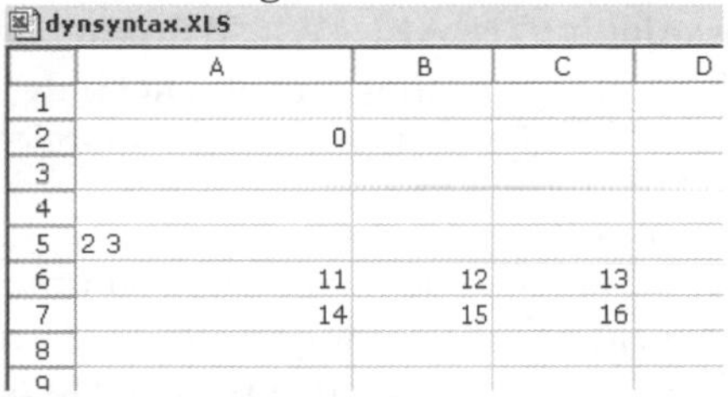   **Figure 5-7 Results view II**

Some aspects of the APL function *DynSyntax* are noteworthy:

• In line [1], the 'Value' of the argument R, passed from **Excel**, is queried—no objects are created within this function.

• The transpose ⍉ in line [2] is necessary because **Excel** is row major whereas APL is column major.

• The ↑¨ in line [3] is necessary to cope with empty cells in the range passed across from **Excel**.

• The ⍕ also in line [3] is necessary because **Excel** cannot show more than one number in any cell unless formatted as character.

• The name of the function must be specified exactly in **Excel**. Unlike **Excel**, APL names are case sensitive.

5.7.5 Redirection or enumeration?

If a property of a collection of objects is to be queried, each item in the collection needs to be referenced in turn—this requires a looping solution. An example of a collection is the names of every worksheet in a workbook. Consider the following example:

```
      ∇ Z←RedirEnum;i
[1]    ⍝ Ajay Askoolum
[2]    'objXL' ⎕wi 'Create' 'Excel.Application'
[3]    'objXL' ⎕wi 'xWorkBooks.Add > .wb'
[4]    :for i :in ⍳5
[5]        ('objXL.wb' ⎕wi 'xWorkSheets.Add') ⎕wi 'xName' ('MySheet - ',⍕i)
[6]    :endfor
[7]    i←'objXL.wb' ⎕wi 'xWorkSheets.Count'
[8]    Z←⊂''
[9]    :for i :in +\i/1
[10]       Z←Z,⊂'objXL.wb' ⎕wi 'xWorkSheets(i).Name'
[11]   :endfor
      ∇
```

Lines [1] – [6] create a workbook, adds five sheets to it, and renames each sheet. Equally, neither the number of worksheets nor their respective names in the workbook may be known in advance.

Line [7] establishes the number of worksheets. As the name of each worksheet is required, each worksheet object need to be referenced in turn—hence the looping solution in lines [9]-[11]. Note the specification of the loop counter i as $+\backslash i/1$—this ensures that index origin or option base 1 is used. Line [10] generates an error:

```
⎕WI ERROR: 80020004 Parameter not found.
RedirEnum[10]    Z←Z,⊂'objXL.wb' ⎕wi 'xWorkSheets(i).Name'
                                    ∧
```

The complaint is that the parameter i is not found—it is being sought in **Excel** where it is undefined. This can be resolved in several ways.

5.7.5.1 Exploiting the APL+Win Windows interface

The APL+Win ⎕wi function provides not only the simplest but also the neatest solution:

```
    'objXL.wb' ⎕wi 'xWorkSheets(1).Name'
MySheet - 5
```

The syntax is correct if the worksheet index is hard coded. Therefore, the problem must be that the counter value is not being substituted. As noted, the first element of the right-hand argument of ⎕wi is always a string. Therefore, it can be constructed using the literal value of i and the resulting string is passed to ⎕wi.

The replacement line [10] is as follows:

```
[10]       Z←Z,⊂'objXL.wb' ⎕wi 'xWorkSheets(',(⍕i),').Name'
```

Note that line [10] is evaluated right-to-left consistent with the way APL expressions are usually evaluated. A neater solution is to specify the worksheet index as an argument:

```
[10]       Z←Z,⊂'objXL.wb' ⎕wi 'xWorkSheets().Name' i
```

With the revised line [10], the following result is obtained:

```
    RedirEnum
MySheet - 5 MySheet - 4 MySheet - 3 MySheet - 2 MySheet - 1 Sheet1
```

The result is surprising on two counts. Firstly, there are six worksheet names and only five were added. Secondly, the names are seemingly in reverse order. This is *not* anomalous. On adding a workbook, a default number of worksheets are created within the workbook.

The number of worksheets created depends on the default specified—a workbook must have at least one worksheet—and may be queries thus:

```
      'objXL' ⎕wi 'xSheetsInNewWorkbook'
1
```

This accounts for six worksheets. The names appear to be in reverse order because, by default, **Excel** adds sheets *before* the current sheet—thus, on adding a worksheet, the newly added worksheet is *Worksheest(1)* and the existing worksheet becomes *Worksheest(2)*. Thus, the default worksheet in the workbook ends up being *WorkSheets(6)*. As we only renamed the five sheets added in lines [4]-[6], the original sheet retained its default name.

An understanding of the workings of the ActiveX object is vital in order to be able to explain its behaviour.

5.7.5.2 Re-iterative redirection

The same result is obtained when lines [9] – [11] are replaced by the following:

```
[9]    :for i :in +\ι/1
[10]        'objXL' ⎕wi 'xWorkSheets > .wb'
[11]        'objXL.wb' ⎕wi 'Item > .ws' 1
[12]        Z←Z,⊂'objXL.wb.ws' ⎕wi 'xName'
[13]   :endfor
    RedirEnum
MySheet - 5 MySheet - 4 MySheet - 3 MySheet - 2 MySheet - 1 Sheet1
```

Unlike the previous solution, which did not create any objects, this one creates each of the *objXL*.wb and *objXL.wb.ws* objects six times, requires more lines of code and leaves the last instance of the object in memory. It is untidy.

5.7.5.3 APL+Win version 4.0

Version 4.0 simplifies the syntax for working with ActiveX objects and promotes much cleaner and more economical code:

```
      ∇ Z←RedirEnum2;i;⎕wsclf
  [1]    ⍝ System Building with APL+Win
  [2]    ⎕wself←'objXL' ⎕wi 'Create' 'Excel.Application'
  [3]    0 0ρ⎕wi 'xWorkBooks.Add'
{1[4]    :for i :in ι5
  [5]        ⎕wi 'xWorkSheets.Add.Name' ('MySheet - ',⍕i)
1}[6]    :endfor
  [7]    i←⎕wi 'xWorkSheets.Count'
  [8]    Z←⊂''
{1[9]    :for i :in +\i/1
  [10]       Z←Z,⊂⎕wi 'xWorkSheets().Name' i
1}[11]   :endfor
      ∇
```

This version does not require the workbook object. In line [5], the worksheet is named directly. Line [10] is much simpler, allowing the argument of the worksheet function to be specified outside—the missing arguments are read sequentially.

Consider a more complicated example: on the first worksheet, the first 10 rows and first 10 columns are populated with the numbers 1 to 100. The code is shown in **Figure 5-8 Place holder arguments**. A set of round brackets indicates a single argument; commas within the brackets separate multiple arguments. The final value is the argument to the ⎕wi call after it has been reconstituted with the preceding arguments.

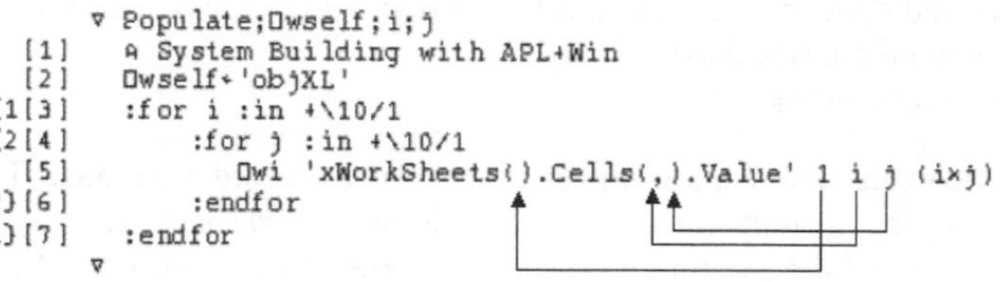

```
      ∇ Populate;⎕wself;i;j
 [1]    ⍝ System Building with APL+Win
 [2]    ⎕wself←'objXL'
{1[3]   :for i :in +\10/1
{2[4]     :for j :in +\10/1
 [5]        ⎕wi 'xWorkSheets().Cells(,).Value' 1 i j (i×j)
2}[6]     :endfor
1}[7]   :endfor
      ∇
```

The arguments are consumed in the manner indicated by the arrows in **Figure 5-8 Place holder arguments**.

Figure 5-8 Place holder arguments

5.7.5.4 APL+Win enumeration

Surprisingly for a language that is so dextrous at handling arrays and one which supports the each construction (¨), APL is very poor at handling collections. This is obvious from the APL function, *APLEnum*, which is comparable to the **Excel** function *GetSheetNames*. The following function illustrates how APL handles collections:

```
      ∇ Z←APLEnum;R;⎕wself
 [1]    ⍝ System Building with APL+Win
 [2]    'objXL' ⎕wi 'xActiveWorkBook.WorkSheets > .SheetsCollection'
 [3]    Z←⊂''
 [4]    0 0⍴'objXL.SheetsCollection' ⎕wi 'EnumStart'
{1[5]   :while 0≠⍴R←'objXL.SheetsCollection' ⎕wi 'EnumNext'
 [6]       ⎕wself←'objXL.IndividualSheet' ⎕wi 'Create' R
 [7]       Z←Z,⊂⎕wi 'xName'
1}[8]   :endwhile
 [9]    'objXL.SheetsCollection' ⎕wi 'EnumEnd'
 [10]   'objXL.SheetsCollection' ⎕wi 'Delete'
 [11]   'objXL.IndividualSheet' ⎕wi 'Delete'
      ∇

      'objXL.SheetsCollection' ⎕wi 'methods'
      Close Create Defer Delete EnumEnd EnumNext EnumStart Event Exec
Modify New Open Ref Send Set SetLinks XAdd XCopy XDelete XFillAcrossSheets
XMove XPrintOut XPrintPreview Xselect
```

The object created in line [2] includes methods handled by APL—these are ones without the 'X' prefix. APL+Win handles enumeration with its own methods *EnumStart*, *EnumNext* and *EnumEnd*. The object to which the methods *EnumStart* and *EnumEnd* are applied must exist physically—this is created in line [2]. With every iteration, the method *EnumNext* creates an object, implicitly. The scope of the object pointer appears to extend to multiple statements. It is interesting to note an inconsistency with ⎕wi. Exceptionally, the syntax in lines [5]-[6] works with APL+Win methods and does not work with ActiveX methods: for example:

```
      R←'objXL' ⎕wi 'xActiveWorkBook'
      'objXL.Sheets' ⎕wi 'Create' R
⎕WI CREATION ERROR: Invalid object pointer
      'objXL.Sheets' ⎕wi 'Create' R
                    ^
```

It is also irritating that *EnumEnd* is required to end the iteration explicitly. The *:while* loop started in line [5] terminates when R inherits a value of 0, which signals that no more items are available in the collection.

This approach produces inelegant code with unnecessary overheads—several intermediate objects are created and need to be deleted at the end. Nonetheless, it produces the expected result, as shown:

```
      APLEnum
MySheet - 5 MySheet - 4 MySheet - 3 MySheet - 2 MySheet - 1 Sheet1
```

This is an example of a situation where APL fails to provide a concise and neat solution—the APL code is 11 lines compared with 7 lines of **Excel**!

5.7.5.5 Interrogating Excel

APL can add and run **Excel** functions and subroutines; functions return results but subroutines do not. For this example, two functions will be added to an instance of **Excel**. *GetSheetNames* will return the names of all worksheets and *GetUsedRange* will return the values contained in the used range within **Excel**.

The following APL function will create an instance of **Excel**, add a workbook, add five sheets to the workbook, add the two functions, and make **Excel** visible:

```
       ∇ HandlingExcelMacros R;⎕wself
[1]    ⍝ System Building with APL+Win
[2]    ⎕wself←'objXL' ⎕wi 'Create' 'Excel.Application' ('visible' 1)
[3]    (⎕wi 'xWorkBooks.Add.VBProject.VBComponents.Add(1).CodeModule') ⎕wi
       'XAddFromString'  R
[4]    :for i :in ⍳5
[5]        ('objXL' ⎕wi 'xActiveWorkBook.WorkSheets.Add') ⎕wi 'xName'
       ('MySheet - ',⍕i)
[6]    :endfor
       ∇
```

The right-hand argument of *HandlingExcelMacros* is a string vector comprising the definition of the *GetSheetNames* and *GetUsedRange* functions. Some values have been manually added to sheet 6, named MySheet-5 and shown in **Figure 5-9 Excel macros,** in order to facilitate the demonstration of the *GetUsedRange* function. A variety of data types are added; note that a number of cells are left empty and that some sheet names are visible.

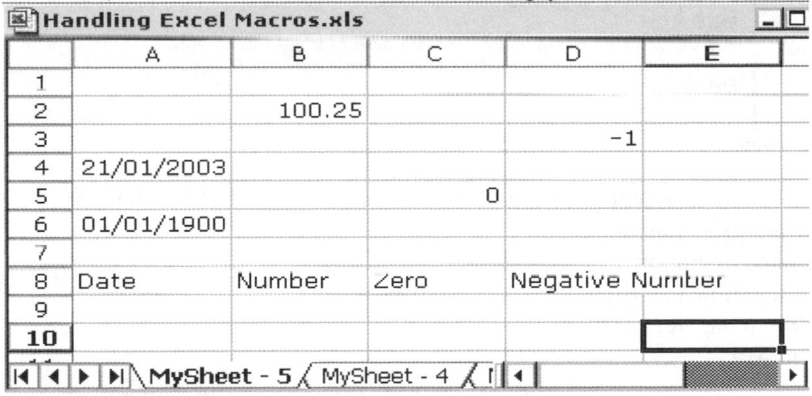

Figure 5-9 Excel macros

There are alternative methods for managing VBA automation code:

• It is expedient to develop and debug such functions within **Excel** and to save them in the APL workspace using **copy** + **paste** in a string variable. In **Excel**, the functions are added to a code module named Module1 by default.

• The code in any module may be written to a text file using the **File | Export File** option in the **Excel VB** Editor (**Alt + F11**) menu.

APL can load such a file into **Excel** just as easily as from an APL variable—in HandlingExcelMacros [2], substitute *XAddFromString* by *XAddFromFile* and the right-hand argument of this function, R, is the name of the file containing the code instead of an APL variable or a literal containing the code.

There are several advantages in using a file exported directly from **Excel:**

● All the functions and subroutines may be debugged within the **Excel** environment before being ported to APL or files.

● The file may contain any number of functions and subroutines.

● This avoids transcription and coding errors.

APL can query the names of the sheets using **Excel** VBA code.	```⊃'objXL' ⎕wi 'Run' 'GetSheetNames'```
	 ```MySheet - 5```
	 ```MySheet - 4```
	 ```MySheet - 3```
	 ```MySheet - 2```
	 ```MySheet - 1```
	 ```Sheet1```

Refer to the code shown in **Figure 5-10 Excel macros code: Excel** returns the names of the worksheets in a variable of type Variant—which comes to APL as a nested vector, which is disclosed (⊃). The **Excel** result is, as to be expected, the same as that of the APL solutions except that the APL result can be displayed more easily.

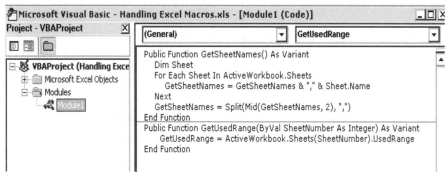

Figure 5-10 Excel macros code

Two functions were added to **Excel**— to demonstrate that multiple functions can be added in the same operation.	```⎕←A←⍴'objXL' ⎕wi 'Run' 'GetUsedRange'```

The GetUsedRange function identifies the range containing data, reads its value and returns it to APL.	``` 100.25```
	 ``` ¯1```
	 ```37642```
	 ```1 0```
	 ```Date Number Zero Negative Number```

APL can identify the empty cells. A zero indicates the position at which the cell is empty.	``` ρ¨∊¨A```
	 ```0 1 0 0```
	 ```0 0 0 1```
	 ```1 0 0 0```
	 ```1 0 1 0```
	 ```0 0 0 0```
	 ```0 0 0 0```
	 ```4 6 4 15```

The used range is rows 2–8 and columns 1-4, inclusive. Thus the range has 7 rows and 4 columns.	``` ρA``` ```7 4```
In APL, the variable is nested 6 levels deep.	``` ≡A``` ```6```

The integrity of cell data type is retained.	A[1;2]×10 1002.5

The latter statement is supported by numeric operation on element A[1;2]. Minus 1 is correctly translated as negative 1, with a high minus. Empty cells come across as empty elements. The one point of contention is that the dates 01/01/1900 and 21/01/2003 are shown as 1 and 37642 respectively. Although this might seem anomalous, in fact, the correct values have been transferred—APL does not understand dates causing the formatting to be lost. **Excel** stores dates as a number internally and just displays it formatted. It is the internal number that is transferred. In this case, 21/01/2003 represents 37642 days from 01/01/1900, which is day 1. Note that **Excel** incorrectly recognises 29/02/1900 as a valid date.

5.8 ActiveX typed parameters

There is a critical semantic difference in the APL and, say, the **VB** environments as regards instances of objects. In APL, an instance of an object is an unassigned string in Windows memory and in **VB** it is an assigned variable holding an object pointer.

Consider the following **VB** code illustrating the Open method of the ActiveX Data Object's (**ADO**) record object:

```
1   Dim cnn1 As ADODB.Connection
2   Dim rstEmployees As ADODB.Recordset
3   Dim strCnn As String
4   Dim varDate As Variant
5   ' Open connection.
6   strCnn = "Provider=sqloledb;" & _
7           "Data Source=MyServer;Initial Catalog=Pubs;User Id=sa;Password=;
        "
8   Set cnn1 = New ADODB.Connection
9   cnn1.Open strCnn
10  ' Open employee table.
11  Set rstEmployees = New ADODB.Recordset
12  rstEmployees.CursorType = adOpenKeyset
13  rstEmployees.LockType = adLockOptimistic
14  rstEmployees.Open "employee", cnn1, , , adCmdTable
```

This code raises these four issues for APL, namely:

• Lines [1] – [4] declare variables as typed objects—lines [1] and [2]—and as string and variant—lines [3] and [4] respectively. APL does not support these types of variables.

• Lines [12] and [13] assigns properties of objects using named constants which have not been declared in the code—in **VB**, these constants are read only properties of the ADO control and are available when the appropriate object is created using *early* binding. APL uses *late* binding and the constants are unavailable as variables.

• In line [14], the connection object created in line [8] becomes an argument or parameter—the object pointer is *cnn1*. In APL+Win, an object exists as a string and not an object pointer.

• In addition, the syntax used in line [14] omits the third and fourth arguments.

5.8.1 Boolean parameters

Depending on the component, Boolean may mean 1 and 0. In most contexts, APL will substitute 1 for **TRUE** although **TRUE** may have a value of ‾1 in the component. However, in circumstances where 1 or 0 result in a runtime error, an explicit Boolean value must be specified:

```
        ('#' ⎕wi 'VT' 1 11) or ('#' ⎕wi 'VT' 0 11)
```

5.8.2 Positional parameters

The syntax for the expression in line [14] is:

```
        recordset.Open Source, ActiveConnection, CursorType, LockType, Options
```

The syntax dictates the ordinal position of each parameters; **VB** affords the luxury of being able to omit one or more of these parameters, as shown in line [14] of the **VB** code:
`rstEmployees.Open "employee", cnn1, , , adCmdTable.`

Any omitted parameter inherits some system driven default value. With APL, it is not be possible to omit a positional parameter. It is good practice to specify the syntax in full as an aid to debugging—the code implicitly documents its construction—and this is not too onerous a task.

5.8.3 Passing object pointers

The critical issue in translating in [14] is being able to pass an object pointer as a parameter. The equivalent code in APL using fully qualified syntax is:

```
      ∇ Resolve
[1]    ⍝ System Building with APL+Win
[2]    0 0⍴'cnn1' ⎕wi 'Create' 'ADODB.Connection' ⍝ Lines [1] and [8]
[3]    0 0⍴'rstEmployees' ⎕wi 'Create' 'ADODB.Recordset' ⍝ Lines [2] and
       [11]
[4]    strcnn←"Provider=sqloledb;Driver={SQL Server};Server=AJAY
       ASKOOLUM;Initial Catalog=Pubs;User Id=sa;Password=;" ⍝ Lines [6]
       and [7]
[5]    'cnn1' ⎕wi 'XOpen' strcnn
[6]    'rstEmployees' ⎕wi 'xCursorType' (⎕wi '=adOpenKeySet')
[7]    'rstEmployees' ⎕wi 'xLockType' (⎕wi '=adLockOptimistic')
[8]    ⎕wself←'rstEmployees'
[9]    ⎕wi 'XOpen' "employee" ('cnn1' ⎕wi 'obj') (⎕wi 'xCursorType') (⎕wi
       'xLockType') 2 ⍝ Line [14]
[10]   → 0
[11]   ⍝ XOpen Syntax = recordset.Open Source, ActiveConnection,
       CursorType, LockType, Options
[12]   array←'rstEmployees' ⎕wi 'GetRows' ‾1 1 ('#' ⎕wi 'VT' ('emp_id' 1
       'minit') 8204)
      ∇
```

This function omits variable declarations, refers to constants by name, and specifies all parameters—and the connection string is adapted, as it is machine specific.

5.8.4 Typed data parameters

APL does not have typed data as other languages. Although this is not usually a problem—passing data parameters, as distinguished from object parameters, by value resolves the data type issue—there are instances when an ActiveX method requires strongly typed data.

An example of an ActiveX method that requires strongly typed data is the GetRows method of the record object. The GetRows method is documented as follows:
`array = recordset.GetRows( Rows, Start, Fields)`

The documentation of the Fields parameter is:
`"Optional. A Variant that represents a single field name or ordinal position,`
`     or an array of field names or ordinal position numbers. ADO returns`
`     only the data in these fields."`

The **VB** code for retrieving the first three fields 'emp_id', 'fname' and 'minit' from the rstEmployees object is:
`array=rstEmployees.GetRows(,,Array("emp_id","fname","minit"))`

Intuitively, the corresponding APL, specifying all positional parameters, should be:
`     array←'rstEmployees' ⎕wi 'XGetRows' ‾1 1 ('emp_id' 'fname' 'minit')`
`⎕WI ERROR: ADODB.Recordset exception 800A0CC1 Item cannot be found in the`
`collection corresponding to the requested name or ordinal.`
`     ◊  array←'rstEmployees' ⎕wi 'XGetRows' ‾1 1 ('emp_id' 'fname'`
`'minit')`

It does not work! However, the following expressions work:

```
ρarray←'rstEmployees' ⎕wi 'XGetRows' ¯1 1 ('emp_id')
```

43 1

```
        ρarray←'rstEmployees' ⎕wi 'XGetRows' ¯1 1 ('emp_id' 0)
43 2
```

In the latter example, the two fields are, respectively, specified by name and ordinal position. In fact, specifying a single name or ordinal works as does a single name and one ordinal. The argument *Fields* gets coerced to type variant implicitly when two or fewer fields are included—it requires explicit coercion when there are more than two fields. The syntax using the field names is:

```
        ρarray←'rstEmployees' ⎕wi 'GetRows' ¯1 1 ('#' ⎕wi 'VT' ('emp_id'
'fname' 'minit') 8204)
43 3
⊃array[⎕io;]                                  PMA42628M
                                              Paolo
                                              M
```

The syntax using the ordinal position of fields—always specified in index origin 0—is:

```
        ρarray←'rstEmployees' ⎕wi 'GetRows' ¯1 1 ('#' ⎕wi 'VT' (0 1 2) 8204)
43 3
        ⊃array[⎕io;]                          PMA42628M
                                              Paolo
                                              M
```

The same technique converts a combination of field names and ordinal positions—that is, string and numeric data—to type variant:

```
        ρarray←'rstEmployees' ⎕wi 'GetRows' ¯1 1 ('#' ⎕wi 'VT' ('emp_id' 1
'minit') 8204)
43 3
        ⊃array[⎕io;]                          PMA42628M
                                              Paolo
                                              M
```

In VB, *TypeName(Variable)* returns the type name and *VarType(variable)* returns the array type. APL+Win supports data types in **Table 5-3 APL+Win external data type**

VB Type Name	Scalar Type	Array Type
Empty	0	Not Applicable
Null	1	Not Applicable
Integer	2	8194
Long	3	8195
Single	4	8196
Double	5	8197
Currency	6	8198
Date	7	8199
String	8	8200
Object	9	8201
Error	10	8202
Boolean	11	8203
Variant	12	8204
DataObject	13	8205
Byte	17	8209

Table 5-3 APL+Win external data type

5.8.5 Passing selective named parameters

Quite often, ActiveX methods take several named arguments. It can be very laborious to specify all the arguments in the predefined order, more so when the default values of the arguments are used. The syntax of the SaveAs method of the **Excel** object is:

```
expression.SaveAs(Filename, FileFormat, Password, WriteResPassword,
      ReadOnlyRecommended, CreateBackup, AddToMru, TextCodePage,
      TextVisualLayout)
```

Of the nine arguments, Filename is the only mandatory argument—all the others are optional. With APL+Win versions prior to 4.0, all the arguments need to be specified:

```
'objXL' ⎕wi 'xActiveWorkBook > .wb'
'objXL.wb' ⎕wi 'XSaveAs' "FromAPL",⍬,⍬,⍬,⍬,⍬,1,⍬,⍬
```

A zero length argument is specified for values where the **Excel** default is used and all the arguments are specified in the required order. This code is cumbersome and prone to errors, and quite unreadable in that there are no visual clues as to the names of the arguments being specified. The **Excel** documentation illustrates that the method may be called using selective arguments only, in any order.

```
ActiveWorkbook.SaveAs AddToMRU:=False, Filename:="FromAPL"
```

In this example, the argument name is specified, selective arguments only are mentioned, and the arguments are specified in a random order. Version 4.0 enables this 'friendly' syntax as follows:

```
'objXL' ⎕wi 'xActiveWorkBook.SaveAs(AddToMRU,Filename)' 0 "FromAPL"
```
1

This syntax closely corresponds to the VBA syntax making it easy to translate native **Excel** examples into APL. A further subtle point is that the second and subsequent levels in the hierarchical syntax do not require the APL prefix: for example `SaveAs` does not need the prefix 'X' in an expression such as `xActiveWorkBook.SaveAs`. In fact, specifying the prefix causes an error:

```
'XL' ⎕wi 'xActiveWorkBook.XSaveAs(AddToMRU,Filename)' 0 "FromAPL"
⎕WI ACTION ERROR: Action "XSaveAs(AddToMRU,Filename)" not found
```

The hierarchical syntax forces the context of the property, method, or event to its parent—there is no scope for confusing it with a like named APL+Win property, method, or event and, therefore, a prefix is not necessary.

5.8.6 Rogue objects

Under some circumstances, ActiveX objects may display behaviour that is best described as roguish—there is no error yet the behaviour is not as expected. Normally, this will be obvious during development or testing. An example will illustrate this. First, create the **Excel** object:

```
⎕wself←'objXL' ⎕wi 'Create' 'Excel.Application'
```

Next, create the ActiveWorkbook object:

```
⎕wi 'xActiveWorkBook > .wb'
```

The properties of this object are:

```
⎕wi 'properties'
```
```
children class clsid data def description events instance nterface links
methods modified modifystop name obj opened pending progid properties self
state suppress version
```

Usually, native ActiveX properties are enumerated with an 'x' prefix within APL+Win and those tagged by APL+Win have no prefix. There are 23 properties; none has the prefix 'x'. What is amiss?

```
ρ'objXL.wb' ⎕wi 'properties'
```
23

In contrast to starting **Excel** directly, when a new workbook is automatically opened, creating an instance of **Excel** as an ActiveX object does not automatically open a default workbook. Since there is no workbook open, there cannot be an active workbook. Hence there are no properties with an 'x' prefix. A runtime error would have helpful in that it would force an investigation—without an error, it is difficult to highlight this exception.

A workbook can be added to the existing instance of **Excel**:

```
⎕wi 'xWorkBooks.Add'
```

The value returned is an object pointer, which can be instantiated using either redirection or dynamic syntax, as follows:

```
⎕wi 'xWorkBooks.Add > .nwb'
```

Or,

```
'objXL.nwb' ⎕wi 'Create' (⎕wi 'xWorkBooks.Add')
objXL.nwb
```

Now, there is an open workbook, and therefore there is an active workbook. What of its properties? Explore the following:

```
ρ'objXL.wb' ⎕wi 'properties'
23
```

Still only 23! It is safe to assume that APL is not seeing the properties of the active workbook. Why? The active workbook object must be re-instantiated:

```
'objXL' ⎕wi 'xActiveWorkBook > .wb'
```

Now APL sees the properties:

```
ρ'objXL.wb' ⎕wi 'properties'
89
```

With ActiveX objects that expose a hierarchy of objects, an instance of a lower level object fixed within APL does *not* refresh dynamically with subsequent changes to those lower level objects. In this example, the object *objXL.wb* did not expose the correct properties count until re-initialised. The lesson is to use the hierarchical syntax, which always refers to an object in its current state.

The dynamic syntax is valid *only* during the execution of an expression. On completion of one expression and the start of the next, the object created by the first expression ceases to exist unless explicitly created. In this context, a statement is any one line of code without the diamond separator.

The dynamic syntax allows an object to be passed as an argument from one context to another in the same environment as well as across environments. When an object is passed as an argument—in fact, an object pointer is being passed—it needs to be used without reassignment. On reassignment, the pointer regresses to being simply a number. The APL+Win prefix must be omitted with all but the first level when using the hierarchical syntax.

5.9 Development environment features

A key advantage that APL confers on its developers is the ability to enumerate the properties, methods, and events of ActiveX Objects using simple interactive code. The general syntax is:

```
'obj' ⎕wi 'properties' ⍝ enumerate all properties of obj
'obj' ⎕wi 'methods' ⍝ enumerate all methods of obj
'obj' ⎕wi 'events' ⍝ enumerate all events
```

5.9.1 User-defined properties

APLX and APL+Win have a unique facility—the ability to defined any number of user-defined properties for any object, including ActiveX objects. The procedure for creating custom properties is as follows:

```
⎕wself←'objWRD' ⎕wi 'Create' 'Word.Application'    ⎕wi '∆MyPropertyOne'
⎕wi '∆MyPropertyOne' (⍳5)                           1 2 3 4 5
⎕wi '∆MyPropertyTwo' (2 2ρ1 0 1)                    ⎕wi '∆MyPropertyTwo'
                                                     1 0
                                                     1 1
```

Any number of user-defined properties can be added and the value of a property can be any value, character or numeric, and any type, simple or nested scalar, vector, matrix except scalar 0. The scalar 0 value assigned to a user-defined property has the effect of deleting it. Scalar 0 is also the default value of an undefined user-defined property. This is quite convenient in that the alternative would be to generate an error; an error would halt processing and would be fatal in that the ⎕wi function does not support procedural code for error trapping. Although this is tidy, it is somewhat incongruous: an uninitialised variable in a workspace does not have a default value.

Once defined, a user-defined property is indistinguishable from other intrinsic properties, except that the user-defined property name always begins with ∆. User-defined properties enable imaginative subtleties with a user interface. For example, an edit box object can have two such properties: one holding its default value and the other its 'current' value. On the value being changed and the new value failing validation, the two properties can be used to restore either the default or 'current' value—the latter acting like an 'undo' feature. User-defined properties have rich potentials in some circumstances; these circumstances emerge naturally during development.

VB has one 'tag' property, which allows arbitrary values to be assigned—the name of the 'tag' property is tag, it cannot be renamed.

5.10 Using ActiveX asynchronously

On the face of it, this is an anomalous concept. The very idea of an application (a client) using an ActiveX component (a server) is that the client and server link is synchronous and the server automatically terminates as soon as the client terminates.

Application components servers, like **Excel**, **Word**, and APL+Win, are different—they can survive the client. The idea of using a component asynchronously is that it can be started by an application and left running to complete any time-consuming task and then terminate itself. For example, **Excel** can be started with the following code:

```
'objXL' ⎕wi 'Create' 'Excel.Application'
```

Likewise, the APL Grid object, say, can be started as follows:

```
      ∇ Grid
[1]     'MyForm' ⎕wi 'Create' 'Form'
[2]     'MyForm.Grid' ⎕wi 'New' 'APL.Grid'
      ∇
```

However, unlike `'MyForm' ⎕wi 'Delete'`, which terminates the instance of APL Grid, `'objXL' ⎕wi 'Delete'` does not terminate the instance of **Excel**. This can be verified via the **Task Manager**—the **Processes** tab will include EXCEL.EXE as a running process. In order to terminate the session of **Excel**, it is necessary to issue `'objXL'  ⎕wi 'xApplication.Quit'` *before* deleting the objXL object.

This behaviour enables APL+Win to create a visible or hidden instance of an application component, load a file with an auto running macro and leave the component to finish a long running task and to terminate itself. Examples of this type of functionality are a **Word** mail merge application or an **Excel** workbook with external links.

The use of an application component asynchronously is useful in that it frees the client application and allows it to continue. The alternative is to run a macro in the server via the client—`'objXL'  ⎕wi 'Run' 'MyMacro'`—and to wait until the macro finishes. It is convenient not to have to wait, especially since the component cannot pass back any custom property either.

This raises an interesting question: why use the application component as an ActiveX server? Why not simply start the application and let it run the same code? There are several reasons, namely:

• Appearance—an application may use an application component to give the impression of creating information is custom file formats, such as **Excel** workbooks, seamlessly and achieves this by keeping **Excel** hidden.

• Timing—the application may create runtime data that needs to be saved, incrementally, in a custom file format and then discarded. It is necessary to start a hidden application component to archive the data as it is created during the application session.

• Dynamic context—the application may create dynamic data, which needs to be passed to the application component. For example, an application may create the data file for a mail merge application and needs to pass associated information, such as the name and location of that file to the component.

5.10.1 Custom properties

There is no provision for ActiveX components to be instantiated with bespoke arguments. APL and APLX allow the creation of a user-defined property for an object holding a pointer to an ActiveX component; however, the component itself will be oblivious of that user-defined property. In the previous example, objWRD will not be aware of ΔMyPropertyOne or ΔMyPropertyTwo.

This raises the question of how an application can pass runtime data to an application component.

5.10.1.1 Passing parameters to Excel

Excel has a 'Names' collection that allows arbitrary variables and string values to be added— if necessary, values can be converted to other types using the usual built-in conversion functions. Consider this example:

```
Application.Names.Add "NEW" ,100   ' Assign value
Application.Names("NEW").Value ' Retrieve value
val(mid(Application.Names("NEW").Value,2) ) ' Convert to numeric
```

The equivalent APL+Win code is:

```
'objXL' ⎕wi 'xActiveWorkbook.Names.Add("New",100)'
```

Or

```
'objXL' ⎕wi 'xActiveWorkbook.Names.Add' 'New' 100
```

The values can be retrieved as follows:

```
'objXL' ⎕wi 'xActiveWorkbook.Names("NEW").Value'
=100
```

Although a numeric value was assigned from APL+Win, **Excel** converts the assignment to character, in the following manner:

```
100×'objXL' ⎕wi 'xActiveWorkbook.Names("NEW").Value'
DOMAIN ERROR
      100×'objXL' ⎕wi 'xActiveWorkbook.Names("NEW").Value'
          ^
```

APL+Win can convert the value to numeric:

```
100×⎕fi 1↓'objXL' ⎕wi 'xActiveWorkbook.Names("NEW").Value'
10000
```

Variables created in this manner do not appear in any cell although they are associated with the active workbook.

5.10.1.2 Passing parameters to Word

Word has a 'Variables' collection, which is equivalent to **Excel**'s 'Names' collection. This emphasizes the fact that ActiveX objects may implement similar functions in different ways. Consider this example:

```
application.ActiveDocument.Variables.Add "Myvar" ,100
?100*application.activedocument.Variables("Myvar").Value
 10000
```

Note that the value is numeric. The equivalent APL+Win code is:

```
   100× 'objWRD' ⎕wi 'ActiveDocument.variables("Myvar").value'
DOMAIN ERROR
     100×  'objWRD' ⎕wi 'ActiveDocument.variables("Myvar").value'
       ^
```

Although a numeric operation is possible in **Word**, APL does not see the value as numeric. Variables created in this manner do not appear in the document and are associated with the active workbook.

5.11 Debugging components

In general, a component delivers either a result or an exception; in other words, an error message to signal that a method could not be concluded. With industry standard components, an error condition is fatal—it causes processing to halt.

However, with custom components, it is possible to trace the flow of execution and, if necessary, to make code changes to remove coding flaws. A third party black box component does not allow tracing because its code is hidden away in a binary file.

5.11.1 VB for Application code

VB for Application (**VBA**) is used to write code in any of the Microsoft Office applications. **Excel** code—that is, code in **Excel** macros or in **Excel** Add-Ins—can be debugged interactively. APL+Win can create an instance of **Excel**, open a workbook, and install an Add-In. If that **Excel** session is made visible, pressing **Alt + F11** in **Excel** will show its code. Functions and subroutine can have breakpoints added by locating the cursor on any line and pressing **F9**. When APL+Win makes a call to a function or subroutine with a breakpoint, **Excel** will stop at the break point allowing the intermediate variables to be reviewed from the immediate execution window of the VBA Integrated Development Environment (**IDE**). Automatic execution can be resumed by pressing **F5**. Alternatively, code can be stepped through one statement at a time by pressing **F8**—including subsequent nested calls to other functions and subroutines.

This procedure applies to all VBA code, including **Word** templates. Thus, code written within a VBA environment for interaction with APL+Win can be tested and debugged interactively. This makes it very easy to develop sophisticated routines in VBA—had this not been possible, it would have been very difficult to provide custom functionality.

5.11.2 VB code

In **Chapter 6** *Mixed Language Programming*, **VB** is used to create an ActiveX DLL, which can be used by APL+Win. A DLL is a binary file containing compiled code. Usually, it is impossible to trace code execution in a binary file.

5.11.2.1 Tracing execution in a VB ActiveX DLL/EXE

It is possible to debug a **VB** ActiveX DLL or ActiveX EXE code line by line, as follows:

● Load the ActiveX project and compile it using **File | Make ***, if this has not been done already. Compilation creates the binary DLL file and registers it as an ActiveX component.

• Click **Debug | Toggle Breakpoint** or press **F9** on the first line of the target subroutine or function to add a breakpoint.

• Click **Run | Start With Full Compile** or press **Ctrl + F5** to start running the **VB** project.

• Start APL, create an instance of the component in the usual way, and call the subroutine or function that will be traced. The syntax for creating an instance of an ActiveX DLL is:

```
Object ⎕wi ActiveXDLL.NameOfClass
```

If the ActiveX project is running, achieved by **Ctrl + F5**, APL uses the registry information but creates an instance of the project currently loaded in memory.

A call to any function or subroutine within the DLL, from APL+Win, will now run within the loaded ActiveX project in **VB**. Execution will stop on any breakpoint within the code and it can be resumed using **F5** or **Run | Start,** or **F8**, or **Debug | Step Into**. **F5** will run the code up to the next breakpoint or the end of the code and **F8** will step through each line of code. **Figure 5-11 Debug session** shows a debug session in progress.

Any value within the session may be queried via the Immediate Window; this is not shown but can be made visible with **View | Immediate Window**. A debugging session in progress, discussed in **Chapter 6 Mixed Language Programming**, is shown in **Figure 5-11 Debug session**. The project is running, as indicated by the greyed out arrowhead on the Debug toolbar. Execution stopped on the breakpoint line 'If IsNumeric...' and **F8** has been pressed once, allowing execution to proceed to the 'Result(1) = 1' line.

This debugging technique also works with VBA code although this code cannot be compiled:

• Create the instance of **Excel** or **Word**.

• Open the desired workbook or document.

• Click **Tools | Macros**, then **VB** Editor, or simply press ALT + F11.

• Locate the function or subroutine to debug, and press **F9** on the appropriate line.

• Press **F5** to start the code running.

• Call the function or subroutine from APL+Win.

• Use **F5** or **F8** to trace execution.

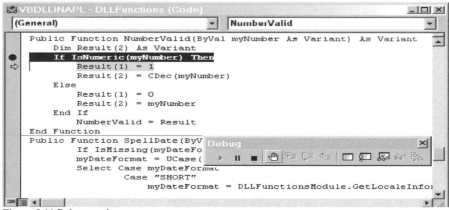

Figure 5-11 Debug session

The ability to debug the source code makes it very easy to test the code using calls from APL+Win. It is fortunate that there is no difference in the interface for debugging **VB** and VBA projects interactively.

This very powerful interactive facility makes it very easy to debug a **VB** ActiveX DLL, EXE project, or a VBA project. Changes to the code are immediately effective without needing to be recompiled, in the case of DLL or EXE, or saved. However, in order to have absolute confidence in the code, compile the project first—this will ensure that the state of the project fulfil all the requirements of the **VB** compiler. It is necessary to pause execution in order to make changes and then execution must be resumed to restore the link to APL. Finally, the changes must be saved to make them permanent and in the case of an ActiveX DLL or EXE, the project must be recompiled. It may be necessary to end the APL+Win session before recompiling as deleting the APL+Win object that is an instance of the ActiveX DLL or EXE or '#' ⎕wi 'ReleaseObjects' does not appear to release it.

Chapter 6

Mixed Language Programming

System building on the Windows platform involves mixed language programming—because it is possible, readily accessible, and reduces development costs. At the very least, standard platform-related tasks can be completed in a routine manner using Windows API calls; these are functions written in a language other than APL+Win. **VB** enables custom ActiveX DLLs, providing bespoke properties, methods, and events, to be written. This approach yields a key advantage, namely, rapid application development.

An APL+Win application may be extended or enhanced by invoking supplemental non-APL code from the workspace. Non-APL code may simplify application building in circumstances where the 'server' language supports features unavailable in APL+Win, such as the comprehensive set of **VB** date functions. APL system development using such functions removes the overhead of writing, testing and debugging them.

The **VB** ActiveX DLLs, Microsoft Script Component, Microsoft Script Host and Microsoft Script Control provide convenient means for integrating *custom* **VB**, VBScript and JavaScript code seamlessly into an APL+Win system to produce high efficiency applications rapidly.

6.1 Application extension trade-offs

From the standpoint of managing APL projects, ActiveX DLLs offer these advantages:

● Suitable ActiveX DLLs may already exist and may be deployed within APL+Win applications.

● If none exists, the ActiveX DLL may be coded by another set of developers: this will reduce the development duration.

● The encapsulation of the code in a black box promotes code reuse in other projects and ensures the availability of a debugged library of functions.

● It is likely that functions within binary or compiled ActiveX DLLs will execute faster than comparable APL+Win functions.

The penalty imposed by ActiveX DLLs is the lack of array handling facilities in such languages. Unlike APL+Win, **VB** and the scripting languages are not array based—array handling must be explicitly coded.

However, the GUI user interface in APL is not array but scalar-based as in most other languages. For instance, validation on data entry forms is invariably scalar-based. Moreover, processing of data retrieved from databases is done one record at a time, thus also making it scalar-based. Therefore, the penalty in using application extension functions written in **VB** or in one of the scripting languages is negligible.

In circumstances where an APL+Win application processes arrays, scalar- based functions may be called re-iteratively by APL+Win or by using the each (¨) construction.

This is discussed below. Depending on the complexity of the functions called, there may not be any performance sacrifice in doing so. In the event that there is a huge performance impediment, some or all of the functions will need to be coded in raw APL, thereby re-introducing the overhead of coding, testing, and debugging.

6.2 VB ActiveX DLLs

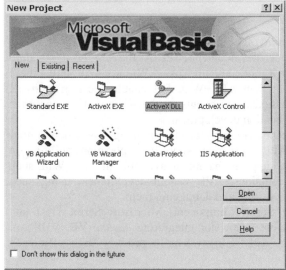

In order to create a **VB ActiveX DLL**, select ActiveX DLL as the project type; this is shown in **Figure 6-1 VB project types**. The project will be called Project1; this may be renamed as appropriate.

VB automatically adds a class named Class1; this may be renamed as appropriate.

Add a Module to the project, renaming it as appropriate. This will contain code that will not be stored in the Class.

Figure 6-1 VB project types

The **VB** menu **File** | **Ma̲ke** *project*.**DLL** will create the ActiveX DLL and register it automatically on the computer where the DLL is created. The DLL may be made available on other computers by copying it to a location other than the WINDOWS\SYSTEM or WINDOWS\SYSTEM32 folder and registering it. The compilation process creates three files—PROJECT.DLL, PROJECT.EXP and PROJECT.LIB. Only the DLL file needs to be distributed together with the **VB** runtime system, which is freely available on the Microsoft web site.

The registration is achieved with **Start** | **Run** and typing the command as shown in **Figure 6-2 Registering DLLs.** The actual location of the DLL must be specified. The same command with the suffix /u will permanently remove the **Registry** entries for the DLL.

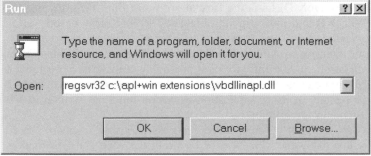

Figure 6-2 Registering DLLs

The target location of the DLL must be chosen with care in order to avoid conflicts with existing, perhaps like-named, DLLs; the operating system stores its DLLs in

Windows\System32. The command REGSVR32.EXE is available with the Windows operating system.

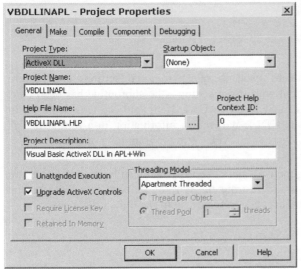

Several properties of the project may be specified on the dialogue shown in **Figure 6-3 Project properties**.

If it is intended to write a help file for the methods in the ActiveX DLL, the name of the help file must be specified.

Figure 6-3 Project properties

A help file (either *.HLP or *.CHM) may be used to document the methods provided together with their respective syntax. APL may use this file to show extended help from the workspace.

6.3 A sample ActiveX DLL project

This project is called VBDLLINAPL; it has one class called *DLLFunctions* and one module called *DLLFunctionsModule*. **VB** takes care of everything except the definition of the functions or methods in the class module. Functions should not contain explicit error messages—these can be returned to the calling environment—and references to arguments must be by value and not by reference. Passing arguments by value avoids the problems with the implicit conversion of data types that must take place during an APL+Win call to the DLL.

The sample ActiveX DLL contains a number of functions to illustrate the means of constructing such a DLL and the mechanism for using it with APL+Win. Comparable functions are not written in APL+Win to illustrate the advantage of using this DLL—the APL+Win programmer almost certainly has an arsenal of similar legacy functions and is invited to perform a comparative analysis of robustness and timing between the ActiveX methods and the raw APL functions.

6.3.1 The DLLFunctionsModule module

The module, with line numbers added manually as an aid to reference and readability, contains the following code:

```
1 Attribute VB_Name = "DLLFunctionsModule"
2 Option Explicit
3 Option Base 1
4 Option Compare Text
5 Declare Function GetLocaleInfo Lib "kernel32" Alias "GetLocaleInfoA"
      (ByVal Locale As Long, ByVal LCType As Long, ByVal lpLCData As String,
      ByVal cchData As Long) As Long
6 Public Function GetLocaleInformation(ByVal Identity As String) As String
```

```
 7     Dim Buffer As String, TruncLength As Long
 8     Buffer = Space(255)
 9     TruncLength = GetLocaleInfo(&H400, Identity, Buffer, 255)
10     If TruncLength = 0 Then
11         GetLocaleInformation = ""
12     Else
13         GetLocaleInformation = Left(Buffer, TruncLength - 1)
14     End If
15 End Function
```

The GetLocaleInfo API call will be used to query regional settings and, where appropriate, return information consistent with those setting—the subroutine has been hard coded to return the settings for LOCAL_USER_DEFAULT. If an application is deployed in several countries, the methods within this DLL will return information in a format that is consistent with their regional settings—there is no APL+Win programming overhead.

6.3.2 The DLLFunctions class

The class contains code for properties, methods, and events, which can be accessed from an APL+Win workspace. The intention is to illustrate the type of functionality that may be provided. It is listed below with line numbers added manually to facilitate reference:

```
 1 VERSION 1.0 CLASS
 2 BEGIN
 3    MultiUse = -1    'True
 4    Persistable = 0   'NotPersistable
 5    DataBindingBehavior = 0   'vbNone
 6    DataSourceBehavior  = 0   'vbNone
 7    MTSTransactionMode  = 0   'NotAnMTSObject
 8 END
 9 Attribute VB_Name = "DLLFunctions"
10 Attribute VB_GlobalNameSpace = False
11 Attribute VB_Creatable = True
12 Attribute VB_PredeclaredId = False
13 Attribute VB_Exposed = True
14 Option Explicit
15 Option Base 1
16 Option Compare Binary
```

Each method included in the class is listed and the means of using it in APL+Win is discussed below.

6.4 Using VBDLLINAPL.DLL

This DLL can be invoked in APL. Consider the cover function VBDLL below:

```
      ∇ Z←L VBDLL R;⎕wself
  [1]    ⍝ System Building with APL+Win
{1[2]    :if 0=⍴'VBDLL' ⎕wi 'self' ⍝ Create instance if VBDLLINAPL exists
{2[3]        :if 0∊⍴'#' ⎕wi 'XInfo' 'VBDLLINAPL.DLLFunctions'
  [4]            Z←0 'VBDLLINAPL Object Not Found On This PC'
> [5]            :return
+ [6]        :else
  [7]            ⎕wself←'VBDLL' ⎕wi 'Create' 'VBDLLINAPL.DLLFunctions'
2}[8]        :endif
+ [9]    :else
  [10]       ⎕wself←'VBDLL'
1}[11]   :endif
  [12]       Z←1 ⎕wself
{1[13]   :if 0=⎕nc 'L'
+ [14]       :orif 0=⍴L~' '
> [15]       :return
```

```
1}[16]   :endif
{1[17]   :select L
```

The first call to this function initialises an instance of the ActiveX control, which remains active during the APL session—the control does not have to be invoked explicitly. Thus, a call may be made from anywhere in the workspace without the need for initialising an instance of the DLL. The important syntax is in line [6]. In general, the syntax is:

```
        Object □wi ActiveXDLL.NameOfClass
```

Each class in the ActiveX DLL project will require a different *object*. The sample DLL contains only one class. It can be invoked in two ways:

• The syntax `VBDLL ''` will return a two element result `1 VBDLL`, where `VBDLL` is the object created, in the event of success or `0 VBDLLINAPL Object Not Found On This PC` in the event of failure—the first element is **TRUE** or **FALSE**.

• Alternatively, the syntax `[Method] VBDLL [argument]` will return the *result* of the particular method called—this is recommended.

6.4.1 Syntax issues

The APL function VBDLL must have a right-hand argument specified at all times, even though the ActiveX method called from within VBDLL may not require an argument—the right-hand argument will be ignored. There is another issue with the passing of arguments to DLL functions. Should a date comparison function, say *After*, be implemented in raw APL code, its call would probably be within a dyadic function taking each date as a left- and right-hand argument; it will be called as follows:

```
        Date1 After Date2
```

A comparable function implemented in a **VB** DLL would require the following syntax:

```
        'After' VBDLL Date1,Date2
```

Arguably, the APL syntax is more readable. However, the DLL solution is more readily implemented as **VB** understands dates and its logical operators return the correct result. The XDateCompare method discussed below illustrates how the readability of a DLL call may be improved. The specification of all optional arguments makes the APL+Win code more readable.

For functions that expect two arguments and the second argument is optional, either enclose the first argument when the optional second argument is omitted or modify the VBDLL function. The current line [63] is:

```
[63]          Z←□wi (⊂'XStringCase'),R
```

This line must be modified as follows:

```
[63]          :if 1==R
[63.1]            Z←□wi 'XStringCase' R
[63.2]          :else
[63.3]            Z←□wi (⊂'XStringCase'),R
[63.4]          :endif
```

6.4.2 Querying the events in VBDLLINAPL

Only one event is defined in this DLL:

```
    17 Public Event BadPropAssignment(ByVal BadValue As Variant)
```

The structure of this one line declaration is quite simple: the name of the event is followed by the arguments passed to the event handler. The arguments will be inherited by the APL+Win □warg variable within the event handler, which must also be defined in APL+Win. In general, the events in a DLL can be enumerated as follows:

```
    ('X'=↑¨'VBDLL' □wi 'events')/'VBDLL' □wi 'events' ⍝ Query
XbadPropAssignment
```

The object must also define how the event will be raised; this is discussed in the next section.

6.4.3 Querying the properties in VBDLLINAPL

The definition of properties requires more analysis. All the properties are created within the following lines:

```
20 Public Property Get ROValue() As Variant
21     ROValue = "APL+Win/VB DLL Version 1.0"
22 End Property
23 Public Property Get ValidDateValue() As Variant
24     ValidDateValue = Format(xValidDateValue,
       DLLFunctionsModule.GetLocaleInformation(&H1F))
25 End Property
26 Public Property Let ValidDateValue(ByVal vNewValue As Variant)
27     If IsDate(vNewValue) Then
28         xValidDateValue = Format(vNewValue,
       DLLFunctionsModule.GetLocaleInformation(&H1F))
29     Else
30         RaiseEvent BadPropAssignment(vNewValue)
31     End If
32 End Property
```

The rules are:

● Any variable declared as *private* are visible only within the class—the calling application cannot see them. In this instance, xValidDateValue is visible only to the ValidDateValue *Get* and *Let* properties defined in lines [23] to [32].

● Any variables declared as *public* and without corresponding property *Get, Let,* or *Set* functions are read and write properties to which any value may be assigned. The property *AnyValue* is such a property, declared in line [19]. The keywords *Get* and *Let* are used with the assignment of literal values, whereas the *Set* keyword is used to assign objects.

● Any variables declared as *public* and with a corresponding *Get* property only—that is without a *Let* property—is a read only property. In this instance, *ROValue* is a read only property: note that it is created directly without recourse to a private variable.

● Any variables declared as *public* with corresponding *Get* and *Let* properties are read and write properties; these properties allow stricter control in updating the variables; the code in line [24] within the Get block formats the return value in the short date format of regional settings.

● There is no point in declaring a variable as *public* and specifying a *Let* property only. This would mean that a property may be created or written but cannot be read.

The available properties may be enumerated in the following manner:

```
('x'=↑¨'VBDLL' ⎕wi 'properties')/'VBDLL' ⎕wi 'properties'
xAnyValue xROValue xValidDateValue
```

As expected, the variable xValidDateValue, declared as *private* in line [18], is not listed and is unavailable. Had it been available, it would be listed as xxValidDateValue—APL+Win adds the prefix *x* to all properties. The xValidDateValue listed is defined in lines [23] to [32]. The internal xValidDateValue is used internally in lines [24] and [28].

The *Let* xValidDateProperty verifies that a valid date is specified and, if not, it raises the *BadPropAssignment* event, in line [30]. In order for APL+Win to be aware of this event, it must provide and event handler:

```
'VBDLL' ⎕wi 'onXBadPropAssignment' 'Msg "Bad date value ", ⍕⎕warg'
```

An attempt to assign an invalid date to the ValidDateValue property will trigger the event handler, which shows the message in **Figure 6-4 Property event handler**.

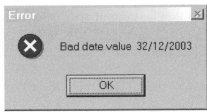

Figure 6-4 Property event handler

```
'VBDLL' ⎕wi 'xValidDateValue'
'32/12/2003'
```

Note that the invalid value is passed into the APL+Win ⎕warg variable. The properties may be read or written in the usual manner: the property is read-only if called without an argument and read/write when called with an argument. This is an example of a read-only property:

```
      'VBDLL' ⎕wi 'xROValue'
APL+Win/DB DLL Version 1.0
```

Where specified, the *Get* and *Let* property code is run implicitly. An example of a method that has a mandatory argument:

```
'VBDLL' ⎕wi 'xValidDateValue'    │  'VBDLL' ⎕wi 'xValidDateValue'
'23/08/2003'                     │  23/08/2003
```

In the absence of any *Get* and *Let* code, a property can be assigned any value, even an APL+Win nested value, such as:

```
'VBDLL' ⎕wi 'AnyValue' (⍳10)     │  'VBDLL' ⎕wi 'AnyValue'
('APL+Win')                      │  1 2 3 4 5 6 7 8 9 10   APL+Win
```

This creates an untidy interface for two reasons:

- Both the read and write operations return the current value of the property.

- The lack of any validation permits rogue values to be assigned to a property—this increase the incidence of runtime errors.

6.4.4 Querying the methods in VBDLLINAPL

APL+Win is almost unique in the facility it provides to query the properties, methods and events of an object after it has been initialised, as seen by:

```
      ⎕wself←'VBDLL' ⎕wi 'Create' 'VBDLLINAPL.Dllfunctions' ⍝ Initialise
      ('X'=↑¨⎕wi 'methods')/⎕wi 'methods' ⍝ Query
XCurrencySymbol XDateCompare XDateScalar XDateValid XDaysOfWeek
XFindReplace XGetAgeAttQ XGetDatePart XGetDateTime XGetLocal XGetTimeStamp
XMonthsOfYear XNumberValid XSpellDate XstringCase
```

Note that every *public* function and subroutine name in the DLLFunctions class is now available as a method in the ActiveX control VBDLLINAPL; those declared *private* will not be enumerated. The names of the methods are returned in ascending order and not in the sequence that the methods were coded.

Although there is no explicit documentation of the syntax of each method within the DLLFunctions class, APL+Win allows the syntax to be queried.

6.4.4.1 Short help on syntax

```
      ⎕wi '?XStringCase'
XStringCase method:
   Result@String ← ⎕wi 'XStringCase' myString@String [myConvFormat]
```

6.4.4.2 Extended help on syntax

APL+Win may show help topics from a help file specified within the ActiveX DLL. Refer to section **Getting help on syntax** for more details. The following code displays the available help text:

```
      'Help' VBDLL 'XcurrencySymbol'
```

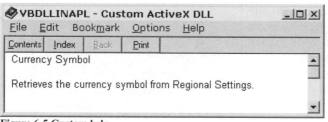

Figure 6-5 Custom help shows the help text that is displayed from the help file.

Figure 6-5 Custom help

The syntax information is extracted from the function header. Arguments shown in square brackets, like [ConvFormat], are optional. The word following the @ character specifies the type of data expected, like myString@String which denotes that the first argument *myString* is expected to be a string. The **VB** variant data type is not reported.

The APL+Win cover function that invokes a custom built ActiveX DLL needs to document each method—this can take the form of inline comment within APL code or a help file or both.

6.4.5 XCurrencySymbol

This method allows the currency symbol specified in regional settings to be queried. The **VB** code for this method is:

```
33 Public Function CurrencySymbol() As String
34     CurrencySymbol = DLLFunctionsModule.GetLocaleInformation(&H14)
35 End Function
```

The APL+Win code to invoke this method:

```
★ [18]        :case 'XCurrencySymbol'
  [19]            ⍝ Result@String←⎕wi 'XCurrencySymbol'
  [20]            Z←⎕wi 'XCurrencySymbol'
```

The APL+Win syntax is:

```
    'XCurrencySymbol' VBDLL ⊖
```
ú

This highlights a problem with the value returned—it is not £ as expected. ActiveX DLLs use the ANSI character set which is not consistent with the APL+Win character set. The APL+Win ANSI2AV function corrects this anomaly:

```
    ANSI2AV 'XCurrencySymbol' VBDLL ⊖
```
£

6.4.6 XDateCompare

This method returns a Boolean result based on the comparison of two dates. The syntax built-into the cover function is:

```
    'XDateCompare' VBDLL date1, ComparisonString, date2
```

ComparisonString is BEFORE | AFTER | NOTAFTER | NOTBEFORE | SAMEAS. This is equivalent to the comparison operators $<$ | $>$ | $<=$ | $>=$ | $=$. The **VB** code for this method is:

```
36 Public Function DateCompare(ByVal myDate1 As Date, ByVal myCompare As
      String, ByVal myDate2 As Date) As Boolean
37     Select Case UCase(myCompare)
38         Case "BEFORE"
39             DateCompare = myDate1 < myDate2
40         Case "AFTER"
41             DateCompare = myDate1 > myDate2
42         Case "NOTAFTER"
43             DateCompare = myDate1 <= myDate2
44         Case "NOTBEFORE"
```

```
45                    DateCompare = myDate1 >= myDate2
46         Case Else
47                    DateCompare = myDate1 = myDate2
48      End Select
49 End Function
```

The APL+Win code to invoke this method is:

```
★ [21]        :case 'XDateCompare'
  [22]            ⍝ Result@Boolean ← ⎕wi 'XDateCompare' myDate1@Date
         myCompare@String myDate2@Date
  [23]            Z←⎕wi (⊂'XDateCompare'),R
```

The APL+Win syntax is illustrated thus:

```
     'XDateCompare' VBDLL '01/01/2002' 'before' '01 December 1980'
0

     'XDateCompare' VBDLL '01/01/2002' 'after' '01 December 1980'
1

     'XDateCompare' VBDLL '01/01/2002' 'notbefore' '01/01/2002'
1

     'XDateCompare' VBDLL '01/01/2002' 'notafter' '01/01/2002'
1

     'XDateCompare' VBDLL '01/01/2002' 'SAMEAS' '01/01/2002'
1
```

The function is capable of seeing dates in different formats and the default comparison method is 'SameAs'. The comparison string is converted to uppercase internally thereby providing some tolerance in how it is specified. The syntax is:

```
     'XDateCompare' VBDLL '01/01/2002' 'after' '01 December 1980'
```

This is more readable than a typical APL+Win syntax:

```
     'After' VBDLL '01/01/2002','01 December 1980'
```

6.4.7 XDateScalar

This method returns an integer corresponding to the internal representation of the date in **VB**, which is able to render it in any custom format. APL+Win usually relies on a hard coded or implicit date format, for example, dd mm yyyy, and holds dates as a vector of three elements. The **VB** code for this method is:

```
50 Public Function DateScalar(ByVal myDate As Date) As Long
51      DateScalar = CDate(myDate)
52 End Function
```

The APL+Win code to invoke this method is:

```
★ [24]        :case 'XDateScalar'
  [25]            ⍝ Result@Long←⎕wi 'XDateScalar' myDate@Date
  [26]            Z←⎕wi 'XDateScalar' R
```

The APL+Win syntax is:

```
     'XDateScalar'  VBDLL formattedDate
```

The argument *formattedDate* is any valid date where day, month, and year are separated by a valid delimiter. If *formattedDate* is specified without a delimiter, it is treated as an invalid date.

Internally, **VB** stores dates as a number of days from a reference date—dates forward of the reference date are positive integers and those earlier than the reference date are negative. This example return the scalar date from a formatted date:

```
     'XDateScalar' VBDLL '12/09/2002'
37511
```

The reference date is the scalar 1, which represents:

```
     'XGetDateTime' VBDLL 1 'Long'
31 December 1899
```

Thus, 10 October 1899 should be represented by a negative integer. This is confirmed by:

```
      'XDateScalar' VBDLL '10 October 1899'
⁻81
```

This method expects a valid date as its argument; it does not see the usual representation of a date in APL, as a vector of three numbers, as a valid date. Some care is required in that **VB** implicitly sees a date in the format set in regional setting but also coerces a date into another format, if possible. In the following example, month is specified incorrectly for the regional settings format but is accepted as valid:

```
      'XGetLocal' VBDLL '&H1F'
dd/MM/yyyy
      'XSpellDate' VBDLL 'XDateScalar' VBDLL '12/28/2002'
28 December 2002
```

VB understands scalars as dates within context; that is, where a date is expected by a method:

```
      'XDateScalar' VBDLL '19/02/1954'
19774
      'XDateScalar' VBDLL '09/09/2002'
37508
      'XGetAgeAttQ' VBDLL 19774 37508
48.5
      'XGetAgeAttQ' VBDLL '19/02/1954' '09/09/2002'
48.5
```

6.4.8 XDateValid

This method returns **TRUE** or **FALSE** indicating whether the string specified is a valid date or not. The value is expected as a variant—were a *date* type to be specified, the method will fail in the event of the value not being a valid date. A date validation method that fails if an invalid date is encountered rather defeats the purpose of such a method. The **VB** code for this method is:

```
53 Public Function DateValid(ByVal myDate As Variant) As Boolean
54     DateValid = IsDate(myDate)
55 End Function
```

The APL+Win code to invoke this method is:

```
★ [27]       :case 'XDateValid'
  [28]           ⍝ Result@Boolean←⎕wi 'XDateValid' myDate
  [29]           Z←⎕wi 'XDateValid' R
```

The APL+Win syntax is:

```
      'XDateValid' VBDLL  formattedDate | DateScalar
```

Note that the method does not expect a date format as an argument. **VB** will attempt to see the *formattedDate* as a date—this may be unexpected or erroneous in some circumstances. For example,

```
      'XDateValid' VBDLL  '12/25/2002'
1
```

In the latter example, the date is seen as 25 December 2002 but may have been intended to be the twelfth of a month with the month specified in error.

```
      'XDateValid' VBDLL 19774
0
      'XDateValid' VBDLL '19/02/1954'
1
```

VB can interpret dates in a variety of formats.

```
      'XDateValid' VBDLL 'Feb 19, 1954'
1
```

```
      'XDateValid' VBDLL '1954, Feb 19'
1
```

Unlike the XNumberValid method, which returns a two-element result where the first element is **TRUE** (the second element is a number) or **FALSE** (the second element is the argument used to call the method), XDateValid simply returns **TRUE** or **FALSE**. This allows the programmer flexibility in choosing the format in which to store dates. A date stored as a scalar within APL+Win takes less storage and may be formatted in any sequence of day, month, and year dynamically. Ideally, the ISO date format *yyyymmdd* ought to be used.

6.4.9 XDaysOfWeek

This method returns a nested vector consisting of the names of the days of the week as specified in regional settings. The **VB** code for this method:

```
56 Public Function DaysOfWeek(Optional ByVal myDowStyle As Variant) As
       Variant
57     If IsMissing(myDowStyle) Then myDowStyle = "long"
58     myDowStyle = UCase(Left(Trim(myDowStyle), 4))
59     Select Case myDowStyle
60           Case "LONG"
61                 myDowStyle = 0
62                 While myDowStyle <> 7
63                     DaysOfWeek = DaysOfWeek &
       DLLFunctionsModule.GetLocaleInformation(&H2A + myDowStyle) & _
64                     IIf(myDowStyle <> 6, ",", "")
65                     myDowStyle = myDowStyle + 1
66               Wend
67           Case Else
68                 myDowStyle = 0
69                 While myDowStyle <> 7
70                     DaysOfWeek = DaysOfWeek &
       DLLFunctionsModule.GetLocaleInformation(&H31 + myDowStyle) & _
71                     IIf(myDowStyle <> 6, ",", "")
72                     myDowStyle = myDowStyle + 1
73               Wend
74     End Select
75     DaysOfWeek = Split(DaysOfWeek, ",")
76 End Function
```

The APL+Win code to invoke this method:

```
* [30]        :case 'XDaysOfWeek'
  [31]             ⍝ Result←⎕wi 'XDaysOfWeek' [myDowStyle]
  [32]             Z←⎕wi 'XDaysOfWeek' R
```

The APL+Win syntax is:

```
      'XDaysOfWeek' VBDLL ''  | 'Long' ⍝ '' is default
```

The names are collated in an array within the **VB** function and become a nested vector in the APL+Win workspace. By default, 'Short' or abbreviated names are returned.

```
      'XDaysOfWeek' VBDLL ''
Mon Tue Wed Thu Fri Sat Sun
      'XDaysOfWeek' VBDLL 'Long'
Monday Tuesday Wednesday Thursday Friday Saturday Sunday
```

The information returned is in the language specified in regional settings. If the language is set to French,

```
      'XDaysOfWeek' VBDLL ''
lun. mar. mer. jeu. ven. sam. dim.
      'XDaysOfWeek' VBDLL 'Long'
lundi mardi mercredi jeudi vendredi samedi dimanche
```

6.4.10 XFindReplace

This method can return a string having replaced the occurrences of a sub string by another sub string. The sub string to be replaced may be located with or without regard to case; by default, the case of the sub string is used. This method does not have a user interface: the arguments of the call specify all its operands. The **VB** code for this method:

```
77 Public Function FindReplace(ByVal myString As String, ByVal myFind As
      String, ByVal myReplace As String, Optional ByVal myCompare As
      VbCompareMethod) As String
78        If IsMissing(myCompare) Then myCompare = vbBinaryCompare
79        FindReplace = Replace(myString, myFind, myReplace, 1, -1,
      myCompare)
80 End Function
```

The APL+Win code to invoke this method:

```
★ [33]        :case 'XFindReplace'
  [34]            A Result@String←⎕wi 'XFindReplace' myString@String
        myFind@String myReplace@String [myCompare@Long_VbCompareMethod]
  [35]            Z←⎕wi (⊂'XFindReplace'), R
```

The APL+Win syntax is:

```
'XFindReplace' VBDLL 'String' 'target' 'replacement' [0 | 1]
```

The syntax for the 'Replace' function in **VB** is:

```
Replace(expression, find, replace[, start[, count[, compare]]])
```

In the method, the 'start' and 'count' parameters have been set to their respective default values, 1, and –1. The options for the 'compare' parameter are vbBinaryCompare (0) and vbTextCompare (1)—the former treats text as case-sensitive, the default, and the latter disregards the case.

```
   'XFindReplace' VBDLL 'with APL+Win' 'wITH' 'With' 0
With APL+Win
   'XFindReplace' VBDLL 'wITH APL+Win' 'wITH' 'With' 1
With APL+Win
```

The last argument is optional and may be omitted—it will default to 0.

```
   'XFindReplace' VBDLL 'wITH APL+Win' 'wITH' 'With'
With APL+Win
```

6.4.11 XGetAgeAttQ

This method calculates the duration between two dates in complete three months periods and returns the result is years. If the second date is earlier than the first, the result is negative. The **VB** code for this method:

```
81 Public Function GetAgeAttQ(ByVal myStartDate As Date, ByVal myEndDate As
      Date) As Double
82 If Day(myStartDate) > Day(myEndDate) Then myEndDate = DateAdd("m", 1,
      myEndDate)
83      GetAgeAttQ = 0.25 * Int(DateDiff("m", myStartDate, myEndDate) / 3)
84 End Function
```

The APL+Win code to invoke this method:

```
★ [36]        :case 'XGetAgeAttQ'
  [37]            A Result@Double←⎕wi 'XGetAgeAttQ' myStartDate@Date
        myEndDate@Date
  [38]            Z←⎕wi (⊂'XGetAgeAttQ'),R
```

The APL+Win syntax is:

```
'XGetAgeAttQ' VBDLL date1 date2
```

The arguments *date1* and *date2* are valid dates or date scalars and *date1* may be before or after *date2*. Either or both dates may be date scalars.

```
   'XGetAgeAttQ' VBDLL  '09/09/2002' '19/02/1954'
⁻48.75
```

```
      'XGetAgeAttQ' VBDLL   '09/09/2002' '19/02/1954'
⁻48.75
      'XDateScalar' VBDLL   '01/09/2002'
37500
      'XGetAgeAttQ' VBDLL   '19/02/1954' 37500
48.5
```

6.4.12 XGetDatePart

This method returns the day, month or year from its argument—a valid date—irrespective of the format of the date. The **VB** code for this method:

```
85 Public Function GetDatePart(ByVal myDate As Date, ByVal myDatePart As
      String) As Integer
 86    myDatePart = Left(UCase(Trim(myDatePart)), 3)
 87    Select Case myDatePart
 88        Case "DAY"
 89            GetDatePart = Day(myDate)
 90        Case "MON"
 91            GetDatePart = Month(myDate)
 92        Case Else
 93            GetDatePart = Year(myDate)
 94    End Select
 95 End Function
```

The APL+Win code to invoke this method:

```
★ [39]        :case 'XGetDatePart'
  [40]            ⍝ Result@Integer←⎕wi 'XGetDatePart' myDate@Date
        myDatePart@String
  [41]            Z←⎕wi (⊂'XGetDatePart'),R
```

The APL+Win syntax is:

```
      'XGetDatePart' VBDLL '12/02/2002' 'Day'
12
      'XGetDatePart' VBDLL '12 Feb 2002' 'Month'
2
      'XGetDatePart' VBDLL 'February 12, 2002' 'Year'
2002
```

6.4.13 XGetDateTime

This method returns a string comprising a date and a time, both formatted as specified in regional settings. It is not unlike ⎕ts, which returns a numeric vector comprising year, month, day, hour, minutes, seconds, and milliseconds from the system clock. Unlike ⎕ts, XGetDateTime can return a string from any valid argument. The **VB** code for this method:

```
 96 Public Function GetDateTime(ByVal myDateTime As Variant, Optional ByVal
      myReturnAs As String) As String
 97    If IsMissing(myReturnAs) Then myReturnAs = "DateTime"
 98    myReturnAs = UCase(Trim(myReturnAs))
 99    Select Case myReturnAs
100        Case "SHORT"
101            myReturnAs =
      DLLFunctionsModule.GetLocaleInformation(&H1F)
102        Case "LONG"
103            myReturnAs =
      DLLFunctionsModule.GetLocaleInformation(&H20)
104        Case "TIME"
105            myReturnAs =
      DLLFunctionsModule.GetLocaleInformation(&H1003)
106        Case "DATETIME"
107            myReturnAs =
      DLLFunctionsModule.GetLocaleInformation(&H1F) & " " & _
108            DLLFunctionsModule.GetLocaleInformation(&H1003)
109    End Select
```

```
110     GetDateTime = Format(myDateTime, myReturnAs)
111 End Function
```

The APL+Win code to invoke this method:

```
★ [42]        :case 'XGetDateTime'
  [43]             A Result@String←⎕wi 'XGetDateTime' myDateTime
        [ReturnAs@String]
  [44]                Z←⎕wi (⊂'XGetDateTime'),R
```

The APL+Win syntax is:

```
    'XGetDateTime' VBDLL 'formattedDate | floating point number' 'long
| short | datetime'
```

VB does not have a *DateTime* data type. In order to keep this method versatile, its first argument is a formatted date, a scalar, or a floating-point number where the fractional part represents elapsed hours, minutes and seconds. Thus, it will *not* detect an invalid date:

```
    'XGetDateTime' VBDLL '50/90/2002' 'Long'
50/90/2002
```

The second argument may be 'SHORT', 'LONG', 'TIME', or 'DATETIME', respectively returning a date in short date format, or a date in long date format, just the time or a date in short date format and time, all in the format set in regional settings. Some examples:

```
    'XGetDateTime' VBDLL '12 September 2002' 'Short'
12/09/2002
    'XGetDateTime' VBDLL  37511 'Long'
12 September 2002
    'XGetDateTime' VBDLL  37511.123 'Time'
02:57:07
    'XGetDateTime' VBDLL  37511.123 'DateTime'
12/09/2002 02:57:07
```

The latter example shows how easily internal DateTime values can be converted to an appropriate format. Such values may arise from external data sources such as **Excel** or industry databases—some adjustment may be necessary if the reference date in the source is not the same as in **VB**. **VB** handles dates in the range January 1, 100 A.D. through December 31, 9999 A.D. with December 31, 1899 as the reference date. In addition, a user-defined format may be used to return a date in custom format.

```
    'XGetDateTime' VBDLL  37511.123 'yyyymmdd'
20020912
    'XGetDateTime' VBDLL  '12/02/1986' 'mmddyyyy'
02121986
    'XGetDateTime' VBDLL  '12/02/1986' 'dd-mm-yyyy'
12-02-1986
```

This method may be used to render date data from databases, which are returned as floating point numbers, into sensible dates, times or timestamps.

6.4.14 XGetLocal

This method allows a call to the GetLocaleInfo API call with any valid argument and returns an element of user regional setting. The **VB** code for this method:

```
112 Public Function GetLocal(ByVal myIdentity As Variant) As String
113     GetLocal = DLLFunctionsModule.GetLocaleInformation(myIdentity)
114 End Function
```

The APL+Win code to invoke this method:

```
★ [45]        :case 'XGetLocal'
  [46]             A Result@String←⎕wi 'XGetLocal' myIdentity
  [47]             Z←⎕wi 'XGetLocal' R
```

The APL+Win syntax is:

```
'XGetLocal' VBDLL '*'
```

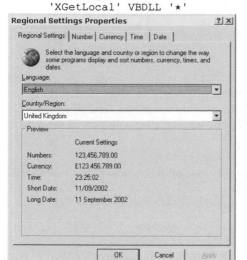

The right-hand argument * represents any valid number. For instance, in order to return the short date format from regional setting, the value is 31. The right-hand argument of the VBDLL function may be specified as either an integer or a hexadecimal string—the latter is recognised by the prefix '&H'. All properties may be queried, including ones that are implicitly set by the system.

The dialogue for specifying/querying regional settings manually is shown in **Figure 6-6 Regional settings.**

Figure 6-6 Regional settings

The examples illustrate how the locale settings may be queried programmatically.

```
'XGetLocal' VBDLL 4    'XGetLocal' VBDLL 2              'XGetLocal' VBDLL '&H1F'
English                English (United Kingdom)         dd/MM/yyyy
```

6.4.15 XGetTimeStamp

This method returns the APL+Win ⎕ts values without the milliseconds in regional setting format. The **VB** code for this method:

```
115 Public Function GetTimeStamp(Optional ByVal myReturnAs As String) As
        String
116     If IsMissing(myReturnAs) Then myReturnAs = "now"
117     myReturnAs = UCase(Left(Trim(myReturnAs), 5))
118     Select Case myReturnAs
119             Case "SHORT"
120                 myReturnAs =
    DLLFunctionsModule.GetLocaleInformation(&H1F)
121             Case "LONG"
122                 myReturnAs =
    DLLFunctionsModule.GetLocaleInformation(&H20)
123             Case "TIME"
124                 myReturnAs =
    DLLFunctionsModule.GetLocaleInformation(&H1003)
125             Case "NOWL"
126                 myReturnAs =
    DLLFunctionsModule.GetLocaleInformation(&H20) & " " & _
127                 DLLFunctionsModule.GetLocaleInformation(&H1003)
128             Case Else
129                 myReturnAs =
    DLLFunctionsModule.GetLocaleInformation(&H1F) & " " & _
130                 DLLFunctionsModule.GetLocaleInformation(&H1003)
131     End Select
132     GetTimeStamp = Format(Now, myReturnAs)
133 End Function
```

The APL+Win code to invoke this method:

```
* [48]       :case 'XGetTimeStamp'
  [49]           ⍝ Result@String←⎕wi 'XGetTimeStamp' [ReturnAs@String]
```

```
[50]            Z←□wi 'XGetTimeStamp' R
```
The APL+Win syntax is:
```
'XGetTimeStamp' VBDLL 'long | short | time | nowl'
```
The right-hand argument is optional within the method but needs to be specified as "
when calling the VBDLL cover function.
```
'XGetTimeStamp' VBDLL ''
15/09/2002 16:17:53
```
The optional argument for this method is 'Short' (date only in short date format) or
'Long' (date only in long date format) or 'Time' (time only in regional format) or 'NowL'
(date and time with date in long date format) or 'Nowl' (date and time with date in short date
format), the default.

6.4.16 XMonthsOfYear

This method returns a nested vector consisting of the names of the months of the year in the
language set. The **VB** code for this method:
```
134 Public Function MonthsOfYear(Optional ByVal myMthStyle As Variant) As
      Variant
135    If IsMissing(myMthStyle) Then myMthStyle = "long"
136    myMthStyle = UCase(Left(Trim(myMthStyle), 4))
137    Select Case myMthStyle
138         Case "LONG"
139             myMthStyle = 0
140             While myMthStyle <> 12
141                 MonthsOfYear = MonthsOfYear &
      DLLFunctionsModule.GetLocaleInformation(&H38 + myMthStyle) & _
142                 IIf(myMthStyle <> 11, ",", "")
143                 myMthStyle = myMthStyle + 1
144             Wend
145         Case Else
146             myMthStyle = 0
147             While myMthStyle <> 12
148                 MonthsOfYear = MonthsOfYear &
      DLLFunctionsModule.GetLocaleInformation(&H44 + myMthStyle) & _
149                 IIf(myMthStyle <> 11, ",", "")
150                 myMthStyle = myMthStyle + 1
151             Wend
152    End Select
153    MonthsOfYear = Split(MonthsOfYear, ",")
154 End Function
```
The APL+Win code to invoke this method:
```
★ [51]       :case 'XMonthsOfYear'
  [52]           ⍝ Result←□wi 'XMonthsOfYear' [myMthStyle]
  [53]           Z←□wi 'XMonthsOfYear'
```
The APL+Win syntax is:
```
'XMonthsOfYear' VBDLL 'Long | short'
```
By default, 'Short' or abbreviated names are returned.
```
'XMonthsOfYear' VBDLL ''
janv. févr. mars avr. mai juin juil. aoρt sept. oct. nov. dec.
'XMonthsOfYear' VBDLL 'Long'
janvier février mars avril mai juin juillet aoρt septembre octobre
novembre décembre
```

Note the loss of accuracy when the extended character set is involved as in *févr* and
août; this anomaly may be avoided by using the ANSI2AV function.

```
  ANSI2AV ¨'XMonthsOfYear' VBDLL ''
janv. févr. mars avr. mai juin juil. août sept. oct. nov. déc.
```

6.4.17 XNumberValid

This method returned a nested vector of two elements. The first element is 1 if the argument is a valid number or 0 if not. The second element is an unformatted representation of the argument, as a number, when the first element is 1 or the argument itself when the first element is 0. The **VB** code for this method:

```
155 Public Function NumberValid(ByVal myNumber As Variant) As Variant
156      Dim Result(2) As Variant
157      If IsNumeric(myNumber) Then
158          Result(1) = 1
159          Result(2) = CDec(myNumber)
160      Else
161          Result(1) = 0
162          Result(2) = myNumber
163      End If
164          NumberValid = Result
165 End Function
```

The APL+Win code to invoke this method:

```
★ [54]       :case 'XNumberValid'
  [55]            ⍝ Result ← ⎕WI 'XNumberValid' myNumber
  [56]            Z←⎕wi 'XNumberValid' R
```

The APL+Win syntax is:

```
'XNumberValid' VBDLL anystring
```

'*anystring*' is a numeric representation of a number, with or without formatting, that is, with or without currency symbol and thousands separator.

```
      'XNumberValid' VBDLL '1,234.56'
1 1234.56
      10×2⊃'XNumberValid' VBDLL '1,234.56'
12345.6
      'XNumberValid' VBDLL '1,234E2.56'
 0 1,234E2.56
      'XNumberValid' VBDLL AV2ANSI '£1,234.45'
1 1234.45
```

The XNumberValid method strips out the currency symbol and thousands separator, and returns a number that can be subject to numeric operations in APL+Win without requiring (⍎) 'execute'. However the British currency symbol £ must be translated to its ANSI equivalent. The function AV2ANSI converts strings from the APL atomic vector to ANSI. The international British monetary symbol cannot be processed: this is usually an acronym.

```
      'XGetLocal' VBDLL 21
GBP
```

6.4.18 XSpellDate

This method returns it first argument, a valid date, formatted as specified in its second argument, a date or time stamp format, independently of regional settings. The custom format may be specified as long or short to default to the corresponding regional settings. The **VB** code for this method:

```
166 Public Function SpellDate(ByVal myDate As Date, Optional ByVal
      myDateFormat As Variant) As String
167          If IsMissing(myDateFormat) Then myDateFormat = "Long"
168          myDateFormat = UCase(Trim(myDateFormat))
169          Select Case myDateFormat
170                  Case "SHORT"
171                      myDateFormat =
      DLLFunctionsModule.GetLocaleInformation(&H1F)
```

```
172                    Case "LONG"
173                        myDateFormat =
      DLLFunctionsModule.GetLocaleInformation(&H20)
174           End Select
175           SpellDate = Format(myDate, myDateFormat)
176 End Function
```

The APL+Win code to invoke this method:

```
★ [57]        :case 'XSpellDate'
  [58]           ⍝ Result@String← ⎕wi 'XSpellDate' myDate@Date
        [myDateFormat]
  [59]              Z←⎕wi (⊂'XSpellDate'),R
```

The APL+Win syntax is:

```
'XSpellDate' VBDLL validdatestring long | short | custom
```

This method takes two arguments, a date and a format. The format is 'Long' for dates in long date format, 'Short' for dates in short date format or user-defined. User-defined formats are independent of regional settings.

```
      'XSpellDate' VBDLL '12 September 2002' 'dddd, dd mmm yyyy'
Thursday, 12 Sep 2002
```

The second argument may be omitted—it will default to 'Long'—however, the first argument must be enclosed.

```
      'XSpellDate' VBDLL ⊂'12/09/2002'
12 September 2002
```

6.4.19 XStringCase

This method provides string conversion functionality similar to those found in **Word**. In addition leading, trailing, and multiple embedded blanks are removed. Conversion into a 'sentence' adds a full stop unless . or ! or ? appear as the last character. The **VB** code for this method:

```
177 Public Function StringCase(ByVal myString As String, Optional ByVal
      myConvFormat As Variant) As String
178    If IsMissing(myConvFormat) Then myConvFormat = "sen"
179    myConvFormat = UCase(Left(Trim(myConvFormat), 3))
180    myString = Trim(myString)
181    While 0 <> InStr(myString, "  ")
182           myString = Replace(myString, "  ", " ")
183    Wend
184    Select Case myConvFormat
185          Case "UPP"
186                 StringCase = StrConv(myString, vbUpperCase) '
      vbUpperCase=1
187          Case "LOW"
188                 StringCase = StrConv(myString, vbLowerCase)
        'vbLowerCase=2
189          Case "TIT"
190                 StringCase = StrConv(myString, vbProperCase) '
      vbProperCase=3; like Title Case in Word
191          Case "TOG"
192                 While 0 <> Len(myString)
193                    If Mid(myString, 1, 1) = UCase(Mid(myString,
      1, 1)) Then
194                       StringCase = StringCase &
      LCase(Mid(myString, 1, 1))
195                    Else
196                       StringCase = StringCase &
      UCase(Mid(myString, 1, 1))
197                    End If
198                    myString = Mid(myString, 2)
```

```
199                    Wend
200            Case Else
201                    StringCase = StrConv(myString, vbLowerCase)
202                    Mid(StringCase, 1, 1) = UCase(Mid(StringCase, 1, 1))
203                    If 0 = InStr("""'?.!,", Mid(StringCase,
       Len(StringCase), 1)) Then
204                            If 0 <> InStr(StringCase, " ") Then StringCase =
       StringCase & "."
205                    End If
206        End Select
207 End Function
```

The APL+Win code to invoke this method:

```
* [60]      :case 'XStringCase'
  [61]          ⍝ Result@String←⎕wi 'XStringCase' myString@String
       [myConvFormat]
  [62]          Z←⎕wi (⊂'XStringCase'),R
```

The APL+Win syntax is:

```
'XStringCase' VBDLL string  lower | upper | title | toggle |
sentence
```

The default value for the second argument is *sentence*.

```
      'XStringCase' VBDLL 'This is a DLL function' 'Lower'
this is a dll function
      'XStringCase' VBDLL 'This is a DLL function' 'upper'
THIS IS A DLL FUNCTION
      'XStringCase' VBDLL 'This is a DLL function' 'title'
This Is A Dll Function
      'XStringCase' VBDLL 'This is a DLL function' 'toggle'
tHIS IS A dll FUNCTION
      'XStringCase' VBDLL 'This is a DLL function' 'sentence'
This is a dll function.
```

The second argument is optional; if omitted, it will default to 'Sentence'—however; the first argument must be enclosed.

```
      'XStringCase' VBDLL ⊂'This is a DLL function'
This is a dll function.
```

6.4.20 Getting help on syntax

```
* [63]      :case 'Help'
  [64]          Z←⎕wi '??',(,R)~'? '
```

The expected syntax for a method is reported in the workspace session. If a help file has been specified in the project properties of the DLL, the help file is opened. Finally, the following additional lines generate an error if VBDLL is called without a valid method.

```
+ [65]      :else
  [66]          Z←'Unable to interpret command'
1}[67]  :endselect
      ∇
```

6.5 Processing APL+Win arrays

The each operator may be used to process APL arrays. Given *a* as an array of dates,

```
12/09/2001
31/10/1989
03/12/1990
```

XAgeAttQ may be called thus:

```
   ,[θ](⊂'XGetAgeAttQ') VBDLL ¨(⊂¨⊂[⎕io+1]a),¨⊂⊂'01/01/2005'
 3.25
15.25
14
```

This complicates the syntax and makes the code less readable. Another approach with string operations is to ravel the array with a known suitable character that may be used to recreate the array. Given *a* as:

```
processing string arrays
with VBDLL
      ρa
2 24
```

The solution is:

```
    b←'XStringCase' VBDLL (,a,⎕tcnl) 'title'
    b←⊃(b≠⎕tcnl)⊂b
    b
Processing String Arrays
With Vbdll
      ρb
2 24
```

6.6 Deploying ActiveX DLLs

A number of new issues require critical attention with ActiveX DLLs.

● Application DLLs must be installed at a location in the filing system that is unlikely to create a conflict. Ideally, choose the application folder as the destination.

● Verify whether the DLL exists at the selected location. If it does, the accepted wisdom is that only older versions of DLLs are replaced. Should a newer version be found, it is retained.

● Unregister the DLL before overwriting it and register it afterwards.

Failure to observe this conventional practice may damage existing **Registry** entries and cause one or more applications to stop working.

If a DLL is shared, that is, used by multiple users, copy it to a shared network folder: the DLL needs to be replaced in one location only. However, it still needs to be registered on each computer.

The version of a DLL may be verified manually as shown in **Figure 6-7 Checking version of DLL**. This is done as follows:

● Locate and highlight it within **Windows Explorer**.

● Right click.

● Click on the **Properties** menu option.

● Click the **Version** tab.

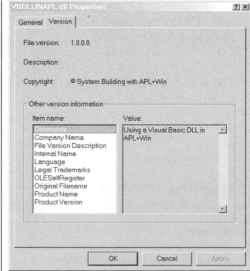

Figure 6-7 Checking version of DLL

If a DLL does not produce the expected behaviour, the first two steps in debugging any errors are:

• First, verify that the DLL is registered correctly. There is no convenient Windows API call to query the existence of an object. However, if an attempt to create an instance in APL+Win fails, the object is either registered incorrectly or not registered at all. If incorrectly registered, remove the **Registry** entries using the optional /u parameter. If the object is registered, the following condition will be **TRUE**.

```
     0≠ρ'#' ⎕wi 'XInfo' objectName.class
```

• Second, verify the version of the DLL. As APL+Win uses late biding for COM objects, it will use the version of the COM object that it finds at runtime rather than a specific version thereof. The version of a DLL can be retrieved programmatically, thus:

```
          ⊃FileVersion 'c:\apl+win application\vbdllinapl.dll'
c:\apl+win application\vbdllinapl.dll
1.0.0.0
     ∇ Z←FileVersion R;⎕wself
[1]    ⍝ System Building with APL+Win
[2]    ⎕wself←'∆FSO' ⎕wi 'Create' 'Scripting.FileSystemObject'
[3]    Z←R (⎕wi 'XGetFileVersion' R)
[4]    ⎕wi 'Delete'
     ∇
```

This function expects the fully qualified name of the DLL as its right-hand argument; it does not verify the existence of the file before querying its version.

6.6.1 Name and location

It is imperative that the name of the DLL is not the same as any existing DLL that could be part of the operating system or belong to another application. This makes the naming of the DLL a rather onerous task.

A safe and recommended practice is to register the DLL from a location other than C:\WINDOWS\SYSTEM or C:\WINDOWS\SYSTEM32. Since Windows looks for DLL references in the application folder first, it is efficient to store DLLs at the same location. This minimises the likelihood of name conflict. The names of the DLL might use a known user-defined prefix such as VBDLL or APLWIN.

6.6.2 General availability

ActiveX DLLs, once registered, become generally available; that is, all other applications can see and use them. APL+Win will see a DLL as *name.class* and may query the availability of all ActiveX DLLs available:

```
     '#' ⎕wi 'XInfo' 'VBDLLINAPL'
VBDLLINAPL.DLLFunctions ActiveObject VBDLLINAPL.DLLFunctions
```

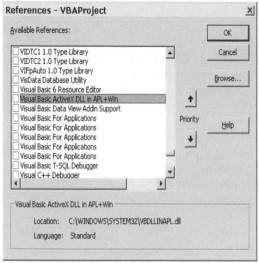

Other applications will see it by the project description property assigned to the DLL. This dialogue, seen in **Figure 6-8 Early binding references**, shows the **Excel** VBA References dialogue: it includes the VBDLLINAPL.DLL ActiveX component.

Figure 6-8 Early binding references

The DLL may be invoked in one of two ways:

```
Sub Method1()
 Dim aa As New VBDLLINAPL.DLLFunctions
End Sub

Sub Method2()
   Set aa = CreateObject("VBDLLINAPL.DLLFunctions")
End Sub
```

The code in the DLL will not be visible but will be available to other COM aware applications.

6.6.3 Updrading/Replacing ActiveX DLLs

The guiding principle of ActiveX DLLs is that an upgrade should not entail the rewriting or recompilation of the applications that deploy it. This is achieved by observing these rules:

• The name of the project and the name of the class should not be changed; new classes may be added to the same project to make a new set of functionality available.

• Properties, methods, and events should not be removed or renamed from a subsequent version.

• The syntax of existing properties, methods, and events should not be changed; this includes the calling convention and the type of parameters required and their order.

• It should be 'safe' to replace a version of an ActiveX DLL by a later version. Newer version should be tested thoroughly to ensure that they work consistently with a preceding version.

This means that the published interface of a DLL must continue to work consistently although the internal code may change in order to fix errors or improve performance and new elements may be added to the interface. If an existing element changes, for whatever reasons, such that its interface changes, add a new element with the new syntax and leave the existing element intact.

In the event of widespread changes that contravene these rules, a new project should be created with one or more new classes, if necessary by renaming the existing project and its existing classes. The new DLL must be registered and the existing DLL must remain unchanged to ensure continued compatibility with previously deployed applications.

When it moved to 32-bit architecture with Windows95, Microsoft renamed core system DLLs by adding the suffix 32 to their names—for example, KERNEL.DLL became KERNEL32.DLL. In itself, the architecture would not require the renaming of the DLLs. The reason they had to be renamed is a consequence of the internal changes to the return values arguments of the methods encapsulated within them: the data type of arguments needed to be changed.

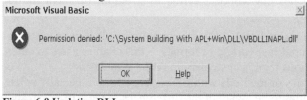

An ActiveX DLL that is in use will fail to update: the error shown in **Figure 6-9 Updating DLLs** arises.

Figure 6-9 Updating DLLs

It may be necessary to terminate the APL+Win session to free the DLL. **VB** automatically registers an ActiveX DLL on the computer on which it is created. If such a DLL is distributed, it will require registration on the target computers. In practice, the registration should be accomplished automatically by the application installation utility. Although the registration process affects the **Registry** of the computer on which it is carried out, it is sensible to locate the DLL on a network drive, as its replacement needs to be carried out in just one location.

6.6.4 ActiveX DLL coding for APL

VB ActiveX DLLs hold great potential for simplifying the process of APL +Win application building in that powerful functions can be readily accessed and reused. However, APL can use ActiveX DLLs as COM objects only; the ⎕wcall and ⎕na function cannot be used to define functions contained in them. These system functions work with DLLs conforming to the Win32 standard and which are usually written in the C language. APL will work with a Win32 DLLs but calls can be quite complicated because of the differences in the respective data types in APL and C. Consider the GetVersionEX API call in KERNEL32.DLL. The documented syntax for this function is:

```
Declare Function GetVersionEx Lib "kernel32" Alias "GetVersionExA" (ByVal
     lpVersionInformation As OSVERSIONINFO) As Long
```

Its APL translation. using name association is:

```
'DLL I4←Kernel32.GetVersionExA(⊂*C1←)' ⎕na 'GetVer'
```

The argument OSVERSIONINFO is a data type that does not exist in APL+Win, albeit it can be constructed should adequate documentation exist to allow interpretation in APL+Win—in general, it does not. In contrast, ActiveX DLLs on a computer are readily accessible by APL. Moreover, APL usually handles the transformation of arguments transparently. In order to avoid surprising results, all **VB** environmental settings must be explicitly declared at the start of every module and class.

• A method that requires a string as an argument, such as XStringCase, can easily be coded such that it can accept a case-insensitive string that may be abbreviated—this makes the APL+Win call programmer friendly.

• An API call that is unavailable in APL+Win—because it has not been added to the ADF file—may be made available from within a DLL. This process is simpler than the modification of the APLW.INI file to add the API call manually.

• Unlike APL+Win, **VB** processes arrays only when a loop is expressly created to do so. However, arrays are returned to APL+Win as nested vectors, as in the XDaysOfWeek function. If a result is returned as a collection from **VB**, it is returned as an object that will require further processing within APL+Win.

• Lack of array facilities and the occasional character set problem curtail the efficiency of ActiveX DLLs; however, they are eminently suitable for the manipulation of dates.

• The validation of input, where array processing is usually not required, may be speeded up if handled within a DLL and will keep the workspace free of a clutter of small utility functions.

• Dates and numbers, as in the functions in VBDLLINAPL, may be handled in accordance with regional settings and may be entered in a variety of formats—this enables users to enter numbers and dates with fewer restrictions than is customary in APL+Win design.

On returning from calls to properties or methods within ActiveX DLLs, especially when validating data entry forms, it is essential to be aware of the differences in the APL+Win and ActiveX environments. For example, if the *DateValid* method is used to validate entries in a data field within the APL+Win GUI and then stored in the workspace, it is necessary to ensure that the value stored in the workspace variable has the correct format. A date such as '11/25/2002' will be valid but may require conversion to '25/11/2003' or some other format before being stored.

6.7 Building a DLL for APL using C# Express 2005

C# is the flagship product of the .NET platform. It is deceptively easy to build a prototype class library (DLL) using C# for use with APL. Start C# Express, click **File | New Project,** and select the **Class Library** template. This creates a default assembly named **ClassLibrary1** and a class named **Class1;** in order to facilitate this demonstration, do not rename the files. Click **Project | ClassLibrary1 properties,** then the **Application** option in the left pane and then the button labelled **Assembly Information**: check the **Make Assembly COM visible** option. Click the **Build** option in the left pane and check the **Register for COM Interop** option. Click the button labelled **Advanced** and ensure that the option **Do not reference mscorlib.dll** is *not* checked. Click the **Debug** option in the left pane and set **Configuration** to **Release**. Click the **Signing** option in the left pane and ensure that **Sign the assembly** is *not* checked: the DLL is *not* strongly named but it will use the codebase option. The shell for the DLL is ready for customisation.

For demonstration, the following code held in the file CLASS1.CS provides a template for defining properties, methods, and events that are accessible by APL. Of all the files in the project, CLASS1.CS is the only one that requires customisation Compile the project using F6 (**Build | Build Solution**) or **Build | Rebuild Solution** menu options and *not* from the command line. The file CLASS1.SC is shown below with comments—line numbers have been added for reference:

```
 1 /// <Author> Ajay Askoolum </Author>  ///
 2 using System;
 3 using System.Collections.Generic;
 4 using System.Text;
 5 /* Next lone is vital for COM */
 6 using System.Runtime.InteropServices;
 7
 8 /* Namespace id the first part of the name of the DLL object */
 9 namespace ClassLibrary1
10 {
11 /* Refer to help files; delegates are necessary for adding events */
12      public delegate void NoArgEvent();
13      public delegate void SingleArgEvent(string arg1);
14      public delegate void MultiArgEvent(int arg1, int arg2);
```

```
15        [InterfaceType(ComInterfaceType.InterfaceIsIDispatch)]
16 /* Builds custom events */
17      public interface UserEvents
18      {
19 /* DispId is a unique integer within the namespace */
20          [DispId(6)]
21          void MyEventNil();
22          [DispId(7)]
23          void MyEventOne();
24          [DispId(8)]
25          void MyEventMulti(int arg1, int arg2);
26      }
27      [InterfaceType(ComInterfaceType.InterfaceIsIDispatch)]
28 /* Class name is the second part of the name of the DLL object */
29      public interface _Class1
30      {
31          [DispId(1)]
32          int MyMethod(int numarg1, int numarg2);
33          [DispId(2)]
34          string MMethod();
35          [DispId(3)]
36          int MyRWProperty { get; set;}
37          [DispId(4)]
38          string MyROProperty { get; }
39          [DispId(5)]
40          void eventm();
41      }
42      [ClassInterface(ClassInterfaceType.None)]
43 /* ProgId attribute is the first column of Xinfo */
44      [ProgId("ASKOOLUM.AJAY")]
45      [ComSourceInterfaces(typeof(UserEvents))]
46      public class Class1 : _Class1
47      {
48          public Class1()
49          {
50          }
51 /* Three events are declared */
52          public event NoArgEvent MyEventNil;
53          public event SingleArgEvent MyEventOne;
54          public event MultiArgEvent MyEventMulti;
55          private int myVarRW;
56 /* A read / write property, accepts integer only */
57          public int MyRWProperty
58          {
59              get { return myVarRW; }
60              set
61              {
62 /* Raises an event before changing value of property */
63                  MyEventNil();
64                  myVarRW = value;
65              }
66          }
67 /* A read-only property */
68          public string MyROProperty
69          {
70              get { return @"Demonstrates C# DLL for APL."; }
71          }
72 /* First method, raises an event if its latter argument is 0 */
73          public int MyMethod(int numarg1, int numarg2)
74          {
75              if (0 == numarg2)
76              {
77                  MyEventMulti(numarg1, numarg2);
78              }
79              return numarg1 + numarg2;
```

```
80          }
81  /* Second method, just returns a string like a read-only property*/
82          public string MMethod()
83          {
84              return @"This is a C# DLL to be called by APL.";
85          }
86  /* Third method, simply raises an event */
87          public void eventm()
88          {
89              MyEventNil();
90              MyEventOne(@"This method just raises an event.");
91          }
92      }
93  }
```

6.7.1 Using the C# DLL

On the computer where compiled, the ClassLibrary1.DLL is available for immediate use. Create an instance of the DLL using either the *ProgId* or the *NAMESPACE.CLASS* convention, as seen in **Table 6-1 Using a C# DLL with APL**.

Create Instance	
APL+Win, APLX	`'xx' ⎕wi 'Create' 'ASKOOLUM.AJAY'`
APL/W	`'xx' ⎕wc 'OLEClient' 'ClassLibrary1.Class1'`
Properties	
APL+Win, APLX	`('x'=↑¨'xx' ⎕wi 'properties')/'xx' ⎕wi 'properties'` `xMyROProperty xMyRWProperty`
APL/W	`((⊂'My')≡¨2↑¨xx.PropList)/xx.PropList` `MyRWProperty  MyROProperty`
Methods	
APL+Win, APLX	`('X'=↑¨'xx' ⎕wi 'methods')/'xx' ⎕wi 'methods'` `XMMethod XMyMethod Xeventm`
APL/W	`xx.MethodList` `MyMethod  MMethod  eventm`
Events	
APL+Win	`('X'=↑¨'xx' ⎕wi 'events')/'xx' ⎕wi 'events'` `XMyEventMulti XMyEventNil XmyEventOne`
APLX	`('X'¹¨'xx' Œwi 'events')/'xx' Œwi 'events'` `onXMyEventMulti onXMyEventNil onXMyEventOne`
APL/W	Like VB, `xx.EventList` does not expose any events directly.
Event Error	
APL/W	DISP_E_EXCEPTION: ClassLibrary1: Object reference not set to an instance of an object.
APLX	Domain Error
VB	Error 80004003: Object reference not set to an instance of an object.

Table 6-1 Using a C# DLL with APL

The error associated with the events relates to enabling event handlers with APL/W, APLX, and VB. APL+Win works more directly.

```
    'xx' ⎕wi 'MyRWProperty' 198        ⍝ Assign, no event handler
    'xx' ⎕wi 'MyRWProperty'            ⍝ Query value
198
    'xx' ⎕wi 'onMyEventNil' '⍳10'      ⍝ Set handler
    'xx' ⎕wi 'MyRWProperty' 100        ⍝ Assign, event handler set
1 2 3 4 5 6 7 8 9 10
    'xx' ⎕wi 'MyRWProperty'            ⍝ Query value
100
```

```
      'xx' ⎕wi 'MyMethod' 9 0                  ⍝ No event Handler
9
      'xx' ⎕wi 'onMyEventMulti' 'a←⎕warg'  ⍝ Event Handler set
      'xx' ⎕wi 'MyMethod' 9 0                  ⍝ Should assign argument
9
      a
9 0
      ⎕wi 'onMyEventOne' 'a←⎕wevent'       ⍝ Another event handler set
      ⎕wi 'eventm'                             ⍝ Should assign event raised
      a
MyEventOne
```

If an event is not handled, APL is unaware that the event has fired. Note that an event can pass back values to APL+Win; in the example above, ⎕warg returns values from the DLL and these would be available to any handler function in the workspace.

6.7.2 Deploying C# DLLs

The technique for deploying a C# DLL is rather elusive. Deployment *requires* that the same version of the .NET Framework is available on both the source and target computers. Several versions of the .NET Framework can exist on the same computer. Ideally, use a deployment wizard in Visual Studio 2005 and avoid direct changes to the registry as this may damage the computer irreversibly. Failing that, ensure that the DLL is *not* in use and try one of the following hands-on approaches at your own risk; the first approach is:

• Copy the DLL to the desired location on the target computer.

• Switch to the command prompt and change directory to the same location and run the following command: **regasm /codebase classlibrary1.dll**

The second approach is:

• On the source computer, create a folder corresponding to the deployment folder on the target computer and copy the DLL to the same location.

• Run the following command: regasm /codebase /regfile:classlibrary.reg classlibrary1.dll

This creates the file CLASSLIBRARY1.REG. Copy the DLL and REG files to the target computer at the same location and import the REG file: use REGEDIT or simply double-click on the REG file from within Windows Explorer. If the DLL registers successfully, create an instance of the DLL using either of the following lines:

```
'CSDLLPID' ⎕wi 'Create' 'ASKOOLUM.AJAY'      ⍝ Uses ProgID
'CSDLLNC' ⎕wi 'Create' 'ClassLibrary1.Class1' ⍝ Uses NAMESPACE.CLASS
```

The ProgID is in line [44], the NAMESPACE is in line [9] and the CLASS is in line [46] of CLASS1.CS. After initial the registration, copy replacement DLLs in place on the target computer without any changes to the registry.

Mixed language programming is a reality: APL can use Scripting directly as well as components built with other languages, including the most recent, C#.

Chapter 7

Application Extension using Scripting

The Microsoft Script Control is an ActiveX server than can be dropped on a form or instantiated in code. APL requires an explicit instance of the server. This component is unlike any other. It is neither a black box nor a white box component: black box or binary components do not expose their internal code, whereas the code of white box components is visible and alterable. The Script Control has the following features:

• It permits new properties, methods, events, and collections to be added at runtime. However, the new configuration of the Script Control cannot be saved and must be recreated with every invocation.

• It can establish a two-way data sharing with APL. That is, it can create data that APL can use and vice versa.

• It can call functions within an APL workspace.

• It can run code in several scripting languages, including VBScript and JavaScript; this may be used to provide APL the ability to run user-specified code in a scripting language of their choice at runtime.

• The Script Control does **not** allow calls to any Windows API.

• The Script Control does not fully understand the APL atomic vector—especially such things as the APL high minus, the currency symbol, etc.

Subject to licensing agreements, all the required software is available for free downloading from Microsoft: the Script Control is in SCT10EN.EXE. The files MSSCRIPT.OCX, DISPEX.DLL, JSCRIPT.DLL, and VBSCRIPT.DLL will be found in the Windows SYSTEM or SYSTEM32 folder and MSSCRIPT.HLP and MSSCRIPT.CNT will be in the Windows HELP folder, if the Script Control is installed.

Although the deployment of the Script Control raises computer security issues, it is widely used; it leaves the computer or network vulnerable to virus attacks

7.1 The APL/VBScript affinity

VBScript has two things in common with APL:

• Both are independent of the version of any automation server they might use—that is, they both use late binding.

• Both have just one data type and, unless the statement *Option Explicit* is declared with VBScript, both can create and reassign variables, mixing implicit data types, at any time.

VBScript uses the variant data type for everything—numbers, strings, dates, objects, collections, etc.; therefore, like APL, variables do not need to be declared. In contrast, VBScript *always* uses index origin 0, whereas the default is 1 for APL and can be set to 0.

VBScript has a rich set of built-in functions that APL may use, notably, date arithmetic functions. VBScript and JavaScript use the normal keyboard and are easy to learn; in fact, advanced users will very likely be familiar with VBScript from their experience with VBA.

Scripting extends the runtime functionality of any application. For APL applications, this is especially valuable because it eliminates all APL keyboard issues:

● APL can expose its variables at runtime for processing with custom user-specified code.

● There is no need to write an equation parser from scratch or to grapple with the quirks of the APL language. Users can use any scripting language supported by the Script Control using the normal keyboard.

● Users may write their own code—that is, the application provides runtime facilities that are not built-into its design. This code can be single lines of code or complete procedures using the predefined functions in the scripting language.

● Moreover, the APL developer can deploy the Script Control to access the built-in functionality of a scripting language.

7.1.1 The VBScript built-in functions

Of particular interest are the VBScript date arithmetic functions that are unavailable in APL; these functions provide a set of debugged functions that can be readily deployed via the Script Control.

7.1.2 Adding the Script Control

Add the Script Control as an automation server and not as an object on an APL form. **Figure 7-1 Adding Scripting** illustrates the process. Initialise the Script Control as follows:

```
     ∇ ScriptShare;⎕wself
[1]    ⍝ System Building with APL+Win
[2]    ⎕wself←'∆ScrCtl' ⎕wi 'Create' 'MSScriptControl.ScriptControl'
[3]    ⎕wi 'xLanguage' ''
[4]    ⎕wi 'xLanguage' 'VBScript'
```

On initialisation in line [2], the Script Control scripting language is null. Lines [3] and [4] add the language of choice: with APL+Win, it is necessary to set the language to null first—this is possibly a bug. Any given instance of the Script Control object may use only one language at a time.

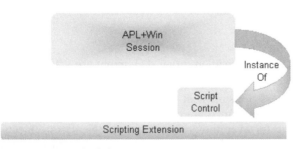

The Script Control serves as an extension to APL+Win and can return results to the APL+Win application.

The Script Control can evaluate any valid expression written in any scripting language; this chapter uses VBScript.

Figure 7-1 Adding Scripting

APLX	`''ScrCtl' ⍒wi 'Create' 'MSScriptControl.ScriptControl' ('xLanguage' 'VBScript')`
APL/W	`'∆ScrCtl'⎕WC'OLEClient' 'MSScriptControl.ScriptControl' ∆ScrCtl.Language←'VBScript'`

7.1.3 An algebraic expression evaluator

The APL syntax is far from ideal for this purpose since neither the APL keyboard (unified or APL) nor the 'right to left' evaluation of APL expressions nor the inherent grammar differences in APL (like * is power not times) is intuitive. It is inefficient to write an algebraic expression parser in APL quite simply because it is not possible to do so without compromise. On the other hand, the Script Control requires just a few lines of code to deploy and users can readily enter algebraic expressions. At this point, the Script Control is functional and all the facilities of VBScript are accessible, as shown by:

```
      ⎕wi 'XEval' 'IsDate("01/25/2000")'
1
      ⎕wi 'XEval' 'FormatDateTime("01/25/2000")'
25/01/2000

      ⎕wi 'XEval' 'IsNumeric("£1,234,56")'
0
      ⎕wi 'XEval' 'IsNumeric("ú1,234,56")'
1
```

From these examples, three things are apparent:

● Expressions require to be constructed as one string: arguments cannot be passed separately.

● There may occasionally be a problem with the APL character set—as with the currency symbol. With APL+Win, this is the familiar ⎕av to ANSI translation problem.

● The scripting expressions deal with scalar data only.

In this mode of deployment, the Script Control is of limited use with APL. At best, it is an integral algebraic VBScript expression evaluator allowing users to specify formulae involving scalar data items.

Figure 7-2 Using the InputBox dialogue

The expression:
```
      '∆ScrCtl' VBInput ''
25/01/2003
```
Invokes the built-in InputBox dialogue and allows the user to specify an expression; this may include any VBScript keywords. **Figure 7-2 Using the InputBox dialogue** shows an example.

The VBInput function creates the dialogue shown in **Figure 7-3 APL+Win InputBox**.

```
      ∇ Z←L VBInput R;⎕wself
 [1]    ⍝ System Building with APL+Win
 [2]    ⍝ L is Script Control object
 [3]    ⍝ R is [Prompt],[Caption],[Expr]
 [4]    ⎕wself←L
{1[5]   :if 1==≡R
 [6]        R←⊂R
1}[7]   :endif
 [8]    R←(3↑R)~¨' '
 [9]    R[⎕io]←⊂((0=⍴(⎕io+1)⊃R)/'Enter Expression')
 [10]   R[⎕io+1]←⊂((0=⍴(⎕io+1)⊃R)/'APL+Win')
 [11]   L←'InputBox("',(⎕io⊃R),'","',((⎕io+1)⊃R),'","', ((⎕io+2)⊃R),'")'
 [12]   R[⎕io+2]←⊂⎕wi 'XEval' L
{1[13]  :if 0≠⍴(⎕io+2)⊃R
```

```
    [14]        Z←□wi 'XEval' ((□io+2)⊃R)
1}[15]    :endif
       ▽
```

Alternatively, use an APL Edit box—such as the one shown in **Figure 7-3 APL+Win InputBox**—to specify the expression. If an expression contains an error, write it back into the Edit box for editing.

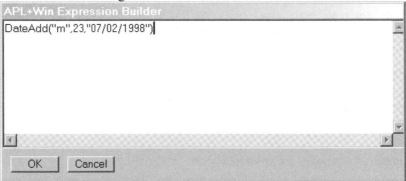

Figure 7-3 APL+Win InputBox

The following function creates a simple form with an Edit box:

```
    ▽ Z←APLInput;□wself
[1]    ⍝ System Building with APL+Win
[2]    □wself←'VBScrInp' □wi 'Create' 'Form' ('visible' 0) ('border' 0)
[3]    □wi 'caption' 'APL+Win Expression Builder'
[4]    □wi 'extent' 10 50
[5]    0 0⍴'.expr' □wi 'New' 'Edit' ('where' (0 0 ,¯2 0.5+□wi 'extent'))
           ('style' 4 16 64 8192 16384)
[6]    0 0⍴'.OK' □wi 'New' 'Button' ('where' ((↑¯1.5+□wi 'extent'), 1 1 6))
           ('onClick' 'Z←"OK" ProcInput "VBScrInp.expr" □wi "text"')
[7]    0 0⍴'.Cancel' □wi 'New' 'Button' ('where' ((↑¯1.5+□wi 'extent'), 8 1
           6)) ('onClick' 'Z←"Cancel" ProcInput θ')
[8]    □wi 'Wait'
       ▽
```

In line [6], an event handler is specified for validating user input; the event handler function is defined as follows:

```
       ▽ Z←L ProcInput R;□elx
  [1]    ⍝ System Building with APL+Win
  [2]    □elx←'→ 0'
  [3]    Z←θ
{1[4]    :select L
★ [5]        :case 'OK'
  [6]            Z←'∆ScrCtl' □wi 'XEval' R
★ [7]        :case 'Cancel'
  [8]            0 0⍴'∆VBScrInp' □wi 'Delete'
  [9]            Z←θ
1}[10]   :endselect
       ▽
```

The **OK** button evaluates the expression entered and returns the result to the APL+Win session; the form remains active thus allowing other expressions to be entered. The **Cancel** button destroys the form.

7.2 Error trapping

It is naïve to expect that expressions submitted to the Script Control for evaluation would never cause a runtime error. Therefore, an error handler is mandatory. Ideally, the handler should provide clues for resolving any runtime error encountered. Consider the following expression:

```
'∆ScrCtl' ⎕wi 'XEval' 'DateAdd("m",23,"07/02/2003"'
```

A runtime error should arise because the expression in incomplete: a closing bracket is missing from the right-hand argument. **Figure 7-4 Runtime diagnosis** shows the diagnosis of the error as reported by the Script Control.

Both the cause of the error and its possible resolution are fully identified.

Detailed information of this type will lead to the rapid resolution of the error.

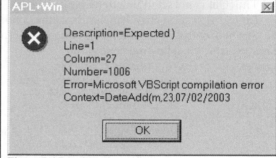

Figure 7-4 Runtime diagnosis

7.3 Exploring the Script Control

In order to deploy the Script Control efficiently, it must be configured to create the desired environment; one such environment is configured by:

```
      ∇ L Script R;⎕wself;⎕elx
 [1]   ⍝ System Building with APL+Win
 [2]   ⍝⎕elx←'"APLError" Script ⎕dm'
{1[3]  :if 0=⍴'∆ScrCtl' ⎕wi 'self' ⍝ Verify if Script Control exists &
       create instance
{2[4]     :if 0∊⍴'#' ⎕wi 'XInfo' 'MSScriptControl.ScriptControl'
 [5]        Msg 'Microsoft Script Control Object is unavailable.'
> [6]         :return
+ [7]      :else
 [8]          ⎕wself←'∆ScrCtl' ⎕wi 'Create'
       'MSScriptControl.ScriptControl'
 [9]          0 0⍴⎕wi 'xLanguage' ''
 [10]         0 0⍴⎕wi 'xLanguage' 'VBScript'
 [11]         0 0⍴⎕wi 'xSiteHwnd' ('#' ⎕wi 'hwndmain')
 [12]         0 0⍴⎕wi 'xTimeOut' '=NoTimeOut'
 [13]         0 0⍴⎕wi 'xAllowUI' 1
 [14]         0 0⍴⎕wi 'onXError' '"XError" Script θ'
 [15]         Z←⎕wself
2}[16]     :endif
+ [17]  :else
 [18]      ⎕wself←'∆ScrCtl'
1}[19]  :endif
{1[20]  :if 0=⎕nc 'L'
> [21]      :return
1}[22]  :endif
{1[23]  :select L
```

In line [11], the APL+Win system object is specified as the window in which the Script Control may display dialogue boxes and other GUI objects. The default value of this property is 0, the **Desktop** window. Line [12] sets the maximum time that the Script Control is allowed to run.

In line [12], the timeout value is set to indefinite; the constant NoTimeOut has value ‾1. Line [13] enables the AllowUI property; this property controls whether the Script Control is able to display GUI objects, such as InputBox, MsgBox and the timeout dialogue. If this property is set to **FALSE**, the Script Control suppresses its internal GUI dialogues; it will be necessary to provide a custom onXTimeOut event handler when AllowUI is set to **FALSE** and TimeOut is not set to indefinite.

Line [14] sets the error handler. Note the conditional statement in line [3]: an instance of the Script Control is *not* recreated when the error handler is called. On encountering a runtime error, the following code is run:

```
★ [24]        :case 'XError'
  [25]            R←▼¨(⊂⎕wi 'XError') ⎕wi¨ 'xDescription'  'xLine' 'xColumn'
          'xNumber' 'xSource' 'xText'
  [26]            R←('Description=' 'Line=' 'Column=' 'Number=' 'Error='
          'Context='),¨'R~¨⊂'"'''
  [27]            R←∊('"',¨'R,¨'"'),¨((‾1+⍴R)/⊂'& vbCR &'),⊂''
  [28]  A         ⎕wi 'xError.Clear'
  [29]        0 0⍴⎕wi 'XExecuteStatement' ( 'MsgBox
          ',R,',vbOKOnly+vbCritical,"APL+Win"')
```

By default, the timeout value is 10000 milliseconds. If the Script Control takes longer than the specified duration, the dialogue shown in **Figure 7-5 Timeout options** appears.

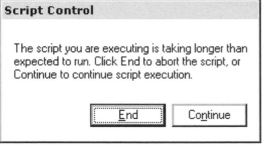

The code can be aborted or allowed to continue running by clicking the appropriate button.

Figure 7-5 Timeout options

Note that the Script Control itself provides the diagnosis and reports it using its own MsgBox dialogue. Note the following:

• The error handler will not display its message if AllowUI is set to **FALSE**.

• The MsgBox dialogue does not understand the unattended run mode discussed in Chapter 2 *Advanced APL+Win Techniques*.

If the function Msg is used instead, substitute line [29] by the following line:

```
  [29]            0 0⍴Msg R
```

The MsgBox function can be called either with or without its arguments enclosed in round brackets: the use of round brackets forces MsgBox to return a result. Use the Script Control's *ExecuteStatement* method when the arguments are not enclosed in round brackets,—otherwise use the *Eval* method. With some errors, such as error number 500, line [29] fails to clear the error condition and reports:

```
        ⎕WI RESULT ERROR: Unable to link next action: codeobject
```

Strictly speaking, it is unnecessary to clear the error as the Script Control routinely clears the properties of the Error object when one of the *Reset*, *AddCode*, *Eval*, or *ExecuteStatement* methods are called; line [30] calls the ExecuteStatement method. However, the latent expression must be reset, if necessary to ⎕dm, in order to avoid an indefinite loop within the error handler routine:

```
★ [30]        :case 'APLError'
  [31]            ⎕elx←'⎕dm'
  [32]          0 0ρMsg R
```

The *Reset* method clears all code and data from the Script Control. The *EVAL* method is used to evaluate an expression that returns an explicit result whereas the *ExecuteStatement* method is used when the expression does not return a result.

7.3.1 The Eval method

This method may also be used to create an APL+Win object using redirection. APL+Win does not have the facility to use a specific and existing instance of an application server: for instance, it cannot use an existing **Excel** session as an automation server. In some circumstances, it might be desirable to be able to do so. For example, **Outlook** may be running and APL+Win may need to send an email. All other things being equal, APL has to create a new instance of **Outlook**.

The GetObject function in **VB** and VBA exist precisely for grabbing hold of existing instances of automation servers. The syntax is:

```
    GetObject([pathname] [, class])
```

Although APL does not have the GetObject functionality, it can emulate it using the Script Control:

```
     ∇ L GetObject R;⎕wself;⎕elx
 [1]   ⍝ System Building with APL+Win
{1[2]   :if 0≠ρL ⎕wi 'self'
 [3]        L ⎕wi 'Delete'
1}[4]   :endif
 [5]   ⎕elx←"⎕elx←'→ 0' ◇ '∆sc' ⎕wi 'Delete' ◇ 0 0ρL ⎕wi 'Create' R ◇ Z←1"
 [6]   ⎕wself←'∆sc' ⎕wi 'Create' 'MSScriptControl.ScriptControl'
 [7]   ⎕wi 'xLanguage' ''
 [8]   ⎕wi 'xLanguage' 'VBScript'
 [9]   ⎕wi ('XEval>',L) ('GetObject(,"',R,'")')
 [10]  '∆sc' ⎕wi 'Delete'
 [11]  Z←1
     ∇
```

The left-hand argument of this function is the name of an APL+Win object; the right-hand argument is the class of the automation server. The syntax for grabbing an existing session of **Outlook** is shown in **Table 7-1 GetObject pathways**. Should a session of the class exists, GetObject uses it, otherwise it creates a new instance. In line [9], redirection is used with the XEval method to create an instance of **Outlook**. The code is:

```
    'olk' GetObject 'Outlook.Application'
```

Uses existing session	Creates a new session

`'olk' ⎕wi 'class'`
`ActiveObject`
`'olk' ⎕wi 'xName'`
`Outlook`
	`'olk' ⎕wi 'class'`
	`ActiveObject Outlook.Application`
	`'olk' ⎕wi 'xName'`
	`Outlook`

Table 7-1 GetObject pathways

Now, this instance may be used to send an email.

The function SendEmail constructs and
sends an email, silently.

The right-hand argument of the function
is the name of the object created using
redirection that is the result of the GetObject
function.

The result of running the SendEmail
function is an item in the Sent folder: it is
shown in **Figure 7-6 Email using Outlook**.

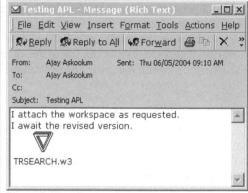

Figure 7-6 Email using Outlook

```
      ∇ SendEmail R
[1]   ⍝ System Building with APL+Win
[2]   ⎕wself←'ml' ⎕wi 'Create' ('ol' ⎕wi 'XCreateItem' (R ⎕wi
        '=olMailItem'))
[3]   ⎕wi 'To' 'aa2e72e@lycos.co.uk'
[4]   ⎕wi 'Subject' 'Testing APL'
[5]   ⎕wi 'xBody' ('I attach the workspace as requested.',⎕tcnl,'I await
        the revised version.',⎕tcnl)
[6]   ⎕wi 'xAttachments.Add' (⎕wsid,'.w3')
[7]   ⎕wi 'XSend'
[8]   'ml' ⎕wi 'Delete'
      ∇
```

On the subject of emails, another route is to use the ShellExecute API call and the mailto
protocol.

ShellExecute=H(HW hwnd,*C lpOperation,*C lpFile,*C lpParameters,*C lpDirectory,I
iShowCmd) ALIAS ShellExecuteA LIB Shell32.DLL

An example of the APL+Win syntax is encapsulated in this function.

```
      ∇ Email R
[1]   ⍝ System Building with APL+Win
[2]   R←'mailto:',R
[3]   ⎕wcall 'ShellExecute' ('#' ⎕wi 'hwndmain') ⍬ R ⍬ ⍬ 'SW_SHOWNORMAL'
      ∇
```

This uses the mailto protocol for activating the email programme found on the
computer. It does not despatch the email. The right-hand argument may comprise of an
email id or several other items of information. For example,

```
'support@Ltd.com&cc=legal@own.net&bcc=market@own.net&subject=Maintenance&b
ody=This is a reminder.'
```

Using this example as the right-hand argument, the message shown in **Figure 7-7 Email Mailto** is created. This opens the 'New Mail Message' window allowing users to compose and send a message.

In the right-hand argument, items of information are delimited by &. Any or all items of information may be eliminated.

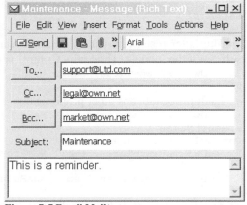

Figure 7-7 Email Mailto

An APL+Win application may use this facility for generating a standard message from a client's computer and destined for the software provider in the event of an unexpected error. Since the application creates the message, it can use the *@body=* item to pinpoint the context of the error as an aid to rapid resolution.

7.4 Extending the Script Control

Upon initialisation, the Script Control does not have any custom properties or methods. That is, it is simply a powerful calculator capable of executing single lines of code that may include the built-in functionality of the scripting language specified: it does not have any custom properties or methods:

```
      ⎕wself←'∆ScrCtl'
      ρ('X'=↑¨(⎕wi 'xCodeObject') ⎕wi 'properties')/(⎕wi 'xCodeObject')
⎕wi 'properties'
0
      ρ('X'=↑¨(⎕wi 'xCodeObject') ⎕wi 'methods')/(⎕wi 'xCodeObject') ⎕wi
'methods'
0
```

The AddCode method allows custom properties (variables or user-defined data types) and methods (subroutines and functions) to be added—there is no requirement for any kind of registration following the addition of custom properties and methods:

```
      ⎕wi 'XAddCode' SampleCode
```

By default, the code is added to the default module named Global; one module is added automatically and is referred to implicitly by the AddCode method:

```
      ⎕wi 'xModules.Count'
1
      ⎕wi 'xModules().Name' 1
Global
```

The full syntax of the AddCode methods is:

```
      ⎕wi 'xModules().AddCode' 'Global' SampleCode
```

New code modules may be added, as:

```
      0 0ρ⎕wi 'xModules.Add' "MyModule"
```

However, individual modules cannot be deleted by reference—the Reset method removes *all* modules. After the code in SampleCode is added, new properties and methods are exposed:

```
      ρ('x'=↑¨(⎕wi 'xCodeObject') ⎕wi 'properties')/(⎕wi 'xCodeObject')
⎕wi 'properties'
3
```

```
    ρ('X'=↑¨(⎕wi 'xCodeObject') ⎕wi 'methods')/(⎕wi 'xCodeObject') ⎕wi
'methods'
10
```

The latter result can also be obtained via the *Procedures* property of the Script Control itself—not the module. The Script Control is referring to the Global module implicitly:

```
    ⎕wi 'xProcedures.Count'
10
```

Each procedure has other properties.

```
    ⎕wi 'xProcedures.Item().Name' 1
GetMonthNames
ⁿ or ⎕wi 'xProcedures().Name' 1
    ⎕wi 'xProcedures.Item().NumArgs' 1
2
    ⎕wi 'xProcedures.Item().HasReturnValue' 1
1
```

The Script Control has a collection of modules and each module has a collection of methods (procedures) and properties (variables). Note that for enumerating the procedures, the index origin is always 1. The *Name* property returns the name of the procedure, the *NumArgs* property returns a count of the number of arguments expected by it, and the *HasReturnValue* property is either **TRUE** (procedure is a function) or **FALSE** (procedure is a subroutine). The names of the arguments may also be queried, thus:

```
    (⎕wi 'xCodeObject') ⎕wi '?GetMonthNames'
XGetMonthNames method:
  Result ← ⎕WI 'XGetMonthNames' GivenMonth Abbreviated
```

Since VBScript has only one data type, namely Variant, the type of the arguments is not reported. The generic syntax for calling procedures is:

```
    ⎕wi 'xModules().CodeObject.GetMonthNames(,)' 'Global' 0 1
 Jan Feb Mar Apr May Jun Jul Aug Sep Oct Nov Dec
```

The result is a nested vector. The short syntax is:

```
    ⎕wi 'xCodeObject.GetMonthNames(,)' 1 0
January
```

The procedure may also be run with the Eval method, thus:

```
    ⎕wi 'XEval' 'GetMonthNames(1,0)'
January
```

However, this is unnecessarily restrictive in that the name of the procedure and its arguments need to be specified as one string. The Run method also allows a named procedure to be executed; however, this requires the syntax of this method's argument to be constructed in manner that is problematic with APL+Win when multiple arguments are involved. The generic syntax of the Run method is:

```
    ⎕wi 'XRun(procedure, arg1, arg2,..., argn)
```

When a procedure requiring one argument is run, both the result and argument are returned—the result is a nested vector:

```
    ⎕wi 'XRun("ValidDate","01/12/2003")'
 1 01/12/2003
    ⎕wi 'XRun("GetDateSerial", "01/01/2003") '
 37622 01/01/2003
    ⎕wi 'xModulesXCodeObject.GetDateSerial' '01/01/2003'
37622
```

However, this syntax fails with APL+Win when multiple arguments are involved, as shown by:

```
    ⎕wi 'XRun("GetMonthNames",1,0)'
⎕WI ERROR: exception 800A01C2
```

Within the APL+Win environment, it is easier to write code that specifies arguments separately as a list.

7.4.1 What is in SampleCode?

SampleCode is a workspace variable, a vector, created within the APL+Win editor. It may be stored in a native or component file or in a database and retrieved at runtime. This variable contains VBScript code. The Script Control does not allow its code to be browsed—even when the code is user-defined—and cannot be saved with the added code. Thus, a permanent copy of the code is necessary and it must be added at runtime.

The variable *SampleCode* is listed and discussed next; line numbers have been added manually to aid reference:

```
1 Option Explicit
```

Line [1] overrides the default VBScript setting; with Options Explicit, all variables must be declared before being referenced. Any reference to an undeclared variable will generate a runtime error, equivalent to APL+Win's *Value Error*.

7.4.1.1 A collection object

A collection is an associative vector comprising of name/value pairs; their construction is simpler than in APL+Win and support for them is integral in VBScript. However, a collection is static upon definition; that is, elements cannot be removed or added dynamically:

```
2 Class MyType
3       Public ID
4       Public Name
5       Public Access
6 End Class
```

Lines [2] to [6] create a user-defined *collection* comprising of three variables. As all variables are Variants in VBScript, it is not possible to define the types of the variables in the collection:

```
7 Dim ClassInstance
8 Set ClassInstance = New MyType
```

The class is exposed in the public interface; an instance, created in lines [7] to [8] of it is:

```
     ⎕wi 'xCodeObject.ClassInstance.Id' '10875'
     ⎕wi 'xCodeObject.ClassInstance.Name' 'System Building'
     ⎕wi 'xCodeObject.ClassInstance.Access' 'APL+Win'
```

7.4.1.2 A read/write property

```
9 Public PropVariable
```

Line [9] defined as property without recourse to *Property Let* and *Property Get*. This is the simplest means of defining a property; any value may be assigned to this property, including an APL+Win nested value, thus:

```
     ⎕wi 'xCodeObject.PropVariable' (⍳10) ('Mixed Vector')
```

However, the Script Control is unable to make *complete* sense of the APL+Win nested vector. On the one hand, it sees it as a nested array:

```
     ⎕wi 'XEval' 'VarType(PropVariable(0))'
8195
     ⎕wi 'XEval' 'TypeName(PropVariable(0))'
Long()
     ⎕wi 'XEval' 'VarType(PropVariable(1))'
8
     ⎕wi 'XEval' 'Typename(PropVariable(1))'
String
```

However, the first element (⍳10) is not seen as a variant array, type 8204. Although it is possible to assign an APL+Win nested value to a VBScript property, this is of limited use, as

VBScript cannot use it although it will return it to APL+Win correctly. This property is simply a variable and may be created arbitrarily at runtime:

```
        ⎕wi 'XExecuteStatement' 'NewProp="Created at runtime"'
        ⎕wi 'xCodeObject.NewProp'
Created at runtime
```

VBScript always use double quotes to enclose string values.

7.4.1.3 A read only property

```
10 Class NewSampleVersion
11      Private xSampleVersion
12      Private Sub Class_Initialize
13          xSampleVersion="APL+Win/Script Control Sample Code 1.0"
14      end Sub
15      Public Property Get SampleVersion
16          SampleVersion=xSampleVersion
17      End Property
18 End Class
```

Lines [10] to [18] create a read only property. The property is read only because there is no *Property Let* code. The value of the property is initialised via the Class_Initialize event: this event runs as soon as a new instance of the class is created. The corresponding Class_Terminate event runs when that instance is destroyed: an instance of a class is destroyed when it is set to *Nothing*.

This property is *not* exposed in the public interface. Knowledge of its existence allows an instance to be created—this exposes it:

```
        ⎕wi 'XExecuteStatement' 'Set Version=New NewSampleVersion'
        ⎕wi 'xCodeObject.Version.SampleVersion'
APL+Win/Script Control Sample Code 1.0
```

Any attempt to write to this property will fail:

```
LENGTH ERROR: Wrong number of arguments
```

7.4.1.4 A read/write property with validation

```
19 Class UDDate
20      Private sUDDate
21      Public Property Get UDDate
22          UDDate = sUDDate
23      End Property
24      Public Property Let UDDate(newvalue)
25          If IsDate(newvalue) then
26              sUDDate = newvalue
27          Else
28              MsgBox "Failed to update UDDate. Invalid date (" & newvalue &
")",vbCritical,"APL+Win"
29          End if
30      End Property
31 End Class
```

Lines [19] to [31] create a class that will validate values assigned to the property within it. This offers the mechanism for creating typed property values. This class is not exposed until an instance of it is created; several instances of the class may be created at runtime:

```
        ⎕wi 'XExecuteStatement' 'Set StartDate=New UDDate'
        ⎕wi 'XExecuteStatement' 'Set ValnDate=New UDDate'
```

The *Property Get* code retrieves the value of the property and the *Property Let* code assigns new values; the property does not have a default value. Should it be desirable to assign a default value, the *Get* code may be amended as follows:

```
[21]      Public Property Get UDDate
[22.1]        If IsEmpty(sUDDate) then
[22.2]            UDDate="01/01/2000"
[22.3]        Else
[22.4]            UDDate = sUDDate
```

```
[22.5]      End If
[23]    End Property
```
With this code, all instances of the class will inherit the default value of 01/01/2000; the default value cannot be changed from outside the class.

A quirk in the APL+Win interface prevents an instance of the class from being updated in the usual manner, as seen below:
```
☐wi 'xCodeObject.ValnDate.UDDate' "31/12/2003"
LENGTH ERROR: Wrong number of arguments
```
It appears that APL+Win is able to see *only* the first property defined within the class— it may be either *Get* or *Let*—and in this instance, it sees the *Get* property only as it is defined first. A way round this is to use the Script Control itself to update the value:
```
☐wi 'XExecuteStatement' 'StartDate.UDDate="31/12/2003"'
☐wi 'xCodeObject.StartDate.UDDate'
31/12/2003
```
The other instance of the class has the default date:
```
☐wi 'xCodeObject.ValnDate.UDDate'
01/01/2000
```

7.4.1.5 Enumerating names of months
```
32 Function GetMonthNames(GivenMonth,Abbreviated)
33     Dim Result(11),i
34     GivenMonth=Int(Abs(GivenMonth))
35     If Abbreviated <>0 then Abbreviated = 1
36     If GivenMonth=0 Then
37         For i = 0 To 11
38             Result(i)=MonthName(i+1,Abbreviated)
39         Next
40         GetMonthNames=Result
41     Else
42         GetMonthNames=MonthName(GivenMonth MOD 12,Abbreviated)
43     End If
44 End Function
```
This method requires two arguments; *GivenMonth* may be 0, representing all months, or any integer between 1 and 12 and *Abbreviated* may be **TRUE** or **FALSE**. If *GivenMonth* is 0, the method returns a nested vector of all months of the year. Otherwise, a simple vector of the name of the month is returned. The argument *Abbreviated* controls whether the full or abbreviated names are returned, as shown by:
```
☐wi 'xCodeObject.GetMonthNames(,)' 0 1
Jan Feb Mar Apr May Jun Jul Aug Sep Oct Nov Dec
☐wi 'xCodeObject.GetMonthNames(,)' 9 0
September
```
The result of this method automatically returns its result in line with regional settings.

7.4.1.6 Enumerating names of days
```
45 Function GetDayNames(GivenDay,Abbreviated)
46     Dim Result(6),i
47     GivenDay=Int(Abs(GivenDay))
48     If Abbreviated <>0 then Abbreviated = 1
49     If GivenDay=0 Then
50         For i = 0 To 6
51             Result(i)=WeekdayName(i+1,Abbreviated)
52         Next
53         GetDayNames=Result
54     Else
55         GetDayNames=WeekdayName(GivenDay MOD 7,Abbreviated)
56     End If
57 End Function
```

This method requires two arguments; *GivenDay* may be 0, representing all months, or any integer between 1 and 7 and *Abbreviated* may be **TRUE** or **FALSE**. If *GivenDay* is 0, the method returns a nested vector of all days of the week. Otherwise, a simple vector of the name of the day is returned. The argument *Abbreviated* controls whether the full or abbreviated names are returned:

```
      ⎕wi 'xCodeObject.GetDayNames(,)' 0 1
Sun Mon Tue Wed Thu Fri Sat
      ⎕wi 'xCodeObject.GetDayNames(,)' 6 0
Friday
```

The result of this method automatically returns its result in the language set in regional settings.

7.4.1.7 Date validation

```
58 Function ValidDate(DateVariable)
59     ValidDate=IsDate(DateVariable)
60 End Function
```

The VBScript *IsDate* function returns **TRUE** if its argument is a valid date and **FALSE** otherwise. However, *IsDate* returns **FALSE** only when its argument is invalid in any date format that it understands, not just the regional long or short date formats, and it does not indicate which date format it has used, as seen by:

```
      ⎕wi ¨(⊂⊂'xCodeObject.ValidDate'),¨'2003/01/25' '2003/25/01'
'25/01/2003' '01/25/2003'
1 0 1 1
```

If this method is used to validate dates entered in the user interface, it will need to be translated into one of the regional date formats, usually the short date format, before being stored in order that other date functions may work correctly.

7.4.1.8 Number validation

```
61 Function ValidNumber(NumericVariable)
62     ValidNumber=IsNumeric(Replace(NumericVariable,"£","ú"))
63 End Function
```

The *IsNumeric* function returns **TRUE** if its argument is numeric and **FALSE** otherwise. The argument may contain a currency symbol, thousand separators and negative numbers formatting. This method has been coded to avoid the APL+Win to ANSI character translation problem for the UK. However, its result is independent of regional settings:

```
      ⎕wi ¨(⊂⊂'xCodeObject.ValidNumber'),¨ '(£22,988.89)' '(£22.988,89)'
'£22988.89'
1 1 1
```

If this method is used to validate user input, the value needs to be translated into a valid number that APL+Win can understand. A general solution is:

```
      ⎕wi 'XEval' 'CDbl(Replace("(£22,988.89)","£",""))'
¯22988.89
```

If dates and numbers are written back to the user interface in a format consistent with regional settings, it will be necessary to call the VBScript built-in *GetFormatNumber*, *GetFormatCurrency,* or *GetFormatDateTime* as appropriate. The *GetFormatDateTime* function returns dates in the short data format specified in regional settings.

7.4.1.9 Formating numbers

```
64 Function GetFormatNumber(ValidNumber)
65     GetFormatNumber=FormatNumber(ValidNumber)
66 End Function
```

This function returns its argument, a valid number with or without formatting, in regional format:

```
      ⎕wi¨ (⊂⊂'xCodeObject.GetFormatNumber'),¨"22988.889" "22,988.889"
"22.988,889" "-22,988.89"
 22,988.89 22,988.89 22.99 (22,988.89)
```

7.4.1.10 Formating currency

```
67 Function GetFormatCurrency(ValidNumber)
68    GetFormatCurrency=Replace(FormatCurrency(ValidNumber),"ú","£")
69 End Function
```

This method works like the GetFormatNumber method except that it appends the local currency symbol to its result.

7.4.1.11 Formating percentages

```
70 Function GetFormatPercent(ValidNumber)
71    GetFormatPercent=FormatPercent(ValidNumber)
72 End Function
```

This method returns a number formatted as a percentage:

```
      ⎕wi¨(⊂'xCodeObject.GetFormatPercent'),¨ "0.089" "-0.089"
8.90% (8.90%)
```

7.4.1.12 Formating dates as a scalar

```
73 Function GetDateSerial(ValidDate)
74    GetDateSerial =
CDbl(DateSerial(Year(ValidDate),Month(ValidDate),Day(ValidDate)))
75 End Function
```

This method relies on several VBScript built-in date functions. It returns its argument as a scalar representing the number of days elapsed from a reference date:

```
      ⎕wi 'xCodeObject.GetDateSerial'  "01/01/2003"
37622
```

This result is consistent with the way the APL+Win Grid object stores dates internally: This function is defined as follows:

```
      ∇ Z←GRIDDate R
[1]    ⍝ System Building with APL+Win
[2]    ⎕wself←'∆APLGrid' ⎕wi 'Create' 'APL.Grid'
[3]    ⎕wi 'Set' ('Rows' 1) ('Cols' 1)
[4]    ⎕wi 'xText' 1 1 R
[5]    Z←⎕wi 'xDate' 1 1
[6]    ⎕wi 'Delete'
      ∇
```

```
GRIDDate "01/01/2003"
37622
```

7.4.1.13 Date arithmetic I

```
76 Function GetDateAdd(Duration,Interval,ValidDate)
77    GetDateAdd=FormatDateTime(DateAdd(Duration,Interval,ValidDate))
78 End Function
```

This method adds a specified duration to a valid date and returns the result in the short date regional format. The valid date may be in any format, for example:

```
      ⎕wi¨(⊂'xCodeObject.GetDateAdd'),¨( "m" 10 "01/12/2003") ("m" 10
"2003/12/01")
01/10/2004 01/10/2004
```

7.4.1.14 Date arithmetic II

```
79 Function GetDateDiff(Interval,ValidDate1,ValidDate2)
80    Dim Adj
81    If Year(ValidDate1)<Year(ValidDate2) Then
82       Adj=1
83    Else
84       Adj=-1
85    End If
86    Select Case Interval
87       Case "CompleteMonths"
88          GetDateDiff = DateDiff("m",ValidDate1, DateAdd("m",(Day(ValidDate1)
> Day(ValidDate2)) * Adj,ValidDate2))
89       Case "AgeAttQ"
```

```
90              . GetDateDiff = 0.25 *
Int(GetDateDiff("CompleteMonths",ValidDate1,Validdate2)/3)
91         Case "AgeNext"
92            If Month(ValidDate2) < Month(ValidDate1) Then
93               ValidDate2=DateAdd("yyyy",Adj,ValidDate2)
94            Elseif Month(ValidDate2) = Month(ValidDate1) and
Day(ValidDate1)<=Day(ValidDate2) Then
95               ValidDate2=DateAdd("yyyy",Adj,ValidDate2)
96            End If
97            GetDateDiff=DateDiff("yyyy",ValidDate1,ValidDate2)
98         Case Else
99            GetDateDiff=DateDiff(Interval,ValidDate1,ValidDate2)
100    End Select
101 End Function
```

This method takes the difference between two valid dates and returns its result as a of the duration. Although VBScript date arithmetic functionality simplify calculations, it has a serious anomaly in that it ignores day of month when calculating duration in months, month when calculating duration in years, etc. This makes raw calculations unsuitable in some circumstances. For example,

```
    ⎕wi 'xCodeObject.GetDateDiff' "m" "10/01/2000" "12/06/2003"
41
    ⎕wi 'xCodeObject.GetDateDiff' "m" "12/06/2003" "10/01/2000"
⁻41
```

This method has been coded such that it can recognise custom intervals, such as *CompleteMonths*, *AgeAttQ* (age attained in quarter years) and *AgeNext* (age next birthday)— these are typical APL+Win date arithmetic calculations—and appropriate adjustments are made to the result. The syntax is consistent irrespective of whether custom intervals or built-in intervals are used:

```
    ⎕wi 'xCodeObject.GetDateDiff' "CompleteMonths" "10/01/2000"
"12/06/2003"
41
    ⎕wi 'xCodeObject.GetDateDiff' "CompleteMonths" "12/06/2003"
"10/01/2000"
⁻40
```

Note the difference in the last result. The other two custom methods are called in the same manner:

```
    ⎕wi 'xCodeObject.GetDateDiff' "AgeAttQ" "10/01/2000" "12/06/2003"
3.25
    ⎕wi 'xCodeObject.GetDateDiff' "AgeAttQ" "12/06/2003" "10/01/2000"
⁻3.5
    ⎕wi 'xCodeObject.GetDateDiff' "AgeNext"  "10/01/2000" "12/06/2003"
3
    ⎕wi 'xCodeObject.GetDateDiff' "AgeNext"  "12/06/2003" "10/01/2000"
⁻4
```

In these examples, using the each operator circumvents the need for writing looping solutions in APL+Win.

7.5 Multi-language programming

As demonstrated, the *AddCode* method of the Script Control allows multi-language programming from within the APL+Win workspace; complete procedures can be executed and the results are returned to the workspace. However, so far the workspace variables have been passed to the Script Control as arguments; that is, by value. Likewise, variables or objects within the Script Control have been retrieved into the workspace by value.

7.5.1 Sharing an APL+Win object

In order to capitalize on the flexibility afforded by the Script Control, it is necessary to be able to retrieve variables and objects by reference. The *AddObject* method enables the Script Control to share variables and run functions within an APL+Win ActiveX server session created by APL+Win. Since APL+Win creates the server session, it is able to do the same, see below:

```
★ [33]          :case 'ShareWSEngine'
  [34]              0 0ρ'∆APL' ⎕wi 'Create' 'APLW.WSEngine'
  [35]                  ⎕wi 'XAddObject' 'APLEngine' ('#' ⎕wi 'UseObject' ('∆APL'
         ⎕wi 'obj'))
         'ShareWSEngine' Script θ
```

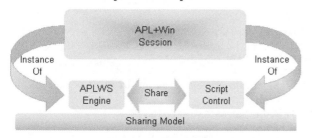

In this scenario, the Script Control is able to read *and* write variables in the WS Engine session as well as run functions within it.

Figure 7-8 Sharing server with Script Control

At this point, the APL+Win server session has a clear workspace. The Script Control may add a variable to it:

```
    ⎕wi 'xCodeObject.APLEngine.Variable' 'SCtl' 100.23
```

In addition, the APL+Win session may read it, as a numeric value:

```
    10×'∆APL' ⎕wi 'Variable' 'SCtl'
1002.3
```

This new value may be written to the server workspace, thus:

```
    NewSCtl←'∆APL' ⎕wi 'Variable' 'SCtl' (10×'∆APL' ⎕wi 'Variable'
'SCtl')
```

In addition, the Script Control may see the new value from within:

```
    ⎕wi 'XExecuteStatement' 'MsgBox
APLEngine.Variable("SCtl"),vbOkOnly,"From APL WSEngine session"'
```

This simple demonstration illustrates that APL+Win and the Script Control may share variables via an APL+Win server session: refer to **Figure 7-9 Sharing variables**.

Ordinarily, the server session will be hidden. Note that the integrity of the numeric value is retained; as VBScript stores dates as formatted strings, it will understand dates if stored as strings in the workspace—in a regional setting format.

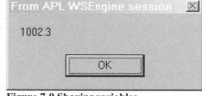

Figure 7-9 Sharing variables

7.5.2 Creating own instance of APL+Win object

An alternative to sharing the instance of the APL+Win ActiveX server, the Script Control may create its own instance of the ActiveX server session created by APL+Win. The code that needs to be added to the Script Control is:

```
1 Dim APLSession
2 Sub GetAPLSession(SessionHandle)
3     Set APLSession = SessionHandle
4 End Sub
    ⎕wi 'xCodeObject.GetAPLSession' ('∆APL' ⎕wi 'obj')
```

The Script variable *APLSession* is the same as the *APLEngine* reference from APL+Win. Either approach may be used as they deliver identical functionality, but not both, as this will increase the overhead of managing the code.

7.5.3 Processing simple numeric arrays

A crucial test of usability is the ability by the Script Control to handle APL+Win arrays internally. Consider the following example:

```
⎕wself←'∆APL' ⎕wi 'Variable' 'NumArg' (+\10/1)
```

The Script Control is able to return the variable intact, as shown by:

```
(+\10/1)≡⎕wi 'XEval' 'APLEngine.Variable("NumArg")'
```
1

How does it see the variable internally?

```
⎕wi 'XEval' 'TypeName(APLEngine.Variable("NumArg"))'
Long()
        ⎕wi 'XEval' 'VarType(APLEngine.Variable("NumArg"))'
8195
```

It sees it a variable of type *Long*—that is numeric—and as an array. It is possible to query the lower and upper indices of the variable:

```
⎕wi 'XEval' 'LBound(APLEngine.Variable("NumArg"))'
0
        ⎕wi 'XEval' 'UBound(APLEngine.Variable("NumArg"))'
9
```

VBScript is able to see the lower and upper bounds of the one-dimensional array correctly. VBScript does not have a function that will return the rank of an array; ordinarily, the rank must be known in advance in order to code the indexing of the array correctly—VBScript always works in index origin 0. The dimensions or shape of the array may be established by the *LBound* and *UBound* functions. The syntax of these functions is:

LBound(arrayname[, dimension])
UBound(arrayname[, dimension])

The optional argument, *dimension*, specifies the dimension of an array. By default, all VBScript arrays are one-dimensional. Since VBScript sees the variable *NumArg* as a numeric array, it ought to be possible to get it to return the sum of the array. The VBScript code to compute and display the sum of *NumArg* is held in a variable *Code2Sum*:

```
 1 Sub SumNumArg()
 2      Dim element
 3      Dim TotalNumArg
 4      Dim NumArg
 5      NumArg = APLEngine.Variable("NumArg")
 6      For each elememt in NumArg
 7          TotalNumArg = TotalNumArg + element
 8      Next
 9      MsgBox TotalNumArg,vbOkOnly,"From Script Control code"
10 End Sub
```

In line [5], the APL+Win Server variable is read into a VBScript variable in order to avoid repeated reference to the server object. In lines [6] to [8], each element is added to a counter. The VBScript/VB **For Each** construction has two advantages:

● It circumvents the necessity for knowing the dimension of the array.

● It neatly avoids the need for indexing an array and any errors relating to the base in which those indices are counted.

The disadvantage is that the elements referred to in the **For Each** loop cannot be reassigned.

Unfortunately, the code does not work. An error is reported; see the error reported in **Figure 7-10 Script error report I**.

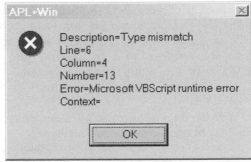

```
APL+Win                                    ×
     ⊗      Description=Type mismatch
            Line=6
            Column=4
            Number=13
            Error=Microsoft VBScript runtime error
            Context=

                         OK
```

Figure 7-10 Script error report I

The key to the problem is in the description, *Type mismatch.* With APL,

• *Type mismatch* usually means DOMAIN ERROR: this happens when numeric operations are attempted on character data.

• A runtime indexing error is usually reported as RANK ERROR or INDEX ERROR.

VBScript is unable to index the variable *NumArg*:
```
⎕wi 'XEval' 'APLEngine.Variable("NumArg")(0)'
```

The direct reference to an index makes the nature of the error more explicit—note the (...) in the description of the error, shown in **Figure 7-11 Script error report** .

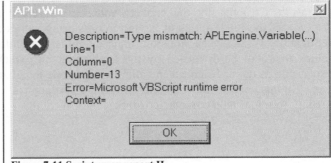

```
APL+Win                                    ×
     ⊗      Description=Type mismatch: APLEngine.Variable(...)
            Line=1
            Column=0
            Number=13
            Error=Microsoft VBScript runtime error
            Context=

                         OK
```

Figure 7-11 Script error report II

Yet, APL+Win can index the values returned:
```
(⎕wi 'XEval' 'APLEngine.Variable("NumArg")')[⎕io]
```
1

What if the variable is passed by value? In order to pass the variable by value, the subroutine requires the following modifications:
```
1 Sub SumNumArg(NumArg)
4   ' Dim NumArg
5   ' NumArg = APLEngine.Variable("NumArg")
```

A single quotation mark in lines [4] and [5] comments the code; these lines will not be executed. The revised syntax for calling the code is:
```
⎕wi 'xCodeObject.SumNumArg()' ('∆APL' ⎕wi 'Variable("NumArg")')
```

Unfortunately, this does not overcome the problem. The suspicion is that VBScript expects the argument as a variant array and the implicit translation of variable types is failing. The argument may be passed as a variant array:
```
⎕wi 'xCodeObject.SumNumArg()' ('#' ⎕wi 'VT' ('∆APL' ⎕wi
'Variable("NumArg")') 8204)
```

As shown in **Figure 7-12 VBScript feedback**, this works! The variable *NumArg* is stored as a simple vector in the server workspace, it is read from the application workspace, converted to a variant array, and passed to the VBScript code within the Script Control.

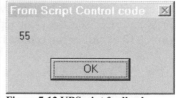

Figure 7-12 VBScript feedback

Will the original *SumNumArg* subroutine work if the server workspace is sent *NumArg* already converted to the variant data type?

```
0 0ρ'∆APL' ⎕wi 'Variable' 'NumArg' ('#' ⎕wi 'VT'(+\10/2) 8204)
```

Then the VBScript code is called with the converted data type being passed:

```
⎕wi 'xCodeObject.SumNumArg()' ('∆APL' ⎕wi 'Variable("NumArg")')
```

Unfortunately, it does not work. On examining the server workspace, *NumArg* is a simple APL+Win vector; see **Figure 7-13 Server session**. In one sense, this is preferable in that it avoids the overhead of transferring variables from the application workspace to the server workspace.

Thus, it is necessary to pass arguments directly from the application workspace—rather than from a server workspace—to a procedure in the Script Control. VBScript does not support the *Cvar* function that converts values to a variant type.

One of the implications of having to pass array values directly to VBScript is that the application workspace cannot store transient values dynamically in the server workspace—values that will be discarded by the application but which may be required for further processing at the end—and then call VBScript code to collate those values in a meaningful way. An instance of this scenario might be the application running in trace mode and needing to report all intermediate values.

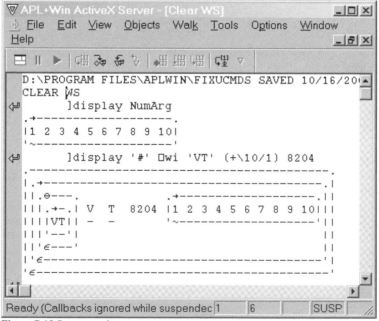

Figure 7-13 Server session

7.5.3.1 Is it APL+Win or VBScript that is failing?

The code in *Code2Sum* can be tried in **Excel** VBA with slight modification.
As shown in **Figure 7-14 Excel VBA feedback**, it works in
Excel VBA: it is able to see the APL+Win server variable as
an array and is able to index each element, perhaps because it
supports the *Cvar* function. This function converts other
types to the Variant type—it is unnecessary because Variant
is the default data type.

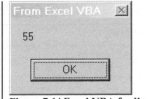

Figure 7-14 Excel VBA feedback

The steps to executing the code in **Excel** are:

● Create an **Excel** session, using:
```
⎕wself←'⍙XL' ⎕wi 'Create' 'Excel.Application' ('visible' 0)
```

● Add the code held in the APL+Win variable XLCode2Sum to a module in **Excel**:
```
      (⎕wi 'xWorkBooks.Add.xVBProject.VBComponents.Add(1). CodeModule')
⎕wi 'XAddFromString'  XLCode2Sum
```

● Initialise an instance of the APL+Win server session within **Excel**:
```
⎕wi 'XRun' 'GetSession' ('⍙APL' ⎕wi 'obj')
```

● Run the subroutine:
```
⎕wi 'XRun' 'SumNumArg'
```

As **Excel** does *not* have an *AddObject* method, an instance to the APL+Win server is
created by the *GetSession* subroutine. The actual function, *SumNumArg* is identical to the
one executed in VBScript.

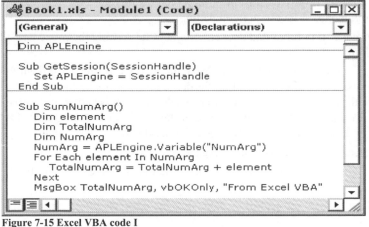

The contents of the
APL+Win variable
XLCode2Sum is
shown in the Excel
IDE in **Figure 7-15
Excel VBA code I**.

Figure 7-15 Excel VBA code I

Excel VBA returns the expected result when the APL+Win variable *NumArg* is passed
by reference also. The same modifications are made to the subroutine—to lines [1], [4] and
[5]. The modifications are shown within the editor in **Figure 7-16 Excel VBA code II**; this
was not possible to do with VBScript.

Essentially, the logic as with VBScript is involved. The difference is that the subroutine
is called directly from APL+Win, with an argument. A call to the *SumNumArg* procedure is
shown in trace mode **Figure 7-16 Excel VBA code II**. Using the **Excel** object variable, ⍙XL,
the syntax is:

● In this example, a variable from the server workspace is passed:
```
⎕wi 'XRun' 'SumNumArg' ('⍙APL' ⎕wi 'Variable( "NumArg")')
```

- In this second example, the argument is passed and by value:

```
⎕wi 'XRun' 'SumNumArg' (+\10/1)
```

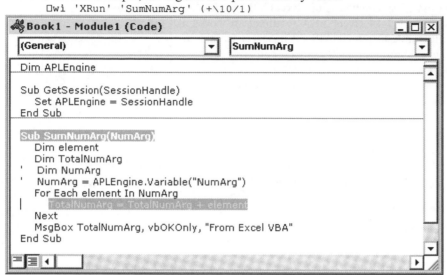

Figure 7-16 Excel VBA code II

Therefore, to answer the question, it is not APL+Win that fails but VBScript.

7.5.3.2 Excel VBA for scripting?

This investigation has been useful if only because it raises an interesting question: Use **Excel** VBA for scripting? The arguments in favour are compelling:

- The **VB** Editor (**VBE**) provides a familiar end user environment for writing the procedures; it has built-in syntax checking and colouring.

- The code can be debugged directly from APL+Win: make the **Excel** session visible, switch to the VBE (**Alt + F11**), press **F9** on the starting line for debugging and make the call from APL+Win. The editor will allow code to be traced via F8. Such a session is shown in **Figure 7-16 Excel VBA code II**. **F9** was pressed on the first line of the *SumNumArg* subroutine, **F8** was pressed several times and the current execution line is within the *for ... next* block.

- During development, the code can be tested and debugged interactively.

- Most of the VBScript functionality is available, as well as the rich set of **Excel** workbook functions.

- **Excel** has a ready-made interface for handling data; this includes viewing, formatting, printing, etc.

- **Excel** workbooks may be saved, with the code, and reused.

7.5.4 Passing arguments to built-in functions

Earlier it was noted that it was desirable to be able to pass arguments directly to VBScript; however, this applies to defined procedures only. Direct execution of built-in functions requires the name of the function and its argument to be passed as a string. With APL+Win, this can be unduly messy. For example, the syntax for the *Ucase* function is:

```
⎕wi 'XEval' 'Ucase("String to be converted")'
STRING TO BE CONVERTED
```

If the argument of the function is already held as an APL+Win variable, in the application workspace, it cannot be passed without pre-processing. One solution is:

```
Str←"String to be converted"
```

```
Fn←'Ucase("*")'
(('*'=Fn)/Fn)←⊂Str
⎕wi 'XEval' (∈Fn)
```
STRING TO BE CONVERTED

However, this will not work with functions that expect one or more of their arguments as arrays, like the Filter function. Moreover, functions that work with scalar values might need to be applied to APL+Win arrays. Consider, for example, a requirement to convert an APL+Win array to uppercase. Two approaches may be used:

• First, scalar arguments may be written to the server workspace and referred to directly.
```
'∆APL' ⎕wi 'Variable("Str")' Str
⎕wi 'XEval' 'Ucase(APLEngine.Variable("Str"))'
```
STRING TO BE CONVERTED

The problem with array variables with this syntax has already been noted.

• Second, a robust solution is to write a cover VBScript function that can return the result to the application workspace. The cover function for *Ucase* is:
```
 1 Function UpperCase(Str)
 2     If IsArray(Str) then
 3         Dim i
 4         For i = 0 to ubound(Str)
 5             Str(i) = Ucase(Str(i))
 6         Next
 7     Else
 8         Str = Ucase(Str)
 9     End If
10     UpperCase = Str
11 End Function
```
It is vital that a VBScript keyword is not used as the cover function name: in this example, *UpperCase* is used in place of *Ucase*. Note also that indexing is used to loop through the elements instead of a *for –next* construction, see lines [3] – [6]. The code is added and tested thus:
```
⎕wi 'XAddCode' UCASE
⎕wi 'xCodeObject.UpperCase' ('#' ⎕wi 'VT' ('mon' 'tue') 8204)
```
MON TUE
```
⎕wi 'xCodeObject.UpperCase' 'mon'
```
MON

If the argument is an array, as with the *Filter* function, the cover function does not require verification for the rank of the argument as in line [2] of *UpperCase*. The code for a filter cover function is:
```
1 Function RetFilter(StrArray,SearchStr,IncludeFlg)
2     RetFilter=Filter(StrArray,SearchStr,IncludeFlg,vbTextCompare)
3 End Function
```
```
⎕wi 'XAddCode' FILTER
APLArray←'System' 'Building' 'With' 'APL+Win'
⎕wi 'xCodeObject.RetFilter' ('#' ⎕wi 'VT' APLArray 8204) 'system' 1
```
System
```
⎕wi 'xCodeObject.RetFilter' ('#' ⎕wi 'VT' APLArray 8204) 'system' 0
```
Building With APL+Win

The *SearchStr* argument to this function may be a sub-string:
```
⎕wi 'xCodeObject.RetFilter' ('#' ⎕wi 'VT' APLArray  8204) 'st' 0
```
Building With APL+Win

A cover function adds flexibility in these respects, since:

• It allows the argument to be specified independently of the VBScript built-in function that it is using.

• It can be written such that it may sense whether its argument is an array and control flow accordingly—this is shown in *UpperCase* line [2].

• It allows some arguments of the built-in function to be omitted from the syntax of the cover function—this is shown in *RetFilter* where the *Filter* function's optional argument *Compare* argument is omitted and allowed to take the default value of the constant *vbTextCompare*.

7.5.4.1 Processing simple character arrays

In the previous section, the string array argument of the *RetFilter* function is passed as a nested vector converted to the Variant data type. This is consistent with passing the numeric vector *NumArg*: nesting is required with strings values in order to separate the individual values.

Excel VBA does not require the string array argument converted to a variant data type, as shown:

```
      TestInXL
 Building With      APL+Win
      ρTestInXL
3
      ⊃TestInXL
Building
With
APL+Win
```

The TestInXL function is:

```
   ∇ TestInXL
[1]   ⍝ System Building with APL+Win
[2]   ⎕wself←'⍋XL' ⎕wi 'Create' 'Excel.Application' ('visible' 0)
[3]   (⎕wi 'xWorkBooks.Add.xVBProject.VBComponents.Add(1). CodeModule')
          ⎕wi 'XAddFromString'  XLFilterCode
[4]   ⎕wi 'XRun' 'GetSession' ('⍋APL' ⎕wi 'obj')
[5]   ⎕wi 'XRun' 'RetFilter' (⊃APLArray) 'system' 0
   ∇
```

The new code added to **Excel** is shown in **Figure 7-17 Excel 'Filter' function**; it is identical to the version used with VBScript. However, it does not require explicit conversion of its arguments to the variant data type. In line [5], the array is passed as a simple two-dimensional array.

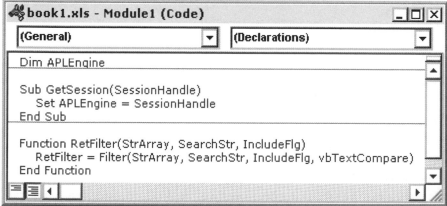

Figure 7-17 Excel 'Filter' function

What about two-dimensional arrays in VBScript?

```
APLArray←2 2ρAPLArray
```

The code shown in **Table 7-2 Runtime code I** is added to the Script Control to convert a two-dimensional nested array.

Note that the planes of a multi-dimensional array are indexed in origin 1 with the *Ubound* and *Lbound* functions.

```
1 Function UpperCase(Str)
2     Dim i,j
3     For i = 0 to Ubound(Str,1)
4         For j = 0 to Ubound(Str,2)
5             Str(i,j) = Ucase(Str(i,j))
6         Next
7     Next
8     UpperCase = Str
9 End Function
```

Table 7-2 Runtime code I

```
      □wi 'xCodeObject.UpperCase' ('#' □wi 'VT' APLArray 8204)
SYSTEM BUILDING
WITH    APL+WIN
      ρ□wi 'xCodeObject.UpperCase' ('#' □wi 'VT' APLArray 8204)
2 2
      ≡□wi 'xCodeObject.UpperCase' ('#' □wi 'VT' APLArray 8204)
2
```

The function works but hides a subtlety.

```
APLArray←4 1ρAPLArray
```

Now *APLArray* is an array with 4 rows. The code has been changed as shown in **Table 7-3 Runtime code II**—note the row indexing is hard coded in index origin 0.

```
1 Function UpperCase(Str)
2     Dim i,j
3     For i = 0 to 3
4         Str(i,0)=Ucase(Str(i,0))
5     Next
6     UpperCase = Str
7 End Function
```

Table 7-3 Runtime code II

Does this version of the UpperCase function work?

```
      □wi 'xCodeObject.UpperCase' ('#' □wi 'VT' APLArray 8204)
```

In fact, it fails because of an index error—see **Figure 7-18 Transposition error**.

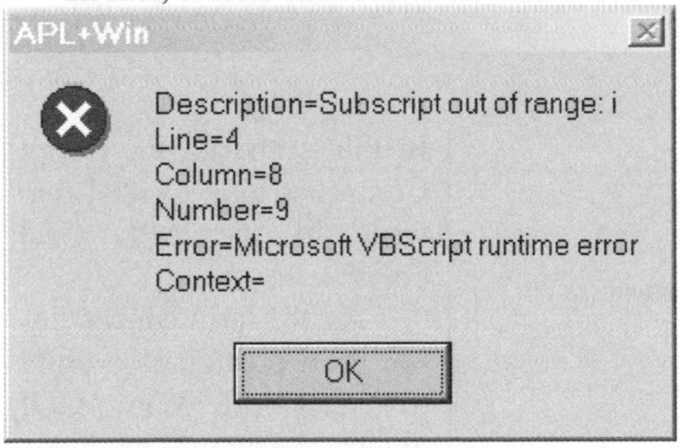

Description=Subscript out of range: i
Line=4
Column=8
Number=9
Error=Microsoft VBScript runtime error
Context=

OK

Figure 7-18 Transposition error

The reason is that VBScript sees APLArray having shape 1 4 although it has shape 4 1 in APL+Win.

The reason why the previous version worked is that VBScript and APL+Win transparently transpose array arguments to suit their respective internal representations.

The hard coded solution will work if *APLArray* is transposed explicitly:

```
      □wi 'xCodeObject.UpperCase' ('#' □wi 'VT' (⍉APLArray) 8204)
SYSTEM BUILDING WITH APL+WIN
      ρ□wi 'xCodeObject.UpperCase' ('#' □wi 'VT' (⍉APLArray) 8204)
1 4
      ⍉□wi 'xCodeObject.UpperCase' ('#' □wi 'VT' (⍉APLArray) 8204)
SYSTEM
BUILDING
WITH
APL+WIN
```

An identical function called within **Excel** does not require the explicit conversion of *APLArray* to the variant data type but does require the transposition. Both VBScript and VBA are column major, whereas APL+Win is row major.

7.5.4.2 VBScript writing to APL+Win server workspace

Since VBScript returns its native vectors as APL nested vectors or APL nested arrays to the application workspace, it is reasonable to assume that if it were to write the results to the server workspace it will write them likewise. Insert the following line to the function UpperCase, shown in **Table 7-2 Runtime code I**:

```
8.5 APLEngine.Variable("UppercaseResult") = APLArray
```

For a variation in the result, APLArray is a nested matrix of dimension 2 2. At this point, the revised function is in the Script Control. Given ⎕wself←'⍙APL':

```
     ⎕wi 'Variable("UpperCaseResult")'
 SYSTEM BUILDING
 WITH   APL+WIN
ρ⎕wi 'Variable("UpperCaseResult")'   │  ≡⎕wi 'Variable("UpperCaseResult")'
 2 2                                  │  2
```

```
      The result is always nested, even when the argument is a nested
vector. Thus, when the application workspace retrieves values added by the
Script Control to the server workspace, some conversion will be necessary
unless those values are scalar.
```

7.6 Sharing with the APL Grid object

The *XADDObject* method of the Script Control requires an object pointer. The APL Grid object can be shared just as the server workspace object. The advantage is that the APL Grid object is an object within the application workspace itself. The code is:

```
★ [36]        :case 'ShareAPLGrid'
  [37]           0 0ρ'⍙Grid' ⎕wi 'Create' 'APL.Grid'
  [38]              '⍙Grid' ⎕wi 'Set' ('xRows' 15) ('xCols' 10)
  [39]                 ⎕wi 'XAddObject' 'APLGrid' ('#' ⎕wi 'UseObject' ('⍙Grid'
      ⎕wi 'obj'))
1}[40]   :endselect
      ▽
      'ShareAPLGrid' Script θ
```

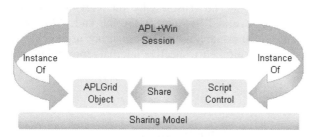

In this scenario, the Script Control is able to read *and* write to the APL Grid object.

As the grid object has not been created as a child of an APL+Win GUI object, it cannot be made visible.

Figure 7-19 Sharing the Grid with Script Control

Line [37] may be modified to create the grid object as a child of an APL+Win GUI object; this will be especially desirable if the grid object is part of the application interface and the Script Control is called upon to manipulate or validate its contents. If the grid object is part of an APL+Win GUI form or MDI form elsewhere, lines [37] -[38] must be modified:

```
      0 0ρ'AppFrm' ⎕wi 'Create' 'Form'
      0 0ρ'AppFrm.Grid' ⎕wi 'Create' 'APL.Grid'
```

Line [39] needs to be changed to use this instance of the grid object:

```
[39]    ⎕wi 'XAddObject' 'APLGrid' ('#' ⎕wi 'UseObject' ('AppFrm.Grid' ⎕wi
      'obj'))
```

The syntax of the alternative lines [39] is such that the visible and non-visible instances of the grid object are referred to by the same object pointer name, *APLGrid*, within the Script Control.

7.7 Concurrent sharing with the Script Control

This example establishes **Excel** as an automation server for two automation clients, APL+Win and the Script Control. Note the code pathway:

- APL+Win creates an instance of the Script Control—the Script Control is an automation server to APL+Win.
- The Script Control creates an instance of **Excel**—**Excel** is an automation server to the Script Control.
- APL+Win the same instance of **Excel** as an automation server—**Excel** becomes an automation server to APL+Win at the *same* time.

This model relies on the Script Control's ability to pass the object pointer of its instance of **Excel** to APL+Win.

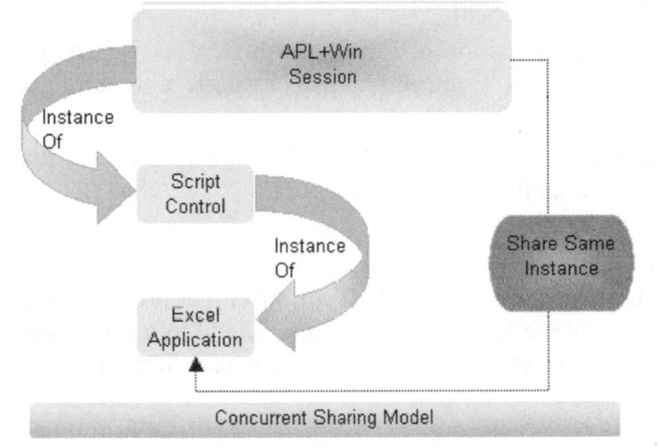

Figure 7-20 Concurrent Sharing with Script Control

APL+Win then uses this pointer to create an object of the same instance of **Excel**.

Now, **Excel** can respond to both APL+Win and the Script Control.

```
     ∇ ConcurrentShare;⎕wself
[1]    ⍝ System Building with APL+Win
[2]    ⎕wself←'∆ScrCtl' ⎕wi 'Create' 'MSScriptControl.ScriptControl'
[3]    0 0⍴⎕wi 'xLanguage' ''
[4]    0 0⍴⎕wi 'xLanguage' 'VBScript'
[5]    ⎕wi 'XExecuteStatement' 'Set xl = CreateObject("Excel.Application")'
[6]    'ShareXL' ⎕wi 'Create' (⎕wi 'XEval' 'xl')
[7]    'ShareXL' ⎕wi 'XWorkbooks.Add'
[8]    'ShareXL' ⎕wi 'xActiveWorkbook.Worksheets().Range().Value' 1 'A1:E5'
         (⍉5 5⍴50?1000)
[9]    'ShareXL' ⎕wi 'xActiveWorkbook.Worksheets().Range().Value' 1 'A7'
         (10×'ShareXL' ⎕wi 'xActiveWorkbook.Worksheets().Range().Value' 1
         'A2')
[10]   ⎕wi 'XExecuteStatement'
         'xl.ActiveWorkbook.WorkSheets(1).Range("A6").Value = 10 *
         xl.Range("A1")'
[11]   ⎕wi 'XExecuteStatement'
         'xl.ActiveWorkbook.WorkSheets(1).Range("E10").Value = 10 *
         xl.Range("E5")'
     ∇
```

The result is shown in **Figure 7-21 Excel concurrent sharing**.

The Script Control creates the instance of **Excel** in line [5]. APL+Win grabs this instance in line [6], opens a new workbook in line [7], and populates range A1:E5 with random numbers.

The Script Control changes the active workbook in lines [10] and [11].

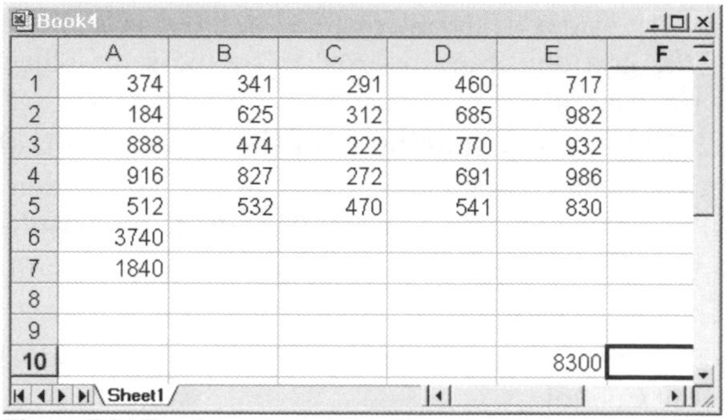

Figure 7-21 Excel concurrent sharing

Therefore, there is a choice of the syntax used to populate the **Excel** workbook. Some subtleties are worth noting:

● Whereas APL+Win can work with arrays, the Script Control works at cell level. Code needs to be added to process every cell of arrays.

● The APL+Win syntax requires the full specification of the first argument as well as the second argument—refer to line [9]. The Script Control requires a single and abridged argument for the same purpose—refer to lines [10] - [11].

● The Script Control can execute **Excel** code exactly as in **Excel** VBA with a prefix that represents the **Excel** object; APL+Win requires the code to be adapted.

7.8 APL+Win and HTML

Hyper Text Mark-up Language (**HTML**) is used to render web pages. JavaScript was designed primarily for web pages; it may be included in HTML to provide a programming capability. This is an example of the use of JavaScript on a web page, saved in a file named AUTHORS.HTM; line numbers are shown for reference and do not form part of the file:

```
1 <html>
2 <head>
3 <script language="JavaScript">
4 var html = '<table width="100%" border="1">';
```

The browser initialises the WEB page with JavaScript and creates a variable *html* that will be built up in memory and then be written to the page in a single operation. This allows the page to be updated at once rather than on a piecemeal basis.

In the example below, SQL Server is used:

```
 5 function GetTable() {
 6     var Cnn = 'Driver={SQL Server};Server=AJAY
ASKOOLUM;Database=pubs;UID=sa;PWD=;';
 7     document.write(Cnn + "<BR>");
 8     var Sql  = "SELECT * FROM AUTHORS";
 9     document.write(Sql+"<BR>");
10     var RS = new ActiveXObject("ADODB.Recordset");
11     RS.Open(Sql,Cnn);
12     RecordSet= RS.GetRows().toArray();
13     RecordSet.columns = RS.Fields.Count;
14     RecordSet.rows = RecordSet.length / RecordSet.columns;
15     document.write("Selected " + RecordSet.rows + " records with " +
RecordSet.columns + " fields.");
16     html += '<tr>\n';
17     for (i = 0;i < RS.Fields.Count;i++){
18     html += '<td>' + RS.Fields(i).Name + '</td>\n';
19     }
20     html += '</tr>';
```

```
21    RS.Close;
22    RS = null;
23    return RecordSet;
24 }
25 </script>
```

Lines [5] – [24] create a function *GetTable*; this connects to the SQL Server pubs database, retrieves the AUTHORS table, and returns a record set object. Lines [16] - [20] add the names of all column names found in the AUTHORS table to the variable *html*.

JavaScript arrays are *always* one-dimensional—its dimension being its *length* property—and are objects rather than data. Record set objects are two-dimensional; the record object is coerced into an array object in line [12]. The dimensions are simulated in lines [13] and [14], they also add the number of columns and rows in the record set as properties of the record object. These properties are used in lines [31] and [33].

Next, the JavaScript code is added:

```
26 </head>
27 <body>
28 <script language="JavaScript">
29 var RecordSet = GetTable();
30 document.write("<hr>");
31 for (var i = 0; i < RecordSet.rows; i++ ) {
32    html += '<tr>\n';
33    for (var j = 0; j < RecordSet.columns; j++) {
34      html += '<td>' + RecordSet[ i + j] + '</td>\n';
35    }
36    html += '</tr>';
37 }
38 html += '</table>\n';
39 RecordSet.Close;
40 RecordSet = null;
41 document.write(html);
42 </script>
43 </body>
44 </html>
```

Within lines [26] – [44], the function GetTable is called and each row of the record set object is written to an HTML table; the code is tagged to the variable *html*. This variable is written to the browser in line [41]. Note the following:

● The code in AUTHORS.HTM is independent of APL+Win—that is, APL+Win has no control.

● The WEB browser interprets the code in AUTHORS.HTM at runtime; that is whenever AUTHORS.HTM is opened.

● The result produced by the code is dynamic: any changes in the AUTHORS table will be reflected in the browser page whenever it is refreshed.

The result is shown in **Figure 7-22 AUTHORS table**.

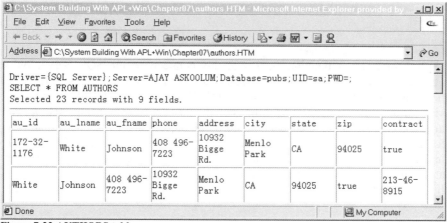

Figure 7-22 AUTHORS table

7.8.1 Creating/Displaying HTML file from APL+Win

One method for APL+Win to provide browser-based extensions is to build an HTML page in a native file and then to call on the browser to display that file. APL+Win can create a native file in which the HTML elements can be written and then the browser can be invoked to display the file, as follows:

```
         ∇ APLtoHTM;FileHwnd;html;⎕wself
    [1]    ⍝ System Building with APL+Win
{1[2]    :if 0≠⎕wcall 'PathFileExists' 'c:\authorsapl.htm'
   [3]       0 0⍴⎕wcall 'DeleteFile' 'c:\authorsapl.htm'
1}[4]    :endif
   [5]    FileHwnd←¯1+⍳/0,⎕nnums,⎕xnnums
   [6]    Cnn←'Driver={SQL Server};Server=AJAY
          ASKOOLUM;Database=pubs;UID=sa;PWD=;'
   [7]    Sql←"SELECT * FROM SALES;"
   [8]    ⎕wself←'RS' ⎕wi 'Create' 'ADODB.Recordset'
   [9]    ⎕wi 'XOpen' Sql Cnn
   [10]   'c:\authorsapl.htm' ⎕xncreate FileHwnd
   [11]   '<html><head>' ⎕nappend FileHwnd
   [12]   ('<text>',Cnn,'<br></text>') ⎕nappend FileHwnd
   [13]   ('<text>',Sql,'<br></text>') ⎕nappend FileHwnd
   [14]   '<table width="100%" border="1">' ⎕nappend FileHwnd
   [15]   '<tr>' ⎕nappend FileHwnd
   [16]   html←∊(⊂'<td>'),¨(⎕wi ¨(⊂⊂'Fields().name'),¨ (¯1++\(⎕wi
          'Fields.Count')/1)),¨⊂'</td>'
   [17]   html←html,'</tr>'
   [18]   html ⎕nappend FileHwnd
{1[19]   :while ~⎕wi 'xEOF'
   [20]      '<tr>' ⎕nappend FileHwnd
   [21]      html←⍕¨⎕wi 'XGetRows' 1
   [22]      html←∊(⊂'<td>'),¨html,¨⊂'</td>'
   [23]      html ⎕nappend FileHwnd
   [24]      '</tr>' ⎕nappend ¯1
1}[25]   :endwhile
   [26]   '</head></html>' ⎕nappend FileHwnd
   [27]   ⎕nuntie FileHwnd
   [28]   ⎕wi 'XClose'
```

```
[29]   ⎕wself←'IE' ⎕wi 'Create' 'InternetExplorer.Application'
[30]   ⎕wi 'Δhwnd' (⎕wcall 'FindWindow' 'IEFRAME' 0)
[31]   0 0ρ⎕wi 'XNavigate2' 'c:\authorsapl.htm'
[32]   ⎕wi 'visible' 1
[33]   0 0ρ⎕wcall 'DeleteFile' 'c:\authorsapl.htm'
     ∇
```

Warning: line [3] of this function is liable to fail if the file is tied within any active APL+Win session or by another application. The file is deleted in line [33]. In other words, the native file is temporary. Should it be necessary to retain the file, another convention for creating a unique file name dynamically is necessary—a common approach is to use a file name with a numeric suffix that is incremented each time the file is created.

In order to avoid multiple instances of **Internet Explorer**, it is necessary to terminate any instance thereof as soon as the user returns to the APL+Win application:

```
     ∇ FormIE;⎕wself
[1]    ⍝ System Building with APL+Win
[2]    ⎕wself←'MyFrm' ⎕wi 'Create' 'Form' 'Close'
[3]    ⎕wi 'caption' 'Component Exe'
[4]    ⎕wi 'where' 9.6 25.6 7 23.4
[5]    ⎕wi 'onFocus' '(⎕wcall ''PostMessage'' (''ΔIE'' ⎕wi ''Δhwnd'')
          ''WM_CLOSE'' 0 0)'
[6]    ⎕wi 'onWait' '⎕wi ''onFocus'' "(⎕wcall ''PostMessage'' (''ΔIE'' ⎕wi
          ''Δhwnd'') ''WM_CLOSE'' 0 0)"'
[7]    ⎕wself←'MyFrm.bn1' ⎕wi 'New' 'Button'
[8]    ⎕wi 'caption' 'Show File'
[9]    ⎕wi 'where' 1.7 6 1.5 10
[10]   ⎕wi 'onClick' 'APLtoHTM'
[11]   'MyFrm' ⎕wi 'Wait'
     ∇
```

This function creates the dialogue shown in **Figure 7-23 APL+Win form to trigger Internet Explorer**.

The Explorer session is created on clicking the **Show File** button shown in **Figure 7-23 APL+Win form to trigger Internet Explorer.**

Figure 7-23 APL+Win form to trigger Internet Explorer

The form is destroyed as soon as the user focuses on the APL form; this implies that the APL form is left visible when the event handler in line [10] creates the browser session.

The result of clicking the button is shown in **Figure 7-24 Triggered Internet Explorer.** APL+Win does not support OLE object embedding—the browser cannot be contained as a child of the APL+Win form.

Note the different handling of the record loop within AUTHORS.HTM and APLtoHTM: neither relies on the *RecordCount* property of record object, as it may not be available. The number of records is calculated as the number of values in the record object divided by the number of fields/values per record.

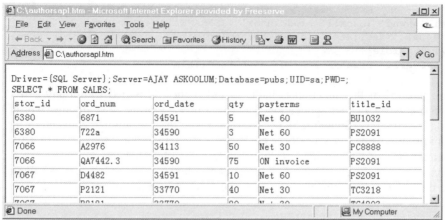

Figure 7-24 Triggered Internet Explorer

As demonstrated, basic HTML content is easily produced by or from APL+Win; some HTML content is vital to provide an APL+Win application with Windows pedigree. An implementation such as the one encapsulated within the function APLtoHTM is preferable as it is contained within the APL+Win workspace. However, two refinements might be in order, namely:

• Currently, the record set object is hard coded; it may be supplied as an argument to a similar function.

• The covering GUI element, in *FormIE*, needs to be smarter: either the HTML file can be generated in advance and then shown on the button being clicked or the button needs to be disabled as soon as it has been clicked. The generation of the HTML file introduces a delay of variable length—it depends on the size of the record object—and may trap the user into clicking the button repeatedly.

7.8.2 Taking control of HTML content

The HTML table tag provides a convenient means for APL+Win to render arrays or columnar data. Although HTML printed output is not 'friendly'—there is no control over pagination and headings—it does make the browsing of APL+Win arrays possible. The recurring issues with APL+Win need to be addressed explicitly. The APL issues are:

• Other environments, which also misrepresent dates from the APL+Win environment even when their source is external, do not readily understand APL vectors and arrays: for example, dates in ADO record sets.

• Other international settings such as currency symbols, etc., do not translate correctly because the APL+Win atomic vector does not map correctly or seamlessly to the ANSI character set.

• The APL minus is specific to APL; it needs to be translated explicitly.

• APL numeric vectors present other problems: loops and transposition may be required in the target environment.

The issue is not whether APL+Win has the techniques for overcoming these problems but whether the developer remembers that they are necessary. If these APL+Win specific issues are addressed, the Script Control enables APL+Win to appear to the user as any other application

7.8.3 APL+Win and XML

Extensible Mark-up Language (**XML**) is a mark-up language with strong similarities to HTML. Like HTML files, an XML file is a text file, that is, human-readable. However, unlike HTML files, which have a consistent layout, XML files are structured files intended primarily for data portability. The structure varies from file to file. For example, XML files containing ADO persisted record sets are different from the XML representation of the APL Grid. XML is easily manageable by APL+Win:

• As XML files are text files, the APL+Win native file functions are adequate to create them. However, APL+Win native file functions can but read XML files sequentially. This is a severe limitation since XML files can be reading a much more structure manner.

• The Microsoft XML parser is an ActiveX control that APL+Win can use to read and write XML files. This ActiveX supports XML **DOM** (Document Object Model) documents and free threaded XML DOM documents.

• There is an emerging XML:SQL standard for accessing data stored in XML format.

The Microsoft XML parser can be downloaded from the Microsoft web site. The APL+Win code to use the parser is:

```
     ⎕wself←'DOM' ⎕wi 'Create' 'MSXML.DOMDocument'
     ¯5↑⎕wself ⎕wi 'properties'
xspecified xtext xurl xvalidateOnParse xxml
     ⎕wself←'DOMF' ⎕wi 'Create' 'MSXML.FreeThreadedDOMDocument'
     ¯5↑⎕wself ⎕wi 'methods'
 XselectSingleNode XsetProperty XtransformNode XtransformNodeToObject
Xvalidate
```

This ActiveX is used in chapter 4 *Working with Windows,* and chapter 12 *Working with ActiveX, Data Object (ADO).* Chapter 2, *Advanced APL+Win Techniques* used the XML format for created persisted record objects.

Chapter 8

Windows Script Components

The Microsoft Script technology may be downloaded, subject to end user licence agreement, from Microsoft's web site: SCRIPTEN.EXE (for Windows 2000), SCR56EN.EXE (for ME and NT). The file SCRDOC56.EXE contains the documentation, and WZ10EN.EXE contains the component wizard.

Script technology opens the gateway to building systems using multiple languages. Core functions of scripting languages—especially those that are absent in APL+Win, like VBScript's date functions—may be used for rapid application development.

The Windows Script Component (**WSC**) wizard makes it very easy to construct a COM object based on a scripting language. A WSC may communicate with any language, which is COM compliant, such as APL. The wizard guides the developer with six interactive steps:

- Creates a text file containing basic XML elements; this file has extension .WSC.

- Prompts for and creates registration information.

- Prompts for the type of interface handling the script component will be using, see **Figure 8-3 Component language**. This requires the selection of a language, web, and debugging options. The web options are irrelevant for APL+Win, and the debugging option is relevant only when the Script Editor is installed.

- Prompts for the *properties* that the component will expose at runtime and creates the necessary elements in the selected scripting language.

- Prompts for the *methods* that the component will expose at runtime and creates the necessary elements in the selected scripting language.

- Prompts for the *events* that the component can fire—the calling environment provides handlers.

The recommended way to build a Script Component is to use the wizard to create the shell WSC file and then to modify its methods, properties, and events. In other words, let the wizard takes care of everything except the actual code corresponding to methods, properties, and events. Unlike usual COM components, which are binary files, the wizard creates a structured text file.

There are some crucial differences between the two types of components:

- COM components are *black box* components that do not expose their contents. These components have source files that are compiled to create a binary file. Only the binary file is distributed. Thus, a COM component (the binary file) is not editable.

- WSC components are *white box* components; their entire content is visible; as text files, they are readily editable using any text editor. These components are not compiled and the text files are distributed.

Equally, there are some similarities between the two types of components:

• Both types of components require registration on the local computer before they are available.

• Both types of components expose properties, methods, and events to any automation client, including APL+Win, found on the local computer.

8.1 Building a Script Component using JavaScript

By all accounts, JavaScript is quite easy to learn and has sound support on the Internet. It provides a readily available library of functions, which, if packaged in a component, can be readily deployed in APL+Win.

A simple example will illustrate the process of building a JavaScript component and its use in APL+Win.

Figure 8-1 Script Wizard

If installed, the wizard shows up in the program menu, shown in **Figure 8-1 Script Wizard**. Click to launch it.

Figure 8-2 Component name shows the first dialogue.

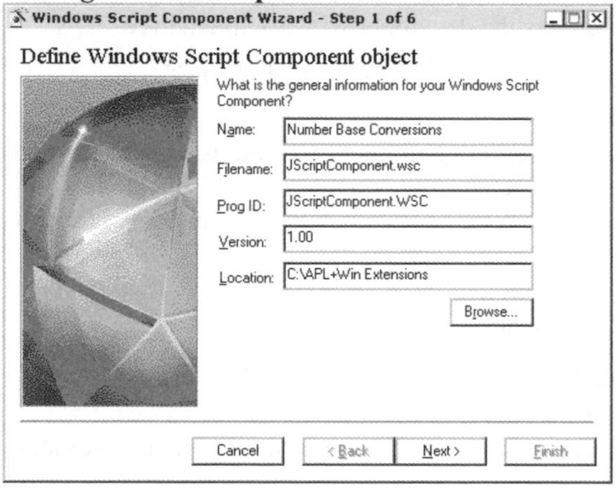

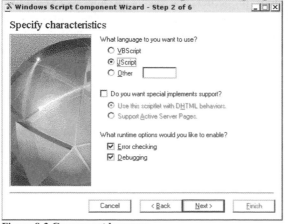

Figure 8-2 Component name

Specify the name of the component, the name of the file in which it is to be stored, its program id, version, and the location of the file. For the sake of simplicity, retain the file extension as WSC and the make the program id the same as the file name.

Specify any meaningful text as the name of the component; at runtime, the name becomes the description of the component.

Click Next. **Figure 8-3 Component language** shows the next dialogue.

Specify the scripting language. For this example, select JScript.

For APL+Win, any type of special implement support is not necessary.

Enable the runtime error checking and debugging options—this facilitates the investigation of runtime errors during development.

Figure 8-3 Component language

Click **Next**. **Figure 8-4 Component properties** shows the properties dialogue. For this initial example, properties are not required.

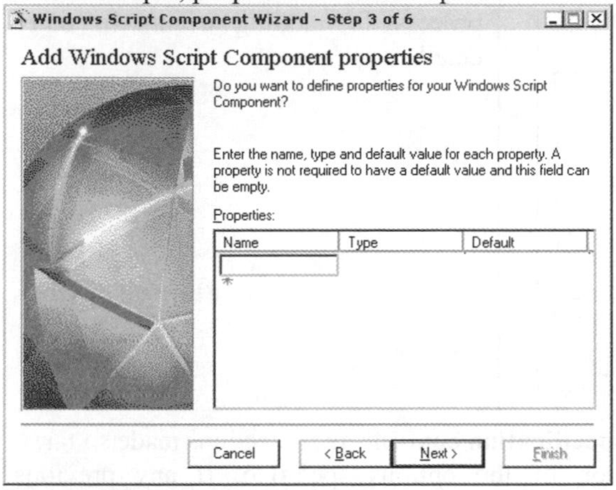

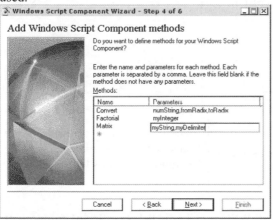

The *Name* column holds the name of a property; on receiving focus, the *Type* column transforms into a combo box with options *Read/Write, Read Only* and *Write Only* and, *Default* column holds the default value of the property, if any.

A later example discusses the process for adding properties: see Table 8-3 VBScript component shell.

Figure 8-4 Component properties

Click **Next**. **Figure 8-5 Component methods** shows the dialogue for adding methods; specify the three methods, Convert, Factorial, and Matrix:

• **Convert** takes three arguments: *numString* is the string representation of the number to be converted, *fromRadix* is the base in which *numString* is expressed, and *toRadix* is the new base in which *numString* should be expressed.

• **Factorial** has a single argument: *myInteger* is a positive integer.

• **Matrix** has two arguments: *myString* is a delimited string and *myDelimiter* is the delimiter used.

The list of methods and their corresponding parameters are listed for visual verification.

Click Next. **Figure 8-6 Component events** shows the dialogue for adding events.

Figure 8-5 Component methods

Should it be necessary to review or correct previous entries, the **Back** button shows the previous dialogue and can be used to navigate all the way to the very first dialogue.

It is necessary to ensure that the component contains all the necessary elements as the wizard only allows new components to be constructed: it is not possible to invoke the wizard with an existing component for remedial changes. Thus, if any structural aspect of a component needs to be changed, it needs to be created afresh if the wizard is used or edited manually—this requires expert knowledge.

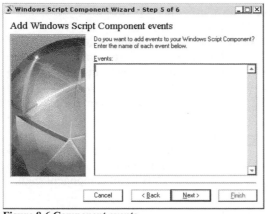

For this example, no events are required. A later example discuses the process for adding events in more detail.

Figure 8-6 Component events

Click <u>N</u>ext. **Figure 8-7 Component specification** confirms the selections made so far.

This final step provides confirmation of the options specified. If any previous specification requires change, click the **Back** button, several times if necessary, make the desired changes, and click **Next**.

Note that the wizard generates the class id for the component

Click <u>F</u>inish. **Figure 8-8 Component creation** shows the outcome of the wizard's attempt to create the component.

Figure 8-7 Component specification

The wizard confirms the successful creation of the component and its location.

Figure 8-8 Component creation

The Wizard creates this NUMBERBASE.WSC file. **Table 8-1 JavaScript component shell** shows the shell code contained in this file:

```
1 <?xml version="1.0"?>
2 <component>
3 <?component error="true" debug="true"?>
4 <registration
5      description="Number Base Conversions"
6      progid="JScriptComponent.WSC"
7      version="1.00"
8      classid="{74da0096-570e-4c75-a967-e13add178088}"
9 >
```

```
10 </registration>
11 <public>
12       <method name="Convert">
13                <PARAMETER name="numString"/>
14                <PARAMETER name="fromRadix"/>
15                <PARAMETER name="toRadix"/>
16       </method>
17       <method name="Factorial">
18                <PARAMETER name="myInteger"/>
19       </method>
20       <method name="Matrix">
21                <PARAMETER name="myString"/>
22                <PARAMETER name="myDelimiter"/>
23       </method>
24 </public>
25 <script language="JScript">
26 <![CDATA[
27 var description = new NumberBase;
28 function NumberBase()
29 {
30       this.Convert = Convert;
31       this.Factorial = Factorial;
32       this.Matrix = Matrix;
33 }
34 function Convert(numString,fromRadix,toRadix)
35 {
36       return "Temporary Value";
37 }
38 function Factorial(myInteger)
39 {
40       return "Temporary Value";
41 }
42 function Matrix(myString,myDelimiter)
43 {
44       return "Temporary Value";
45 }
46 ]]>
47 </script>
48 </component>
```

Table 8-1 JavaScript component shell

The line numbers shown are not integral to the file—they are manually included to aid reference—continuation lines are self-evident. Line [25] specifies the language as JScript—optionally, change this to JavaScript.

The wizard adds lines [27] – [33] automatically; as these lines are unnecessary, remove them from the file when defining its contents—properties, methods, and events. The final step is to register the component.

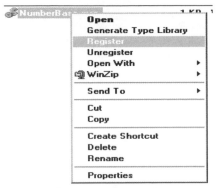

Figure 8-9 Using the Register option

Using Windows Explorer, locate the file; right click on it and the dialogue shown in **Figure 8-9 Using the Register option** appears.

The **Register** option will register the WSC component.

Subsequently, provided that only the code within the <script> and </script> block is altered—to define the Convert function—the component does not need to be Unregistered and Registered again.

However, reference to the component needs to be recreated in APL+Win after changes. The 'Register' option executes the following command:

```
regsvr32 "C:\APL+Win Extensions\NumberBase.WSC"
```

The 'Unregister' option executes the following command:

```
regsvr32 "C:\APL+Win Extensions\NumberBAse.WSC /u
```

A message box dialogue confirms the status of the registration process, shown in **Figure 8-10 Registration status.**

Figure 8-10 Registration status

In this case, **Figure 8-10 Registration status** confirms that the registration was successful. At this stage, the component is ready for APL+Win to use, as shown:

```
'JS' ⎕wi 'Create' 'NumberBase.WSC'
JS
```

The methods within the component are:

```
'JS' ⎕wi 'methods'
Close Create Defer Delete EnumEnd EnumNext EnumStart Event Exec Modify New
Open Ref Send Set SetLinks XConvert XFactorial XMatrix
```

Note that only the methods with the X prefix are available within the component—the others are APL+Win GUI methods:

```
'JS' ⎕wi '??XConvert'
XConvert method:   Result ← ⎕WI 'XConvert' numString fromRadix toRadix
[OptionalArgs]
```

8.1.1 Coding the methods

At this point, the component holds the bare shell of the methods. If invoked, the methods achieve very little, for example:

```
'JS' ⎕wi 'XConvert' 'FF' 16 10
Temporary Value
```

As expected, the function does not return a sensible value. The *description* and *progid* properties return sensible values as expected. The *Name* and *ProgID* assigned in the very first step are respectively exposed as *description* and *progid*:

```
      'JS' □wi 'description'
Number Base Conversions
      'JS' □wi 'progid'
Numberbase.WSC
```

The next step is to code the methods. Although fully coded algorithms are abundant on the Internet, a good reference on JavaScript is essential in this endeavour. The contents of this file is listed in **Table 8-2 JavaScript fully coded component**.

The file, NUMBERBASE.WSC, is a text file; it can be edited with any text editor such as NOTEPAD.EXE. It is critical that the information within the tag <registration> and </registration> is NOT altered manually: the information in this block is already written to the **Registry** and will be required when the component is used on another computer where it will also need to be registered. The line numbers are not contained in the file: they are shown as an aid to readability:

```
 1 <?xml version="1.0"?>
 2 <component>
 3 <?component error="true" debug="true"?>
 4 <registration
 5      description="Number Base Conversioons"
 6      progid="Numberbase.WSC"
 7      version="1.00"
 8      classid="{74da0096-570e-4c75-a967-e13add178088}"
 9 >
10 </registration>
11 <public>
12      <method name="Convert">
13              <PARAMETER name="numString"/>
14              <PARAMETER name="fromRadix"/>
15              <PARAMETER name="toRadix"/>
16      </method>
17      <method name="Factorial">
18              <PARAMETER name="myInteger"/>
19      </method>
20      <method name="Matrix">
21              <PARAMETER name="myString"/>
22              <PARAMETER name="myDelimiter"/>
23      </method>
24 </public>
25 <script language="JavaScript">
26 <![CDATA[
27 function Convert(numString,fromRadix,toRadix)
28 {
29   // Build 3 element Boolean string (fromRadix in range),(toRadix in
range),(numString valid in base fromRadix)
30   var ConvValid="";
31   // Is fromRadix within range 2-36 and integer
32
ConvValid=ConvValid.concat(Math.abs(fromRadix==Math.min(36,Math.max(2,Math.floor(
fromRadix)))));
33   // Is toRadix within range 2-36 and integer
34
ConvValid=ConvValid.concat(Math.abs(toRadix==Math.min(36,Math.max(2,Math.floor(to
Radix)))));
35   // Proceed if both radices are within range
36   if (3==parseInt(ConvValid,2))
37      {
38      // Check if numString is valid in base fromRadix
```

```
39
ConvValid=ConvValid.concat(Math.abs(numString==parseInt(numString,fromRadix).toSt
ring(fromRadix)));
40      // Proceed if valid
41      if (7==parseInt(ConvValid,2))
42        {
43          return parseInt(numString,fromRadix).toString(toRadix);
44        }
45      else
46        {
47          return "Error Code ".concat(parseInt(ConvValid,2));
48        }
49    }
50    else
51    {
52      return "Error Code ".concat(parseInt(ConvValid,2));
53    }
54 }
55 function Factorial(myInteger)
56 {
57 myInteger = Math.floor(myInteger);  // If the number is not an integer, round
it down.
58 if (myInteger < 0)
59    {
60    return 0;  // If the number is less than zero, reject it.
61    }
62      if (myInteger == 0)
63        {
64          return 1;  // If the number is 0, its factorial is 1.
65        }
66        else return (myInteger * Factorial(myInteger - 1));  // Otherwise,
recurse until done.
67 }
68 function Matrix(myString,myDelimiter)
69 {
70   return myString.split(myDelimiter);
71 }
72 ]]>
73 </script>
74 </component>
```

Table 8-2 JavaScript fully coded component

Now, the function *Convert* ought to convert a numeric string, *numString*, from base *fromRadix* to base *toRadix*:

```
    'JS' ⎕wi 'XConvert' 'FF' 16 10
Temporary Value
```

This is unexpected—the revised component does <u>not</u> need to be re-registered but APL+Win must create a new instance of it:

```
    'JS' ⎕wi 'Create' 'numberbase.wsc'
    'JS' ⎕wi 'Convert' 'FF' 16 10
Error Code 6
```

The function fails—this is because JavaScript uses lowercase for *numString*. Like APL+Win, JavaScript is case sensitive. Try:

```
    'JS' ⎕wi 'Convert' 'ff' 16 10
255
```

The function Convert will convert *numString* expressed in base *fromRadix* to base *toRadix* if *numString* is valid in base *fromRadix* and both *fromRadix* and *toRadix* are integer and within range 2 to 36—*numString* may be positive or negative. Some examples illustrate this:

```
    'JS' ⎕wi 'XConvert' '3eff' 16 2
11111011111111
```

```
      'JS' ⎕wi 'XConvert' '-3eff' 16 2
‾11111011111111
      'JS' ⎕wi 'XConvert' '-11111011111111' 2 16
‾3eff
      'JS' ⎕wi 'XConvert' '-11111011111111' 2 10
‾16127
```

The file SCRDOC56.EXE provides JScript documentation; JScript is Microsoft's implementation of NetScape's JavaScript. While JavaScript does not have the array processing facilities of APL+Win, there are clear affinities between the languages and the APL+Win programmer should find it easy to learn. For example, the APL+Win equivalent of 36⌊2⌈⌊fromRadix is:

```
Math.abs(fromRadix==Math.min(36,Math.max(2,Math.floor(fromRadix))))
```

JavaScript returns –1 for **TRUE**, hence the need for Math.abs. In addition, a number of functions take vectors as arguments, like APL:

```
Math.min([number1[, number2[. . . [,numberN]]]])
```

The deployment of JavaScript components has the potential to minimise the overhead in building APL+Win applications because it supports features that have to be coded from scratch in APL+Win. For instance, the one JavaScript *Convert* function saves 35*2 or 1,225 APL functions —assuming a coding convention of HEXtoDEC and DECtoHEX. *Convert* converts any base to another base within its specification; this is the range 2 to 36.

The Factorial method also works as expected. Although this is simply !*n* in APL+Win, the coding of this method illustrates the looping structure in JavaScript and the way in which the calling application can be oblivious of it:

```
      (⊂'JS') ⎕wi ¨(⊂⊂'XFactorial'),¨0 5 10
1 120 3628800
```

The matrix method produces a surprise: the result is an object rather than an array. JavaScript arrays are objects!

```
      'JS' ⎕wi 'XMatrix' 'System Building With APL+Win' ' '
15620496
```

At first sight, the result seems odd: it is an object pointer—arrays are objects in JavaScript. In order to see the result, the properties of the object need to be queried:

```
      ⊃('JS' ⎕wi 'XMatrix' 'System Building With APL+Win' ' ') ⎕wi¨ 'x0'
'x1' 'x2' 'x3'
System
Building
With
APL+Win
```

8.2 Building a Script Component using VBScript

Once again, it is advisable to use the wizard to create the WSC shell file. As it is possible to use several scripting languages, it may be advisable to define a read only property to hold the language specified. For this example, one property, *Language* and three methods will be defined—*DateValid*, *Factorial* and *Matrix*. The raw WSC file, with line numbers added manually, is shown in **Table 8-3 VBScript component shell**.

```
 1 <?xml version="1.0"?>
 2 <component>
 3 <?component error="true" debug="true"?>
 4 <registration
 5     description="VBScriptComponent"
 6     progid="VBScriptComponent.WSC"
 7     version="1.00"
 8     classid="{1a2700ee-7bce-453a-8d2d-92b62908c3a2}"
 9 >
10 </registration>
11 <public>
```

```
12        <property name="Language">
13                <get/>
14        </property>
15        <method name="DateValid">
16                <PARAMETER name="myString"/>
17        </method>
18        <method name="Factorial">
19                <PARAMETER name="myInteger"/>
20        </method>
21        <method name="Matrix">
22                <PARAMETER name="myString"/>
23                <PARAMETER name="myDelimiter"/>
24        </method>
25 </public>
26 <script language="VBScript">
27 <![CDATA[
28 dim Language
29 Language = "VBScript"
30 function get_Language()
31     get_Language = Language
32 end function
33 function DateValid(myString)
34     DateValid = "Temporary Value"
35 end function
36 function Factorial(myInteger)
37     Factorial = "Temporary Value"
38 end function
39 function Matrix(myString,myDelimiter)
40     Matrix = "Temporary Value"
41 end function
42 ]]>
43 </script>
44 </component>
```

Table 8-3 VBScript component shell

APL+Win can deploy this component only after it is registered. An instance of the component is created in the usual manner:

```
'VBS' ⎕wi 'Create' 'VBScriptComponent.wsc'
```

The list of properties includes the x*Language* property:

```
'VBS' ⎕wi 'properties'
 children class clsid data def description events instance interface links
methods modified modifystop name obj opened pending progid properties self
state suppress version xLanguage
        'VBS' ⎕wi 'xLanguage'
VBScript
```

The x*Language* property is a read only property, as intended:

```
'VBS' ⎕wi 'xLanguage' 'JavaScript'
LENGTH ERROR: Wrong number of arguments
        'VBS' ⎕wi 'xLanguage' 'JavaScript'
             ∧
```

The methods available in the component include both the *DateValid, Factorial,* and *Matrix* methods, as expected:

```
'VBS' ⎕wi 'methods'
Close Create Defer Delete EnumEnd EnumNext EnumStart Event Exec Modify New
Open Ref Send Set SetLinks XDateValid XFactorial XMatrix
```

A call to the *DateValid* method should return Temporary Value—as appearing in the code and as the method, *Convert* did in the JavaScript component. The syntax of the call can be easily verified:

```
'VBS' ⎕wi '??DateValid'
XDateValid method:
```

```
Result ← □WI 'XDateValid' MyDateString [OptionalArgs]
```
The basic syntax requires one argument:
```
    'VBS' □wi 'XDateValid' '01/01/1980'
□WI ERROR: 800A01C2 Unknown
```
This fails, unexpectedly. The Microsoft Knowledge Base asserts that this error, 800A01C2, is due to "Wrong Number of Arguments or Invalid Property Assignment". This does not explain the error. As VBSCRIPTCOMPONENT.WSC is a COM, it is deployable with any COM compliant language—does it work with VBScript?

The file DATEVALID.VBS contains the following code:

Figure 8-11 Value at runtime

```
Set vbs=CreateObject("vbscriptcomponent.wsc")
MsgBox vbs.datevalid("01/01/1980")
```

Double clicking the file name in **Windows Explorer** executes its code. It produces the expected result; see **Figure 8-11 Value at runtime**. The 800A01C2 error has plagued developers using other languages too; that is, this is not a problem specific to APL+Win. It transpires that each function or subroutine needs to be specified with a dummy argument not within the <method> and </method> block but within the <script> and </script> block for the component to work correctly with APL+Win. Therefore, a modification is necessary in line [36]:
```
36 function Factorial(MyInteger,Dummy)
       'VBS' □wi '??DateValid'
XDateValid method:
   Result ← □WI 'XDateValid' MyDateString [OptionalArgs]
```
Similar changes are necessary to the headers of the other two methods.

After these corrections and the code for *DateValid*, *Factorial*, and *Matrix*, VBSCRIPTCOMPONENT.WSC is shown in **Table 8-4 VBScript fully coded component**.
```
 1 <?xml version="1.0"?>
 2 <component>
 3 <?component error="true" debug="true"?>
 4 <registration
 5      description="VBScriptComponent"
 6      progid="VBScriptComponent.WSC"
 7      version="1.00"
 8      classid="{1a2700ee-7bce-453a-8d2d-92b62908c3a2}"
 9 >
10 </registration>
11 <public>
12      <property name="Language">
13              <get/>
14      </property>
15      <method name="DateValid">
16              <PARAMETER name="myString"/>
17      </method>
18      <method name="Factorial">
19              <PARAMETER name="myInteger"/>
20      </method>
21      <method name="Matrix">
22              <PARAMETER name="myString"/>
23              <PARAMETER name="myDelimiter"/>
24      </method>
25 </public>
26 <script language="VBScript">
```

```
27 <![CDATA[
28 dim Language
29 Language = "VBScript"
30 function get_Language()
31     get_Language = Language
32 end function
33 function DateValid(MyDateString,Dummy)
34     DateValid = IsDate(MyDateString)
35 end function
36 function Factorial(MyInteger,Dummy)
37         If IsNumeric(MyInteger) Then
38             If MyInteger <= 1 Then
39                 Factorial = 1
40             Else
41                 Factorial = MyInteger*Factorial(MyInteger-1,0)
42             End If
43         Else
44             Factorial = 0    ' Error code.
45         End If
46 end function
47 function Matrix(myString,myDelimiter,Dummy)
48             Matrix = Split(myString,myDelimiter)
49 end function
50 ]]>
51 </script>
52 </component>
```

Table 8-4 VBScript fully coded component

```
        'VBS' ⎕wi 'Create' 'vbscriptcomponent.wsc'
```

The *DateValid* method works as expected:

```
        'VBS' ⎕wi 'XDateValid' '01/01/1980'
1
        'VBS' ⎕wi 'XDateValid' '1 January 1980'
1
        'VBS' ⎕wi 'XDateValid' '30 February 1980'
0
```

So does the *Factorial* method:

```
        'VBS' ⎕wi 'XFactorial' 5
120
```

The method Factorial has been included to illustrate recursion in scripting languages—and to explain the necessity of coping with dummy variables when using VBScript. In APL, factorial is simply *!n*. The advantage of having recursive functions in a component is that the system overheads involved do not affect workspace memory.

As a consequence of this additional dummy argument being specified as the last item in the list of arguments, any recursive calls to the function found within the <script> and </script> block must, of necessity, specify a dummy value. VBScript does not allow optional arguments. However, a call to the method must specify only the arguments listed within the <PARAMETER> block.

Thus, a call to the Factorial method specifies a single argument:

```
        'VBS' ⎕wi 'XFactorial' 6
720
```

However, the recursion call must specify the dummy argument as well—refer to line [41] in **Table 8-4 VBScript fully coded component**.

8.3 About the VBS file

In the previous section, the investigation of the VBScript components used the DATEVALID.VBS file. A VBS file is an executable file; its content is executed by:

```
        ↑↑/⎕wcall 'FindExecutable' 'datevalid.vbs' 'c:\temp' (1024⍴⎕tcnul)
```

```
C:\WINDOWS\WScript.exe
```

When a file with extension VBS is double clicked in **Windows Explorer**, WSCRIPT.EXE executes the code within it. CSCRIPT.EXE executes its code if the file name if entered at the command prompt.

8.4 Runtime errors in script components

It is a matter of preference whether error trapping is included within the methods in a component, or whether data is validated within APL+Win *before* calling the methods in components. Note that the JavaScript version of *Factorial* ensures that its argument is integer and the VBScript version does not. Data validation and error trapping within APL will help to keep code in components simple. This approach lends itself to system users being able to write their own code in WSC files—error trapping will make the task more complex.

However, error checking and debugging options must be switched off before the component is deployed to users—this will help prevent accidental changes to the component.

8.5 Which Scripting language?

Clearly, it is possible to use VBScript and JScript or JavaScript as illustrated. In fact, script components may use several scripting languages and multiple scripting languages at once. The choice of language must depend on the ease with which a language delivers the required functionality to the APL+Win programmer and, realistically, on the APL+Win programmer's competency with that language.

Consider the Matrix function in each of the components using VBScript and JavaScript. First, consider the VBScript component:

```
      'VBS' ⎕wi 'XMatrix' 'jan,feb' ','
 jan feb
      a←'VBS' ⎕wi 'XMatrix' 'jan,feb' ','
      ≡a
2
      1⊃a
jan
      ⊃a
jan
feb
```

Next, consider the JavaScript component:

```
      'JS' ⎕wi 'XMatrix' 'jan,feb' ','
27359968
      'JS.Obj' ⎕wi 'Create' ('JS' ⎕wi 'XMatrix' 'jan,feb' ',')
JS.Obj
      'JS.Obj' ⎕wi 'properties'
 children class clsid data def description events instance interface links
methods modified modifystop name obj opened pending progid properties self
state suppress version x0 x1
      'JS.Obj' ⎕wi 'x0'
jan
      'JS.Obj' ⎕wi 'x1'
feb
```

The VBScript component returns the result directly to the workspace, whereas the JavaScript component returns an object; the result is available as properties of the object. This is an unnecessary overhead. In this case, the VBScript solution is preferable.

JavaScript methods are functions in the component—a function will return an explicit result only if contains a return (variable) statement within its definition. VBScript methods are either functions or subs (subroutines). The distinction is that functions may return an explicit result to the host whereas subroutines never return a result; both functions and subroutines may require arguments.

8.5.1 A wise choice?

From the point of view of the developer's self-interest, experience of JavaScript is a more portable skill but is harder than VBScript. The trade-off is between prospects in the long term and immediate productivity.

Scripting languages are used for web programming. Internet Explorer supports both VBScript and JavaScript, but a number of other browsers, such as Netscape, support JavaScript only.

8.6 Multi-language Script component

Script components come very close to enabling system building using multiple programming languages. In this example, the script component will expose the properties, methods, and events in **Table 8-5 Custom Script component:**

• The JSProp property implemented in JavaScript will fire the JSPropEvt event also written in JavaScript.

• The JSEval method written in JavaScript will fire the JSEvalEvt event upon an error being encountered.

• The JSEval method will evaluate any valid JavaScript expression and return the result—it is based on the eval() function: **eval(**codeString**)**

	JavaScript	VBScript
Property	JSProp	VSProp
Property_Event	JSPropEvt	VBSPropEvt
Method	JSEval	VBSMsg
Method_Event	JSEvalEvt	VBSMsgEvt
Property—Read only	Copyright	

Table 8-5 Custom Script component

The VBSProp property implemented in VBScript will fire the VBSPropEvt event also written in VBScript. The VBSMsg method written in VBScript will fire the VBSMsgEvt event should an error be encountered. The VBSMsg method calls the MsgBox function:

```
MsgBox(mesage[, buttons][, title][, helpfile, context])
```

Table 8-6 Multi-language fully coded component shows the content of the WSC file, after coding the properties, methods, and events.

```
 1 <?xml version="1.0"?>
 2 <component>
 3 <?component error="true" debug="true"?>
 4 <registration
 5     description="Multi-Language Script Compoment"
 6     progid="MultiLanguage.WSC"
 7     version="1.00"
 8     classid="{34f28cc2-be27-4272-823c-68777a445d0a}"
 9 >
10 </registration>
11 <public>
12     <property name="Copyright">
13         <get/>
14     </property>
15     <property name="JSProp">
16         <get/>
17         <put/>
18     </property>
```

```
19        <property name="VBSProp">
20                <get/>
21                <put/>
22        </property>
23        <method name="JSEval">
24                <PARAMETER name="myExpr"/>
25        </method>
26        <method name="VBSMsg">
27                <PARAMETER name="myMessage"/>
28        </method>
29           <event name="JSPropEvt"/>
30           <event name="JSEvalEvt"/>
31           <event name="VBSPropEvt"/>
32           <event name="VBSMsgEvt"/>
33 </public>
34 <script language="JavaScript">
35 <![CDATA[
36 var Copyright="APL+Win System Building With Components.";
37 var JSProp = "Managed by JavaScript";
38 function get_Copyright()
39 {
40      return Copyright;
41 }
42 function get_JSProp()
43 {
44      return JSProp;
45 }
46 function put_JSProp(newValue)
47 {
48      JSProp = newValue;
49         fireEvent("JSPropEvt");
50 }
51 function JSEval(myExpr)
52 {
53   try {
54        var result=eval(myExpr);
55        }
56      catch(e)
57         {
58        fireEvent("JSEvalEvt");
59           result=0;
60        }
61      finally
62         {
63           return(result);
64        }
65 }
66 ]]>
67 </script>
68 <script language="VBScript">
69 <![CDATA[
70 dim VBSProp
71 VBSProp = "Managed By VBScript"
72 function get_VBSProp()
73      get_VBSProp = VBSProp
74 end function
75 function put_VBSProp(newValue)
76      VBSProp = newValue
77         fireEvent("VBSPropEvt")
78 end function
79 function VBSMsg(myMessage,Dummy)
80         fireEvent("VBSMsgEvt")
81      VBSMsg = MsgBox(myMessage)
82 end function
83 ]]>
```

```
84 </script>
85 </component>
```

Table 8-6 Multi-language fully coded component

8.7 What is in MULTILANGUAGE.WSC?

First, register the component as illustrated in **Figure 8-9 Using the Register option** and create an instance of it, as follows:

```
'MLC' ⎕wi 'Create' 'multilanguage.wsc'
```

The properties of the component are verifiable:

```
'MLC' ⎕wi 'properties'
children class clsid data def description events instance interface links
methods modified modifystop name obj opened pending progid properties self
state suppress version xCopyright xJSProp xVBSProp
```

As are its methods:

```
'MLC' ⎕wi 'methods'
Close Create Defer Delete EnumEnd EnumNext EnumStart Event Exec Modify
New Open Ref Send Set SetLinks XJSEval XVBSMsg
```

Moreover, its events are:

```
'MLC' ⎕wi 'events'
Close Delete Modified Open Reopen Send XJSEvalEvt XJSPropEvt XVBSMsgEvt
XVBSProPEvt
```

As expected, the custom properties, methods and events are available. In the spirit of COM objects, APL+Win need not know the implementation of any properties, methods, or events.

8.7.1 Is a property read only or read/write?

The *xCopyright* property is read only as intended:

```
'MLC' ⎕wi 'Copyright'
APL+Win System Building With Components.
'MLC' ⎕wi 'Copyright' 'Plagiarise copyright'
LENGTH ERROR: Wrong number of arguments
'MLC' ⎕wi '??xCopyright'
xCopyright : ↑
  Ref: Variant
'MLC' ⎕wi '??xJSProp'
xJSProp : ↑
  Ref: Variant
  Set: Variant
```

Note that the *xCopyright* property, being read only, has only a *Ref* argument whereas the *JSProp* property, being read and write, has both a *Ref* and a *Set* argument.

8.7.2 Firing an event associated with a property

The properties JSProp and VBSProp have been coded such that they fire an event on being written—the events are JSPropEvt and VBSPropEvt, respectively—but not on being read:

```
'MLC' ⎕wi 'xJSProp' ⍝ Expect default value
Managed by JavaScript
'MLC' ⎕wi 'xVBSProp' ⍝ Expect default value
Managed By VBScript
'MLC' ⎕wi 'xJSProp' 'Implemented in JavaScript'
'MLC' ⎕wi 'xVBSProp' 'Implemented in VBScript'
```

Seemingly, the events appear to fail to fire when executing the last two statements. The reason is simple: the visibility of events is not obvious unless the host application creates handlers for them. In fact, the events did fire but there were no visible feedback since handlers were not set. In APL+Win, the events require the prefix 'on', namely:

```
     'MLC' ⎕wi 'onXJSPropEvt' "'Event handler for JSPropEvt. Can execute
APL code' (⍳5)"
     'MLC' ⎕wi 'onXVBSPropEvt' "'Event handler for VBSPropEvt. Can
execute APL code' (2×⍳5)"
```

Now the handlers have been set. An assignment of the JSProp property should fire its event handler:

```
     'MLC' ⎕wi 'xJSProp' 'Implemented in JavaScript'
 Event handler for JSPropEvt. Can execute APL code  1 2 3 4 5
     'MLC' ⎕wi 'xVBSProp' 'Implemented in VBScript'
 Event handler for VBSPropEvt. Can execute APL code  2 4 6 8 10
```

Moreover, the events do fire on writing the properties with new values. Properties and events associated with properties were set for both JavaScript and VBScript in the component MULTILANGUAGE.WSC. The VBScript and JavaScript code defining the properties and their events are identical. This example provides a powerful demonstration of the use of JavaScript, VBScript, and APL in the same context.

8.7.3 Firing an event associated with a method

The method JSEval has an associated event, JSEvalEvt, which will fire in the event of an error but not otherwise. In contrast, the method VBSMsg will trigger its associated event, VBSMsgEvt, every time. The method JSEval will execute any valid JavaScript expression and return its result. Should an error be encountered, it returns 0 AND fires its event—provided the handler has been defined:

```
     'MLC' ⎕wi 'XJSEval' 'Math.min(1,4,8,0.5,90)'
0.5
     'MLC' ⎕wi 'XJSEval' 'Math.Min(1,4,8,0.5,90)'
0
```

The latter expression encountered an error, returned 0, but did not fire its event. The expression is incorrect because of the letter *m* in the word *min* is in uppercase—as noted, JavaScript is case sensitive. The event handler can be used to verify this:

```
     'MLC' ⎕wi 'onXJSEvalEvt' "'JavaScript is case sensitive. Math.Min
should be Math.min'"
     'MLC' ⎕wi 'XJSEval' 'Math.min(1,4)'
1
     'MLC' ⎕wi 'XJSEval' 'Math.Min(1,4)'
JavaScript is case sensitive. Math.Min should be Math.min
0
```

Without an event handler, the event VBSMsgEvt does fire but its effect in undetectable:

```
     'MLC' ⎕wi 'XVBSMsg' 'Greetings APL+Win. Nice to talk to you. from
VBScript'
```

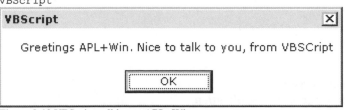

The call to the XVBMsg method generates the message shown in **Figure 8-12 VBScript talking to APL+Win**.

Figure 8-12 VBScript talking to APL+Win

The message appears in the APL+Win session.

```
     'MLC' ⎕wi 'onXVBSMsgEvt' "'Hello VBScript. Nice to hear from you.'"
     'MLC' ⎕wi 'XVBSMsg' 'Calling APL+Win. VBScript calling APL+Win → '
```

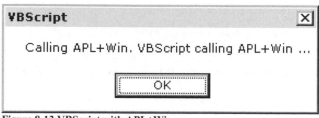

This call generates a new message shown in **Figure 8-13 VBScript with APL+Win**.

Figure 8-13 VBScript with APL+Win

The following is returned to the APL+Win session:

```
Hello VBScript. Nice to hear from you.
1
```

The message arises from the APL+Win event handler and the 1 is the value of the **OK** button, which is returned.

8.8 Finally, just because it is possible...

APL+Win is clearly COM compliant as a host. The system documentation stresses that it is also compliant as a COM server. Therefore, it must be possible to create a script component using APL+Win as a COM server, to create an instance of the script component in APL+Win or VBScript and have the APL+Win COM server return results to its host—APL+Win or VBScript. Just for fun!

The script component creates an instance of APL+Win COM server, uses VBScript to define a function, which will call on APL+Win to evaluate an expression and return its result. **Table 8-7 APL+Win Server component with corrections** shows the component.

```
 1 <?xml version="1.0"?>
 2 <component>
 3 <?component error="true" debug="true"?>
 4 <registration
 5     description="aplserver"
 6     progid="aplserver.WSC"
 7     version="1.00"
 8     classid="{6b07d128-d2b7-42f2-a89e-c8a8b5d241b9}"
 9 >
10 </registration>
11 <object id="objAPLServer" progid="APLW.WSEngine"/>
12 <public>
13     <method name="Evaluate">
14         <PARAMETER name="MyExpr"/>
15     </method>
16 </public>
17 <script language="VBScript">
18 <![CDATA[
19 function Evaluate(MyExpr,Dummy)
20     Evaluate = objAPLServer.Exec(MyExpr)
21 end function
22 ]]>
23 </script>
24 </component>
```

Table 8-7 APL+Win Server component with corrections

8.8.1 Testing with APL+Win as the server application

```
        'APL' ⎕wi 'Create' 'aplserver.wsc'
APL
        'APL' ⎕wi 'Evaluate' '(2++\(⍳10)÷2)' ⍝ ⎕io is 1
2.5 3.5 5 7 9.5 12.5 16 20 24.5 29.5
```

It works! Note that the APL+Win symbols, including ⍳, reach the component without compromise.

8.8.2 Testing with VBScript as the client application

The simplest way is to create a freestanding VBS file: a VBS file is directly executable.

Consider the VBS file, shown in **Table 8-8 APLSERVER.VBS**.

```
1 Set APL =CreateObject("aplserver.wsc")
2 MsgBox APL.Evaluate("+/+\10/1")
3 set APL=Nothing
```
Table 8-8 APLSERVER.VBS

It is not possible to specify APL characters within a VBS file; however, APL expressions constructed with ASCII characters should not present any problems—hence, 10/1, rather than ι10, is specified. An unexpected error arises; this is because VBScript too expects the dummy argument. **Figure 8-14 Script error trapping** shows the error message.

Figure 8-14 Script error trapping

Change line [2] as follows:
```
MsgBox APL.Evaluate("+/+\10/1",0)
```

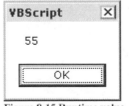

Figure 8-15 Runtime value

This cures the problem as shown in **Figure 8-15 Runtime value**: the expected result is returned.

However, having to specify a dummy argument is unsatisfactory.

The necessity of a dummy argument makes the code confusing. Therefore, the deployment of the APL WSEngine server in a component using VBScript is unviable. In contrast, a directly executable VBS file does not require a dummy argument; it executes successfully as confirmed in **Figure 8-16 VBS+APL+Win Server**. **Table 8-9 APLWIN.VBS** lists the code.

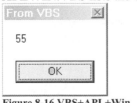

Figure 8-16 VBS+APL+Win Server

```
Set apl = CreateObject("APLW.WSEngine")
Function Evaluate(expr)
     Evaluate=apl.exec(expr)
end function
MsgBox Evaluate("+/+\10/1"),vbOK, "From VBS"
Set apl = nothing
```
Table 8-9 APLWIN.VBS

8.8.3 JavaScript as the client application

The corresponding extension for an executable file containing JavaScript code is JS. **Table 8-10 JAVA.JS** shows the content of the file JAVA.JS.

```
1 var apl = new ActiveXObject('APLW.WSEngine');
2 function JSwithAPL()
3 {
4   apl.Exec('"JavaScript using APL+Win"');
5   apl.SysCommand("XLOAD D:\\PROGRAM FILES\\APLWIN\\EXAMPLES\\DEMOAPL");
6  return(apl.Exec("+/PRIMES 500"));
7  }
8 WScript.Echo( JSwithAPL());
```
Table 8-10 JAVA.JS

Rather than use the WSC component, this code uses APL+Win directly as an ActiveX server. **Figure 8-17 JavaScript, APL+Win in the browser** shows the results.

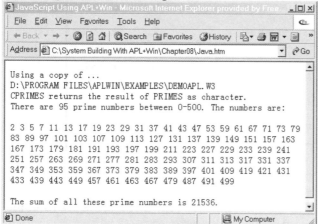

In the following HTML example, JavaScript creates an instance of APL+Win, loads a workspace, and updates the browser page.

Although this works as expected, the difficulty is in being able to write APL code in the HTML document—this is a text file.

Figure 8-17 JavaScript, APL+Win in the browser

Table 8-11 JAVA.HTM shows the code that creates the browser page:

```
 1 <HTML>
 2 <TITLE>JavaScript Using APL+Win</TITLE>
 3 <HEAD>
 4 Using a copy of ...<BR>
 5 D:\PROGRAM FILES\APLWIN\EXAMPLES\DEMOAPL.W3<BR>
 6 CPRIMES returns the result of PRIMES as character.<BR>
 7 <SCRIPT LANGUAGE="JavaScript">
 8 var apl = new ActiveXObject("APLW.WSEngine");
 9 apl.SysCommand("XLOAD C:\\DEMOAPL");
10 function JSwithAPL()
11 {
12   var sumprimes = 0;
13   var allprimes = apl.Exec("CPRIMES 500").split(' ');
14   document.write("There are " + allprimes.length + " prime numbers between 0-
500. The numbers are: <BR><BR>");
15   for (i=0;i<allprimes.length;i++){
16     sumprimes = sumprimes + parseInt(allprimes[i]);
17     document.write(allprimes[i] + " ");
18   }
19   document.write("<BR><BR>The sum of all these prime numbers is " + sumprimes +
".");
20  }
21 </SCRIPT>
22 </HEAD>
23 <BODY>
24 <SCRIPT>
```

```
25 JSwithAPL();
26 </SCRIPT>
27 </BODY>
28 </HTML>
```
Table 8-11 JAVA.HTM

Line [13] retrieves the prime numbers from APL+Win as a character vector and converts it to an array using space as a delimiter.

8.8.4 Comparing VBScript and JavaScript

JavaScript is unable to receive the array of all prime numbers directly when invoked in a JS or HTM file.

There appears to be some subtlety in the way JavaScript works in different contexts.

When APL+Win invokes JavaScript—or VBScript—directly there are no problems passing arrays.

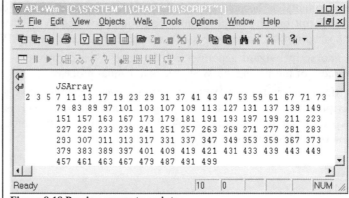

Figure 8-18 Passing arrays to scripts

Figure 8-18 Passing arrays to scripts shows the results of using the function JSArray to pass the array of all primes.

```
    ∇ Z←JSArray;⎕wself
[1]   ⍝ System Building with APL+Win
[2]   ⎕wself←'∆ScCtl' ⎕wi 'Create' 'MSScriptControl.ScriptControl'
[3]   ⎕wi 'xLanguage' ''
[4]   ⎕wi 'xLanguage' 'JavaScript'
[5]   ⎕wi 'XExecuteStatement' 'var apl = new
        ActiveXObject("APLW.WSEngine");'
[6]   ⎕wi 'XExecuteStatement' 'apl.SysCommand("XLOAD D:\\Program
        Files\\APLWIN\\Examples\\DEMOAPL");'
[7]   ⎕wi 'XExecuteStatement' 'var allprimes = apl.Exec("PRIMES 500");'
[8]   Z←⎕wi 'XEval' 'allprimes'
[9]   ⎕wi 'XExecuteStatement' 'apl.SysCommand("OFF");'
[10]  ⎕wi 'XExecuteStatement' 'apl = null'
[11]  '∆ScCtl' ⎕wi 'Delete'
    ∇
```

The equivalent function using VBScript is:

```
    ∇ Z←VBArray;⎕wself
[1]   ⍝ System Building with APL+Win
[2]   ⎕wself←'∆ScCtl' ⎕wi 'Create' 'MSScriptControl.ScriptControl'
[3]   ⎕wi 'xLanguage' ''
[4]   ⎕wi 'xLanguage' 'VBScript'
[5]   ⎕wi 'XExecuteStatement' 'set apl = CreateObject("APLW.WSEngine")'
[6]   ⎕wi 'XExecuteStatement' 'apl.SysCommand("XLOAD D:\Program
        Files\APLWIN\Examples\DEMOAPL")'
[7]   ⎕wi 'XExecuteStatement' 'allprimes = apl.Exec("PRIMES 500")'
[8]   Z←⎕wi 'XEval' 'allprimes'
```

```
[9]    ⎕wi 'XExecuteStatement' 'set apl=nothing'
[10]   'ΔScCtl' ⎕wi 'Delete'
     ∇
```

A comparison of these two functions, line by line, shows some of the differences in the two script languages.

8.9 The way forward with script components

In this chapter, the examples illustrate the use of script components—these tend to be very small—in APL+Win system building. The advantages must be self-evident. Rapid progress in building components is possible in two ways: either find scripting experts to build the components or become one! It is outside the realm of this book to teach VBScript and JavaScript but the scripting documentation provided by Microsoft is adequate for learning the scripting languages, especially for those who have learnt APL.

• The script wizard generates all the code for a component in the language selected; the process of modifying the WSC file is a manual process.

• Manual alterations to WSC files require knowledge of XML and the syntax of the selected scripting language. Until the programmer's literacy level is up to scratch, clearly the expedient tactic is to use example WSC files and to restrict modifications to the <script> and </script> block of codes only.

Chapter 9

Working with Excel

Microsoft **Excel** is probably the most popular and powerful desktop application on the Windows platform; this popularity has ensured a substantial user base and accumulated experience. **Excel** is a fully functional development environment in its own right, capable of producing highly sophisticated applications. As a COM application, it makes all its functionality available to a host language such as APL+Win.

At the very least, APL can easily deploy **Excel** to present its data and results, use the sophisticated printing control that **Excel** offers, use the graphical capabilities of **Excel** for visual presentation of data and use it as a data source. Moreover, since **Excel** communicates effectively with other tools and applications such as databases, APL may use it as a gateway for accessing data via **Excel**.

APL development can exploit **Excel**'s intrinsic functionality with little development overheads and take advantage of the large **Excel** installation base. Moreover, the presentational content of an APL application may be developed in parallel in **Excel**. A further tangible benefit in using **Excel** is that APL can make its data and results available in an easily manageable electronic form—this includes distribution via electronic mail—as workbooks.

9.1 Application or automation server

When **Excel** is launched from the menu—either by using the **Excel** shortcut or by double clicking on an XLS file—it is an application. It is visible and appears as an application within the **Task manager**. When APL creates an instance of **Excel**, **Excel** is an automation server and, by default, is not visible.

It is desirable to be able to establish the mode in which **Excel** is working and perhaps to switch the mode. For example, if APL wants to leave **Excel** running for user interaction, it needs to make it visible and switch it to run as an application. The **Excel** UserControl read/write property is **TRUE** when **Excel** is running as an application, and **FALSE**—irrespective of visibility—when running as an automation server:

```
      ⎕wself←'xl' ⎕wi 'Create' 'Excel.Application'
      ⎕wi 'xUserControl'
0
```

However, should the user interact with a visible automation server session, the value of this property is automatically changed to **TRUE** and **Excel** functions as both an application—allowing user interaction—and an automation server—allowing interaction with the automation client.

The value of this property must be noted—upon creation of the session—if the instance of **Excel** is to toggle between application and automation server mode, as it is volatile.

This can be done using the 'Names' collection in **Excel**: the names collection is available only when **Excel** has an active workbook. The active workbook may be a new workbook or an existing workbook that has been opened:

```
0 0½Œwi 'xWorkbooks.Add'                © Create
    0 0ρ⎕wi 'xWorkbooks.Open' 'd:\ajay\savedwbk.xls' ⍝ Open
```

The property may be recorded in a custom hidden name within the active workbook, thus:

```
    ⎕wi 'xNames.Add(Name,RefersTo,Visible)' 'Session' (⎕wi
'xUserControl' ) 0
```

Names within workbooks are unique; they follow the usual rules for variable names, and case insensitivity, irrespective of visibility. Unless the visible property of a name is set to **FALSE**, it is visible in the dialogue box invoked by **Insert | Name** followed by any option from the sub menu displayed. The name is saved in the workbook, and is not visible in the workbook; that is, it is not shown in a cell and therefore it is non-intrusive. Thus, the name in which the UserControl property is recorded must be unique in order to avoid logic errors. This can act as a flag and used to control programme flow within **Excel** VBA code; the state of 'User Control' may be recorded thus:

```
0 0ρ⎕wi 'xNames.Add(Name,RefersTo,Visible)' '_APL_' (⎕wi 'xUserControl') 0
```

The value of an existing name or variable may be queried:

```
    ⎕wi 'xNames("_APL_").Value'
=0
```

This returns a formula. The actual value, retaining its type, may be returned thus:

```
    ⎕wi 'xEvaluate' (⎕wi 'xNames("_APL_").Value')
0

    10+⎕wi 'xEvaluate' (⎕wi 'xNames("_APL_").Value')
10
```

Any reference to a name that does not exist causes a runtime error; therefore, unless a name is known to exist, its presence must be queried first. The value of an existing name may be changed from character to numeric and vice versa by reassignment. This is achieved either by recreating the variable or by reassigning its value—the new value must be enclosed within quotation marks, as shown:

```
    ⎕wi 'xNames.Item("_APL_").Value' 1
    ⎕wi 'xNames.Item("_APL_").Value'
="§"

    ⎕wi 'xNames.Item("_APL_").Value' "1"
    ⎕wi 'xNames.Item("_APL_").Value'
=1
```

Given the name of an APL's **Excel** object that has an active workbook, the following function can return the value of a name within it. If the name does not exist, it is created with a default value of zero, thus:

```
    '_APL_' GetXLName 'xl'
0

    ∇ Z←L GetXLName R;⎕wself
[1]   ⍝ System Buidling With APL+Win
[2]   ⎕wself←R ⍝ L=Name, R=Excel object
[3]   R←((⊂⎕wself) ⎕wi¨ (⊂⊂'xNames.Item'),¨+\(⎕wi 'xNames.Count')/1) ⎕wi¨
        ⊂'xName'
[4]   :if ~(⊂L)∊R
[5]       0 0ρ⎕wi 'xNames.Add' L 0 0
[6]   :endif
[7]   Z←⎕wi 'xEvaluate' ((⎕wi 'xNames.Item' L)⎕wi 'xValue')
    ∇
```

Note line [3]—the names collection indices are based on index origin 1. An existing name can be deleted or reassigned (made visible) by the following statements:

```
(⎕wi 'xNames.Item' '_APL_') ⎕wi 'XDelete'
0 0ρ(⎕wi 'xNames.Item' '_APL_') ⎕wi 'xVisible' 1
```

The VB/VBA version of the code is:

```
1 Function GetXLName(ByVal UsrName As String) As Variant
2     Dim i As Long
3     For i = 1 To Names.Count
4         If Names.Item(i).Name = UsrName Then
5             GetXLName = Evaluate(Names.Item(i).Value)
6             Exit For
7         End If
8     Next
9     If IsMissing(GetXLName) Then Names.Add UsrName, Str(UserControl), 0
10 End Function
```

Both the APL+Win and VB/VBA versions of this function create the name '_APL_' with a default value of 0, if it is found missing.

An **Excel** workbook may be saved, saved to another name, or changes may be abandoned—use the SaveAs method to save a new workbook, as follows:

```
0 0ρ⎕wi 'xActiveWorkbook.SaveAs(FileName)' 'c:\text.xls'
0 0ρ⎕wi 'xActiveDocument.Save'
0 0ρ⎕wi 'ActiveWorkbook.Close(SaveChanges)' 0
```

9.1.1 Using the automation flag

The UserControl property of the **Excel** application object may be used directly. However, this property is volatile in that it changes automatically on user interaction. This is a workbook and not an application property. If APL+Win creates a session and allows user interaction, the property no longer indicates that the session was created as an automation server. Therefore, it is desirable to assign the value of this property to a variable as soon as a workbook is opened. The variable may be used to restore the initial state of the property. The automation flag—UserControl or the variable that holds its initial value—can be used for two purposes: controlling program flow and controlling user interaction.

9.1.1.1 Controlling program flow

The code in **Table 9-1 Tracking user control** illustrate how a variable holding the state of the UserControl property may be used to control programme flow within APL+Win and VBA. The variable or name must be created on opening the workbook and the workbook must be saved in order to retain the value assigned. The saved state of UserControl may be used to control programme flow as follows:

APL+Win	VBA
∇ UsingFlag	1 Private Sub Workbook_Open()
[1] A System Building with APL+Win	2 ' Workbook open event
{1[2] :select '_APL_' GetXLName 'xl'	3 Select Case GetXLName("_APL_")
★ [3] :case 0	4 Case 0
[4] A automation mode	5 ' automation mode
★ [5] :case 1	6 Case 1
[6] A application mode	7 ' application mode
+ [7] :else	8 Case Else
[8] A unknown mode	9 ' unknown mode
1}[9] :endselect	10 End Select
∇	11 End Sub

Table 9-1 Tracking user control

Note the difference in the syntax of the *Select* structure in the VBA code in **Table 9-1
Tracking user control**.

9.1.1.2 Controlling event driven code

An alternative method for controlling event driven code within **Excel** is also available; this
method is suitable when there is no requirement to switch the instance of **Excel** between
application and automation server mode.

An **Excel** Add-In or workbook may contain a WORKBOOK_OPEN routine that will
run on a workbook being opened—this may be necessary to prime the workbook for a
particular purpose. It may be desirable to prevent such a routine from running—because
either the code needs to be debugged, or the routine may be time consuming— when the
workbook is opened by APL + Win. The following code will prevent the
WORKBOOK_OPEN and other event driven routines from running:

```
     ∇ DisableExcelEvents;⎕wself
[1]    ⍝ System Building with APL+Win
[2]    ⎕wself←'xl' ⎕wi 'Create' 'Excel.Application'
[3]    ⎕wi 'xApplication.EnableEvents' 0
[4]    ⎕wi 'xWorkbooks.Open' 'c:\my workbook.xls'
[5]    ⍝ ...
     ∇
```

Unless events are raised, all event handler code lay dormant. Events may be switched
off or permitted to fire by re-setting the property, thus:

```
     ⎕wi 'xApplication.EnableEvents' 1 ⍝ True or False
```

9.1.1.3 Controlling user interaction

When APL creates an **Excel** session and opens a workbook, it may make the **Excel** session
visible. Should the user interact with the visible session, the automation flag, UserControl, is
automatically set to **TRUE**. The session may be made visible for demonstration purposes or
for providing visual feedback to the user: the progress of APL+Win updating and populating
the workbook will be observable. This is achieved by making the automation server visible,
disabling user interaction and by bringing into focus the portion of the worksheet being
worked on. The code is shown below:

```
     ⎕wi 'xVisible' 1
     ⎕wi 'xInteractive' 0 ⍝ Disable interaction
     0 0⍴⎕wi 'Range("F10:H30").Select'
     ⎕wi 'xActiveWindow.Zoom' ('#' ⎕wi 'VT' ¯1 8203)
```

These lines of code have the following effects:

• The first line makes the **Excel** session visible; this can be made invisible by setting the
property to 0.

• The second line disables user interaction on the visible **Excel** session—the mouse pointer
turns to an hourglass: the session can be made interactive by setting the property to 1. This
should be used with caution since a session that is not interactive can be terminated either by
the automation client or via the **Task Manager**—the **Excel** menu becomes inaccessible.

• The third line selects a range—any contiguous range can be selected—on the active sheet.
This may be a range that APL will change programmatically. If the **Excel** session is made
visible to receive user input in a range, the interactive flag must be set to **TRUE** and, in
order to prevent the user from straying outside the permitted range, the scroll area property
needs to be confined to this range, thus:

```
     ⎕wi 'xActiveSheet.ScrollArea' (⎕wi 'xSelection.Address')
```

• The fourth line attempts to fit the selected range into the visible window by adjusting the size of each cell: the bottom right corner of the selection area is visible. Note that the default argument to the zoom method is 100, it can range from 100 to 400: a value equal to the **VB TRUE** automatically calculates the zoom factor outside the range and needs to be passed as a **VB** Boolean data type. The visual effect depends on the size of the selected range and the user's screen.

It is advisable to set the 'DisplayAlerts' property to **FALSE** in automation mode in order to prevent **Excel** from providing user feedback via independent application messages that halt **Excel** and await user interaction.

9.2 The basic structure of Excel

Excel creates *workbooks* which contains *worksheets* comprising of 65,536 *rows* and 256 *columns*, this is the system limit on worksheet size irrespective of available computer resources; each intersection is a *cell* which can contain data, either keyed in directly, derived from a formula or a link to an external data source. Rectangular clusters of cells are referred to as a *range* that may be named. Each workbook can contain code or macros in one or more modules as can each worksheet—the code is written in **VB** for Applications (**VBA**). **Excel** may also create workbook independent code modules, which may be saved in **Excel** Add-Ins. The workbooks, worksheets, ranges, cells and Add-Ins expose properties, methods and events, which APL may use to exchange data.

Excel's cell content corresponds to an APL+Win scalar value, a range may correspond to a vector, or multi-dimensional nested matrix.

9.2.1 Orphan Excel sessions

A session of **Excel** created by APL+Win as an automation server, by default, will have its visible property set to **FALSE**. If the APL+Win session is abruptly terminated, the existing automation server session still exists and is invisible. This is an orphan server session; it continues to use system resources and may have indeterminate crossover effects on an active automation session.

The **Excel** application object is the top-level object. An orphan session also arises when APL creates an instance of a lower level object in the application object hierarchy and the APL object being an instance of the application object is terminated, as shown below:

```
      ⎕wself←'xl' ⎕wi 'Create' 'Excel.application'
      ⎕wcall 'FindWindow' 'XLMAIN' 0
3496
      ⎕wi 'Workbooks>xlwb'
      ⎕wi 'Quit'
      ⎕wi 'Delete'
      ⎕wcall 'IsWindow' 3496
1
      ↑↑/⎕wcall 'GetWindowText' 3496 (256⍴⎕tcnul) 256
Microsoft Excel
```

Although the *Quit* method of the *xl* object is invoked and the *xl* object itself is deleted, the *xlwb* object continues to exist; the workbook object can only exist as an object of the hierarchy of the application object. Thus, the application object continues to exist until the *Close* method of the workbook object is invoked.

Consider the following APL+Win session started after **Excel** :

```
      ⎕wcall 'FindWindow' 'XLMAIN' 0 ⍝ Existing
852224
      ⎕wself←'xl' ⎕wi 'Create' 'Excel.Application' ⍝ First session
      ⎕wcall 'FindWindow' 'XLMAIN' 0
```

```
983280
      Owcall 'IsWindow' 852224
1
      Owcall 'IsWindow' 983280
1
      Owself←'xl' Owi 'Create' 'Excel.Application' ⍝ 2nd xl session
      Owcall 'FindWindow' 'XLMAIN' 0
1048816
      Owcall 'IsWindow' 852224
1
      Owcall 'IsWindow' 983280
0
      Owcall 'IsWindow' 1048816
1
```

After the first 'xl' object is created, there are two **Excel** sessions; the API call IsWindow confirms this. When the same object is recreated, a new handle is returned and the first instance is implicitly terminated. The statement `'xl' Owi 'Create'` implicitly deletes an existing `'xl'` object. The IsWindow API call is:

IsWindow=U(HW hwnd) LIB USER32.DLL

9.2.2 Excel Data representation

A constant source of frustration with **Excel**/APL data interchange is the difference is the way they store data. Consider a range—or an APL+Win matrix—containing the following data:

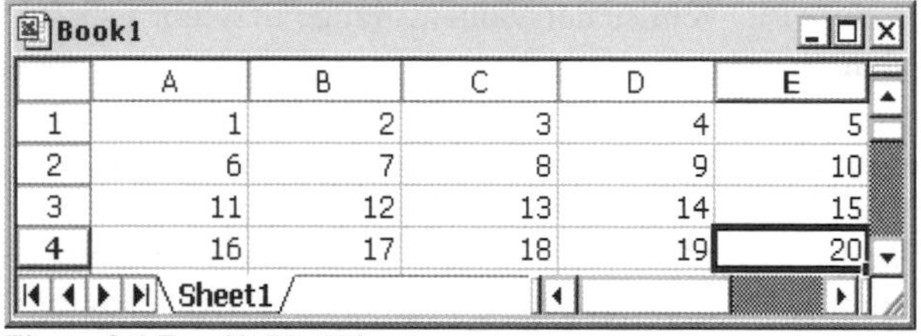

```
          O←A←4 5ρι20
      1  2  3  4  5
      6  7  8  9 10
     11 12 13 14 15
     16 17 18 19 20
```

Figure 9-1 Excel/APL+Win view

In memory, both **Excel** and APL+Win store the array in consecutive locations; however, there is a crucial difference.

9.2.2.1 Column/Row order

Conceptually, **Excel** stores this in the following order:
{1,6,11,16},{2,7,12,17},{3,8,13,18},{4,9,14,19},{5,10,15,15,20}

APL+Win, on the other hand, stores it in the following order:
{1,2,3,4,5},{6,7,8,9,10},{11,12,13,14,15},{16,17,18,19,20}

Note that the braces have been used to emphasize that **Excel** stores its data in column order whereas APL+Win stores it row order. The APL+Win transpose (⍉) primitive changes data from column order to row order and vice versa, and must be used when data is read from or written to **Excel** in order to ensure that the data is presented correctly across the two environments.

Excel is said to be column major and APL+Win is row major; a different looping construction is required for efficient access of each order, element by element. The code to populate a range or matrix of m rows by n columns in **Excel** and APL+Win is shown in **Table 9-2 Row/Column major**. Note that the range or matrix indexing is *always* by row and column, regardless of storage order. With column major storage the column index changes slower, within the loop, than the row index and vice versa for the row major order, is shown below:

Excel	APL+Win
| ```
For j = 1 to n
   For i=1 to m
      Cell(i,j).value=0
   Next j
Next i
``` | ```
:for i :in ιm
   :for j :in ιn
      matrix[i;j]←0
   :next j
:next i
``` |

Table 9-2 Row/Column major

9.2.2.2 Data shape limitations

A further point to note is that **Excel** does not understand APL+Win's multi-dimensional arrays or nested data structures. Data being written to **Excel** must be sent as simple scalars, vectors or two-dimensional arrays. Likewise, data read from **Excel** need to be converted to APL multi-dimensional arrays or nested data structures in the APL workspace, where appropriate.

9.3 APL arrays and Excel ranges

A very simplistic APL view of **Excel** is that it is a four dimensional array:

- The first dimension is the workbook index.

- The second dimension is the worksheets index.

- The third dimension is the row index.

- The fourth dimension is the column index.

Any cell may be manipulated by reference to the four coordinates. A short form notation for the workbook is *ActiveWorkbook*, which for the row and column indices is *ActiveCell*. Note that the *Sheets* count includes charts shown on a separate sheet rather than in a range on a worksheet; the *WorkSheets* count excludes chart sheets.

A range is a subset of these coordinates and like any rectangular collection of cells, in an **Excel** Workbook, is like an APL mixed array. The term array or matrix denotes a set of scalar values or elements held together in one or more dimensions:

```
      □wself←'xl' □wi 'Create' 'Excel.Application'
      □wi 'xWorkbooks.Add'
      □wi 'xActiveWorkbook.Sheets().Cells(,).Value' 1 1 1 100
      □wi 'xActiveWorkbook.Sheets().Cells(,).Value' 1 1 1
100
      □wi 'xWorkbooks().WorkSheets().Cells(,).Value' 1 1 1 1
100
```

In the present context, the latter two expressions are identical; the first uses the short form reference to the first workbook, since:

- The short form notation has the potential to make code less reliable—any workbook may be active and there may be chart sheets in any workbook; therefore, the full notation is recommended.

- The full notation is convenient for APL since any APL variable may be specified as the arguments.

Although this approach to receiving or sending data to and from **Excel** will work, it implies a looping solution, which may be time consuming. It might be desirable to work with arrays and ranges directly, because:

- An array whose dimension is, say, 10 2 in APL+Win is an array of dimension 2 10 in **Excel**. If the dimensions are required to be identical in both environments, the array needs to be transposed (⍉) on leaving and entering the workspace. This is not necessary with APLX.

• From APL+Win, the **Excel** xlR1C1 reference style is ideal for determining an array range in **Excel**. However, since **Excel** works with reference style xlA1 by default, the method ConvertFormula may be used to translate the xlR1C1 reference to xlA1 reference.

The code for transposing arrays and converting array references is:

```
     ∇ ShowTranspose;⎕wself;xlAddress
[1]    ⍝ System Building with APL+Win
[2]    ⎕wself←'∆xl' ⎕wi 'Create' 'Excel.Application'
[3]    0 0⍴⎕wi 'xWorkbooks.Add'
[4]    APLArray←2 3⍴6?1000
[5]    xlAddress←⊂5 4 GetRangeAddress ⍴APLArray
[6]    xlAddress←⎕wi (⊂'XConvertformula'),xlAddress,(⎕wi ¨'=xlR1C1' '=xlA1'
          '=xlRelative')
[7]    ⎕wi 'xWorkbooks().WorkSheets().Range().Value' 1 1 xlAddress
          (⍉APLArray)
[8]    ExcelArray←⍉⎕wi 'xWorkbooks().WorkSheets().Range().Value' 1 1
          xlAddress
[9]    ⎕wi 'xActiveworkbook.Saved' 1
[10]   ⎕wi 'XQuit'
[11]   ⎕wi 'Delete'
[12]   APLArray≡ExcelArray
     ∇
```

Line [5] calls another function to construct the xlR1C1 reference—the objective is to send an APL array of shape 2 3 to **Excel** starting at row 5 column 4 of the first worksheet in the first workbook. The xR1C1 reference is returned by the following function:

```
     ∇ Z←L GetRangeAddress R
[1]    ⍝ System Building with APL+Win
[2]    Z←(∊'R' 'C' ,¨⍕¨L),':',∊∊'R' 'C' ,¨⍕¨L+¯1+R
     ∇
```

In line [7] the array is transposed before being sent to **Excel**; in line [8] the array is transposed on being returned to APL+Win. Note that i line [9], the saved property of the active workbook is set to **TRUE**: this ensures that **Excel** does not prompt for the workbook to be saved in line [10] when the session is terminated. An alternative way of accomplishing the same effect is to close the workbook with an additional argument of **FALSE** before invoking the Quit method of the application object. Different expressions are required depending on whether there is a single, multiple or no workbooks open, as follows:

```
       ⎕wi 'xActiveWorkbook.Close(SaveChanges)' 0 ⍝ Single
       0 0⍴(⊂⎕wself) ⎕wi ¨((⊂⊂'xWorkbooks().Close(Savechanges)') ,¨(+\(⎕wi
'xWorkbooks.Count')/1)),¨0 ⍝ Multiple or none
```

The first expression will fail if there is no workbook open whereas the second expression will work regardless.

When the ShowTranspose function is run, line [12] returns **TRUE**, indicating that the APL array and the one retrieved from **Excel** are identical. The two arrays are shown in **Table 9-3 APL/Excel arrays**

ShowTranspose

APLArray	ExcelArray
322 321 66	322 321 66
615 479 69	615 479 169

Table 9-3 APL/Excel arrays

9.3.1 Writing to multiple sheets

It may be necessary to export APL+Win arrays to several worksheets in a workbook. The following function demonstrates how an array is sent to two worksheets:

```
     ∇ ShowTranspose2;⎕wself;xlAddress
[1]    ⍝ System Building with APL+Win
```

```
[2]    ⎕wself←'∆xl' ⎕wi 'Create' 'Excel.Application'
[3]    0 0⍴⎕wi 'xWorkbooks.Add'
[4]    APLArray←2 3⍴12?1000
[5]    xlAddress←⊂5 4 GetRangeAddress ⍴APLArray
[6]    xlAddress←⎕wi (⊂'XConvertformula'),xlAddress,(⎕wi ¨'=xlR1C1' '=xlA1'
       '=xlRelative')
[7]    ⎕wi 'xWorkbooks().WorkSheets().Select' 1 ('#' ⎕wi 'VT' ('Sheet1'
       'Sheet2') 8204)
[8]    ⎕wi 'xSelection.Range().Select' xlAddress
[9]    ⎕wi 'xSelection.Value' (⍉APLArray)
[10]   ExcelArray1←⍉⎕wi 'xWorkbooks().WorkSheets().Range().Value' 1
       'Sheet1' xlAddress
[11]   ExcelArray2←⍉⎕wi 'xWorkbooks().WorkSheets().Range().Value' 1
       'Sheet2' xlAddress
[12]   ⎕wi 'xActiveworkbook.Saved' 1
[13]   ⎕wi 'XQuit'
[14]   ⎕wi 'Delete'
       ∇
```

In line [7], two sheets are selected and line [8] selects the same range on both sheets— this technique is useful when sending common text to multiple sheets in a workbook. Line [9] sends an APL⁺Win array to the selected range on both sheets.

The same ranges are read in different arrays in lines [10] – [11]. The three arrays are shown **Table 9-4 Array views**.

APLArray	ExcelArray1	ExcelArray2
592 858 435 746 510 534	592 858 435 746 510 534	592 858 435 746 510 534

Table 9-4 Array views

In general, the *Select* and *Activate* methods of the **Excel** object hierarchy are time consuming and ought to be avoided. The workbook and worksheet index may be specified either by an ordinal index or by name: any reference to an index or name that is missing results in a runtime error. All range objects expose *rows* and *columns* properties. A special range in a worksheet is the area that is in use, as shown below:

```
      ⎕wi 'xActiveWorkbook.Worksheets().Cells(,).Value' 1 3 4 100
      ⎕wi 'xActiveWorkbook.Worksheets().Cells(,).Value' 1 13 24 2
      ⎕wi 'xActiveWorkbook.Worksheets().UsedRange.Address' 1
$D$3:$X$13
      ⎕wi 'xActiveWorkbook.Worksheets().UsedRange.Rows.Count' 1
11
      ⎕wi 'xActiveWorkbook.Worksheets().UsedRange.Columns.Count' 1
21
```

The UsedAddress property has some important characteristics, namely:

• It spans a rectangular collection of cells— only two cells are populated in the worksheet in the previous example but the used range is D3:X13.

• It is available when one or more workbooks are open; it is *never* empty.

• It does *not* always begin with the first physical cell, A1 or R1C1.

• The address of the used range may be used to read the entire worksheet.

• The *rows* and *columns* properties may be used to construct a looping solution, if necessary.

Excel has array formulae: highlighting a range of cells, entering a formula and pressing **Ctrl + Shift + Enter** enters array formulae. An automation client may also enter array formula in **Excel**, thus:

```
      ⎕wi 'xWorkbooks().Worksheets().Range().FormulaArray' 1 1 'A1:C2'
'={1,2,8;3,4,9}'
      ⍀⎕wi 'xWorkbooks().Worksheets().Range().Value' 1 1 'A1:C2'
 1 2 8
 3 4 9
      ⍀⎕wi 'xWorkbooks().Worksheets().Cells(,).HasArray' 1 1 1 1
1
      ⍀⎕wi 'xWorkbooks().Worksheets().Cells(,).CurrentArray.Value' 1 1 1 1
 1 2 8
 3 4 9
```

Any attempt to use array formula in overlapping ranges or a subset of the range containing an array formula result in a runtime error.

Cells in a range in **Excel** may contain data of different types—character, numeric, date or null. A *simple* APL array contains data of the same type—either numeric or character. A *nested or mixed* APL array may contain null values and indeed each cell or element may be other arrays—**Excel** does not support mixed arrays at the cell level.

Consider a simple array:

```
                                    ⎕←A←4 5ρι20
                                  1   2   3   4   5
                                  6   7   8   9  10
                                 11  12  13  14  15
                                 16  17  18  19  20
```

Individual elements of this array can be reassigned as follows:

```
      A[2;2]←⎕tcnul
      A[4;5]←⊂2 2ρ1 2 3 'System Building with APL+Win'
```

After these reassignments, the variable *A* becomes a nested array. The array can be viewed using the user command]display A; this is shown in **Figure 9-2 Nested data representation**. APLX has a system function, ⎕display, that allows nested arrays can be printed.

```
      ]display A
.-+-----------------------------------------------------------.
↓ 1    2    3    4    5                                        |
| 6    _    8    9   10                                        |
| 11   12   13   14   15                                       |
|                       .-+-----------------------------------.|
| 16   17   18   19  ↓ 1    2                                 ||
|                      |                                      ||
|                      |    .-+----------------------------.  ||
|                      | 3  |System Building with APL+Win|  ||
|                      |    '----------------------------'  ||
|                      '∊----------------------------------'|
'∊-----------------------------------------------------------'
```

Figure 9-2 Nested data representation

The integrity of the individual elements of *A* is retained after the reassignments as is confirmed by numeric operation on the elements or sub elements:

```
      A[4;5]
                                                            2
 1
 3 System Building with APL+Win
      100× (⊂1 2)⊃ (⊂4 5)⊃A
200
```

A structure such as the one at A[4;5] cannot be held in the corresponding cell E4 of the **Excel** range. Although the reference to any cell is the (*column, row*) coordinate such as E4, the standard convention for referring to a cell or element is by the (*row,column*) coordinate. In **Excel**, the standard reference can be restored with **Tools + Options** and, within the **General** tab, ticking **R1C1 Style Reference** in the **Settings options** group.

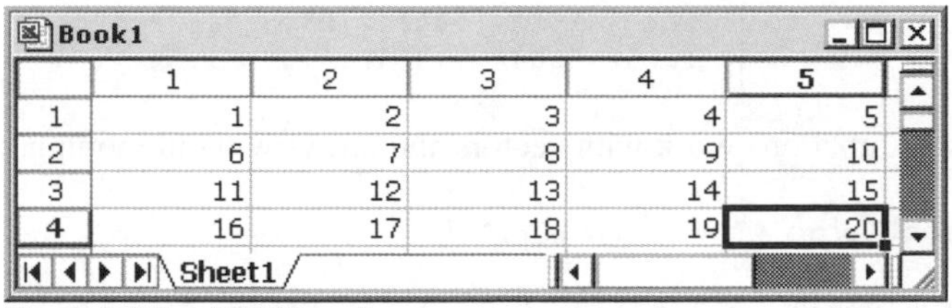

```
┌──────────────────────────────────────────────────┐
│ 🖳 Book1                               _│□│x│ ▲    │
│       1       2       3       4       5    │      │
│  1       1       2       3       4       5 │      │
│  2       6       7       8       9      10 │      │
│  3      11      12      13      14      15 │      │
│  4      16      17      18      19      20 │  ▼   │
│ │◀│◀│▶│▶│\Sheet1/              │◀│        │▶│     │
└──────────────────────────────────────────────────┘
```
Figure 9-3 Excel R1C1 mode

With the **Formulas** option ticked in the **Tools | Options**, **View** tab, the (*row,column*) reference is more apparent.

In **Excel**, a cell may be empty or hold a null value. Elements of an array or range are accessible through row and column indices, although the syntax of reference is different—in APL it is *variable*[*row;column*] whereas in **Excel** it is R*n*C*n*. **Excel** VBA always uses the (*row,column*) reference for arrays created programmatically. In summary, the following scenarios may be problematic in **Excel** to APL and APL to **Excel** data transfer:

• An **Excel** range containing mixed data types needs to be read into an APL nested array.

• An **Excel** range containing character data need to be read into an APL nested array. A simple APL character array treats each character as a column.

• An **Excel** range containing numeric (or date) data may be read into a simple array. Although **Excel** displays formatted dates, the internal representation of the dates, which are integers counting the number of days from an internal reference date, is transferred. The integer values will then require translation into dates in the workspace.

• A direct mapping of an **Excel** cell to APL array element may *not always* be possible with APL nested arrays, such as *A* shown in **Figure 9-2 Nested data representation**, after the reassignments.

• Nonetheless, in practice, it would be unnecessary to replicate arrays from one environment into the other. This makes the transfer process viable.

9.3.2 Excel Worksheet functions

Excel has a rich set of functions available. APL+Win may simply use **Excel** as a presentational environment—much like a grid object—or use **Excel** to full advantage by deploying the built-in functionality, including worksheet functions, of **Excel**:

```
     ⎕wself←'xl' ◇ A←(⎕wi 'XWorkSheetFunction') ⎕wi 'methods'
     ρ('X'=↑¨A)/A
175
     ⎕wi 'XApplication.Version'
9.0
```

Version 9.0 has 175 built-in functions; all these functions are very well documented in the help files. The advantage with these functions is that they may take APL variables as arguments. Consider a vector of numbers:

```
     ⎕←A←10?1000
869 245 541 466 778 949 320 319 65 611
```

The *n*th largest or smallest number in the vector may be established using worksheet functions:

```
⎕wi 'XWorksheetFunction.Large' A 5      ⎕wi 'XWorksheetFunction.Small' A 5
541                                     466
```

Consider the following array:

```
        ⎕←A←5 5ρ25?19992
 7345    4870     180 12687 15051
  349    5543    4930  6965 18531
 7361    2304   14532 11260  4665
```

```
10434 15227 14484 12799 16648
10256  2044  1814 18760  1904
```

A number of built-in worksheet functions work with vectors and arrays without looping. Some examples:

```
     □wi 'XWorksheetFunction.Large' A 6 ⍴ (,A)[⍒,A][6]
14532
     □wi 'XWorksheetFunction.Small' A 9 ⍴ (,A)[⍋,A][9]
4930
     □wi 'XWorksheetFunction.Roman' (□←A←10?1092)
289 300  1 122 835  755 705 529 444 167
 CCLXXXIX CCC I CXXII DCCCXXXV DCCLV DCCV DXXIX CDXLIV CLXVII
     a←90 23 78 28 90 28 17 23 45 78 90 19
     b←¯1↓,□wi 'XWorksheetFunction.Frequency'  (a,0) (a)
```

Clearly, the worksheet functions work with data from the workspace:

• The real benefit of this is that APL+Win can use **Excel**'s functionality to retrieve computed data and send this data rather than the formulae to **Excel**.

• This minimises the number of interactions with **Excel.**

In the latter example, neither of the two arguments needs to be sent to **Excel** to compute the result—instead, an APL result may be sent directly to **Excel**:

```
     □(×b)/¨a b          │ □wi 'ActiveSheet.Range().Value' 'A1:G2'
 90 23 78 28 17 45 19    │ (⍉⊃(×b)/¨a b)
  3  2  2  2  2  1  1    │
```

Of course, the worksheet functions may be inserted into **Excel**, if necessary:

```
     □wi 'xWorkbooks().Worksheets().Cells(,).FormulaR1C1' 1 1 10 5
'=Sum(10,20,30)'
     □wi 'xWorkbooks().Worksheets().Cells(,).Value' 1 1 10 5
60
     □wi 'xWorkbooks().Worksheets().Cells(,).Font.Size' 1 1 10 5 14
```

9.3.3 Excel user-defined functions

Excel allows user-defined functions to be used just as built-in functions on worksheets. The code must be added to module; the code is saved as part of the workbook unless it is explicitly removed. The code to be added may be held in a workspace variable or contained in a file. The workspace variable InRange contains **Excel** VBA code that is added in line [7] of the following function:

```
     ∇ Z←ExcelUDF;□wself
[1]    ⍝ System Building with APL+Win
[2]    □wself←'⍙xl' □wi 'Create' 'Excel.Application'
[3]    0 0⍴□wi 'xWorkbooks.Add'
[4]    □wi 'xWorkbooks().Worksheets().Range().Value' 1 1 'A1' (1?50)
[5]    □wi 'xWorkbooks().Worksheets().Range().Value' 1 1 'A2' (1?100)
[6]    □wi 'xWorkbooks().Worksheets().Range().Value' 1 1 'B1:B4' (⍴4
          1⍴4?150)
[7]    'String' AddCode InRange
[8]    □wi 'xWorkbooks().Worksheets().Range().Value' 1 1 'C1'
          '=InRange(B1,A1,A2)'
[9]    □wi 'xWorkbooks().Worksheets().Range().Value' 1 1 'C2'
          '=InRange(B2,A1,A2)'
[10]   □wi 'xWorkbooks().Worksheets().Range().Value' 1 1 'C3'
          '=InRange(B3,A1,A2)'
[11]   □wi 'xWorkbooks().Worksheets().Range().Value' 1 1 'C4'
          '=InRange(B4,A1,A2)'
[12]   Z←⍉□wi 'xWorkbooks().Worksheets().Range().Value' 1 1 'C1:C4'
```

```
[13]    ⎕wi 'xActiveWorkbook.Saved' 1
[14]    ⎕wi 'xDisplayAlerts' 0
[15]    ⎕wi 'XQuit'
[16]    '#' ⎕wi 'ReleaseObjects'
[17]    ⎕wi 'Delete'
      ∇
```

This function creates an instance of **Excel**, puts a minimum value in A1, a maximum value in A2, populates B1:B4 with random numbers based on the minimum and maximum values, adds a user-defined function, adds formulae in C1:C4 and returns the results in range C1:C4. Line [10] adds the code, from a variable or file, using the following function:

```
        ∇ L AddCode R
   [1]    ⍝ System Building with APL+Win
   [2]    0 0⍴⎕wi 'xApplication.VBE.ActiveVBProject.VBComponents.Add' 1 ⍝
          'vbext_ct_StdModule'
   [3]    ⎕wi
          'xApplication.VBE.ActiveVBProject.VBComponents.Item().Activate'
          'Module1'
{1[4]    :select L
★ [5]        :case 'String'
{2[6]            :if 2=⍴⍴R
   [7]              R←,R,⎕tcnl
2}[8]            :endif
   [9]            ⎕wi
          'xApplication.VBE.SelectedVBComponent.CodeModule.AddFromString' R
★ [10]       :case 'File'
   [11]           ⎕wi
          'xApplication.VBE.SelectedVBComponent.CodeModule.AddFromFile' R
1}[12]   :endselect
      ∇
```

The code held in the workspace variable *InRange* is:

```
1 Option Explicit
2 Function InRange(ByVal CellValue As Long, _
3               ByVal MinValue As Long, _
4               MaxValue As Long) As String
5     Select Case Sgn(MinValue - CellValue) + Sgn(MaxValue - CellValue)
6         Case -2
7             InRange = CellValue & " is greater than the maximum value in"
8         Case -1
9             InRange = CellValue & " is equal to the maximum value in"
10        Case 0
11            InRange = CellValue & " is within"
12        Case 1
13            InRange = CellValue & " is equal to the minimum value in"
14        Case 2
15            InRange = CellValue & " is lower than the minimum value in"
16    End Select
17    InRange = InRange & " range(" & MinValue & "," & MaxValue & ")"
18 End Function
```

The workspace variable must be simple and of rank 1; a new line character must delimit each line. The ability to add user-defined code depends on the **Tools | Macro | Security + Security Level** setting. Running the function **Excel**UDF produces the following output:

```
      ExcelUDF
63 is greater than the maximum value in range(15,57)
37 is within range(15,57)
2 is lower than the minimum value in range(15,57)
45 is within range(15,57)
```

9.3.4 Excel dialogues

Excel has many built-in dialogues that have the potential for saving a great deal of development time to accomplish the same functionality. The number of built-in dialogues is returned by the following expression:

```
      ⎕wi 'xApplication.Dialogs.Count'
706
      ⎕wi 'xApplication.Dialogs().Show' (⎕wi '=xlDialogFileDelete')
```

Some of these dialogues cannot be invoked from the **Excel** user interface but are available only via VBA code and to automation clients. An example is shown in **Figure 9-4 Excel delete file dialogue**.

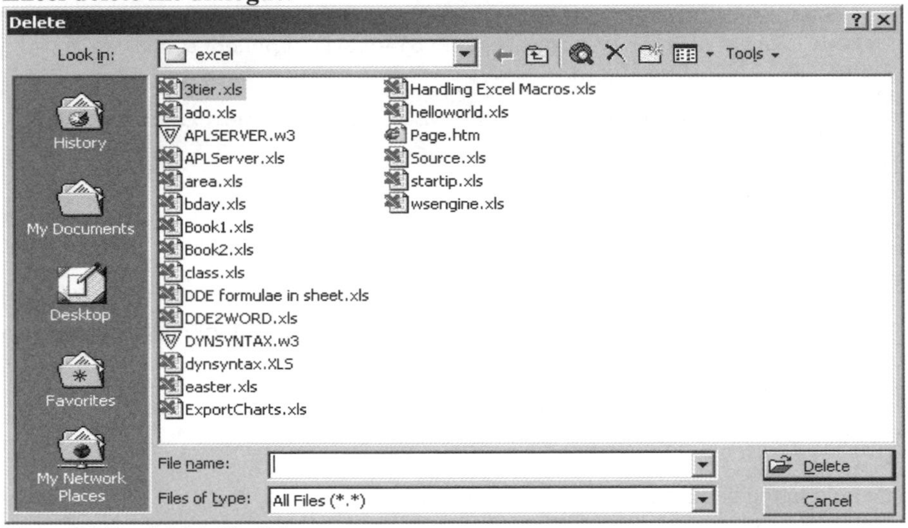

Figure 9-4 Excel delete file dialogue

Some dialogues such as xlDialogFileDelete provide all the functionality for selecting a file for deletion, including options to cancel and to confirm the operation. Unless the automation server session is visible, a dialogue may not be the top window—some care is required when deploying applications that use **Excel** dialogues.

9.3.5 Excel charts

Another feature of **Excel** is its ability to produce sophisticated charts, which update dynamically when the data underlying them changes. If it is desirable to keep historical versions of charts or to embed the charts in other applications, it is convenient to archive each chart to an individual file in a common picture format such as GIF or JPG. **Excel** stores charts individually in chart sheets or can store charts in chart sheets in the same workbook. All charts can be saved independently of the workbook, in separate files using the following function:

```
      ∇ ExportCharts;⎕wself;path;ExpFile;name;i;j
 [1]    ⍝ System Building with APL+win
 [2]    ⎕wself←'∆xl' ⎕wi 'Create' 'Excel.Application'
 [3]    0 0⍴⎕wi 'xWorkbooks.Open' 'd:\ajay\more picture exports.xls'
 [4]    path←1⌽'\',⎕wi 'xActiveWorkbook.Path'
 [5]    name←⎕wi 'xActiveWorkbook.Name'
{1[6]   :for i :in +\(⎕wi 'xActiveworkbook.Sheets.Count')/1
```

```
{2[7]        :for j :in +\(□wi
        'xActiveWorkbook.Sheets().ChartObjects.Count' i)/1
  [8]           ExpFile←path,(∊⍕¨name '-Chart' i j),'.GIF'
  [9]           0 0ρ□wi
        'xActiveWorkbook.Sheets().ChartObjects().Chart.Export' i j
        ExpFile 'GIF'
2}[10]        :endfor
1}[11]    :endfor
{1[12]    :for i :in +\(□wi 'xActiveWorkbook.Charts.Count')/1
  [13]          ExpFile←path,(∊⍕¨name '-Chart' i),'.GIF'
  [14]          0 0ρ□wi 'xActiveWorkbook.Charts().Export' i ExpFile 'GIF'
1}[15]    :endfor
  [16]    0 0ρ□wi 'xActiveWorkbook.Close(SaveChanges)' 0
  [17]    □wi 'XQuit'
  [18]    □wi 'Delete'
     ∇
```

This function creates an instance of **Excel** and opens a workbook in line [3]; the objective in using a saved workbook is that the workbook has a path and a name. The exported picture files are stored in the same path. The names of the files start with the name of the workbook followed by 'Chart': for charts held as objects in worksheets, the worksheet index and the chart index are added to the name—the file extension is GIF. Such charts are exported in lines [6] – [10]. Charts held in chart sheets are exported in lines [12] – [15]; the worksheet index is omitted from the name of the file. The workbook opened in line [3] has three charts. During the export process, **Excel** displays the dialogue shown in **Figure 9-5 Exporting Excel charts** even though the **Excel** session itself is not visible—the name of each file being created is shown dynamically, but there is no hint that **Excel** is doing all the work!

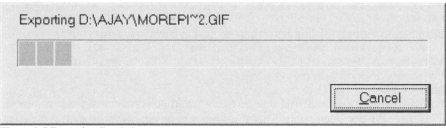

Figure 9-5 Exporting Excel charts

If the naming convention produces a file name that exists already, it is automatically deleted. The names of the files that are created are:

```
Charts in worksheets
D:\ajay\more picture exports.xls-Chart21.GIF
D:\ajay\more picture exports.xls-Chart22.GIF
Charts in chart sheets
D:\ajay\more picture exports.xls-Chart1.GIF
```

The APL+Win picture control supports BMP format pictures only—this is not an option supported within **Excel**; this control may be used to display the exported pictures after they have been converted to BMP format.

9.3.6 Excel ad hoc

The process of APL+Win populating an **Excel** workbook is an essentially ad hoc process and as such, it is fraught with practical problems. Therefore, what APL variable goes where, and how this is presented, needs to be well documented. All of the **Excel** object hierarchy is available for this purpose—this is a vast topic, outside the scope of this chapter.

9.3.6.1 Excel timer event

An **Excel** application using APL+Win as an automation server relies on the server staying 'active' for the duration of the **Excel** session. This requires that **Excel** periodically verifies the existence of the server and start it if not found. There are several contexts where it might be desirable to carry out a task periodically; for instance, it might be desirable to save the workbook at fixed intervals. **Excel** offers the onTime facility:

```
Application.OnTime Now + TimeValue("00:00:15"), "my_Procedure"
```

In order to force **Excel** into a time loop, it is necessary to bury this code in a subroutine. For example, in order to make **Excel** verify the presence of the APL server—in the APLSERVER.XLA Add-In—every ten minutes, call the CheckIt subroutine in the Workbook_Open event handler. This event handler runs as soon as the Add-In is installed: the user interaction is not halted. Note that the CheckIt subroutine sets a call to itself every ten minutes, verifies the presence of the APL+Win server and conditionally starts it, thus:

```
Sub CheckIt()
    Application.OnTime Now + TimeValue("00:10:00"), "CheckIt"
    If Not ExcelWithAPL.APLServer.IsAPLServerActive Then
        ExcelWithAPL.APLServer.StartAPLServer
    End If
End Sub
```

This type of timed calls works well if the code that is called works quickly and 'silently' and does not halt the user or show intrusive messages thereby stopping user interaction.

9.3.6.2 Hiding a sheet

Occasionally, it is desirable to hold information in a sheet and to hide it from the user in order to safeguard sensitive information. A hidden sheet may be made visible from the user interface with **Format | Sheet | Unhide**. The following hides a sheet that cannot be made visible except by recourse to VBA code:

```
    ⎕wi 'xActiveWorkbook.Worksheets().Visible' 1 (⎕wi '=xlveryhidden') ⍝
hides worksheet 1
```

In this example, worksheet 1 is hidden. It can be made visible using this code:

```
    ⎕wi 'xActiveWorkbook.Worksheets().Visible' 1 1 ⍝ From APL
    ActiveWorkbook.Worksheet(1").Visible = True ' From Immediate window
of VB Editor
```

9.3.6.3 Navigation links

An application that creates a large workbook or many workbooks must make the available information visible with documentation. This introduces maintenance overhead. An easier option if to add hyperlinks to the workbooks—these allow navigation to the correct area of a workbook or to the correct workbook via a single click of a mouse button.

Two navigation shortcuts or hyperlinks are shown in **Figure 9-6 Excel hyperlinks**.

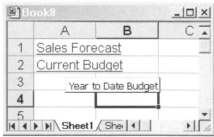

Figure 9-6 Excel hyperlinks

The hyperlinks are created by the following function:

```
    ∇ AddLink;Anchor;Address;SubAddress;ScreenTip;TextToDisplay
[1]    ⍝ System Building with APL+Win
```

```
[2]    ⎕wself←'∆xl' ⎕wi 'Create' 'Excel.Application'
[3]    0 0⍴⎕wi 'xWorkbooks.Add'
[4]    0 0⍴⎕wi 'XActiveworkbook.Worksheets().Activate' 1
[5]    Anchor←(⎕wi 'xActiveSheet.Cells(,)' 1 1) ⎕wi 'obj'
[6]    Address←''
[7]    SubAddress←'Sheet4!A1'
[8]    ScreenTip←'Sales forecast for Quarter 4'
[9]    TextToDisplay←'Sales Forecast'
[10]   0 0⍴⎕wi
          'xActivesheet.Hyperlinks.Add(Anchor,Address,SubAddress,ScreenTip,
          TextToDisplay)' Anchor Address SubAddress ScreenTip TextToDisplay
[11]   Anchor←(⎕wi 'xActiveSheet.Cells(,)' 2 1) ⎕wi 'obj'
[12]   Address←'d:\ajay\formchart.xls'
[13]   SubAddress←'Sheet1!A1'
[14]   ScreenTip←'Year to Date Budget'
[15]   TextToDisplay←'Current Budget'
[16]   0 0⍴⎕wi
          'xActivesheet.Hyperlinks.Add(Anchor,Address,SubAddress,ScreenTip,
          TextToDisplay)' Anchor Address SubAddress ScreenTip TextToDisplay
[17]   ⎕wi 'visible' 1
     ∇
```

The first link is set in lines [5] to [10]—this refers to the active workbook itself. The second link refers to another workbook; if that workbook is not open, **Excel** opens it automatically.

9.3.6.3.1 The ID property

Another, simpler, method for adding hyperlinks involves the assignment of the ID property of a cell or a range:

VBA	`ActiveCell.ID = "MyLink"` `ActiveSheet.Range("A50").ID = "Ratio"`
APL+Win	`⎕wi 'ActiveCell.ID' 'MyLink'` `⎕wi 'ActiveSheet.Range().ID' 'A50' 'Ratio'`

In order to create the hyperlinks:

• Use **File | SaveAs** and save the workbook in HTM format; that is as a WEB page.

• Manually insert code like `<A HREF="#MyLink">Quarterly earnings</A>` immediately before the <body> tag begins within the HTM file. In this example, 'Quarterly earnings' will appear as a hyperlink when the file is opened in the browser.

That the HTM file must be edited manually is a nuisance. However, this does have an advantage: the ID property may be used to insert information in a worksheet that is not visible.

9.3.6.4 Formulae without audit

APL can build the contents of a workbook in two ways:

• All computations are done in APL and sent to **Excel**: the **Excel** workbook does not contain any formulae.

• APL sends basic data to **Excel** and writes formulae in **Excel** to compute further results from the basic data. **Excel** automatically recalculates results if the basic data is changed manually or inadvertently.

Either option is viable depending on the sensitivity of the application, especially when the used range is made read only via password protection. Another option is to send the basic data to **Excel** and to get **Excel** to evaluate formulae based thereon and the results are sent back to **Excel** as raw data; this approach merits consideration if formulae are easily evaluated using **Excel** functionality:

```
□wi 'xActiveSheet.Range("A1:A10").Value' (0.325×⍳10 1⍴10?1000)
```

In the immediate window of the **VB** Editor, it is possible to write a formula that sends its results as data to the active sheet; this expression may equally be included in a subroutine— note the square brackets on the right-hand side.

```
Range("B1:B10")=[A1:A10 / Sum(A1:A10)]
```

Note that the intention is to populate 10 cells in and from **Excel**, B1:B10, without a loop. The result of this operation is seen in APL+Win:

```
2⍕¨□wi 'xActiveSheet.Range("B1:B10").Value'
0.04   0.10   0.17   0.07   0.08   0.16   0.08   0.11   0.14
```

However, this range does not have a formula although its content is based on a formula:

```
□wi 'xActiveSheet.Range("B1:B10").HasFormula'
0
```

This operation may also be carried from APL+Win: in order to make the result visibly different, it is incremented by 1:

```
□wi 'xActiveSheet.Range("B1:B10").Value' (□wi 'XEvaluate' '1+ A1:A10
/ Sum(A1:A10)')
    2⍕¨□wi 'xACtiveSheet.Range("B1:B10").Value'
  1.04   1.04   1.10   1.17   1.07   1.08   1.16   1.08   1.11   1.14
    □wi 'xActiveSheet.Range("B1:B10").HasFormula'
0
```

The target cells do not have formulae even when populated from APL+Win—note the omission of the square brackets. The approach to cell computations without loops or explicit formulae in those cells may be implemented either in VBA or directly from within APL+Win.

9.3.6.5 Visibly non-interactive

A basic requirement with any application is that the user needs feedback about progress. If APL+Win is busy populating a workbook, the workbook may be made visible so that changes are apparent to the user. This makes the process more time consuming but a more serious problem is that the user might interact with the **Excel** session thereby introducing the distinct possibility of runtime errors. Setting a property disables user interaction: this must be re-enabled on completion and before saving the workbook:

```
□wi 'xApplication.Interactive' 0
```

It is necessary to bring the appropriate range of worksheet into view. This is achieved by making the top-left-hand cell of the range the top left-hand cell in the visible window—note that the xlR1C1 reference style is required. A range is brought into focus thus:

```
□wi 'xApplication.Goto(Reference,Scroll)' 'R324C5' 1
1
```

User feedback can be provided by reiteratively making the top left-hand cell the actual cell about to be populated. The status bar can be used to provide feedback:

```
    ∇ Progress
[1]    ⍝ System Building with APL+win
[2]    □wi 'xApplication.DisplayStatusBar' 1
[3]    □wi 'visible' 1
[4]    □wi 'xApplication.Interactive' 0
[5]    □wi 'xApplication.StatusBar' 'Populating R324C5'
[6]    □wi 'xApplication.Goto(Reference,Scroll)' 'R324C5' 1
[7]    □wi 'xActiveCell.Value' 90
[8]    0 0⍴□dl 1
[9]    □wi 'xApplication.StatusBar' 'Populating R34C10'
```

```
[10]   ⎕wi 'xApplication.Goto(Reference,Scroll)' 'R34C10' 1
[11]   ⎕wi 'xActiveCell.Value' 9.89
[12]   ⎕wi 'xApplication.Interactive' 1
[13]   ⎕wi 'xApplication.StatusBar' ''
   ∇
```

The user can see the window change each time and a message appears on the application status bar. This sort of ruse is quite impressive during demonstrations but inadvisable in a deployed application.

If APL+Win creates a workbook that contains sensitive information—such as life rating or underwriting factors—it is advisable to store such information out of view and reach. Simple protection, even with a password, leaves the whole worksheet visible. This is achieved by setting a scroll area:

```
       ⎕wi 'xActivesheet.ScrollArea' 'B50:C100'
```

In this case, cell B50 needs to be the top left-hand cell in the visible **Excel** window:

```
       ⎕wi 'xApplication.Goto(Reference,Scroll)' 'R50C2' 1
1
```

Another technique is to zoom in on an area of the worksheet:

```
       ⎕wi 'xRange().Select' (⎕wi 'xActivesheet.ScrollArea')
1

       ⎕wi 'xApplication.ActiveWindow.Zoom' ('#' ⎕wi 'VT' 1 11)
```

Usually, the argument for the Zoom method is an integer. The latter example shows a special case where the application object is left to decide the scaling factor. In VBA, this is **TRUE**, that is Boolean, and APL+Win needs to supply a Boolean argument explicitly.

9.4 Object syntax

An APL+Win application is likely to update a workbook intermittently; that is, as it progresses. In this context, the chaining object syntax—as used in this chapter—is appropriate as there is only one way to work with the object hierarchy. Object creation using redirection is inadvisable as it makes it very difficult to track whether a lower level exists or not. This becomes a serious problem in a large application or in an application that creates objects within conditional code segments.

Some Windows APIs are useful for managing APL+Win/**Excel** automation.

9.4.1 The FindWindow API

This API requires either the class name of the application or its exact caption and returns the application handle:

FindWindow=H(*C lpClassName, *C lpWindowName) ALIAS FindWindowA LIB USER32.DLL

Unlike **Excel**, whose class name is *XLMAIN* regardless of version, or **Word**, whose class name is *OpusApp* regardless of version, the APL+Win class name is *not* static. The caption of an APL+Win session varies depending on the workspace name and whether it is an ActiveX server or not. Clearly, neither the class name nor the caption may be used to establish the handles of APL sessions with total confidence, especially in a distributed application. Besides, if the class name is used, only the first handle is returned—there may be multiple sessions.

Fortunately, APL+Win has a system property, which contains the dynamic windows handle:

```
       '#' ⎕wi 'hwndmain'
2404
```

The APL syntax for the FindWindow and GetClassName calls is shown below:

```
⎕wcall 'Findwindow' 'Afx:400000:8:9fe:0:3927' 0 | ⍝ using classname
2404
```

```
↑↑/⎕wcall 'GetClassName' 2404 (256ρ⎕tcnul) 256    ⍝ confirm classname
Afx:400000:8:9fe:0:3927
⎕wcall 'FindWindow' 0 'APL+Win - [Clear WS]'       ⍝ using the caption
3212
```

9.4.1.1 Closing server sessions

The problem in losing the server is not only that it may not be possible to restart it unless it is the development version. In a runtime version, the server must be re-initialised and the session bound state—at the point of failure—of the workspace, if required, can neither be saved and retrieved nor recreated easily. In addition, a runtime version of the server *may* leave a hidden and orphaned server session. Such a session will continue to hold Windows resources and application files until Windows is restarted; it is not visible.

An application whose handle is known may be terminated using either the *PostMessage* or *SendMessage* API call. These calls are identical except in one respect—they both place the message in the message queue, but unlike *SendMessage*, *PostMessage* does *not* wait for the message to be processed. The API is defined as:

PostMessage=B(HW hwnd,U wMsg,U wParam,L lParam) ALIAS PostMessageA LIB
      USER32.DLL

SendMessage=B(HW hwnd,U wMsg,U wParam,L lParam) ALIAS SendMessageA LIB
      USER32.DLL

The APL syntax for this API is:

```
⎕wcall 'SendMessage' 1852 16 0 0 ⍝ WM_CLOSE = 16
```

Where the first argument is the window handle, 1852 in this example, the second argument is WM_CLOSE (in decimal), and the remaining arguments are zeros. This call is tolerant in that if the window handle no longer existed, it does not generate an error.

Note that if the target *development* APL+Win session has the **Options** + **Confirm Exit** option switched on, i.e. is ticked, APL+Win will await user interaction prior to terminating, as shown in **Figure 9-7 End APL+Win session**.

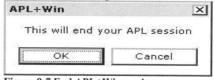

Figure 9-7 End APL+Win session

This option may be switched off in the APLW.INI or APLWR.INI file, thus:

```
[Session]
Confirm Exit=0
```

The following function will close existing **Excel** sessions if APL+Win has created instances of the application object only:

```
     ∇ CloseExcel;Z
 [1]     ⍝ System Building with APL+Win
{1[2]    :while 0≠Z←⎕wcall 'Findwindow' 'XLMAIN' 0
 [3]         Z←⎕wcall 'SendMessage' Z 16 0 0 ⍝ WM_CLOSE = 16
1}[4]    :endwhile
 [5]     ⍝ WARNING: This may loop indefinitely.
     ∇
```

If APL+Win has created an instance of an object within the hierarchy of an instance of **Excel**, this function is liable to loop indefinitely. Its use is only recommended in the development workspace where it can be readily stopped with **Ctrl + Break**. The recommendation is that the chaining syntax is used and no hierarchical object be created in APL+Win.

9.4.2 The GetWindowText API call

There were two concurrent APL+Win sessions—note their different class name. Their captions may be returned by the *GetWindowText* API call using their respective handles. The API is defined as:

```
GetWindowText=I(HW hwnd,>C lpString,I cch) ALIAS GetWindowTextA LIB USER32.DLL
      ↑↑/⎕wcall 'GetWindowText' 3212 (256⍴⎕tcnul) 256
APL+Win - [Clear WS]
      ↑↑/ ⎕wcall 'GetWindowText' 2404 (256⍴⎕tcnul) 256
APL+Win ActiveX Server - [D:\AJAY ASKOOLUM\EXCEL\APLSERVER]
```

9.4.3 Does the ActiveX server still exist?

Therefore, given the APL+Win session handle, the *GetClassName* API provides a full proof method for **Excel** to establish whether the APL+Win ActiveX server still exists. The API is defined as:

```
GetClassName=I(HW hwnd,>C lpClassName,I nMaxCount) ALIAS GetClassNameA LIB
      USER32.DLL
```

The *GetClassName* API requires a window handle, a string buffer, and the length of the string buffer as arguments. The APL syntax of the call is:

```
      ↑↑/⎕wcall 'GetClassName' ('#' ⎕wi 'hwndmain')(256⍴⎕tcnul) 256
Afx:400000:8:9fe:0:401f
```

If the APL+Win handle no longer exists, *GetClassName* returns null. If the APL+Win does not exist, **Excel** can restart it. This allows an application to establish the state of an ActiveX control dynamically at runtime.

The client, **Excel**, may call an APL function in the ActiveX server session to acquire its handle on first initialising the server:

```
   ∇ Z←GetHandle
[1]   ⍝ Return handle of current APL+Win session
[2]   Z←'#' ⎕wi 'hwndmain'
   ∇
```

The *GetClassName* API will return null if the server is no longer active.

9.5 Excel using APL+Win to retrieve APL data

So far, the approach has been for APL+Win to use **Excel** as an automation server. Equally, **Excel** may use APL+Win as an automation server. However, because of the near impossibility of sending APL+Win characters from **Excel**, and the vagaries of arrays in the two environments, a simple **Excel** With APL (**EWA**) model can be constructed as follows:

• An **Excel** Add-In is required to manage the APL+Win server.

• The server APL+Win session is required to respond to **Excel** requests. In theory, the APL+Win server is capable of processing any APL instruction.

• In order to minimize problems with the APL character set in **Excel**, it is expedient to have predefined APL functions, whose names comprise entirely of keyboard characters, which can be called from **Excel**.

9.5.1 Usability models

The success of the EWA model depends on the simplicity of its implementation. Considering the array of skills of users—from novices to **Excel** and/or APL experts—this requires particular attention for the following reasons:

• Any configuration changes to **Excel** must be non-intrusive to eliminate all risk of alienating users; **Excel** is an application in its own right.

• **Excel** users must be able to use the ActiveX server from the **Excel** user interface.

● Users must be able to use EWA without any mandatory knowledge of APL programming. Of necessity, this requires the development of supporting APL workspaces to satisfy users' requirements.

9.5.1.1 Usability model I

One approach is for APL+Win to create an instance of **Excel**, to open a 'special' workbook, to check-in the Add-In, and then to make **Excel** visible for users' to manage. This model, depicted in **Figure 9-8 Restricted automation model**, can control which template workbooks can be opened. The template workbook may contain all visual presentational aspects such as fonts, colours, borders, print ranges, page headers, footers, etc. as well as non-APL static and dynamic data.

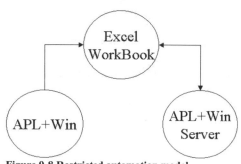

The APL data may be calculated dynamically or be stored in a workspace, an APL component file, or in a native file.

Figure 9-8 Restricted automation model

```
        ∇ AutomationModelI R;⎕wself
 [1]    ⍝ System Building with APL+Win
 [2]    ⎕wself←'∆xl' ⎕wi 'Create' 'Excel.Application'
{1[3]   :if 0=⎕wcall 'PathFileExists' (⍕R)
 [4]        0 0⍴⎕wi 'xWorkbooks.Add'
+[5]    :else
 [6]        0 0⍴⎕wi 'xWorkbooks.Open' R
1}[7]   :endif
 [8]    ⎕wi 'xApplication.Addins().Installed' 'Excel With APL+Win Server'
        0
{1[9]   :if 0=⎕wcall 'PathFileExists' 'C:\System Building With
        APL+Win\Chapter09\APLServer.xla'
 [10]       0 0⍴Msg 'Unable to find add-in',⎕tcnl,'Excel With APL+Win
        Server'
+[11]   :else
 [12]       ⎕wi 'xApplication.Addins.Add(Filename)' 'C:\System Building
        With APL+Win\Chapter09\APLServer.xla'
 [13]       ⎕wi 'xApplication.Addins().Installed' 'Excel With APL+Win
        Server' 1
1}[14]  :endif
{1[15]  :if 0=⎕wcall 'PathFileExists' (⍕R)
{2[16]      :if 0=⎕wi 'xApplication.Dialogs().Show' (⎕wi
        '=xlDialogSaveAs')
 [17]           0 0⍴Msg 'You failed to save the workbook.'
2}[18]      :endif
1}[19]  :endif
 [20]   ⎕wi 'xApplication.UserControl' 1
 [21]   ⎕wi 'visible' 1
        ∇
```

In lines [3] - [7], the workbook specified as the right-hand argument is opened if it exists otherwise a new workbook is created. Line [8] removes the Add-In and it is added anew in line [13]—this ensures that the events in the Add-In are triggered. If the Add-In is checked in and an existing workbook is opened, the events in the Add-In do not file by default. In lines [20] – [23], the APL+Win session returns control to **Excel** and terminates its instance of it, leaving the user to manage the **Excel** session. Note that the instance of **Excel**, ∆xl, still exists for subsequent use. The sequential processes are:

• APL+Win starts the **Excel** session.

• APL+Win opens the desired **Excel** workbook.

• APL+Win can check the Add-In

• APL+Win updates all links, replaces underlying APL+Win based and other calculation formulae, and saves the workbooks under new names:

```
'∆xl' ⎕wi 'xApplication.CalculateFull'
'∆xl' ⎕wi 'XRun' 'CopyValues'
'∆xl'⎕wi 'xActiveWorkbook.SaveAs(FileName)' filename
```

Alternatively, APL+Win may leave the workbooks open and users are responsible for managing the workbooks.

This mode of use is ideal for refreshing an existing   application vendor supplied—workbook that already contains all the links to APL+Win as well as native **Excel** logic such as charts etc. In other words, APL+Win uses **Excel** as a reporting tool to present its data and results.

9.5.1.2 Usability model II

A more general-purpose approach, depicted in **Figure 9-9 EWA Model**, is to allow users autonomy in starting **Excel** and on whether they want to use the APL+Win Add-In.

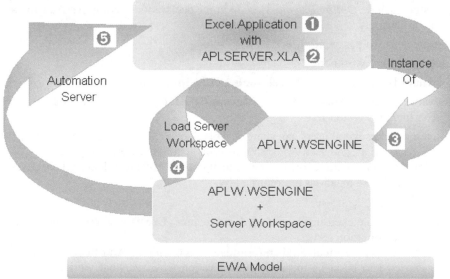

Figure 9-9 EWA Model

In this mode, users are responsible for all aspects of interaction:

- User starts **Excel**.

- User opens an existing workbook or chooses to work in the active workbook.

- User must check-in the APSERVER.XLA Add-In; this should start the APL+Win server session and load the appropriate workspace.

- User must recalculate workbook.

- User must save workbook.

- Users must maintain a library of reference workbooks that include formulae, formatting, printer settings, etc.

- The reference workbooks must be updated and saved under different names and without formulae.

In this mode of use, the vital responsibility for workbook version control procedures rests with users.

9.6 The Excel Add-In

A custom **Excel** Add-In provides the bridge between **Excel** as an automation client using APL+Win as an automation server. An Add-In is a workbook that does not expose its workbook or its code—macros—to the user in **Tools | Macro | Macros**. A macro is a function or subroutine.

The difference between a function and a subroutine is that the former returns an explicit result and the latter does not—they can both have zero or more arguments. If selected functions or subroutines are to be hidden or their scope is to be restricted to the module in which they occur, use *Private* rather than *Public* when declaring them; by default, all declarations are *Public*. An alternative mechanism for hiding subroutines is to declare them with arguments. A global method of hiding all macros is to declare the module in which they appear as private—include *Option Private Module* in the declaration section. An Add-In has several advantages, namely:

- It may be protected by a password.

- It is independent of any workbook.

- It may be switched on or off via the **Excel Tools** + **Add-Ins** menu.

- It hides its names from the **Tools | Macro | Macros** (or **ALT** + **F8**) dialogue box, thereby keeping the user interface clutter free.

- If stored on a network, several users can share it. This also simplifies its maintenance.
An Add-In may be saved at, and used from, any location including a network drive.

A convenient *default* location for storing Add-Ins is one of the folders where **Excel** stores other Add-Ins. These locations are returned by:

```
Application.LibraryPath
Application.UserLibraryPath
```

An Add-In that is checked-in is re-loaded every time **Excel** is started either as an application or as an automation server.

When a workbook is saved with Add-In formulae—even with the Add-In checked out—the Add-In is expected to be at its initial/original location whenever that workbook is re-opened and the Add-In is checked in.

Use the **Browse** button to locate the Add-In file, see **Figure 9-10 Add-In dialogue**. However, do not re-locate an Add-In after it has been used.

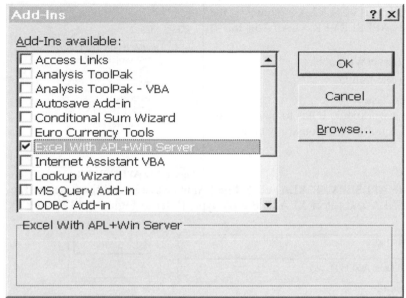

Figure 9-10 Add-In dialogue

9.6.1 Add-In visiblility

In order to prevent the Add-In from interfering with workbooks that are not **Excel +
APL+Win** ActiveX server workbooks, the Add-In must be smart to stay switched on when
required and to switch off when not.

At a minimum, one module containing at least **one *public*** function is required. **Excel**
will expose such functions as user-defined functions when the Add-In is switched on.

The list of user-defined functions is visible when **Insert + Function...** are clicked and
the **User-defined** category is highlighted. A tooltip for each function in the *Function name*
box is shown upon the function being highlighted; this comprises of the function's syntax.
Refer to **Figure 9-11 Excel user-defined functions**.

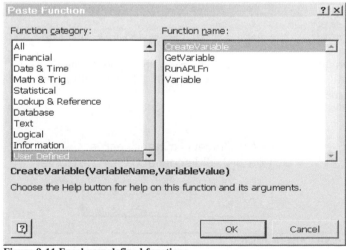

Unfortunately, this tip
fails to indicate the type
of the arguments required
or whether any argument
is optional. The **Help**
button is usually
inoperative for user-
defined functions—unless
a custom help file is
specified in the project
properties. Such a help
file can be used to
document user-defined
functions.

Figure 9-11 Excel user-defined functions

Figure 9-12 APLSERVER.XLA shows the components of the APLSERVER.XLA Add-In. It contains:

Figure 9-12 APLSERVER.XLA

- A workbook, APLServerWorkbook
- A module, APLServerModule
- A class, APLServerClass

As an Add-In is a workbook, it has to have at least one worksheet—this is not used to house any code.

In order to create APLSERVER.XLA, click **File | New | SaveAs** and then use the file name APLSERVER.XLA and select XLA as the file type. Refer to **Figure 9-13 Saving as Add-In**.

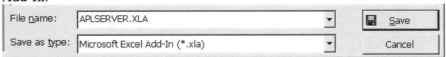

Figure 9-13 Saving as Add-In

This creates the shell. Rename the workbook module, then add and rename the module and class. The Add-In code can be viewed and edited by pressing **Alt+F11**.

The code in the Add-In is discussed next. As an aid to reference, line numbers have been added to all the code as if it were a continuous listing. In reality, there are three separate code segments.

9.6.2 APL server workbook code

The environment settings and global variables are declared in this section—the scope of these settings extends to the whole module and the module only, thus:

```
1 Option Explicit
2 Option Base 1
3 Option Compare Text
```

9.6.2.1 Add-In open event handler

This function is the event handler that is triggered when the Add-In is checked in:

```
4 Private Sub WorkBook_Open()
5     APLServerModule.Initialize
6     Application.OnKey "^%~"
7     Application.OnKey "^%~", AddInPath & "!APLServerWorkbook.IsActive"
8 End Sub
```

It calls the *Initialize* subroutine in the module and sets up keyboard shortcuts for activating or deactivating the Add-In. The short cut is **Ctrl + Alt + Enter** and may be used at any time after the Add-In is checked in.

This invokes the dialogue shown in **Figure 9-14 Add-In manager**

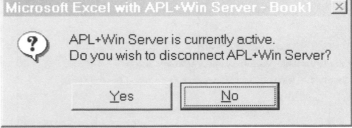

Figure 9-14 Add-In manager

A similar dialogue is shown if the server is inactive: it prompts for connection and the default button is **Yes**.

9.6.2.2 Is the APL Server active?

This subroutine is run when the keyboard short cut is used. Different dialogues are presented depending on whether the APL server is active, line [12], or inactive, line [25]:

```
 9 Public Sub IsActive()
10    Select Case ExcelWithAPL.APLServer.IsAPLServerActive
11       Case True
12          Select Case MsgBox("APL+Win Server is currently active." &
      vbCr & _
13                "Do you wish to disconnect APL+Win Server?", _
14                vbQuestion + vbYesNo + vbDefaultButton2,
      Application.Caption)
15             Case vbYes
16                ExcelWithAPL.APLServer.TerminateServer
17                ExcelWithAPL.APLServer.DeleteToolbar
18                MsgBox "APL+Win Server de-activated." & vbCr & _
19                "It may be re-activated by pressing Ctrl+Alt+Enter",
      _
20                vbInformation + vbOKOnly, Application.Caption
21             Case vbNo
22                ExcelWithAPL.APLServer.CreateToolBar
23          End Select
24       Case False
25          Select Case MsgBox("APL+Win Server is currently inactive." &
      vbCr & _
26                "Do you wish to activate it?", _
27                 vbQuestion + vbYesNo, Application.Caption)
28             Case vbYes
29                ExcelWithAPL.APLServer.StartAPLServer
30                ExcelWithAPL.APLServer.CreateToolBar
31                Application.CalculateFull
32             Case vbNo
33                ExcelWithAPL.APLServer.DeleteToolbar
34          End Select
35       End Select
36 End Sub
```

9.6.3 APL server module code

In **Excel**, press **ALT + F11** and then click **Insert + Module** to add a module.

All environment settings and global variables are declared in this section—the scope of these settings and variables extend to the whole module and to the module *only*, thus:

```
37 Option Explicit
38 Option Base 1
39 Option Compare Text
40 Global APL As Object
41 Global APLServer As New APLServerClass
42 Global APLHandle As Long
43 Global AddInPath As Variant
```

9.6.3.1 Server initialization

This section interrogates the Add-In, initialises global variables and calls a routine to start APL+Win as an automation server. Note that the *App* property of the class module is set to the application, thereby allowing the class module to respond to application level events—**Excel** does not define these by default, as shown:

```
44 Public Sub Initialize()
45    Application.Caption = "Microsoft Excel with APL+Win Server"
46    For Each AddInPath In Application.AddIns
```

```
47            If UCase(AddInPath.Name) = "APLSERVER.XLA" Then
48                AddInPath = "'" & AddInPath.FullName & "'"
49                Exit For
50            End If
51        Next
52        Set APLServer.App = Application
53        DoEvents
54        APLServer.StartAPLServer
55 End Sub
```

9.6.3.2 The Variable method

This cover function retrieves the value of a variable from the APL server workspace, without any modifications, that exists in the server workspace to **Excel**. This does not attempt to transpose the APL array to fit the **Excel** range:

```
56 Public Function Variable(ByVal VariableName As String)
57     ' Return value of VariableName from current APL+Win workspace
58     Variable = APL.Variable(VariableName)
59 End Function
```

Its usage is =*Variable("variableName")* in a cell, where *variableName* is the name of the workspace variable, is case sensitive and must be enclosed in double quotes. If this formula is entered in multiple cells, press **Ctrl + Shift + Enter** otherwise press **Enter**.

9.6.3.3 The GetVariable method

This shares the syntax of the *Variable* method and performs the same function but, in an attempt to fit the **Excel** range, it does transpose the value retrieved from the server workspace. The code is:

```
60 Public Function GetVariable(ByVal VariableName As String, Optional
       TargetArray As Variant) As Variant
61     ' Return value of VariableName from current APL+Win workspace,
       adjusting shape to fit Excel destination array range
62     Select Case TypeName(Application.Caller)
63         Case "Range"
64             TargetArray = Application.Caller.Rows.Count + 0.001 *
       Application.Caller.Columns.Count
65         Case Else
66             TargetArray = 0
67     End Select
68   GetVariable = APL.Call("GetVariable", VariableName, TargetArray)
69 End Function
```

9.6.3.4 The ShowAPL method

This function toggles the visibility of the APL server session; that is, hides the APL session or makes it visible. The code is:

```
70 Private Sub ShowAPL()
71     If TypeName(APL) = "WSEngine" Then
72         APL.Visible = Not APL.Visible
73     Else
74         APLServerWorkBook.IsActive
75         If TypeName(APL) = "WSEngine" Then APL.Visible = Not APL.Visible
76     End If
77     If TypeName(APL) = "WSEngine" Then
78         Select Case APL.Visible
79             Case True
80                 Application.CommandBars("Excel With APL+Win
       Server").Controls(6).Caption = "HideAPL"
81             Case False
```

```
82            Application.CommandBars("Excel With APL+Win
        Server").Controls(6).Caption = "ShowAPL"
83         End Select
84      End If
85 End Sub
```

9.6.3.5 The ShowArray method

This function determines if the active cell in part of an array and if so, it highlights the whole array for enhanced visibility. If the active cell is not part of an array, an error message is shown:

```
86 Private Sub ShowArray()
87     If IsValidContext Then
88         Application.Range(ActiveCell.Address).Select
89         If ActiveCell.HasArray Then
90             ActiveCell.CurrentArray.Select
91         Else
92             MsgBox "Current cell is not part of an array.", vbInformation +
        vbOKOnly, Application.Caption
93         End If
94     End If
95 End Sub
```

9.6.3.6 The ShowAllArrays method

This function highlights all arrays—an array is simply a rectangular block of cells that share the same array formula. This is useful when the formula has to be re-entered: **Excel** does not allow any array to be partially modified. The relevant code is:

```
96 Private Sub ShowAllArrays()
97     If IsValidContext Then
98         Application.ScreenUpdating = False
99         Dim CurrentCell As Variant, ArrayRanges As Variant, i As Variant
100        CurrentCell = Application.ActiveCell.Address
101        Application.Range("A1").Select
102        On Error Resume Next ' No arrays found
103        Application.Selection.SpecialCells(xlCellTypeFormulas, 23).Select
104        ArrayRanges = CurrentCell & "," & Application.Selection.Address
105        ArrayRanges = Split(ArrayRanges, ",")
106        For i = 1 + LBound(ArrayRanges) To UBound(ArrayRanges)
107            Application.Range(ArrayRanges(i)).Select
108            If Not Application.Range(ArrayRanges(i)).HasArray Then
        ArrayRanges(i) = "*"
109        Next i
110        ArrayRanges = Replace(Join(ArrayRanges, ","), ",*", "")
111        Application.Range("A1").Select
112        Application.Selection.Range(ArrayRanges).Select
113        Application.ScreenUpdating = True
114     End If
115 End Sub
```

9.6.3.7 The CalculateCurrentArray method

If the active cell is part of an array, this method refreshes the array—if the workspace variable has been changed, its new values are retrieved; otherwise the error message in line [124]. See below:

```
116 Private Sub CalculateCurrentArray()
117     Application.ScreenUpdating = False
118     If IsValidContext Then
119         Application.Range(ActiveCell.Address).Select
120         If ActiveCell.HasArray Then
121             ActiveCell.CurrentArray.Select
122             Selection.FormulaArray = Selection.FormulaArray
123         Else
```

```
124           MsgBox "Unable to Calculate" & vbCr & "Current cell is not part
        of an array.", vbInformation + vbOKOnly, Application.Caption
125         End If
126     End If
127     Application.ScreenUpdating = True
128 End Sub
```

9.6.3.8 Calculate workbook

This function forces a recalculation of the whole workbook irrespective of the current *Application.Calculation* setting; this may be *xlCalculationAutomatic, xlCalculationManual,* or *xlCalculationSemiautomatic.* The code is:

```
129 Private Sub CalculateAll()
130     Application.CalculateFull
131 End Sub
```

Normally, this method is used conditionally depending on the value of *CalculationVersion* property of the *Application,* or *ActiveWorkbook* object, or both.

9.6.3.9 The IsValidContext method

Although the **Excel** application object exposes methods and properties, a number of these are inoperative until there are one or more open workbooks. This function determines if there are any workbooks open:

```
132 Private Function IsValidContext() As Boolean
133     IsValidContext = 0 <> Application.Workbooks.Count
134 End Function
```

This method is for internal use only.

9.6.3.10 Running APL functions

This function calls a predefined APL function in the server workspace, passing arguments from **Excel**, and returns the results of the APL function into the cell where the function is entered as a formula. Enter an array formula if the result spans a range:

```
135 Function RunAPLFn(ByVal APLFn As String, _
136                   Optional ByVal RightArg As Variant, _
137                   Optional ByVal LeftArg As Variant, _
138                   Optional ByVal TransposeFlags As Variant)
139     APL.Call "SetTranspose", IIf(IsMissing(TransposeFlags), "",
        TransposeFlags)
140     If IsMissing(LeftArg) And IsMissing(RightArg) Then
141         RunAPLFn = APL.Call(APLFn) ' Niladic
142     Else
143         If IsMissing(LeftArg) Then
144             RunAPLFn = APL.Call(APLFn, RightArg) ' Monadic
145         Else
146             RunAPLFn = APL.Call(APLFn, RightArg, LeftArg) ' Dyadic
147         End If
148     End If
149 End Function
```

It is possible to run functions from the immediate window of the VBE: press **Alt + F11** and **View | Immediate Window** to display it.

9.6.3.11 The CopyValues method

This function reads each worksheet in the active workbook and removes all formulae, leaving just the values. This method may be run from the toolbar and should be run just before a workbook is saved under a *new* name. The code is:

```
150 Sub CopyValues()
151     Dim Sheet
152     Application.ScreenUpdating = False
153     For Each Sheet In ActiveWorkbook.Worksheets
154         Sheet.Select
```

```
155          With Selection
156              Range("A1").Select
157              Range(Selection, ActiveCell.SpecialCells(xlLastCell)).Select
158              Selection.Copy
159              Selection.PasteSpecial Paste:=xlValues, _
160                                     Operation:=xlNone, _
161                                     SkipBlanks:=False, _
162                                     transpose:=False
163              Range("A1").Select
164              Selection.Copy
165          End With
166      Next
167      Application.ScreenUpdating = True
168 End Sub
```

9.6.4 CreateVariable

The CreateVariable function allows an **Excel** variable to be created:

```
169 Public Function CreateVariable(ByVal VariableName As String, ByVal
        VariableValue As Variant)
170     Select Case TypeName(VariableValue)
171       Case "Range"
172           VariableValue = VariableValue.Value
173     End Select
174     APL.Call "CreateVariable", VariableValue, VariableName
175 End Function
```

Such a variable is saved with the workbook but is not visible.

9.6.5 APLServer class code

In **Excel**, press **ALT+F11** and then click **Insert + Class Module**. The environment settings and global variables are declared in this section   the scope of these settings extend to the whole module in which it appears and <u>not</u> to any other modules that may also exist.

The code in the Workbook class starts as soon as the Add-In is loaded. The code starts the APL+Win ActiveX server and modifies the **Excel** user interface, for the duration that the Add-In remains switched on, to allow the ActiveX server functionality to be used. The code is:

```
176 Private Declare Function GetClassName Lib "user32" Alias "GetClassNameA"
        (ByVal hwnd As Long, ByVal lpClassName As String, ByVal nMaxCount As
        Long) As Long
177 Private Declare Function PathFileExists Lib "shlwapi.dll" Alias
        "PathFileExistsA" (ByVal pszPath As String) As Long
178 Private Declare Function PostMessage Lib "user32" Alias "PostMessageA"
        (ByVal hwnd As Long, ByVal wMsg As Long, ByVal wParam As Long, lParam
        As Any) As Long
179 Private Declare Function IsWindow Lib "user32" (ByVal hwnd As Long) As
        Long
180 Public WithEvents App As Application
```

The event handler code is in a class module within the Add-In. For a discussion of how to work with the arguments of event handler code within APL+Win, refer to section 1.2 in Chapter 5 *The Component Object Model*.

9.6.5.1 New workbook event handler

The NewWorkbook application level event handler runs as soon as a new workbook is opened:

```
181 Private Sub App_NewWorkbook(ByVal Wb As Workbook)
182     If TypeName(APL) <> "WSEngine" Then StartAPLServer
183 End Sub
```

The workbook itself does not have a workbook open event.

9.6.5.2 Workbook activate event handler

The WorkbookActivate application level event handler runs as soon as a workbook is activated via **Windows** | *workbook*.

```
184 Private Sub App_WorkbookActivate(ByVal Wb As Workbook)
185     If TypeName(APL) <> "WSEngine" Then StartAPLServer
186 End Sub
```

9.6.5.3 Add-In install event handler

The WorkbookAddinInstall application level event handler runs as soon as the APL server or any other Add-In is checked in:

```
187 Private Sub App_WorkbookAddinInstall(ByVal Wb As Workbook)
188     Application.OnKey "^%~", AddInPath & "!APLServerWorkbook.IsActive"
189     StartAPLServer
190 End Sub
```

9.6.5.4 Add-In uninstall event handler

The WorkbookAddinUninstall application level event handler runs as soon as the APL server or any other Add-In is checked out:

```
191 Private Sub App_WorkbookAddinUninstall(ByVal Wb As Workbook)
192     Application.OnKey "^%~"
193     TerminateServer
194 End Sub
```

9.6.5.5 Workbook before close event handler

The WorkBookBeforeClose application level event handler runs as soon as a workbook is saved. This checks out the Add-In thereby ensuring that it does not interfere with the normal usage of **Excel**:

```
195 Private Sub App_WorkbookBeforeClose(ByVal Wb As Workbook, Cancel As
        Boolean)
196     Application.AddIns("Excel With APL+Win Server").Installed = False
197     If UCase(Wb.Name) = "APLSERVER.XLA" Then Exit Sub
198     TerminateServer
199 End Sub
```

9.6.5.6 WorkBook open event handler

The WorkBookOpen application level event handler runs whenever another workbook is opened in the existing **Excel** session. The APL server will be activated whether appropriate or not. Users should ensure that they open automation workbooks only:

```
200 Private Sub App_WorkbookOpen(ByVal Wb As Workbook)
201     If UCase(Wb.Name) = "APLSERVER.XLA" Then Exit Sub
202     If IsActive Then SetApplicationResponse
203 End Sub
```

9.6.5.7 Starting the APL server

This function achieves the following:

• It starts the APL server.

• It loads an APL+Win workspace, APLSERVER.W3 from the default location where workbooks are saved.

The APL workspace should contain functions for facilitating data acquisition within **Excel**. The code is:

```
204 Public Sub StartAPLServer()
205     If TypeName(APLServerModule.APLServer) = "APLServerClass" Then
206         If TypeName(APL) = "WSEngine" Then Exit Sub
207         On Error GoTo NoAPLServer
208         If Not IsAPLServerActive Then
209             Set APL = CreateObject("APLW.WSEngine")
210             APL.Visible = False
```

```
211              If PathFileExists(Application.DefaultFilePath &
         "\APLSERVER.W3") Then
212                  APL.Syscommand ("load " & Application.DefaultFilePath &
         "\APLSERVER")
213                  APLHandle = APL.Exec("GetHandle")
214                  DoEvents
215                  CreateToolBar
216                  Application.CalculateFull
217                  MsgBox "APL+Win Server started.", vbInformation +
         vbOKOnly, Application.Caption
218              Else
219                  MsgBox "Unable to find initial workspace." & vbCr &
         UCase(Application.DefaultFilePath & "\APLSERVER.W3"), vbCritical +
         vbOKOnly, Application.Caption
220              End If
221          End It
222      End If
223      Exit Sub
224 NoAPLServer: MsgBox "Unable to find APL+Win Server.", vbCritical +
         vbOKOnly, Application.Caption
225 End Sub
```

9.6.5.8 Setting Excel behaviour

This function temporarily switches off screen updating and error signalling within **Excel**, recalculates the workbook, and restores screen updating and error signalling. Essentially, this speeds up recalculation:

```
226 Private Sub SetApplicationResponse()
227     Application.ScreenUpdating = False
228     Application.DisplayAlerts = False
229     Application.CalculateFull
230     Application.ScreenUpdating = True
231     Application.DisplayAlerts = True
232     DoEvents
233 End Sub
```

9.6.5.9 Is APL server active?

This function verifies whether the APL+Win server session is still active—it is used internally:

```
234 Public Function IsAPLServerActive() As Boolean
235     Dim Classname As String * 256
236     If 0 <> GetClassName(APLHandle, Classname, 256) Then
        IsAPLServerActive = True
237 End Function
```

9.6.5.10 Is APL server toolbar present?

This function verifies whether the APL server toolbar is present—it is used internally:

```
247 Private Function FindToolbar() As Boolean
248     Dim i As Integer
249     For i = 1 To Application.CommandBars.Count
250         If Application.CommandBars.Item(i).Name = "Excel With APL+Win
        Server" Then
251             FindToolbar = True
252             Exit For
253         End If
254     Next i
255 End Function
```

9.6.5.11 Delete APL server toolbar

This function deletes the APL server toolbar—it is used internally:

```
256 Public Sub DeleteToolbar()
257     Dim i As Integer
```

```
258      For i = 1 To Application.CommandBars.Count
259          If Application.CommandBars.Item(i).Name = "Excel With APL+Win
         Server" Then
260              Application.CommandBars("Excel With APL+Win Server").Delete
261              Exit For
262          End If
263      Next i
264      DoEvents
265 End Sub
```

9.6.5.12 Create APL server toolbar

This function creates the APL server toolbar—it is used internally:

```
266 Public Sub CreateToolBar()
267      If FindToolbar Then Exit Sub
268      Dim i As Integer
269      Dim myTB As Office.CommandBar
270      Set myTB = Application.CommandBars.add(Name:="Excel With APL+Win
         Server", Position:=msoBarTop, Temporary:=True)
271      For i = 1 To 6
272          Set MyCtrl = myTB.Controls.add(Type:=msoControlButton)
273              With MyCtrl
274                  .Visible = True
275                  Select Case i
276                      Case 1
277                          .FaceId = 158 ' 1778
278                          .TooltipText = "Show array containing active cell"
279                          .OnAction = "ShowArray"
280                          .Caption = "ShowArray"
281                          .Style = msoButtonCaption
282                      Case 2
283                          .FaceId = 163 ' 2234
284                          .TooltipText = "Show All Arrays on active sheet"
285                          .OnAction = "ShowAllArrays"
286                          .Caption = "ShowAllArrays"
287                          .Style = msoButtonCaption
288                      Case 3
289                          .FaceId = 1062
290                          .TooltipText = "Calculate current array"
291                          .OnAction = "CalculateCurrentArray"
292                          .Caption = "CalculateCurrentArray"
293                          .Style = msoButtonCaption
294                      Case 4
295                          .FaceId = 940
296                          .TooltipText = "Calculate All arrays"
297                          .OnAction = "CalculateAll"
298                          .Caption = "CalculateFull"
299                          .Style = msoButtonCaption
300                      Case 5
301                          .FaceId = 938
302                          .TooltipText = "Replace Formulae with current
         Values"
303                          .OnAction = "CopyValues"
304                          .Caption = "CopyValues"
305                          .Style = msoButtonCaption
306                      Case 6
307                          .FaceId = 609 ' 214 ' 1549
308                          .TooltipText = "Toggle visibility of APL+Win
         Server"
309                          .OnAction = "ShowAPL"
310                          If 0 <> InStr(APL.Call("WhichAPL"), "Development")
         Then
```

```
311                             .Enabled = True
312                         Else
313                             .Enabled = False
314                         End If
315                         .Caption = "ShowAPL"
316                         .Style = msoButtonCaption
317                     End Select
318                 End With
319     Next i
320     Application.CommandBars("Excel With APL+Win Server").Visible = True
321 End Sub
```

9.6.6 The APL server toolbar

The APL server toolbar, shown in **Figure 9-15 APL server toolbar**, is created as a temporary toolbar; its purpose is to allow easy access to some of the functions available in the Add-In.

Figure 9-15 APL server toolbar

Toolbar buttons are a highly productive element of the user interface:

• They make the functionality built-into the Add-In visible.

• They provide an ideal way for providing access to code that does not return an explicit result.

In the CreateToolBar code, face ids are specified: this retrieves **Excel** icons and places them with button captions. However, the style option specified, msoButtonCaption, ignores the icons. The button captions correspond to the function names corresponding to the caption shown on the toolbar.

9.6.7 The initial ActiveX server workspace

The APL+Win ActiveX server—development or runtime version—does not require an initial workspace or one with a latent expression. However, it is almost mandatory to load an initial workspace with functions such as *GetHandle* and *WhichAPL* in order to simplify the process of transferring data and results. It is virtually impossible to specify APL characters from within **Excel**!

A convenient location for this initial workspace is the default file path that **Excel** uses to store workbooks; the path can be queried dynamically using *Application.DefaultFilePath* and appended to a known workspace name. It is quite likely that the default path will vary from user to user. Alternatively, the path of the current workbook may be used—*Application.ActiveWorkbook.Path*. This convention will avoid reference to a hard coded fully qualified workspace name, as this is likely to cause runtime problems.

The workspace loaded by the server may contain any number of functions. APLSERVER.W3 contains these functions.

The CreateVariable function is called by the CreateVariable function in **Excel**; it allows **Excel** to create a variable in the server workspace, as shown:

```
    ∇ L CreateVariable R
[1]   ⍝ System Building with APL+win
[2]   ⍕(,L),'←⍎R'
    ∇
```

The SurfaceArea function is called by the **Excel** RunAPLFn function:

```
    ∇ Z←L SurfaceArea R
[1]   ⍝ Calculate Area of surface bound by coordinates L,R
```

```
{1[2]     :if '#' ⎕wi '∆L'
  [3]          L←⍉L ⎕wi 'value' ⍝ Vector of X Coordinates (Columns)
+ [4]     :else
  [5]          L←L ⎕wi 'value' ⍝ Vector of X Coordinates (Rows)
1}[6]     :endif
{1[7]     :if '#' ⎕wi '∆R'
  [8]          R←⍉R ⎕wi 'value' ⍝ Vector of Y Coordinates (Columns)
+ [9]     :else
  [10]         R←R ⎕wi 'value' ⍝ Vector of Y Coordinates (Rows)
1}[11]    :endif
  [12]   Z←0.5×(L-1⌽L)+.×⍉R+1⌽R
        ▽
```

The SurfacePerimiter function is called by the **Excel** RunAPLFn function:

```
     ▽ Z←L SurfacePerimeter R
  [1]     ⍝ Calculate Perimeter of surface bound by coordinates L,R
{1[2]     :if '#' ⎕wi '∆L'
  [3]          L←⍉L ⎕wi 'value' ⍝ Vector of X Coordinates (Columns)
+ [4]     :else
  [5]          L←L ⎕wi 'value' ⍝ Vector of X Coordinates (Rows)
1}[6]     :endif
{1[7]     :if '#' ⎕wi '∆R'
  [8]          R←⍉R ⎕wi 'value' ⍝ Vector of Y Coordinates (Columns)
+ [9]     :else
  [10]         R←R ⎕wi 'value' ⍝ Vector of Y Coordinates (Rows)
1}[11]    :endif
  [12]   Z←⍉L,[⎕io]R
  [13]   Z←((Z-1⊖Z)+.*2)+.*0.5
         ▽
```

If it is necessary to control the array orientation, that is, row or column major, explicitly, use the following function:

```
     ▽ SetTranspose R
[1]     ⍝ Ajay Askoolum
[2]     R←,⎕wcall 'CharUpper' ((,R)~' ')
[3]     R←'ZLR'∊3↑R
[4]     '#' ⎕wi ¨(⊂¨R/'∆Z' '∆L' '∆R'),¨~R/'#' ⎕wi¨'∆Z' '∆L' '∆R'
        ▽
```

The right-hand argument is 'L', 'R', or 'Z'—for the left- or right-hand argument and the result, respectively—or any combination of these letters. If these properties have not been defined, the following example sets the flag for the left-hand argument to **TRUE**:

```
SetTranspose 'L'                    │ SetTranspose 'ZR'
'#' ⎕wi¨ '∆Z'  '∆L'  '∆R'          │ '#' ⎕wi¨ '∆Z'  '∆L'  '∆R'
0 1 0                               │ 1 1 1
```

This function toggles the value of the specified user-defined property. If a property is **TRUE**, that corresponding argument or result is transposed before it is used. The transposition flag applies to all APL function calls—refer to lines [2]-[6] of SurfacePerimiter. In practice, it is unnecessary to set any flags.

9.7 The EWA model in action

The EWA model requires an understanding of **Excel** user-defined formulae, Add-Ins and array formulae as well as general fluency in usage. The worksheet depicted in **Figure 9-16 Automation view** shows the EWA model in action.

Essentially, a worksheet may include any **Excel** function/calculation and can be formatted freely:

- **Figure 9-16 Automation view** shows the x-y coordinates on rows 3 and 4—the same information is shown in columns 1 and 2 on rows 6 to 15.
- The graph is an x-y plot of either range. Rows 3 to 5 in column 12 show the area of the surface bound by the x-y coordinates and rows 11 to 13 show the perimeter of the surface.
- Each of these six cells may be calculated and formatted individually.

Superficially, the worksheet contains just mundane data.

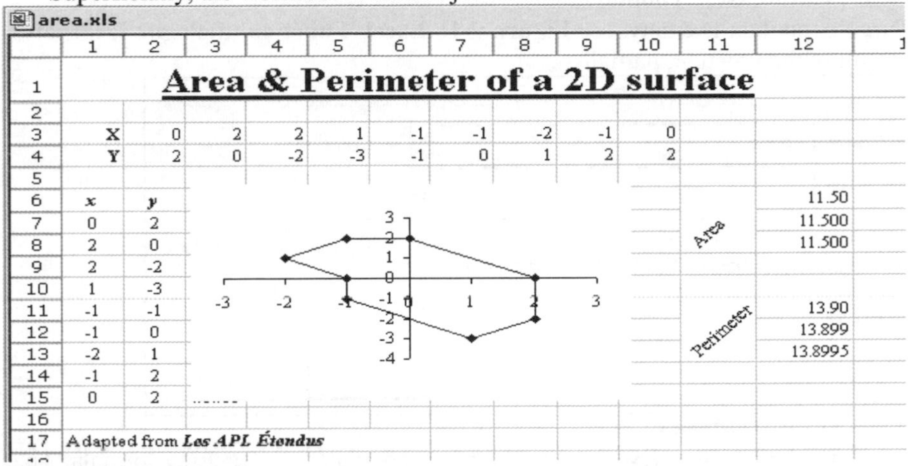

	1	2	3	4	5	6	7	8	9	10	11	12
1			**Area & Perimeter of a 2D surface**									
2												
3	X	0	2	2	1	-1	-1	-2	-1	0		
4	Y	2	0	-2	-3	-1	0	1	2	2		
5												
6	*x*	*y*										11.50
7	0	2									Area	11.500
8	2	0										11.500
9	2	-2										
10	1	-3										
11	-1	-1										13.90
12	-1	0									Perimeter	13.899
13	-2	1										13.8995
14	-1	2										
15	0	2										
16												
17	Adapted from ***Les APL Étendus***											

Figure 9-16 Automation view

Enabling the Formula option in Tools | Options | *General* | *Window Option* in Excel reveals how the worksheet is put together—refer to **Figure 9-17 Automation: formulae view**.

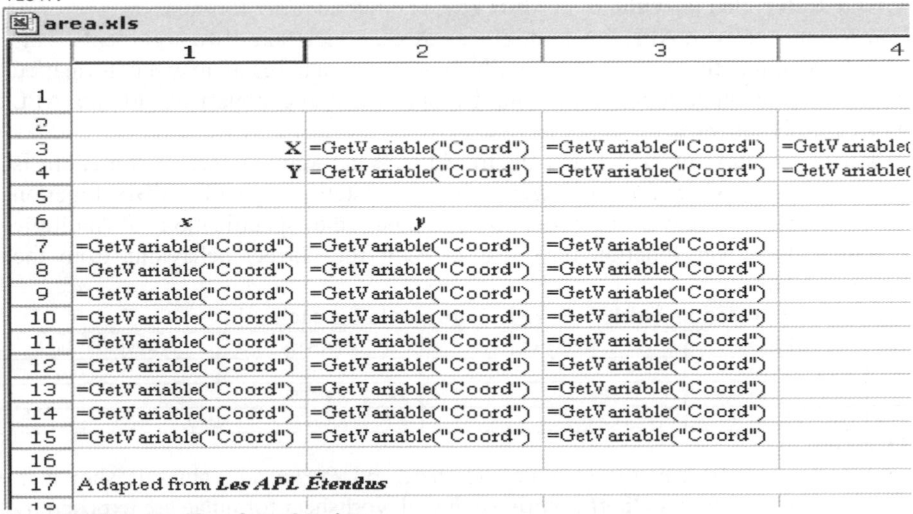

	1	2	3	4
1				
2				
3	X	=GetVariable("Coord")	=GetVariable("Coord")	=GetVariable(
4	Y	=GetVariable("Coord")	=GetVariable("Coord")	=GetVariable(
5				
6	*x*	*y*		
7	=GetVariable("Coord")	=GetVariable("Coord")	=GetVariable("Coord")	
8	=GetVariable("Coord")	=GetVariable("Coord")	=GetVariable("Coord")	
9	=GetVariable("Coord")	=GetVariable("Coord")	=GetVariable("Coord")	
10	=GetVariable("Coord")	=GetVariable("Coord")	=GetVariable("Coord")	
11	=GetVariable("Coord")	=GetVariable("Coord")	=GetVariable("Coord")	
12	=GetVariable("Coord")	=GetVariable("Coord")	=GetVariable("Coord")	
13	=GetVariable("Coord")	=GetVariable("Coord")	=GetVariable("Coord")	
14	=GetVariable("Coord")	=GetVariable("Coord")	=GetVariable("Coord")	
15	=GetVariable("Coord")	=GetVariable("Coord")	=GetVariable("Coord")	
16				
17	Adapted from ***Les APL Étendus***			

Figure 9-17 Automation: formulae view

The x-y coordinates are supplied from the variable *Coord* in an APL+Win ActiveX server workspace. In order to enter the formula, highlight the range of cells and type the formula shown, *=GetVariable("Coord")*, and press **Ctrl + Shift + Enter**. That is, an **Excel** *array formula* is entered manually—**Excel** automatically denotes array formula by enclosing the formula in double braces. Although the braces are not shown when the formula window option is switched on, they are visible when the active cell is within the range *and* the 'Show Formula' option is switched off.

An array formula cannot be entered using **Insert | Function** or the *fx* (Paste Function) button on the **Excel** Standard Toolbar, but can be reviewed by clicking the *fx* button. Highlight the range containing the array formula and click *fx* to display the following dialogue. The *fx* button is found on the **Standard** toolbar: check-in the **Standard** toolbar found at **Tools | Customise + Toolbars**.

The TargetArray box is empty, see **Figure 9-18 Excel 'Function' dialogue I**; this also shows the result of the function, partially.

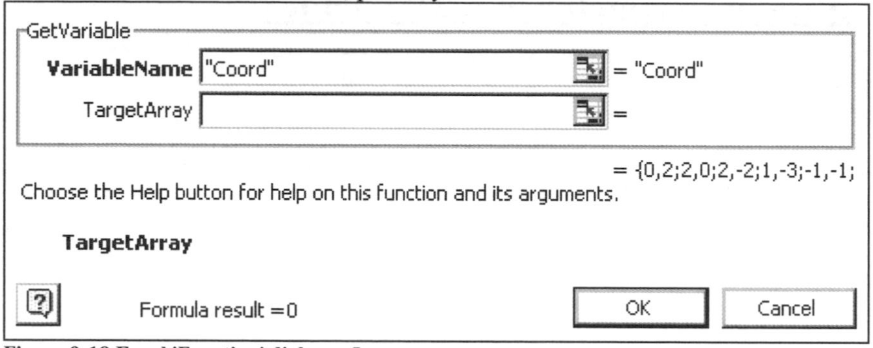

```
┌GetVariable──────────────────────────────────────────────────────┐
│   VariableName │"Coord"                              │ ▦ │ = "Coord"        │
│     TargetArray │                                     │ ▦ │ =                │
│                                                    = {0,2;2,0;2,-2;1,-3;-1,-1; │
│   Choose the Help button for help on this function and its arguments.        │
│      TargetArray                                                             │
│   [?]       Formula result =0                    │  OK  │  │ Cancel │         │
└──────────────────────────────────────────────────────────────────┘
```

Figure 9-18 Excel 'Function' dialogue I

The x-y plot is built within **Excel** using the range populated by the array formula. The *Area* and *Perimeter* cells are also array formulae and must be entered manually in the same manner—first, highlight the target range, type the formula and press **Ctrl + Shift + Enter**. An array formula is *not* the same as the same formula entered in an array: highlight a range and press **Ctrl + Enter**, that is, without the **shift** key, to enter an array formula.

The *Area* formula calls an APL function, SurfaceArea, which takes two arguments. This value is calculated three times, in order to illustrate that each result may be formatted independently in **Excel** and, more important, by specifying the arguments to the APL function in three different ways.

Note that the syntax of a VBA call to APL+Win ActiveX server function is *functionname, right_argument, left_argument*. In this instance, a public VBA function RunAPLFn is used as a cover function—this enables the specification of multiple arguments, which the cover function may use to elicit the correct behaviour from the ActiveX server function.

The arguments are listed in **Table 9-5 APL arguments** for each cell.

Cell	APLFn	RightArg	LeftArg	TransposeFlags
R6C12	SurfaceArea	R7C1:R15C1	R7C2:R15C2	" "
R7C12	SurfaceArea	R3C2:R3C10	R4C2:R4C10	"LR"
R8C12	SurfaceArea	R3C2:R3C10	R7C2:R15C2	"R"

Table 9-5 APL arguments

The specification of the call to calculate the *Area* and *Perimeter* values is shown in **Figure 9-19 Automation: Excel client formulae**. **Excel** worksheet formulae are exposed by **Tools | Options** and checking the **Formulas** option on the **View** tab. The keyboard shortcut **Ctrl + ¬** achieves the same effect: this also acts as a toggle switch.

area.xls				
	10	11	12	
1				
2				
3	=GetVariable("Coord")			
4	=GetVariable("Coord")			
5				
6			=runaplfn("SurfaceArea",R7C1:R15C1,R7C2:R15C2,"")	
7		Area	=runaplfn("SurfaceArea",R3C2:R3C10,R4C2:R4C10,"LR")	
8			=runaplfn("SurfaceArea",R3C2:R3C10,R7C2:R15C2,"R")	
9				
10				
11			=runaplfn("SurfacePerimeter",R7C1:R15C1,R7C2:R15C2,"")	
12		Perimeter	=runaplfn("SurfacePerimeter",R3C2:R3C10,R4C2:R4C10,"LR")	
13			=runaplfn("SurfacePerimeter",R3C2:R3C10,R7C2:R15C2,"R")	
14				
15				
16				
17				

Figure 9-19 Automation: Excel client formulae

If a range containing an array formula is highlighted—this is achieved by making any cell within the array the active cell and pressing **Ctrl + Shift + *** --and the *fx* (Paste Function) button is clicked, it can be examined more easily.

The value of the function and those of its arguments are shown partially, in **Figure 9-20 Excel 'Function' dialogue II**. Note the following:

• Although the formula was specified using the R1C1 reference style, the range addresses referred to in the formula are automatically adjusted to show addresses in the selected style. In this example, the R1C1 style as switched after the formula was specified.

• Arrays may be entered instead of a reference to a range by enclosing the values in double braces and separating each element by a comma.

• Literal arguments must be specified within double quotation marks.

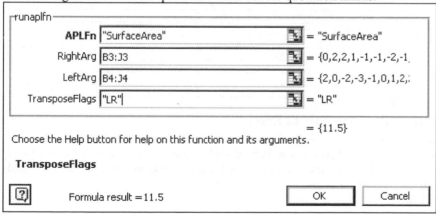

runaplfn
APLFn "SurfaceArea" = "SurfaceArea"
RightArg B3:J3 = {0,2,2,1,-1,-1,-2,-1
LeftArg B4:J4 = {2,0,-2,-3,-1,0,1,2,:
TransposeFlags "LR" = "LR"

= {11.5}
Choose the Help button for help on this function and its arguments.

TransposeFlags

[?]   Formula result = 11.5   OK   Cancel

Figure 9-20 Excel 'Function' dialogue II

Before further investigation of the mechanisms at work, it would be pertinent to validate the *Area* and *Perimeter* formulae:

- The area of the shape, shown in **Figure 9-21 Area of shape**, is confirmed by calculating the shapes of each individual geometric shape, in a clockwise movement starting on the top left-hand quadrant.

- The perimeter may also be calculated by adding the length of the outer sides of each geometric shape starting in the top left quadrant.

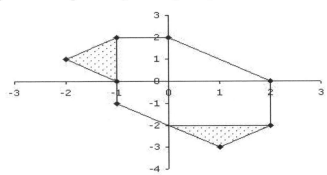

Table 9-6 Formulae audit shows an audit of the calculation of the results of *Area* and *Perimeter*.

Figure 9-21 Area of shape

The area and perimeter is calculated step by step by breaking the surface area shown in **Figure 9-21 Area of shape** into simple geometric shapes. This is used to confirm the results shown in **Figure 9-16 Automation view**, where formulae are used to calculate the same results. This type of debugging exercise is often necessary when proving formulae implemented in both APL+Win and **Excel** VBA.

Area		
The top left shaded triangle	0.5 × 2 × 1	1.00
The trapezium, left of the y axis	0.5 × 1 × (4+3)	3.50
The trapezium, right of the y axis	0.5 × 2 × (4+2)	6.00
The shaded triangle below the y axis	0.5 × 2 × 1	1.00
	Total	11.50
Perimeter		
The two outer sides of the top left shaded triangle	2×((1*2)+1*2)*0.5	2.828427125
The outer side of the adjacent rectangle		1
The hypotenuse of the triangle above the x axis	((2*2)+2*2)*0.5	2.828427125
The outer side of the rectangle below the x axis		2
The two outer sides of the shaded triangle	2×((1*2)+1*2)*0.5	2.828427125
The hypotenuse of the triangle in the trapezium	((1*2)+1*2)*0.5	1.414213562
The outer side of the rectangle in the trapezium		1
	Total	13.89949494

Table 9-6 Formulae audit

9.8 Transferring APL+Win data to Excel

A more robust or tolerant version of the SendTo**Excel** function is presented next. This will send the contents of a record object to a workbook and manage the dimensions of the record object by reference to either user-specified limits or the limits in **Excel**; **Excel** limits are used by default and override user parameters wherever necessary:

```
      ∇ SendToExcel R;⎕wself;UserSheets;PageCount;SheetSize;Fields
 [1]    ⍝ System Building with APL+Win
 [2]    ⎕wself←R
{1[3]   :if ∧/(⊂⎕wself) ⎕wi¨ 'xBOF' 'xEOF'
 [4]       0 0⍴Msg 'Record set is empty'
>[5]       :return
```

```
+ [6]      :else
{2[7]          :if ¯1=⎕wi 'xRecordCount'
  [8]              0 0⍴⎕Msg 'Cursor type must be adOpenKeySet(1) or
         adOpenStatic (3)'
> [9]              :return
+ [10]        :else
  [11]            ⎕wi 'xMoveFirst'
  [12]            ⎕wself←'xl' ⎕wi 'Create' 'Excel.Application'
  [13]            0 0⍴⎕wi 'xWorkbooks.Add' ⍝ To get Rows count
  [14]            ⎕wi ¨('xScreenUpdating' 0) ('xDisplayAlerts' 0)
{3[15]            :if 1=R ⎕wi 'xPageSize'
  [16]                PageCount←⌈(R ⎕wi 'xRecordCount')÷¯1+⎕wi 'xRows.Count'
  [17]                SheetSize←(¯1+⎕wi 'xRows.Count')⌊R ⎕wi 'xRecordCount'
+ [18]            :else
  [19]                PageCount←R ⎕wi 'xPageCount'
  [20]                SheetSize←R ⎕wi 'xPageSize'
3}[21]            :endif
  [22]            ⎕wi 'xWorkbooks.Close'
  [23]            UserSheets←⎕wi 'xSheetsInNewWorkBook'
  [24]            ⎕wi 'xSheetsInNewWorkBook' PageCount
  [25]            0 0⍴⎕wi 'Workbooks.Add'
  [26]            ⎕wi 'xSheetsInNewWorkBook' UserSheets
  [27]            Fields←(⎕wi 'xColumns.Count')⌊R⎕wi 'xFields.Count'
  [28]            (⎕wi 'Sheets.Item' ('#' ⎕wi 'VT' (+\PageCount/1) 8204))
         ⎕wi 'xSelect'
  [29]            0 0⍴⎕wi 'xRange().Select' (⎕wi 'XConvertFormula'
         ('R1C1:R1C',⍕Fields),⎕wi¨ '=xlR1C1' '=xlA1')
  [30]            ⎕wi 'xSelection.FormulaR1C1' ((⊂R) ⎕wi
         ¨(⊂⊂'xFields.Item().Name'),¨¯1++\(R ⎕wi 'xFields.Count')/1)
  [31]            ⎕wi 'xSheets(1).Select'
  [32]            PageCount←+\PageCount/1
{3[33]            :repeat
  [34]                R ⎕wi 'xAbsolutePage' (θ⍴PageCount)
  [35]                0 0⍴⎕wi 'xSheets.Item().Range().CopyFromRecordSet'
         (θ⍴PageCount) "A2" (R ⎕wi 'obj') SheetSize Fields
  [36]                0 0⍴⎕wi 'xSheets.Item().Columns.AutoFit'
         (θ⍴PageCount)
3}[37]            :until 0=⍴PageCount←1↓PageCount
  [38]            0 0⍴⎕wi 'xSheets.Item().Range().Select' 1 "A2"
  [39]            ⎕wi ¨('xScreenUpdating' 1) ('xDisplayAlerts' 1)
  [40]            ⎕wi 'visible' 1
  [41]            ⎕wi 'Delete'
2}[42]        :endif
1}[43]    :endif
         ∇
```

The result of an SQL query is shown **Figure 9-22 Send to Excel**. The first two sheets contain ten records each and the third contains three records—page size is set to 10. The first row, on all sheets, contains the names of the columns.

The following function manages the SQL query: it retrieves all the records from the table AUTHORS in the PUBS database and sends the record set to **Excel**.

```
    ∇ GetAuthors;⎕wself;Cnn;sql
[1]   ⍝ System Building with APL+Win
[2]   Cnn←'Driver={SQL Server};Server=Ajay
      Askoolum;Database=pubs;UID=sa;PWD=;'
```

```
[3]     sql←'Select * from authors'
[4]     ⎕wself←'ADOR' ⎕wi 'Create' 'ADODB.Recordset'
[5]     ⎕wi 'xPageSize' 10
[6]     ⎕wi 'XOpen' sql Cnn,⎕wi ¨'=adOpenStatic' '=adLockReadOnly'
        '=adCmdText'
[7]     SendToExcel 'ADOR'
[8]     'ADOR' ⎕wi 'XClose'
[9]     'ADOR' ⎕wi 'Delete'
    ∇
```

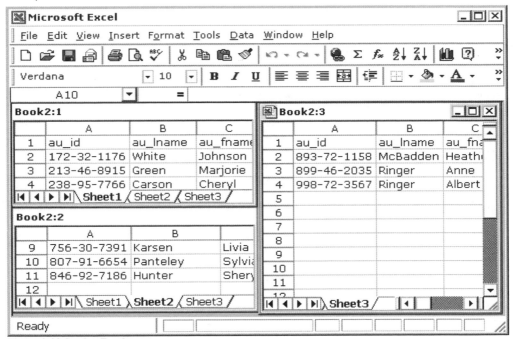

Figure 9-22 Send to Excel

9.9 Automation issues

The **Excel** macro recorder is invaluable in developing the code for APL+Win. Use the recorder to capture any task and then translate it for APL+Win.

The formatting of a Workbook is a very time consuming task with automation. Depending on the size of a workbook, it is necessary to keep formatting to a minimum.

9.9.1 APL+Win issues

Excel has some functionality that can be accessed only via VBA: it is not possible to use the macro recorder. For such functionality, use the immediate window in the IDE to construct the VBA code and then translate it for APL.

Quite often **Excel** uses objects as arguments to functions: in APL+Win, use the *obj* property to do the same.

9.10 Why use Excel with APL?

APL notoriously lacks a means of presenting data in a human readable form and **Excel** is the most widely used personal computing application. The compelling reasons for using **Excel** with APL is that it serves not only as a universally accepted reporting tool but also endows

APL with an intuitive method for presenting its data; the catch is that APL needs to convert its data to rectangular arrays. If Excel is used as a reporting tool, APL needs to use the extensive formatting capability of Excel.

Excel allows APL to present and share its data.

Chapter 10

Working with Word

Microsoft **Word** is the leading industry standard word processor capable of producing sophisticated documents; it is also a COM application, which means that APL+Win can drive it. In contrast, APL+Win is devoid of any word processing features. After the user interface, an application is judged by the quality of its output. The ideal opportunity is to use **Word** to format APL+Win output because:

• The entire document-handling capabilities of **Word** are available, including formatting, printing, and password protection.

• It is possible to produce electronic output for electronic distribution and archives.

APL+Win is capable of launching any feature of **Word**. However, a free-hand approach to **Word** programming is almost certainly detrimental to the objective to hand, namely, formatted output from APL+Win. Unlike **Excel**, **Word** is not a programming environment but a document composition tool that has programming facilities for automating routine tasks.

Any custom reporting tool is expensive—it takes time to create, it incurs an ongoing maintenance cost and involves compromises. Tools such as **Word** are readily accessible for the following reasons:

• Microsoft meets the ongoing cost of development and as users generally have relevant experience of **Word**.

• The document formats supported by **Word**, especially *.DOC, *.RTF and *.HTM, provide flexibility in the deployment of reports as **Word** documents. There are supporting tools such as GhostScript and PDF995 that convert *.DOC format to the PDF format that is readable by Acrobat Reader.

• A **Word** viewer is freely available—this is **Word** with some structural functionality such as editing suppressed; this allows documents to be viewed.

10.1 The Word difference

In the Microsoft Office suite of applications, **Word** is different from the others in two crucial ways. First, all the other applications have predefined place holders: by default, **Excel** has worksheets, ranges, cells, etc., **Access** has tables, queries, etc. In contrast, **Word** has but one predefined placeholder. The cursor position in the active document defines the current paragraph range in the active document; this is automatically selected. The automation client may either fill a document sequentially or it must define specifically its own placeholders. Placeholders in **Word** are sections, pages, bookmarks, paragraphs, tables, text boxes, and several types of fields. In order to fill a specific placeholder, the client must find and select it first and then fill it. This makes the usage of **Word** as an automation client very code-intensive.

Second, **Word** can create a library of document templates that can be used selectively. Each template may customise the **Word** user interface and contain numerous custom styles for controlling the appearance of a document. A style allows parts of a document to be formatted: this includes font including bold, italics, underlining, font size, colour, and justification, auto text entries, language settings, toolbars, etc. Styles represent an efficient way for automation clients to control the appearance of a document, since:

● Document templates may be freely created with **Word;** that is, without using an automation client. In **Word**, click **File | New** and select the **Template** option in the **Create New** box.

● Styles are saved with a document thereby making the document self-sufficient. This is convenient for the distribution of the documents. Ideally, highly specialised documents should be distributed with the templates they are based on.

● Template styles may be applied with a single instruction from the automation client.

10.2 Word templates

A **Word** template is a document that holds the characteristics of the documents it is associated with. In addition, it holds code that can be run from any of the associated documents. The characteristics include: page size and margins, default fonts, styles that define paragraph layout, fonts, indentation, etc. It also holds the last user interface configurations such as layout (**View | Normal,** etc.), **View | Zoom, View | Document Map, Tools | Customize, Tools | Set Language** options, etc.

By default, all documents share the global template called NORMAL.DOT—even documents based on other user-defined templates. If NORMAL.DOT cannot be found, **Word** automatically creates it. **Word** does not allow the global templates path to be altered; this path is determined during installation. Unless a template from another location is attached to the active document, the path for global templates is the same as that for user templates. More precisely, user templates should be stored in an alternative location, perhaps in a folder beneath the global templates path.

The option **Tools | Templates and Add-Ins…** displays the dialogue shown in **Figure 10-1 Word templates**. The current template location and name is shown. A different template can be attached to the document by using the **Attach** button. If the **'Automatically update document styles'** check box is ticked, the document is automatically updated to reflect the styles.

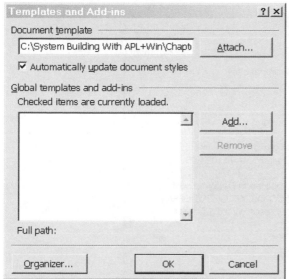

Figure 10-1 Word templates

A document template holds all basic and user-defined styles as well as **Word** interface settings such as page margins, default font, etc.

Thus, a template encapsulates a house style for both the overall appearance of document and those of individual paragraphs in it.

In order to define a style, highlight a paragraph or section of a document, change its appearance and layout as desired, and then define a style using the **Format | Style + New** option. Subsequently, that style can be applied to other selections, paragraphs, and sections. This ensures consistency of appearance both within and across documents using a particular template, especially if the automatic update option is switched on. Failure to define styles implies that each paragraph must be formatted individually.

The default locations used by **Word** can be examined and reset via the **Tools | Options + File Locations** menu option. The dialogue is shown in **Figure 10-2 User paths**. Note the following:

• Each path can be reset either by double clicking on it or by highlighting it and clicking the **Modify** button.

• The location shown in the dialogue box is truncated; use the **Modify** button to see the full path.

• The workgroup templates path is expected to be on a network volume in order that multiple users can share it.

• Any changes to paths take effect immediately; note that any files at the previous locations are not automatically moved to the new location.

10.2.1 Global, user, and workgroup templates

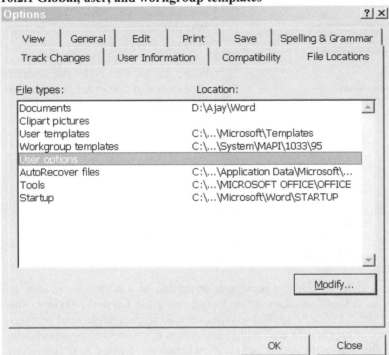

Figure 10-2 User paths

On selecting **File | New**, a set of global templates appears in the **General** tab of the dialogue, shown in **Figure 10-3 Word templates**. Select the **Template** radio button to create a new template. Specialist templates such as ones for APL+Win reporting may be stored in a separate location in order to avoid confusion. If user templates are used, any changes to them must be made on each computer.

Workgroup templates should be stored on a volume that multiple users may share, usually on a network volume. Usually, a workgroup template will be made read only to safeguard from inadvertent changes. Any updates to a workgroup template need only be made at one central location.

Pre-defined document templates promote the efficient use of **Word**, both directly and as an automation server. Imagine that a document contains certain paragraphs that are shown italicised and indented from the left margin. One approach is to format each of those paragraphs manually: this will certainly achieve the desired visual effect. Suppose that the font of those paragraphs need to be changed or that the text should no longer be italicised. If a style is used to define the characteristics of those paragraphs, each one must be reformatted manually. Had a style been used, the alteration of the style will reformat all the paragraphs automatically: this is much more efficient.

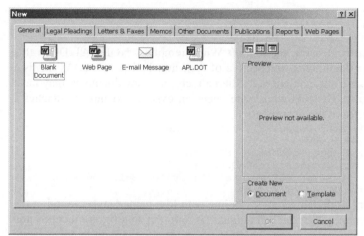

Consider a document that contains the listing of an APL+Win application with text to annotate the purpose of each function. One approach is to include the listing of every function like any other paragraph, with the font changed to an APL font.

Figure 10-3 Word templates

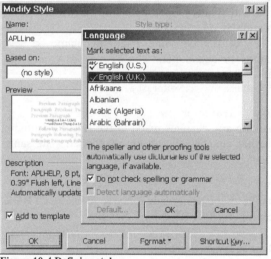

Had a style been used to format the APL code listing, a redefinition of the style would alter the appearance of the paragraphs of APL code. The style may also be defined so that the spell checker omits paragraphs using it. **Figure 10-4 Defining styles** shows the style *APLLine* that is set so that it is not spell checked.

Figure 10-4 Defining styles

- If this document were subject to the spell checker, the APL code will be checked too.
- It would be impossible to use the Find facility to skip from APL listing to APL listing.

The complete functionality for managing styles is available from the **Format | Style …** menu option.

10.3 Starting Word

An automation client can start **Word** quite simply by using:

```
     ∇ StartWord
[1]    ⍝ System Building with APL+win
[2]    ⎕wself←'∆wd' ⎕wi 'Create' 'Word.Application'
     ∇
```

Note that, unlike **Excel**, **Word** has an *Options* property that can be used to configure its behaviour. The next step is to add a new document, perhaps with a preferred template, as shown:

```
      Template←(⎕wi 'xOptions.DefaultFilePath()' (⎕wi
'=wdUserTemplatesPath')),'\NORMAL.DOT'
0 0ρ⎕wi 'xDocuments.Add(Template)' Template
```

This creates a new document using the *default* **Word** template, NORMAL.DOT; this templates applies to all open documents irrespective of their attached template. As this is a new document, it does not have any content yet. Alternatively, the new document may have another template attached to it. The following code opens an existing document, attaches a template, and updates the styles in the document:

```
      FileName←(⎕wi 'xOptions.DefaultFilePath()' (⎕wi
'=wdDocumentsPath')),'\DOC1.DOC'
    0 0ρ⎕wi 'xDocuments.Open(Filename)' FileName
    ⎕wi 'xActiveDocument.UpdateStylesOnOpen' 1
    ⎕wi 'xActiveDocument.AttachedTemplate' 'C:\System Building With
APL+Win\Chapter0\rsp.dot'
```

This forces the document to update its styles in line with like-named styles found in the attached template. In this example, the attached template is found in an arbitrary location and not in the default user templates path. A template that is stored in the default templates path is always visible with **File | Open,** but one stored at another arbitrary location is not. The default file paths **Word** uses can be reviewed with **Tools | Options + File Locations**; these are stored in the template and can vary from template to template.

An active document may be saved under its original name or another name—a new document must be saved under another name—or abandoned. The **Word** VBA help file provides full details of all the options available, thus:

```
    ⎕wi 'xActiveDocument.SaveAs(FileName)' 'c:\example.doc'
    ⎕wi 'xActiveDocument.Save'
    ⎕wi 'xActiveDocument.Close(SaveChanges)' 0
```

The last expression forces **Word** to abandon any changes made to the active document without prompting for a decision as to whether changes should be abandoned. The same effect is achieved with the following code:

```
    ⎕wi 'xActiveDocument.Saved' 1
    0 0ρ⎕wi 'xActiveDocument.Close'
```

The saved property is **FALSE** when changes have been made to a document and **TRUE** if no changes have been made—setting the property to **TRUE** bypasses the prompt to save or abandon the changes.

If it is necessary to store information in an active document to control automation—such information will usually be used by VBA code found within the template attached to the document—it can be done as follows:

```
    0 0ρ⎕wi 'xActiveDocument.Variables.Add(Name,Value)' 'ThisVar' 100
    ⎕wi 'xACtiveDocument.Variables().Value' 'ThisVar'
100
    ⎕wi 'xActiveDocument.Variables().Delete' 'ThisVar'
```

Word is unable to store numeric values—although it accepts them—in document variables; it stores all values as string.

10.4 Word as a report generation component

Word is eminently suitable as a report writing tool because it has the facilities for producing complex and professional documents.

As a COM component, **Word** can be treated as a report generation component of APL+Win applications. The idea is to create a **Word** template to encapsulate a corporate style and then to create shell documents for dynamic information that APL+Win can supply at runtime. In this regime, a small subset of **Word** features becomes the placeholders for information that is to be supplied by APL+Win.

For ad hoc reports, 'copy and paste' and 'drag and drop' are convenient tools for collating information from several sources. The document can be freely formatted or formatted in line with an adopted corporate style.

10.4.1 Tables

Word tables, inserted via **Tables | Insert | Table**, provide a grid of *m* rows by *n* columns with some key features that lend themselves well to formatted output, namely:

● Initial rows, usually the first, from the *m* rows can be set as header rows—header rows are repeated automatically in the next page if a table were to be broken over multiple pages.

● Every cell can be formatted and populated independently, retains its position relative to other cells, and automatically adjusts its size depending on its content in conjunction with adjacent cells. For example, a cell may show a variable number of lines for an address in a cell; its size will vary depending on the number of lines shown.

● The grid lines that delineate each cell may be switched off so that they do not appear on hard copy output.

● Adjacent cells may be merged for presentational purposes.

● Tables imply columnar output—this is the inherent nature of APL array data.

Tables can be deleted from a populated document if found to be empty. This means that APL can produce output that omits inapplicable information. This feature is quite valuable; a user interface cannot anticipate user input and therefore cannot present data from relevant entry forms alone. In contrast, an output generator needs to omit inapplicable information—it keeps the output relevant and makes the application appear intelligent. For example, if an invoicing system needs to show a delivery address—depending on whether delivery is required—the user interface must allow delivery address, expected time of delivery, etc., to be specified. However, if delivery is not required, the output does not need to indicate particulars relating to delivery; the omission of these details implies collection by customers, because:

● **Word** tables enable precise control of columnar output. Columnar output has close affinity with APL array output. Each individual column can be formatted individually.

Tables are ideal for showing APL arrays—they ensure that rows and columns are properly delineated irrespective of the font used. In general, proportional fonts allow less control over the presentation of columns of data than mono spaced fonts as they cause columns to run into each other: tables prevent this.

10.4.2 Building an APL+Win array

The APL+Win array must contain carriage return and line feed (`□tcnl,□tclf`) as the row delimiter and tab (`□tcht`), or any character other than space, as the column delimiter. The final column should not have this delimiter; a delimiter in the final column will cause a superfluous blank column in the table. Following **Excel** conventions, columns showing numeric values are right justified and those showing text values, included formatted dates, are left justified. For illustration, it is assumed that the array sent to **Word** as a table is either numeric or character; dates would be formatted and would be character.

10.4.2.1 Character arrays

Character arrays are probably the most common type of APL+Win array that would be sent to **Word**. They can occur in two forms—homogeneous or mixed.

10.4.2.1.1 Homogeneous character arrays

A homogeneous character array contains only one type of APL+Win data, formatted as character. It may be formatted into multiple columns—each column separated by at least one space—or it may be a single column that would require re-shaping into multiple columns. A single column may be formatted into multiple columns with this function:

```
        ∇ Z←L CharArray1 R
  [1]     ⍝ System Building with APL+Win
  [2]     ⍝ L= Columns[,Orientation], R = single column text
  [3]     (Z L R)←(⍕R) (2↑L) (' '=R)
  [4]     Z←(+/∧\R)⌽Z ⍝ left justify
  [5]     L[⎕io]←L[⎕io]⌊↑⍴Z
  [6]     R←(⌈(↑⍴Z)÷⊖⍴L),⊖⍴L
{1[7]     :if ×0⊥L
  [8]         R←⍉(⌽R)⍴⎕ior(×/R)↑1+⍳↑⍴Z
+ [9]     :else
  [10]        R←R⍴⎕ior(×/R)↑1+⍳↑⍴Z
1}[11]    :endif
  [12]    Z←(' ',[⎕io]Z,⎕tcht)[R;]
  [13]    Z←0 ¯1↓,[1↓⍳⍴⍴Z]Z
  [14]    Z←(~∧⌿' '=Z)/Z ⍝ remove superfluous trailing blanks
        ∇
```

The left-hand argument specifies the number of columns and, optionally, whether the columns are populated row-wise (default) or column wise. The right–hand argument is a single column text array, of rank 2, to be re-formatted as follows:

```
        a←⊃GetLocaleInformation ¨42+0,+\6/1
        3 0 CharArray1 a ⍝ or 3 CharArray1 a
Monday          Tuesday Wednesday
Thursday        Friday  Saturday
Sunday
        3 1 CharArray1 a
Monday          Thursday        Sunday
Tuesday         Friday
Wednesday       Saturday
```

Alternatively, if the APL+Win array contains multiple columns, a column of vertical spaces must delimit the columns, thus:

```
        ∇ Z←CharArray2 R
  [1]     ⍝ System Building with APL+Win
  [2]     ⍝ R = multi column text, each column separated by at least one space
  [3]     Z←⍕R
  [4]     R←∧⌿' '=Z
  [5]     Z←(~(∧\R)∨⌽∧\⌽R)/Z
  [6]     R←∧⌿' '=Z
  [7]     (R/Z)←⎕tcht
        ∇
```

Each row may not have the same number of columns—missing cells contain spaces—but the columns must be delineated vertically. An array may not have the same number of columns on each row. Use CharArray1 unless an array is irregular; that is, it does not have the same number of logical cells on each row. CharArray2 implies the overhead of having to construct the array in the first place.

It would be expected that each column of the table is a sparsely populated array; these arrays can be individually formatted as character and then concatenated to produce the irregular array.

10.4.2.1.2 Mixed character arrays

Unlike character arrays, which contain a single type of data, character, or numeric formatted as character, and have depth 2, mixed arrays contain some character data and some numeric data; some elements may be empty. That is, mixed arrays have ranks higher than or equal to

2. If the depth of the mixed array is greater than 2, one or more of its cells contain another array. In **Word**, this means creating a table and then creating other tables within cells of the original table. The programming overhead for a table within another table is significantly higher as each cell will need to be formatted individually: character cell should be left justified and numeric cells should be right justified. This introduces a significant overhead.

10.4.2.2 Currency arrays

Currency arrays are numeric but show the currency symbol with each element; it is formatted as character before being exported to **Word**.

Although the APL+Win ⎕fmt function may be used to format currency amounts, it has the limitation that an arbitrary format must be specified. If international settings are to be observed, the amounts must be formatted by reference to the computer's regional settings. Use the function CharArray2 may be used to render the result for exporting to **Word**.

Formatting in accordance with regional settings is discussed next.

10.4.2.3 Numeric arrays

Numeric arrays present the same problems as currency amounts as far as regional settings are concerned. A further problem is that there is only one numeric format specified in regional settings. In practice, an application will need to show integer amount and floating-point amounts to differing decimal precision depending on context. This makes a persuasive argument for using the APL+Win ⎕fmt function.

10.4.2.3.1 Regional formats

The Windows GetCurrencyFormat and GetNumberFormat APIs return currency and numeric amounts in accordance with settings found on the computer's regional settings. Both these APIs work with scalar data only and need to be adapted to cope with APL+Win arrays, as shown by:

```
       ∇ Z←L GetLocaleFormat R
  [1]    ⍝ System Building with APL+Win
  [2]    Z←0 1↓⍎,[θ],R
  [3]    ((,'‾'=Z)/,Z)←'-'
  [4]    (,R)←(⊂[⎕io+1]Z)~¨' '
  [5]    Z←~(-0⌷⍴R)↑1
{1[6]    :if 0=⎕nc 'L'
  [7]        L←'Number'
1}[8]    :endif
{1[9]    :select L
★ [10]       :case 'Currency'
  [11]          (,R)←↑¨↑/¨⎕wcall ¨(⊂'GetCurrencyFormat' 1024 0),¨(⊂¨,⍕¨R)
            ,¨⊂0 (256⍴⎕tcnul) 256
+ [12]       :else
  [13]          (,R)←↑¨↑/¨⎕wcall ¨(⊂'GetNumberFormat' 1024 0),¨(⊂¨,⍕¨R)
            ,¨⊂0 (256⍴⎕tcnul) 256
1}[14]   :endselect
  [15]   (Z/R)←(Z/R),¨⎕tcht
  [16]   Z←~∧≠' '=⍕R
  [17]   Z←Z/⍕R
  [18]   ⍝((' ú'=,Z)/,Z)←'£'
       ∇
```

The left-hand is optional and defaults to *Number*; the alternative option is *Currency*. The right-hand argument is a simple APL+Win array of rank 2. Consider the following array:

```
       a←2 2⍴4?19923
       a[1;2]←‾1
```

```
      ⎕rl←2004
      a←2 2⍴4?19923
      a[⎕io;⎕io+1]←¯1×a[⎕io;⎕io+1]
```

The API calls that format amounts into regional formats cope with scalar data only and do not understand the APL+Win high minus. These limitations are overcome in lines [3] – [4], [11] and [13]. Some sample results are shown in **Table 10-1 Sample formatting**:

Numbers	Currency
`GetLocaleFormat a`	`'Currency' GetLocaleFormat a`
`19,611.00     -7,935.00`	`ú19,611.00     -ú7,935.00`
`15,279.00     17,234.00`	`ú15,279.00     ú17,234.00`

Table 10-1 Sample formatting

Although the currency symbol appears incorrectly in the APL+Win session, it translates correctly in **Word** when it is sent directly as seen in the DemoTable function.

10.4.2.4 Blank arrays

It may be necessary to create an empty table that can be populated on an ad hoc basis, as follows:

```
      ∇ Z←BlankArray R
[1]     ⍝ System Building with APL+Win
[2]     Z←(0 ¯1+R)⍴⎕tcht
      ∇
```

This is used in line [7] of DemoTable.

10.4.2.5 Nested arrays

A nested **Word** table is required to show a nested array with depth higher than 2. Nested arrays of depth 2, such as those returned by the XGetRows method of an ADO record object can be formatted and shown in a simple **Word** table, using:

```
      ∇ Z←CharArray3 R
[1]     ⍝ System Building with APL+Win
[2]     Z←0 ¯1↓⍉(⍉¨R),¨⎕tcht
      ∇
```

If the nested array is a record object, it is necessary to add the field names as the first row of the array:

```
      'ADO Record set' '' 1 WordTable CharArray3 RSTable
```

The left-hand argument of WordTable contains the table caption, the alignment constant, as shown in WordTable [18], and a Boolean flag—0 indicates that the first row should be formatted as the table header, repeat on all pages and shown as bold:

```
      ∇ Z←RSTable;Cnn;Sql;⎕wself
[1]     ⍝ System Building with APL+Win
[2]     Cnn←"Provider=SQLOLEDB;Data Source=AJAY ASKOOLUM;Initial
           Catalog=pubs; User ID=sa;Password=;"
[3]     Sql←"SELECT * FROM AUTHORS WHERE au_lname like 'S%'"
[4]     ⎕wself←'ADORS' ⎕wi 'Create' 'ADODB.Recordset'
[5]     ⎕wi 'XOpen' Sql Cnn
[6]     Z←⎕wi 'XGetRows'
[7]     Z←(⎕wi ¨(⊂⊂'xFields().Name'),¨¯1++\(⎕wi 'xFields.Count')/1),[⎕io]Z
      ∇
```

The first row contains the name of each column. The issues relating to mixed character arrays apply to nested arrays as well. An example is shown in **Figure 10-5 ADO record set as table**.

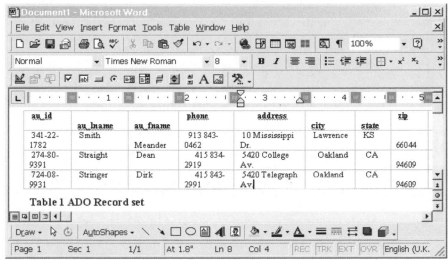

Figure 10-5 ADO record set as table

10.4.2.6 Adding bookmarks in tables

As with the active document, it is possible to add bookmarks to the cells of a table:

```
      ∇ Z←AddTableBookmarks R
[1]     ⍝ System Building with APL+Win
[2]     Z←⊂''
[3]     0 0ρR ⎕wi 'xSelect'
{1[4]   :for i :in +\(R ⎕wi 'xRows.Count')/1
{2[5]       :for j :in +\(R ⎕wi 'xColumns.Count')/1
[6]            R ⎕wi 'XCell(,).Range.Select' i j
[7]            Z←Z,⊂∊'tbl',⍕¨i j
[8]            0 0ρ⎕wi 'xSelection.Bookmarks.Add' (∊'tbl',⍕¨i j)
2}[9]       :endfor
1}[10]  :endfor
      ∇
```

With bookmarks, the cells of a table can be referred to without explicit reference to the table itself; refer to the examples in lines [23] and [24] of DemoTable.

10.4.3 APL+Win array to Word table

The code for working with **Word** tables can be illustrated by the example below.

```
      ∇ DemoTable;A
[1]     ⍝ System Building with APL+Win
[2]     ⎕wself←'∆wrd' ⎕wi 'Create' 'Word.Application'
[3]     0 0ρ⎕wi 'xDocuments.Add'
[4]     ⎕wi 'xSelection.TypeText' 'This is a freely typed paragraph.'
[5]     ⎕wi 'xSelection.TypeParagraph'
[6]     ⎕wi 'xSelection.TypeParagraph'
[7]     '.t1' ⎕wi 'Create'('Word Tables' '' WordTable BlankArray 5 5)
[8]     AddTableBookmarks '.t1'
[9]     '.t1' ⎕wi 'XSelect'
[10]    (⎕wi 'xSelection') ⎕wi ¨(⊂¨(⊂'Font'),¨'.Name' '.Size'
          '.Italic'),¨⊂¨'Arial' 10 1
[11]    '.t1' ⎕wi 'XCell(,).Range.Text' 1 1 'Populating...'
[12]    '.t1' ⎕wi 'XCell(,).Select' 3 3
[13]    A←0.01×4 4ρ16?12345
```

```
[14]    A[⎕io;⎕io]←¯1×A[⎕io;⎕io]
[15]    '.t2' ⎕wi 'Create' ('Numbers' 'Right' WordTable GetLocaleFormat A)
[16]    '.t2' ⎕wi 'XSelect'
[17]    (⎕wi 'xSelection') ⎕wi ¨(⊂¨(⊂'Font'),¨'.Name' '.Size'
            '.Italic'),¨⊂'Verdana' 8 0
[18]    '.t1' ⎕wi 'XCell(,).Select' 3 4
[19]    A←3 CharArray1 ⊃'Mon' 'Tue' 'Wed' 'Thu' 'Fri' 'Sat' 'Sun'
[20]    '.t3' ⎕wi 'Create'('Days' 'JustifyLow' WordTable A)
[21]    '.t3' ⎕wi 'xSelect'
[22]    (⎕wi 'xSelection') ⎕wi ¨(⊂¨(⊂'Font'),¨'.Name' '.Size' '.Italic'
            '.Bold'),¨⊂'Verdana' 8 0 1
[23]    ⎕wi 'xActiveDocument.Bookmarks.Item().Range.Text' 'tbl24' ('Plain
            text',⎕tclf,'Entered directly.')
[24]    ⎕wi 'xDocuments(1).Bookmarks.Item().Range.Style' 'tbl24' ((↑⎕wi
            'xDocuments(1).Styles("List Number")')⎕wi 'obj')
[25]    '.t1' ⎕wi 'XCell(,).Select' 4 1
[26]    '.t4' ⎕wi 'Create' ('Currency' 'Right' WordTable 'Currency'
            GetLocaleFormat 2 2⍴4?12288)
[27]    '.t4' ⎕wi 'XSelect'
[28]    (⎕wi 'xSelection') ⎕wi ¨(⊂¨(⊂'Font'),¨'.Name' '.Size'
            '.Italic'),¨⊂'Verdana' 9 0
[29]    '.t1' ⎕wi 'XCell(,).Select' 4 3
[30]    '.t4' ⎕wi 'Create' ('Text' 'JustifyLow' WordTable CharArray2 2
            15⍴30↑'System Building With APL+Win')
[31]    '.t4' ⎕wi 'XSelect'
[32]    (⎕wi 'xSelection') ⎕wi ¨(⊂¨(⊂'Font'),¨'.Name' '.Size'
            '.Italic'),¨⊂'Times New Roman' 8 0
[33]    '.t1' ⎕wi 'xAutoFitBehavior' (⎕wi '=wdAutoFitContent')
[34]    (⎕wi 'children') ⎕wi ¨⊂'Delete'
[35]    ⎕wi 'xvisible' 1
      ∇
```

Any explanation of this function is liable to be verbose: the **Word** VBA help file contains documentation of the syntax of the object properties, methods, and events. It produces the table shown in **Figure 10-6 Working with tables.** Note that **Word** uses the method Cell, in the singular, rather than the property Cells as **Excel** does, to refer to a row, column intersection.

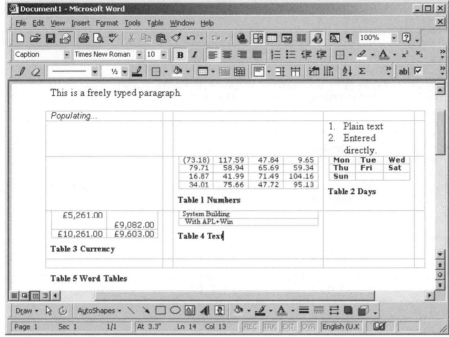

Figure 10-6 Working with tables

The key to understanding DemoTable is to execute each line of code in immediate mode and to observe the effect of each line in the **Word** session: make **Word** visible after line [2]. Some aspects of this function require explanation, as follows:

● Isolated parts of the original table, and of the tables within the original table, are formatted differently to allow the code and result to be verified.

● The basic clue is that it all works by selection of the appropriate placeholder within the active document. As predicted, **Word** is very code intensive.

● 'Selection' is a property of the application object; it refers to the active document. If multiple documents are open, prefix the 'Selection' property with the application's 'xDocuments()' property; its argument is the document index.

● Either the 'xDocuments()' or 'xActiveDocument' prefix is required with properties and methods of the document object—for example, refer to lines [23] and [24].

● Note the creation of objects from the return value of an APL function. For instance, in line [7] an object representing a table is created as ⍙wrd.t1: this facilitates reference to the table object. Always creates such objects as children of the **Word** object so that they are deleted with the **Word** object.

● In line [24], the style property is assigned a new value: the value is an object rather than a literal value.

DemoTable calls WordTable to insert the tables. Predictably, **Word** is very code intensive. Refer to the calls to WordTable in DemoTable and effect illustrated in **Figure 10-6 Working with tables** to trace the effect of each line of code in WordTable:

```
     ∇ Z←L WordTable R;Cols;Rows
[1]    ⍝ System Building with APL+Win
```

```
 [2]    ⍝ L = (Caption ColumnAlignment HeaderFlag), R = Array to populate
        cells
 [3]    R←1⌽⎕tclf,R,⎕tcnl
 [4]    Cols←1++/⎕tcht=((,R)⍳⎕tclf)↑,R
 [5]    Rows←1⌈↑⁻1↓⍴R
 [6]    ⎕wi 'xSelection.TypeText' ((,R)~⎕tcnul)
 [7]    0 0⍴⎕wi 'xSelection.MoveUp(Unit,Count,Extend)' (⎕wi
        '=wdParagraph') (⍫Rows) (⎕wi '=wdExtend')
 [8]    0 0⍴⎕wi
        'xSelection.ConvertToTable(Separator,NumRows,NumColumns,AutoFit)'
        (⎕wi '=wdSeparateByTabs') Rows Cols 1
 [9]    ⎕wi 'xSelection.Tables(1).AutoFitBehavior' (⎕wi
        '=wdAutoFitWindow')
 [10]   ⎕wi 'xSelection.Font.Size' 8
{1[11]  :for R :in ⎕wi ¨(⊂'=wdBorder') ,¨'Left' 'Right' 'Top' 'Bottom'
        'Horizontal' 'Vertical'
 [12]       ⎕wi 'xSelection.Tables(1).Borders().LineStyle' R (⎕wi
        '=wdLineStyleNone')
1}[13]  :endfor
{1[14]  :if 0≠⍴(,⎕io⊃L)~' '
 [15]       ⎕wi
        'xSelection.Tables(1).Range.InsertCaption(Label,Title,Position)'
        (⎕wi '=wdCaptionTable') (' ',,⎕io⊃L) (⎕wi
        '=wdCaptionPositionBelow')
1}[16]  :endif
 [17]   L←1↓L
 [18]
        R←'Center/Distribute/Justify/JustifyHi/JustifyLow/JustifyMed/Righ
        t'
 [19]   R←(R≠'/')⊂R
{1[20]  :if ↑(⎕io⊃L)∊R
 [21]       ⎕wi 'xSelection.ParagraphFormat.Alignment' (⎕wi
        '=wdAlignParagraph',∊(R∊L[⎕io])/R)
1}[22]  :endif
 [23]   L←1↓L
{1[24]  :if 0=↑0⍴∊L
 [25]       ⎕wi 'xSelection.Tables(1).Rows().Select' 1
 [26]       ⎕wi 'xSelection.Rows.HeadingFormat' (⎕wi '=wdToggle')
 [27]       ⎕wi 'xSelection.Font.Bold' ('=wdToggle')
1}[28]  :endif
 [29]   Z←'#' ⎕wi 'UseObject' (⎕wi 'xSelection.Tables(1)')
 [30]   0 0⍴⎕wi 'xSelection.EndKey(Unit)' (⎕wi '=wdStory')
 [31]   ⎕wi 'xSelection.TypeParagraph'
        ∇
```

The simplest and fastest method to export APL+Win arrays to a **Word** table is to send whole arrays in one operation. The optional left-hand argument specifies the caption for the table to be inserted; the right-hand argument is the array used to populate the table. Lines [3] – [4] determine the size of the table. Lines [12] and [14] refer to the first table in the selection range. As this function inserts a single table in the selected range, it will only ever contain one table, albeit this table contains other tables; that is, it is a nested table. The following line returns the number of tables in the active document:

```
      ⎕wi 'xActiveDocument.Tables.Count'
1
```

Table indices are sequential and, therefore, volatile. If another table is inserted between them, the index of the newly added table is 2 and the same line refers to the new table. The tables count property of the active document returns the number of tables within it—it does not include tables within tables:

```
⎕wi 'xActiveDocument.Tables(1).Tables.Count'
4
```

However, the automatic caption numbering counts all the tables. Hence, the caption of the single table created by DemoTable is 5. It is advisable to keep tables as simple as possible, for the following reasons:

● Work with a single table at a time—complete all the processing required within a dedicated function.

● Unless necessary, it is strongly advisable to work with rectangular tables rather than nested tables as the complexity of the code is far less and the information is more readable.

● Use **Word** as an automation server for pre-programmed reports, not for ad hoc reports, that are tested and debugged before distribution. It is virtually impossible to handle all errors from **Word** as the automation server.

The mechanism for sending APL+Win arrays to **Word** is straightforward. However, **Word** takes control of how it shows a table when its content does not permit the table to fit on the page orientation. For large tables, it may be necessary to create a section whose orientation is landscape—this does not resolve the problem but increases the chances of the problem not occurring in the first place:

```
       ∇ Z←L ReOrient R
 [1]    ⍝ System Building with APL+Win
 [2]    Z←⎕wi 'xSelection.PageSetup.Orientation'
 [3]    ⎕wi 'xSelection.InsertBreak(Type)' (⎕wi
        '=wdSectionBreakContinuous')
{⍳[4]   :select L
★ [5]        :case 'Portrait'
{2[6]           :if Z≠⎕wi '=wdOrientPortrait'
 [7]               ⎕wi 'xSelection.PageSetup.Orientation' (⎕wi
        '=wdOrientPortrait')
2}[8]           :endif
★ [9]        :case 'Landscape'
{2[10]          :if Z≠⎕wi '=wdOrientLandscape'
 [11]              0 0⍴⎕wi 'xSelection.PageSetup.Orientation' (⎕wi
        '=wdOrientLandscape')
2}[12]          :endif
1}[13]  :endselect
 [14]   0 0⍴'Section' ''  WordTable R
 [15]   ⎕wi 'xSelection.InsertBreak(Type)' (⎕wi
        '=wdSectionBreakContinuous')
 [16]   ⎕wi 'xSelection.PageSetup.Orientation' Z
       ∇
```

This function inserts a section break, switches the page orientation to landscape, inserts the table, and switches the orientation back to portrait:

```
      'Landscape' ReOrient 3 CharArray1 ⎕nl 2
0
      ⎕wi 'xActiveDocument.Sections.Count'
3
      ⎕wi 'xActiveDocument.Sections().PageSetup.Orientation' 1
0
```

```
        ⎕wi 'xActiveDocument.Sections().PageSetup.Orientation' 2
1
        ⎕wi 'xActiveDocument.Sections().PageSetup.Orientation' 3
0
```

10.4.4 Active document random access

All methods of the active document, such as delete, print, find, etc., act on the currently-selected section of the document: by default, this is the current cursor position with no characters being selected. The active document object provides mobility within the active document by relative reference to paragraphs or sentences; navigation by reference to the relative position of sentences is tortuous:

```
        ⎕wi 'xActiveDocument.Paragraphs.First.Range.Select' ⍝ First
        ⎕wi 'xActiveDocument.Paragraphs.Last.Range.Select' ⍝ Last
        ⎕wi 'xActiveDocument.Paragraphs.Last.Previous(1).Range.Select'
        ⎕wi 'xActiveDocument.Paragraphs.First.Next(1).Range.Select'
```

This allows the first, last or a paragraph offset from either position to be selected. Additionally, a paragraph can be selected randomly, by using:

```
        ⎕wi 'xActiveDocument.Paragraphs().Range.Select' 3
```

In practice, it may be very difficult to be precise about locating the 'correct' paragraph, especially when multiple paragraphs may be added via code.

10.4.4.1 Bookmarks

For total control, it is necessary to define custom placeholders within the active document—**Word** calls these bookmarks. With bookmarks in place, the automation client may work with—that is, populate, format, delete, print, etc.—at random. Bookmarks can be added to the active document with this function:

```
        ∇ Z←L AddBookmarks R
   [1]    ⍝ System Building with APL+Win
{1[2]     :if 0=⎕nc 'L'
   [3]         L←'bmk'
+ [4]     :else
   [5]         L←(,L)~' '
   [6]         L←L,(0=⍴L)/'bmk'
1}[7]     :endif
   [8]         Z←⎕wi 'xActivedocument.Paragraphs.Last.Range.Select'
   [9]         ⎕wi 'xSelection.TypeParagraph'
{1[10]    :for R :in +\R/1
   [11]        ⎕wi 'Selection.TypeText' (L,⍕R)
   [12]        0 0⍴⎕wi 'xSelection.HomeKey(Unit,Extend)' (⎕wi 'wdLine') (⎕wi
             '=wdExtend')
   [13]        Z←⎕wi 'xActiveDocument.Bookmarks.Add(Range,Name)' ((⎕wi
             'Selection.Range')⎕wi 'obj') (L,⍕R)
   [14]        Z←⎕wi 'xSelection.EndKey(Unit)' (⎕wi '=wdLine')
   [15]        ⎕wi 'xSelection.TypeParagraph'
1}[16]    :endfor
   [17]    Z←⎕wi¨(⊂⊂'xActiveDocument.Bookmarks().Name'),¨+\(⎕wi
             'xActiveDocument.Bookmarks.Count')/1
        ∇
```

The optional left-hand argument is a prefix, default *bmk*, for all new bookmark names; the actual name is determined by adding a serial number based on the right-hand argument, which specifies the number of bookmarks to be added.

The function adds the bookmarks at the end of the document and returns *all* the bookmarks that exist in the active document; this may be used by the calling function to track the active document. Any existing bookmark by the same name is deleted implicitly and without warning. The bookmarks contain the names of the bookmarks themselves; they are made visible by **Tools | Options | View** and **Show Bookmarks** is ticked and **Field Shading** is selected as 'Always':

```
      'sys' AddBookmarks 5
code1 code2 code3 sys1 sys2 sys3 sys4 sys5
```

In the latter example, five bookmarks called sys*n* have been added; the other bookmarks existed already. The following procedures are recommended:

- Bookmark names must be valid variable names; the same rules as for APL+Win variable names apply.

- Bookmark names are *not* case sensitive.

- Bookmark names are unique, irrespective of their location, within the active document.

- An existing bookmark name is deleted without warning if a new one with the same name is inserted.

- Bookmark names are returned in alphabetical order.

Bookmarks—either the bookmark name only or both the bookmark name and content may be deliberately deleted. If the content is to be deleted, the bookmark range must be deleted, as shown:

```
      ⎕wi 'xActiveDocument.Bookmarks().Delete' 'code1' ⍝ Name only
      0 0⍴⎕wi 'xActiveDocument.Bookmarks().Range.Delete' 'code1'
```

Available bookmarks can be queried with the following line:

```
      ⎕wi¨(⊂⊂'xActiveDocument.Bookmarks().Name'),¨+\(⎕wi
'xActiveDocument.Bookmarks.Count')/1
 code2 code3 sys1 sys2 sys3 sys4 sys5
```

From an automation client's point of view, an ideal solution is to be able to populate the range of a bookmark, randomly, and for the bookmark name to be deleted from the document—thereby making it unavailable in subsequent operations, thus:

```
      ∇ Z←GetBookmark R
[1]    ⍝ System Building with APL+Win
[2]    Z←⎕wi¨(⊂⊂'xActiveDocument.Bookmarks().Name'),¨+\(⎕wi
        'xActiveDocument.Bookmarks.Count')/1
[3]    Z←(⊂R)∈Z
[1[4]  :if Z
[5]        ⎕wi 'xActiveDocument.Bookmarks().Range.Select' R
+[6]   :else
[7]        Z←0
1}[8]  :endif
      ∇
```

This function returns 1 if the bookmark name specified as the right-hand argument exists: the bookmark range in the active document. Otherwise, it returns 0 and the current selection is the cursor position in the document:

```
      GetBookmark 'code2'
1
      ⎕wi 'xSelection.TypeText' (,⎕tclf,(⊃'This is a new.' 'Two paragraphs
are added'),⎕tcnl)
      ⎕wi¨(⊂⊂'xActiveDocument.Bookmarks().Name'),¨+\(⎕wi
'xActiveDocument.Bookmarks.Count')/1
 code3 sys1 sys2 sys3 sys4 sys5
```

In the example shown in **Figure 10-7 Populating bookmarks**, the bookmark *code2* is selected and its range is replaced by two paragraphs; thereafter, the bookmark name no longer exists. The range remains selected; therefore, formatting may be applied to it:

```
□wi xSelection.Font.Italic' 1
```

NOTE: The functions illustrating the use of bookmarks assume that □wself is assigned an instance of **Word** and that there is a document open.

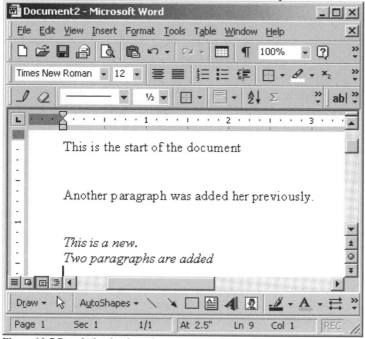

Bookmarks provide APL+Win with the ability to work randomly with named sections of **Word** documents. Any bookmarks that may not have been used may be deleted before saving the completed document.

Figure 10-7 Populating bookmarks

```
     ∇ DeleteBookmarks;i
 [1]    ⍝ System Building with APL+Win
{1[2]   :for i :in □wi¨(⊂⊂'xActiveDocument.Bookmarks().Name'),¨+\(□wi
         'xActiveDocument.Bookmarks.Count')/1
 [3]        □wi 'xActiveDocument.Bookmarks().Delete' i
1}[4]   :endfor
     ∇
```

This right-hand argument specifies an object that is a table in the active document. It adds bookmarks named tbl*ij*—where *i* and *j* represent the row and column—to each cell of the table and returns the collection of these bookmarks. Refer to line [8] and [23] of DemoTable on the usage of bookmarks in tables.

10.4.4.2 Ranges

A range object has a read only start and end property representing a selection or location in the active document; the selection does not need to be named. A predefined bookmark is a named range:

```
     □wi 'xActiveDocument.Range.Start' ⍝ always in index origin 0
0
     □wi 'xActiveDocument.Range.End'
1
```

Any arbitrary selection or bookmark occupies a part of a document: this may be a single character or whole paragraphs. Text can be inserted before or after a range using its *InsertAfter* and *InsertBefore* methods; the *InsertFile* method may be used to insert a complete file.

10.4.4.3 Styles

Format | Styles + New creates new styles; styles are stored in the document template in use and are available with any document that is based on that template. A style defines the visual content of the document range to which it is applied. In all likelihood, an organisation will have a preferred visual layout for its documents and already have a **Word** template. Generally:

● Styles provide a convenient route to the visual formatting of a document; the automation client needs much less code to achieve the visual layout.

● Styles can be re-applied automatically if **Tools | Templates and Add-Ins... + Automatically update document styles** is checked. This ensures that any changes to styles are automatically reflected in a document when it is opened.

● If a visual layout is achieved by the manual selection of font, size, etc., a different visual layout can be achieved only by re-selecting these visual properties; this is very code intensive. Line [17] of DemoTable illustrates the manual method and line [24] shows the application of styles.

● The Normal style is used predominantly in any document. This style is applied by default. Thus, if the visual layout of this style is 'correct', the automation client does not need to apply any formatting or style to most of the document.

10.4.5 Formula

The menu option **Insert | Field...** displays a dialogue that enables fields to be inserted into the document. The syntax for entering formulae is shown in lines [9] – [18] of the IncludePicture function. The variation is in the text argument: in order to find the text argument, insert the desired formula in a document, press **ALT + F9** and examine the formula. This is used to insert formulae, DDE fields, IncludePicture, etc. fields.

Ctrl + F9 inserts { } and the formula is manually specified within the { }.

If { }, double braces, is typed in the document it will not be recognised as a placeholder for a field: it must be inserted using the **Ctrl + F9** keystroke.

Form Fields are usually inserted using the **Forms** toolbar shown in **Figure 10-8 Field formula**. Alternatively, locate the cursor at the desired insertion point and press **Ctrl + F9.**

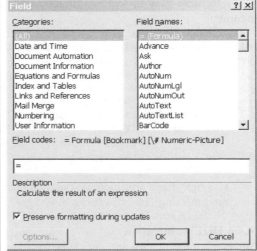

Figure 10-8 Field formula

Specialist formulae may be entered with the Table | Formula… option, shown in **Figure 10-9 Form field formula**.

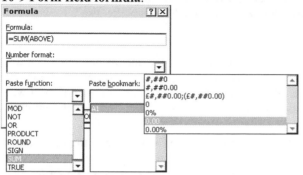

The flexibility of this method is shown in the dialogue to the left.

This method is designed for inserting computational formulae in tables; however, it works equally well outside tables.

Figure 10-9 Form field formula

A '= (Formula)' field, accessible using **Insert | Field**, can use the predefined functions shown below. All functions may take numbers, other formulae, or bookmarks holding numeric values as arguments:

- The functions shown with empty parentheses accept vectors of any length; a comma (,) is used to separate elements.

- Other functions cannot work with vectors.

- The following functions can accept references to table cells as arguments: AVERAGE(), COUNT(), MAX(), MIN(), PRODUCT(), and SUM().

- Table cells are referred to by the intersection of a column, specified by a letter from A to Z, and a row specified as an integer starting from 1; all elements of a column may be specified as 'ABOVE' and all elements of a row as 'LEFT'. These keywords mean all cells in the direction implied up to the topmost or leftmost cell or a non-numeric cell is that direction. Contiguous elements may be specified as *StartCell:EndCell*, for example C2:C7 specifies elements 2 to 7 from the third column.

A formula may be entered directly, by default at the cursor position, into the active document:

```
     ∇ InsertFormula R
[1]    ⍝ System Building with APL+Win
[2]    ⍝ ⎕wself is instance of Word (it has an open document)
[3]    0 0⍴⎕wi 'xSelection.Fields.Add(Range,Type,PreserveFormatting)' ((⎕wi
          'xSelection.Range') ⎕wi 'obj') (⎕wi '=wdFieldEmpty') 0
[4]    ⎕wi 'xSelection.Range.Text' (((('='≠↑⍴R)/'='),R)
[5]    0 0⍴⎕wi 'xSelection.Fields.Update'
[6]    ⎕wi 'xActiveWindow.View.ShowFieldCodes' 0
[7]    0 0⍴⎕wi 'xSelection.EndKey(Unit)' (⎕wi '=wdLine')
[8]    ⎕wi 'xSelection.TypeText' ' '
     ∇
```

The right-hand argument is specified with or without a leading = sign: line [4] adds = if it is found missing: InsertFormula 'Average(3,4,5,9)'
```
⍝ Or InsertFormula '=Average(3,4,5,9)'
```

10.4.6 DDE automation

This is a legacy feature and should be avoided as it works less and less reliably with current versions of Windows, because:

• Microsoft no longer recommends the use of Dynamic Data Exchange (**DDE**); this technology may be removed or discontinued in future versions of Windows.

• **Word** must start the server application before it can retrieve data and it does not terminate the server session on completion. This makes it very time consuming as well as untidy.

• DDE automation works differently between **Excel** and **Word**. APL+Win does not support this method directly but via its GUI objects only.

• DDE links tend to be problematic and unreliable; this has a tendency to destroy user confidence.

The syntax can be illustrated as follows:

```
{ DDEAUTO Excel "c:\\ajay\\office\\Book1.xls" MyRange }
```

This will insert a predefined range called 'MyRange' from the workbook 'C:\AJAY\OFFICE\BOOK1.XLS' as a table in **Word;** it will appear as follows:

```
{ DDEAUTO WINWORD "c:\\ajay\\office\\source.doc" abmk }
```

This will insert the contents of a predefined bookmark 'abmk' from the document SOURCE.DOC into the current document. This is achieved more reliably using the INCLUDETEXT field.

10.4.7 INCLUDEPICTURE

Although pictures within documents may at first sight appear to be a cosmetic gimmick, there are persuasive reasons for their use, especially for automation. In particular:

• The automation client may customise the document headers, continuation headers, footers, and continuation footers to provide a visual identity to the documents it creates.

• The picture file can include company logos or, at a lower level, the identification of branches and teams within the organisation who may have different contact points within the same organisation.

• A further benefit of being able to incorporate headers and footers in a document is that it eliminates the need for pre-printed stationery. Should these headers and footers include colour images, the organisation needs only to organise bulk printing on a colour printer, possibly entrusted to a print bureau.

• Graphical representation of complex information—charts—may be included in the document in order to enhance presentation.

A picture may be incorporated in a document using **Insert | Picture | From File**; this includes a copy of the picture in the document. This APL function achieves the same thing:

```
      ∇ L AddPicture R
 [1]    ⍝ System Buildint with APL+Win
{1[2]   :if 0=⎕nc 'L'
 [3]        L←0
1}[4]   :endif
 [5]    L←↑×L
 [6]    0 0⍴⎕wi
        'xSelection.InlineShapes.AddPicture(FileName,LinkToFile,SaveWithD
        ocument)' R L (~L)
      ∇
```

The left-hand argument default to 0; it controls whether the picture is included as a static image or whether a reference to the picture file is stored. The right-hand argument is the name of the picture file.

An alternative, more powerful, method is to use the INCLUDEPICTURE field; this is accessed via the **Insert | Field...** menu option in **Word**. This enables an external picture file to be incorporated in a document, conditionally if necessary. Most picture formats are supported.

Figure 10-10 INCLUDEPICTURE dialogue shows the dialogue for entering picture fields interactively. This field takes two arguments; its syntax is as follows:

```
INCLUDEPICTURE path\filename switches
```

If the optional switch (\d) is used with this field, the specified picture is not stored in the document, thereby reducing file size:

• Thus, the picture file can be maintained independently of the document.

• If a picture in included with the \d switch using a relative path reference, the picture file must be distributed also. The document can be reconstructed if both files are stored at the same location.

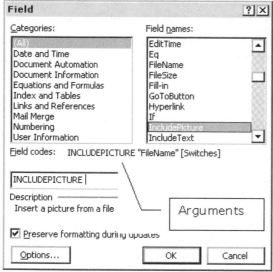

The \d switch causes a reference to the location of the picture file rather than the picture itself to be stored in the document.

If the name of the picture file is specified without a path, its path is relative to the main document. If both path and filename are specified, the picture is expected at the specified path.

In the latter case, the document can be reconstructed on another computer if the picture file exists at the same path name.

Figure 10-10 INCLUDEPICTURE dialogue

This is likely to be problematic for documents that are distributed, for the following reasons:

• It is preferable to omit the \d switch and to force the picture to be stored in the document. The picture is stored as a static object and will not change if the picture file itself is changed.

• If the picture needs to be refreshed in the document—for example, if it is a chart object that is likely to change, use relative reference to its path.

The function IncludePicture can be used to add a picture field to a document:

```
     ∇ Z←L IncludePicture R
[1]    ⍝ System Building with APL+Win
{1[2]   :if 0=⎕wi 'xSelection.Fields.Count'
[3]       R←,R
{2[4]     :if 0≠⎕wcall 'PathFileExists' R
[5]         ((R='\')/R)←c'\\' ⍝ Need to double path separators
[6]         Z←'INCLUDEPICTURE "*"\d'
[7]         ((Z='*')/Z)←c∈R
[8]         R←∈Z
[9]         Z←c((⎕wi 'xSelection.Range') ⎕wi 'obj') ⍝ Range
[10]        Z←Z,c(⎕wi '=wdFieldEmpty')              ⍝ Type
```

```
   [11]              Z←Z,⊂R                                      ⍝ Text
   [12]              Z←Z,⊂1                          ⍝ PreserveFormatting
   [13]              Z←×⎕wi
          (⊂'xSelection.Fields.Add(Range,Type,Text,PreserveFormatting)'),Z
{3[14]                :if 0≠⎕nc 'L'
   [15]                  ⎕wi 'xSelection.MoveLeft(Unit,Count,Extend)' (⎕wi
          '=wdCharacter') 1 (⎕wi '=wdExtend')
   [16]                  ⎕wi 'xSelection.InsertCaption(Label,Title,Position)'
          (⎕wi '=wdCaptionTable') (' ',,L) (⎕wi '=wdCaptionPositionBelow')
3}[17]                :endif
   [18]              0 0⍴⎕wi 'xSelection.TypeParagraph'
+ [19]          :else
   [20]                  Z←0×Msg 'Unable to find file ',R
2}[21]          :endif
+ [22]      :else
   [23]          Z←0×Msg 'Unable to insert picture at current location.'
1}[24]      :endif
        ∇
```

The optional left-hand argument of this function is the caption of the picture object. The right-hand argument specifies a string that represents the *path\filename*. The \d switch is used by default—see line [6]. In immediate execution mode, enter the following expression:

```
      'Picture Caption' IncludePicture 'c:\my documents\my
pictures\iceberg.jpg'
1
```

This includes the picture in the active document; it appearance in the document depends on a **Word** Boolean flag; the flag can be queried thus:

```
      ⎕wi 'xActiveWindow.View.ShowFieldCodes'
0
```

If this flag is **TRUE**, **Figure 10-11 Formula field** illustrates how a field is seen instead of the picture.

{ INCLUDEPICTURE "c:\\my documents\\my pictures\\iceberg.jpg"\d * MERGEFORMAT }

Table { SEQ Table * ARABIC } Picture

Figure 10-11 Formula field

Pictures in **Word** documents interact with the surrounding text; this means that the automation client needs to control the appearance of the pictures. In order to avoid this overhead, include pictures within cells of a table: this confines the picture to the cell and prevents interaction with text—table positions are fixed at the point of insertion in the document. Select the desired cell before using AddPicture or IncludePicture; both use the selected range to insert pictures.

10.4.8 INCLUDETEXT

The INCLUDETEXT field is accessed via the **Insert | Field...** menu option in **Word**. **Figure 10-13 INCLUDETEXT field** shows the dialogue presented when the fields is added interactively.

This enables data from an external source to be incorporated in a document, conditionally if necessary. Unfortunately, **Word** does not understand APL+Win component files or workspace: these sources cannot be used.

This field may be used to include textual information from other **Word** documents. The formatting of the bookmark range in the source document is preserved in its entirety in the active document.

This feature of **Word** facilitates the creation of a document as a collation of bookmarks from other saved documents. Starting with a new document, some text is added and then an INCLUDETEXT field is inserted at the current selected range. This inserts a field in the active document.

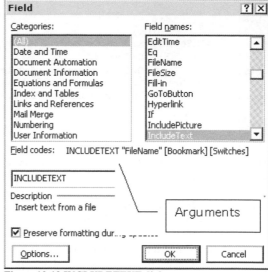

Figure 10-12 INCLUDETEXT dialogue

```
    ⎕wi 'xSelection.TypeText' 'This document has got some text added by
the automation client.'
    ⎕wi 'xSelection.TypeParagraph'
    'LifeCover' IncludeText 'c:\system building with
apl+win\chapter10\life rating.doc'
1
```

The net result is shown in **Figure 10-13 INCLUDETEXT field**. The text that is highlighted has been inserted from another file. The highlighting is achieved by switching field shading on using **Tools | Options + View**. Note the gaps in the text drawn in from another document: this is highlighted in **Figure 10-13 INCLUDETEXT field**. The gaps are placeholders for other information.

The function IncludeText can be used to add a text field to a document:

```
      ∇ Z←L IncludeText R
  [1]      ⍝ System Building with APL+Win
{1[2]      :if 0=⎕wi 'xSelection.Fields.Count'
{2[3]          :if 0≠⎕wcall 'PathFileExists' R
  [4]              R←,R
  [5]              ((R='\')/R)←⊂'\\'  ⍝ Need to double path separators
  [6]              Z←'INCLUDETEXT "*" ? '
  [7]              ((Z='*')/Z)←⊂∊R
  [8]              Z←∊Z
{3[9]              :if 0≠⎕nc 'L'
  [10]                 ((Z='?')/Z)←⊂∊L
+ [11]              :else
  [12]                 Z←(Z≠'?')/Z
3}[13]             :endif
  [14]             R←∊Z
  [15]             Z←⊂((⎕wi 'xSelection.Range') ⎕wi 'obj')  ⍝ Range
  [16]             Z←Z,⊂(⎕wi '=wdFieldEmpty')               ⍝ Type
  [17]             Z←Z,⊂R                                    ⍝ Text
  [18]             Z←Z,⊂1                                    ⍝
      PreserveFormatting
```

```
 [19]            Z←×⎕wi
        (⊂'xSelection.Fields.Add(Range,Type,Text,PreserveFormatting)'),Z
 [20]           0 0ρ⎕wi 'xSelection.TypeParagraph'
+ [21]     :else
 [22]            Z←0×Msg 'Unable to find file ',R
2}[23]    :endif
+ [24]  :else
 [25]        Z←0×Msg 'Unable to insert text at current location.'
1}[26]  :endif
     ▽
```

The optional left-hand argument of this function is the name of a bookmark in the document specified as the right-hand argument, specified as a string that represents the *path\filename*:

```
    'LifeCover' IncludeText 'c:\underwriting\life rating.doc'
```

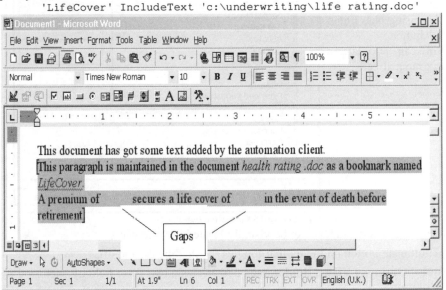

Figure 10-13 INCLUDETEXT field

Thus far, a single field has been inserted into the document:

```
    ⎕wi 'xActiveDocument.Fields.Count'
3
```

Yet, the document has 3 fields; the gaps mark the position of the other two fields defined within the document *life rating.doc,* from which the text has been drawn. The latter two fields are form fields; their definition has been preserved in the new document. A document can be assembled quickly by drawing text (saved as bookmarks) from any number of saved documents.

10.4.9 Form fields

Form fields are placeholders within a document. The original intention for these fields is that the document can be protected, leaving users to fill in the fields: hence the name derivation. An automation client can use the definition of these fields to add their content programmatically using data from the workspace. Several types of form fields are supported.

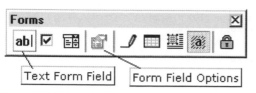

Figure 10-14 Forms toolbar options

Select Tools | Options | Customize; this displays the dialogue shown in **Figure 10-15 Forms toolbar**.

Select the Forms toolbar to display the toolbar shown in **Figure 10-14 Forms toolbar options**.

The **Text Form Field** button, available on the 'Forms' toolbar, inserts a field at the cursor location.

Then, the field may be customised by using the **Form Field Options** button.

APL+Win can populate a form field programmatically.

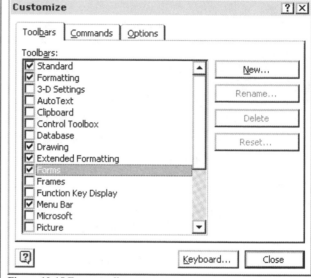

Figure 10-15 Forms toolbar

A text form field in a document can be visually located in three ways:

● First, the penultimate button on the toolbar toggles the shading on and off.

● Second, **Tools | Options | View | Bookmarks** demarcates a text form field by enclosing it within square brackets.

● Third, **Tools | Options | View | Field Codes** toggles the view between the field identification text and the field contents.

A form text field may be highlighted in any or all three ways.

Word text form fields are highly configurable from a single and dynamic dialogue box, shown in **Figure 10-16 Form field (Number) dialogue**.

10.4.9.1 Form field options

Word automatically assigns and names a bookmark for each form field; the pattern of the name is the string 'Text' followed by a number.

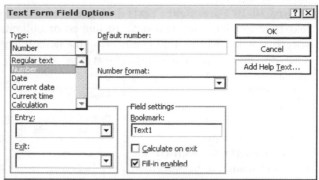

Figure 10-16 Form field (Number) dialogue

The automatic numbering convention starts with 1; any intermediate numbers that may have become available because a field might have been deleted are not reused.

Word 2000 imposes a limit of 40 characters for bookmark names and 16,379 bookmarks per document; a document size is limited to 32 mega bytes. In practice, these limits are not restrictive especially since some form fields can be populated with array data.

The **AddHelp Text** button displays the dialogue shown in **Figure 10-17 Form field custom help**, allowing customised help text for each field and optional **AutoText** entries. The add help text facility provides a convenient means of documenting the contents of a field.

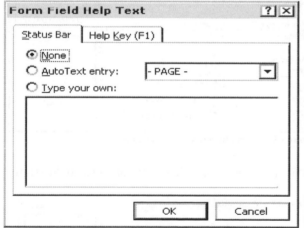

Form fields also boast the facility to specify names of custom macros to run on a form field gaining and losing focus.

In order to facilitate the identification of a form field within APL+Win, a method using ordinal numbers is less reliable than one using names; ordinal numbers adjust dynamically but names are unique.

Figure 10-17 Form field custom help

Word form field supports several field types. The content of a form text can be regular text, numeric, date, current date and current time read from the system clock, and a formula. Each field type is independently configurable. The field configuration dialogue is invoked either by clicking the **Form Field Options** button or by double clicking on the field.

Form fields can be referred to by an ordinal number starting from the top of the document, irrespective of the order in which they were added to the document. In other words, form fields are automatically renumbered; also, they are automatically resized to accommodate their contents.

10.4.9.2 Regular text

Figure 10-18 Form field (Text) dialogue shows the configuration of text fields. A regular text field can have any text data, of limited or unlimited length. The length of the field is specified in the 'Maximum Length' box—the default length is 'Unlimited'.

The specified length and format of a text field is applied automatically.

This means that the automation client need only supply the contents.

If a length is specified, the text is truncated to the specified length.

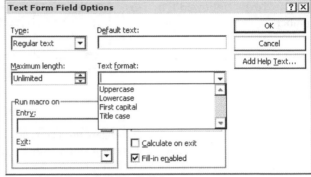

Figure 10-18 Form field (Text) dialogue

10.4.9.3 Number

The configuration options for a number field are shown in **Figure 10-19 Form field (Number) formats**.

A number form field accepts any valid number; its specified format is applied automatically.

A number of predefined formats, including percentage, currency symbol, thousand separators, and negative numbers are available.

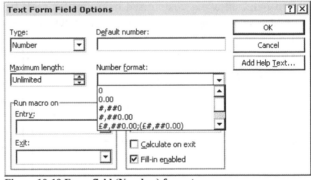

Figure 10-19 Form field (Number) formats

10.4.9.4 Date

The configuration options for a date field are shown in **Figure 10-20 Form field (Date) formats**.

A date form field accepts any valid date and formats its in accordance with the specified format, if any.

A custom format can be specified or a format may be selected from the list of predefined formats.

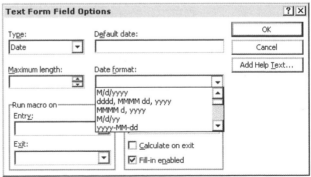

Figure 10-20 Form field (Date) formats

10.4.9.5 Current date/Current time

These fields do not accept arbitrary input; their content is read from the system clock, in the selected format, if any. If either of these fields is included in a document, it provides a reference to the date when the document was updated and saved. Thus, the content of these fields is volatile.

10.4.9.6 Calculation

The configuration options for a date field are shown in **Figure 10-21 Form field (Calculation) dialogue**.

A calculation form field accepts any valid formula whose result is output in accordance with the specified format, if any. The formula can be expressed in terms of other numeric form fields in the same document. This is a numeric field.

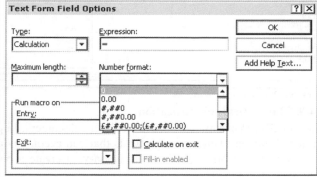

Figure 10-21 Form field (Calculation) dialogue

Theoretically, a **Word** document can have a maximum of 16,379 bookmarks. In practice, a document that exceeds more than a few dozen bookmarks becomes unmanageable.

A bookmark is inserted in a document by highlighting a portion of its content and clicking **Insert | Bookmark...** The dialogue is shown in **Figure 10-22 Bookmark management**.

Bookmark names are unique within a document—note that the dialogue shows the existing bookmark names as an aid to managing the list of bookmarks.

If an existing bookmark is selected, the **Go To** button highlights the section of the document the bookmark spans.

Optionally, hidden bookmarks may be removed from the list by unchecking the **Hidden Bookmarks** check box.

Figure 10-22 Bookmark management

An alternative method of identifying a bookmark is by pressing **F5** (the dialogue in **Figure 10-23 Tracking bookmarks** is displayed) and selecting **Bookmark**.

Any visible bookmark, that is, one that is not hidden, can be selected.

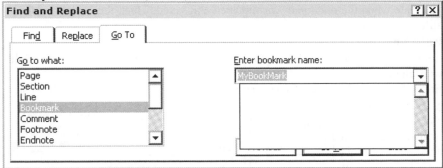

Figure 10-23 Tracking bookmarks

10.4.9.7 Form fields identification

Theoretically, **Word** allows 32,000 fields in a document. In practice, a workable limit is less than a few dozens. Unlike merge fields that can accommodate scalar data only, form fields can accommodate arrays; it is not necessary to create a bookmark for every item of data when using form fields.

Word identifies form fields as type 70, *wdFieldFormTextInput*; this field type contains the sub types shown in **Table 10-2 Form field identification**.

A format for each type may be specified, as can a macro for pre-processing its content and another for post-processing.

Type	Description
0	Regular Text
1	Number
2	Date
3	Current Date
4	Current Time
5	Calculation

Table 10-2 Form field identification

The custom help facility, 'entry', and 'exit' macros work only when the form—document—is protected; the form is protected by clicking the button showing a padlock on the toolbar (see **Figure 10-14 Forms toolbar options).** The form must be unprotected for purposes of automation from APL+Win; therefore, these facilities will not be used except when designing the form. If macros are used, the template attached to the form must be distributed with the form; the macros are stored in the template.

The calculation form field type can contain a formula.

10.5 Populating form fields

Although intended for on-screen forms, form fields provide ideal placeholders for automation. A shell document can be formatted as desired with form fields; the form fields can be populated by the automation client and saved under a different name. This preserves the original document for reuse:

```
⎕wself←'wd' ⎕wi 'Create' 'Word.Application'
⎕wi 'xDocuments.Open' 'c:\ajay\ReportxShell.Doc'
```

This creates an instance of **Word** and opens a document in preparation for filling its form fields using automation: note that ⎕wself is assigned the instance of **Word**.

Form fields can be very sophisticated. For APL+Win, they are particularly useful because they can contain arrays. For trouble-free formatting, always insert fields that will contain arrays in table cells; this prevents the columns from running into each other because:

● Form fields are numbered sequentially from the start of the active document and not in the order in which they were created. The first field index is always 1 irrespective of the APL+Win index origin.

● Any form field may be referred to either by ordinal index or by name; a form field name is not case sensitive.

Form fields may be populated in two ways:

● If the field is to be deleted after it has been populated, use this syntax:
```
⎕wi 'xActiveDocument.FormFields(1).Range.Text' 'Deletes it'
```

● If the field is to be preserved after being populated, use this syntax:
```
⎕wi 'xActiveDocument.FormFields(2).Result' 'Preserves it'
```

The latter method is preferable for APL+Win because it permits a simple loop for completing all the fields. The function FormText is defined as follows:

```
      ∇ FormText;Result
 [1]    ⍝ System Building with APL+Win
{1[2]    :for i :in +\(⎕wi 'xActiveDocument.FormFields.Count')/1
 [3]        Result←GetResult ⎕wi 'xActiveDocument.FormFields(1).Name'
 [4]        ⎕wi 'xActiveDocument.FormFields(1).Result' Result
1}[5]    :endfor
      ∇
```

The GetResult function retrieves the data that is used to populate the field. A simple arrangement is to create a collection of name/value pairs, shown in **Figure 10-24 Form field name/value pairs**.

```
      Name←'Name' 'Salary' 'Address'
      Formula←'Name' 'MyArray[1;]' 'Address[⍳5;]'
      FormDef←'Create' Collection 'Name' 'Formula'
      ]display FormDef
.→------------------------------------------------------------------------------.
|.→--------------.. →-----------------------------------------------------------.|
|| .→---..→------.||.→==============================..→==========================.||
|||Name||Formula|||| .→---..→------..→-------. ||.→---..→-----------..→----------.|||
|||'----''-------'||||Name||Salary||Address||||Name||MyArray[1;]||Address[⍳5;]||||
||'∈--------------'|||'----''------''--------'|||'----''-----------''-----------'|||
|                  ||'∈---------------------'||'∈------------------------------'||
|                  |'∈-----------------------------------------------------------'|
'∈------------------------------------------------------------------------------'
```

Figure 10-24 Form field name/value pairs

This function makes it very straightforward to return the appropriate data; it is defined as:

```
      ∇ Z←L GetResult R
 [1]    ⍝ System Building with APL+Win
 [2]    Z←'Value' Collection L 'Name'
      ∇
```

Any other implementation for the acquisition of the correct data is also viable and may be desirable if more complex data retrieval is necessary.

Name	Formula	Comment
Name	Surname	Returns contents of variable Surname
Sex	⎕fread tie, 3	Returns component 3 of the file tie to *tie*.
Salary	Salary[;9]	Returns an array of salary.

Table 10-3 Field data sources

The essential principle is to match the name of the form field with the Name column in **Table 10-3 Field data sources** and to execute the corresponding entry in the Formula column. The Formula column may be as simple or complex as desired.

In general, the form field definition should include all its desired properties. One exception is that if an array of value is returned, APL+Win needs to transform it into a vector. For example, the Salary[;9] column should be specified thus:

```
,(⍕Salary[;,9]),⎕tcnl
```

Note that the column is extracted as a matrix, formatted as character, a column of carriage returns is added, and then the result is ravelled: the result is a vector. Should it be necessary to have precise control over the way the contents of form fields, a different approach is required. This is coded as follows:

```
       ∇ FormText1;Result
  [1]    ⍝ System Building with APL+Win
{1[2]    :for i :in +\(⎕wi 'xActiveDocument.FormFields.Count')/1
  [3]        Result←GetResult ⎕wi 'xActiveDocument.FormFields(1).Name'
{2[4]        :select ⎕wi 'xActivedocument.FormFields().TextInput.Type' i
★ [5]            :case 0 ⍝ Regular Text
  [6]                ⍝ custom processing for Result
  [7]                ⎕wi 'xActiveDocument.FormFields(1).Result' Result
★ [8]            :case 1 ⍝ Number
  [9]                ⍝ custom processing for Result
  [10]               ⎕wi 'xActiveDocument.FormFields(1).Result' Result
★ [11]           :case 2 ⍝ Date
  [12]               ⍝ custom processing for Result
  [13]               ⎕wi 'xActiveDocument.FormFields(1).Result' Result
★ [14]           :case 3 ⍝ Current Date
  [15]               ⍝ custom processing for Result
  [16]               ⎕wi 'xActiveDocument.FormFields(1).Result' Result
★ [17]           :case 4 ⍝ Current Time
  [18]               ⍝ custom processing for Result
  [19]               ⎕wi 'xActiveDocument.FormFields(1).Result' Result
★ [20]           :case 5 ⍝ Calculation
  [21]               ⍝ custom processing for Result
  [22]               ⎕wi 'xActiveDocument.FormFields(1).Result' Result
2}[23]       :endselect
1}[24]   :endfor
       ∇
```

This function tracks the form field types and custom processing can be inserted where the comment appears.

On completion, the active document needs to be saved to another name:

```
⎕wi 'xActiveDocument.SaveAs(FileName,FileFormat)' 'c:\2004Q1\ReportX.Doc'
(⎕wi '=wdFormatDocument')
```

10.5.1 Error trapping

Word generates a runtime error on any attempt to populate a form field with incompatible data. Note that all fields are populated with character data. APL+Win needs to specify a literal; amounts and dates need to be formatted as character. For example, an attempt to write text to a date form field will cause an error.

Using form fields for generating documents implies that the document is created in advance. The automation process should be tested to ensure that the correct data is used throughout.

10.6 Word vs. Excel for APL+Win automation

Unlike **Excel**, which allows custom user-defined functions on worksheets, **Word** does not allow such functions to be used in the active document. In **Excel**, user-defined functions within cells or ranges can retrieve data from APL+Win—interaction with APL+Win can be invoked from the **Excel** user interface. The predefined functions that can be used within the active document are listed in section **10.4.5 Formula**.

Word is the ideal tool for bulk document creation using its mail merge facility. There are two basic limitations with **Word** 2000 and a further one with mail merge itself, namely:

● There is no mechanism for producing individual documents; the result is either a new document, subject to Windows resource availability, or printed output.

● There is no way to use ADO record objects as the data source. Internally, DDE is used. However, text files may be used as a data source.

● Mail merge inserts scalar data only in the mail merge fields; multiple rows of data cannot be inserted.

As the form letter and the data source are free standing, the implementation of mail merge facilities is not ideal for automation: this type of facility is implemented more efficiently using the **Word** user interface. Mail merge applications tend to be very time consuming, and will lock the automation client in a wait loop for the duration.

10.7 Automation

Although APL+Win creates its own instance of **Word**, the automation of **Word** via APL+Win works better when **Word** is not in use. This avoids system resource problems and saves the user from confusion. Whether or not **Word** is in use can be determined programmatically as follows:

```
      0=⎕wcall 'FindWindow' 'OpusApp' 0
```

The result is **TRUE** if **Word** is not in use and **FALSE** otherwise. An existing **Word** session may be closed forcibly—thereby losing the current session changes—with the following function:

```
      ∇ CloseWord;Z
 [1]    ⍝ System Building with APL+Win
 [2]    Z←⎕wcall 'Findwindow' 'OpusApp' 0
{1[3]   :if 0≠Z
 [4]        ⎕wcall 'SendMessage' Z 16 0 0 ⍝ WM_CLOSE = 16
1}[5]   :endif
      ∇
```

10.7.1 APL+Win automation issues

A common issue with **Word** automation is the formatting of the documents created. Formatting is usually very time consuming and must be avoided wherever possible. This can be achieved in two ways:

● Create and save a document using **Word** and open it with APL+Win for adding runtime information. The shell document can be formatted as required, preferably using a **Word** template. The placeholders in this document may be form fields or IncludeText and IncludePicture fields.

• Create and save a library of **Word** documents; each of these can contain paragraphs that are formatted as desired and contain fields. APL+Win can create a new document and within it, collate the requisite paragraphs from one or more of these documents. Any fields in the new document can then be completed in the usual way.

The technique for automation from APL+Win is the same as for **Excel**. Use the macro recorder to capture the code and translate it to APL+Win. The **Word** macro recorder is less effective than that of **Excel** because some of the functionality, such as those relating to the *Options* and *Information* properties, is not available via the **Word** user interface; therefore, the macro recorder cannot produce sample code that can be adapted for APL.

Chapter 11

Working with Access

Microsoft's **Access** is a complex application; its deployment requires more effort than other applications in the Office suite. **Access** does not have a macro recorder facility, unlike the **Excel and Word** macro recorder, which produces code that can be examined and adapted to APL. If APL, an automation client, is to use **Access**, as an automation server, the APL developer needs detailed knowledge of the **Access** objects hierarchy and their respective properties, methods and events. In practice, this quest may well prove unattainable — the impact on the development time cycle and cost may be unjustifiable.

An alternative approach is required: **Access** is a database application and as such, it can be manipulated with the ActiveX Data Object (**ADO**) and Structured Query Language (**SQL**): refer to Chapter 12 *Working with ActiveX Data Object (ADO)* and Chapter 14 *Structured Query Language*. Both ADO and SQL are much easier to use with APL+Win than the **Access** object. This approach also promises a consistent approach to data sources—after all **Access** is a data source—such as text files, **Excel** files, SQL Server and Oracle which do not have the Component Object Model (**COM**) interface as **Access** does.

11.1 The Access pathways

Access is a complex application that can be used in several ways. Each option exposes a different compromise, namely.

● It is a powerful desktop application: its native object hierarchy and **VB** for Application (VBA) enable detailed control of the 'front end' user interface and easy management of tables, queries, reports, forms. **Access** uses Data **Access** Object (**DAO**).

● It provides a user interface to other data sources. An **Access** database can contain links to tables that exist in most popular data sources, including text files, **Excel** workbooks, Oracle and SQL Server databases, and another **Access** database. In this respect, **Access** is still a file-based MDB data source.

● It provides a user interface to SQL Server (or MSDE) databases when working with data projects (ADP data sources). In this respect, **Access** is a client/server data source.

● It is a COM application and can be used independently of its user interface. In this mode, the automation client uses the **Access** native hierarchy and VBA for managing the database contents.

● **Access** databases represent a data source and, like other data sources, it can be used via ADO. ADO has its own object hierarchy, which is very different from the **Access** object hierarchy.

If ADO is used, there are two routes for connecting to the **Access** data source:

● Use the ODBC driver.

• Use the JET provider.

It is clear that **Access** offers competing strategies for its deployment. In order to decide on the optimal deployment of **Access** as an extension to an APL+Win application, a detailed assessment of the requirements of the application is necessary. There are two propositions:

• Use ADO with the ODBC driver to access the **Access** data source; the tables and queries are built with the **Access** user interface. This option is recommended if the intention is to use the data source from the Access user interface and APL + Win.

• Use ADO and the JET provider to manage the **Access** data source contents, independently of the **Access** user interface. This option is recommended if the **Access** data source is to be used exclusively by APL+Win.

The overriding qualification is that **Access** data sources are used exclusively with native objects; that is, linked tables or data projects are managed at source and not from **Access**. Quite apart from anything else, the performance compromise involved in having **Access** mediate between APL+Win and the data source underlying linked tables or data projects sources is difficult to justify, especially since APL+Win can access these sources directly.

11.1.1 Access smoke and mirrors

The JET provider is a master magician who can stall application development in ways that are nothing short of spectacular. A simple example using the NWIND.MDB database will demonstrate the traps that JET lays silently.

Assuming that a query by the same name does not already exist, the following APL+Win function can create a QUERY in NWIND.MDB:

```
      ∇ CreateQueryJET R;⎕wself
[1]    ⍝ System Building with APL+Win
[2]    ⍝ R is the SQL defining a query; uses a new copy of NWIND.MDB
[3]    ⎕wself←'ADO' ⎕wi 'Create' 'ADODB.Connection'
[4]    ⎕wi 'XOpen' 'Provider=Microsoft.Jet.OLEDB.4.0;Data
          Source=c:\nwind.mdb'
[5]    0 0⍴⎕wi 'XExecute' R
[6]    'ADO' ⎕wi 'Delete'
      ∇
```

```
      CreateQueryJET 'Create View [JetQuery] as select * from customers'
```

Note that the right-hand argument of the function is an SQL statement. Although **Access** shows 'Queries' in its user interface, the SQL word is 'View': the two descriptions are interchangeable. As the function did not generate a runtime error, it can be assumed that the query has been created; running it can prove this:

```
      ExecuteQueryJET 'JetQuery'
91
```

The function ExecuteQueryJET is defined as follows:

```
      ∇ Z←ExecuteQueryJET R
[1]    ⍝ System Building with APL+Win
[2]    ⍝ R is the SQL defining a query; uses a new copy of NWIND.MDB
[3]    ⎕wself←'ADO' ⎕wi 'Create' 'ADODB.Connection'
[4]    ⎕wi 'XOpen' 'Provider=Microsoft.Jet.OLEDB.4.0;Data
          Source=c:\nwind.mdb'
[5]    0 0⍴⎕wi 'XExecute>ADO.Recordset' ('select * from [',R,']' )
[6]    ⎕wself←'ADO.Recordset'
[7]    ⎕wi 'XClose'
[8]    ⎕wi 'xCursorType' 3 ⍝ adOpenStatic
[9]    ⎕wi 'XOpen'
```

```
[10]   Z←□wi 'xRecordCount'
[11]   'ADO' □wi 'Delete'
       ∇
```

The function runs successfully and returns a count of the records in the CUSTOMERS table. Therefore, the **Access** user interface must show it in the Queries dialogue shown in **Figure 11-1 JET hides query**.

Figure 11-1 JET hides query

It does not show! Does the query really exist? Although it was stated that a 'Query' and 'View' are synonymous, there is creeping doubt that this is not true. Will the ODBC driver see the query or view?

```
       ExecuteQueryODBC 'JetQuery'
91
```

This function uses the ODBC driver and returns the same result. It is defined as follows:

```
       ∇ Z←ExecuteQueryODBC R
[1]    ⍝ System Building with APL+Win
[2]    ⍝ R is the SQL defining a query; uses a new copy of NWIND.MDB
[3]    □wself←'ADO' □wi 'Create' 'ADODB.Connection'
[4]    □wi 'XOpen' 'Provider=MSDASQL;Driver={Microsoft Access Driver
          (*.mdb)};DBQ=c:\nwind.mdb;'
[5]    0 0ρ□wi 'XExecute>ADO.Recordset' ('select * from [',R,']' )
[6]    □wself←'ADO.Recordset'
[7]    □wi 'XClose'
[8]    □wi 'xCursorType' 3 ⍝ adOpenStatic
[9]    □wi 'XOpen'
[10]   Z←□wi 'xRecordCount'
[11]   'ADO' □wi 'Delete'
       ∇
```

The ODBC driver can create queries too. The following function achieves this:

```
       ∇ CreateQueryODBC R;□wself
[1]    ⍝ System Building with APL+Win
[2]    ⍝ R is the SQL defining a query; uses a new copy of NWIND.MDB
[3]    □wself←'ADO' □wi 'Create' 'ADODB.Connection'
```

```
[4]    ⎕wi 'XOpen' 'Provider=MSDASQL;Driver={Microsoft Access Driver
        (*.mdb)};DBQ=c:\nwind.mdb;'
[5]    0 0ρ⎕wi 'XExecute' R
[6]    'ADO' ⎕wi 'Delete'
   ∇
      CreateQueryODBC 'CREATE VIEW ODBCQry AS SELECT * FROM CUSTOMERS'
```

The function runs successfully. Will the query be visible?

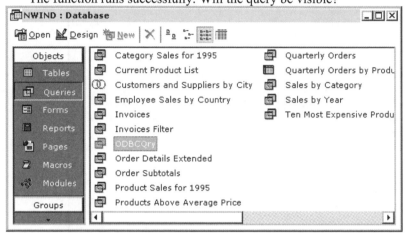

Figure 11-2 ODBC driver exposes query

The query is visible in the Access interface as shown in **Figure 11-2 ODBC driver exposes query**.

Note that the ODBC driver does not allow long names or names with punctuation. Hence, the right-hand argument has a slightly different syntax. The query can be retrieved via ODBC:

```
      ExecuteQueryODBC 'ODBCQry'
91
```

It works equally well with the JET provider:

```
      ExecuteQueryJET 'ODBCQry'
91
```

The original query created by the JET provider is still not visible. This can cause problems with both JET and ODBC. Any attempt to use the name 'JetQuery' causes a runtime error:

```
      CreateQueryODBC 'Create view JetQuery as select * from customers'
⎕WI ERROR: Microsoft OLE DB Provider for ODBC Drivers exception 80040E14
[Microsoft][ODBC Microsoft Access Driver] Object 'JetQuery' already
exists.
CreateQueryODBC[5] 0 0ρ⎕wi 'XExecute' R
                  ∧
      CreateQueryJET 'Create view JetQuery as select * from customers'
⎕WI ERROR: Microsoft JET Database Engine exception 80040E14 Object
'JetQuery' already exists.
CreateQueryJET[5] 0 0ρ⎕wi 'XExecute' R
                 ∧
```

The query 'JetQuery' exists and yet is not visible: this is the stuff of developers' nightmares! Any attempt to use the query name within the **Access** interface will cause a similar error; see **Figure 11-3 Overwrite query**.

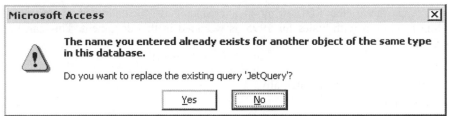

Figure 11-3 Overwrite query

If the **Yes** button is clicked, the hidden query is overwritten and the name becomes visible. This is a foolproof method for destroying the existing query—there is no way of verifying its content within context.

A further manifestation of this problem is more pernicious. The sample query 'JetQuery' uses ANSI query and both the JET and ODBC providers can execute it. However, if JET or ODBC specific SQL keywords are used, only the provider that created the query can execute it.

For the sake of completeness, it is useful to show the code for adding a new query via the **Access** object—the name must not exist for the same type of object:

```
      ∇ AddQuery;⎕wself
[1]   ⍝ System Building with APL+Win
[2]   ⎕wself←'ac' ⎕wi 'Create' 'Access.Application'
[3]   ⎕wi 'XOpenCurrentDatabase' 'c:\nwind.mdb'
[4]   0 0⍴⎕wi 'xCurrentDB.CreateQueryDef' 'NewQuery' 'select * from
      customers'
[5]   ⎕wi 'XQuit'
      ∇
```

The AddQuery function adds the query/view NewQuery to the database. This is a trivial example; it simply creates a copy of the CUSTOMERS table. In practice, views are based on complex SQL Queries.

11.2 The Access object

During development, it is expedient to work with the **Access** object or with the **Access** interface: this promotes rapid application development. It is quicker to design a table using the table wizard than with SQL statements. Likewise, the query wizard enables queries to be constructed via a graphical interface.

11.2.1 Dynamic query definition

A query defined with the aid of the wizard can be retrieved into the APL+Win workspace for incorporation into APL+Win code. The function QueryDef retrieves the SQL associated with the query named 'Category Sales for 1995' in the NWIND.MBD database:

```
      ∇ Z←QueryDef;⎕wself
[1]   ⍝ System Building with APL+Win
[2]   ⎕wself←'ac' ⎕wi 'Create' 'Access.Application'
[3]   ⎕wi 'XOpenCurrentDatabase' 'c:\nwind.mdb'
[4]   Z←(⎕wi 'xCurrentDB.QueryDefs("Category Sales for 1995")') ⎕wi 'xSQL'
[5]   ⎕wi 'XQuit'
      ∇

      ⎕←Z←QueryDef
SELECT DISTINCTROW [Product Sales for 1995].CategoryName, Sum([Product
Sales for 1995].ProductSales) AS CategorySales
FROM [Product Sales for 1995]
GROUP BY [Product Sales for 1995].CategoryName;
```

Despite its appearance, the SQL string is a character *vector* delimited by carriage return—the line beginning with 'Sales for 1995' is wrapped in this document. This can be established easily:

```
ρZ                                            +/Z∊⎕tcnl
199                                        │ 3
```

Access formats SQL statements internally.

11.2.1.1 Reusing a query

Some applications require transient queries; that is, queries that are used for a specific purpose and then discarded. It might be advantageous to create a single query as a placeholder for transient queries and reuse it with a different SQL statement when required. The function AddQuery adds a query to the NWIND database. The current SQL statement associated with 'NewQuery' can be retrieved by the function QueryDef: unfortunately, the syntax of the 'QueryDefs' method is such that it requires the query name to be hard coded. This is resolved by the following function:

```
       ∇ Z←L QueryDef2 R;⎕wself
  [1]    ⍝ System Building with APL+Win
  [2]    ⎕wself←'ac' ⎕wi 'Create' 'Access.Application'
  [3]    ⎕wi 'XOpenCurrentDatabase' 'C:\NWIND.MDB'
  [4]    Z←'xCurrentDB.QueryDefs("#")'
  [5]    Z←∊((Z≠'#')⊂Z),¨R ''
{1[6]    :if 0=⎕nc 'L'
  [7]        Z←(⎕wi Z) ⎕wi 'xSQL'
+ [8]    :else
  [9]        Z←(⎕wi Z) ⎕wi 'xSQL' L
1}[10]   :endif
  [11]   ⎕wi 'XQuit'
       ∇
```

The key element of this function is in line [4]. The character # is used as a placeholder for the name of the query, passed as the right-hand argument, and is replaced in line [5].

```
       QueryDef2 'NewQuery'
SELECT *
FROM customers;
```

If a new SQL statement is required for this query, it is passed as the left-hand argument:

```
       "select * from orders where shipcountry = 'UK'" QueryDef2 'NewQuery'
       QueryDef2 'NewQuery'
SELECT *
FROM orders
WHERE shipcountry = 'UK';
```

The newly defined query can be executed:

```
       ExecuteQueryODBC 'NewQuery'
56
```

It returns the correct result.

11.2.2 Queries based on user-defined functions

Within its user interface, **Access** allows user-defined functions to be used within queries. A simple example can illustrate this.

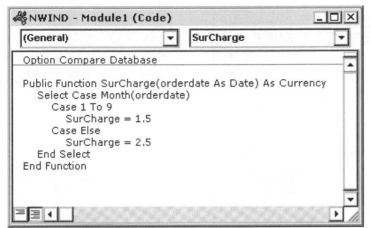

In the NWIND.MDB database, add a module called Module1 and define a function SurCharge, shown in **Figure 11-4 User-defined function I.**

Figure 11-4 User-defined function I

Note that the **Access** case statement allows ranges to be specified; this is not available with the APL+Win case statement. Next, create a new query, ScaleFactors. The SQL view of this query is shown in **Figure 11-5 User-defined function II.**

```
ScaleFactors : Select Query                          _ □ ×
SELECT surcharge(orderdate) AS surcharge, shipcountry, orderdate
FROM Orders
WHERE shipcountry = 'UK';
```

Figure 11-5 User-defined function II

The query contains a calculated column (surcharge) returned by the user-defined function. As before, 56 records are returned: see **Figure 11-6 Query data view**. This query cannot be executed from outside the **Access** user interface—it causes a runtime error:

```
      ExecuteQueryODBC 'ScaleFactors'
⎕WI ERROR: Microsoft OLE DB Provider for ODBC Drivers exception 80040E14
[Microsoft][ODBC Microsoft Access Driver] Undefined function 'surcharge'
in expression.
ExecuteQueryODBC[5] 0 0ρ⎕wi 'XExecute>ADO.Recordset' ('select * from
[',R,']' )
                          ∧
```

The reason for failure is that the ODBC driver cannot see the user-defined function shown in **Figure 11-5 User-defined function II;** the JET provider cannot see it either:

```
      ExecuteQueryJET 'ScaleFactors'
⎕WI ERROR: Microsoft JET Database Engine exception 80040E14 Undefined
function 'surcharge' in expression.
ExecuteQueryJET[5] 0 0ρ⎕wi 'XExecute>ADO.Recordset' ('select * from
[',R,']' )
```

ScaleFactors : Select Query		
surcharge	Ship Countr	Order Date
£1.50 UK		26-Sep-1994
£2.50 UK		27-Oct-1994
£2.50 UK		01-Nov-1994

Figure 11-6 Query data view shows the result of the query in **Figure 11-5 User-defined function II**.

The number of records (56) is shown on the status bar.

surcharge	Ship Countr	Order Date
▶ £1.50	UK	26-Sep-1994
£2.50	UK	27-Oct-1994
£2.50	UK	01-Nov-1994
£2.50	UK	03-Nov-1994
£2.50	UK	16-Dec-1994
£2.50	UK	22-Dec-1994
£2.50	UK	27-Dec-1994
£1.50	UK	09-Jan-1995
£1.50	UK	16-Jan-1995
£1.50	UK	19-Jan-1995
£1.50	UK	01-Feb-1995
£1.50	UK	07-Mar-1995
£1.50	UK	24-Mar-1995
£1.50	UK	03-Apr-1995

ScaleFactors : Select Query

Record: 1 of 56

Figure 11-6 Query data view

11.2.2.1 The UDF workaround

If is unfortunate that user-defined functions are unrecognised by the JET provider and ODBC driver—potentially, such functions can provide powerful functionality at source. The method for retrieving queries such as ScaleFactors is via the **Access** object itself:

```
     ∇ GetQuery;⎕wself
[1]    ⍝ System Building with APL+Win
[2]    ⎕wself←'ac' ⎕wi 'Create' 'Access.Application'
[3]    ⎕wi 'XOpenCurrentDatabase' 'c:\nwind.mdb'
[4]    (⎕wi 'xCurrentDB.QueryDefs("ScaleFactors")') ⎕wi
         'XOpenRecordset>UDFQuery'
[5]    ⍝ ... note that the Access object has not been terminated
     ∇
```

This function creates the object UDFQuery; it contains the required records:

```
     'UDFQuery' ⎕wi 'XMoveFirst'
     a←'UDFQuery' ⎕wi 'XGetRows' 100
```

The argument to the XGetRows method is deliberately higher than the number of records in the query; however, the **Access** record set contains the correct number of records and columns:

```
     ⍴a
56 3
```

Despite appearances, UDFQuery is not an ADO record set object—note the difference in the syntax of the common GetRows method:

```
     'UDFQuery' ⎕wi '?GetRows' ⍝ This is DAO
XGetRows method:
  Result ← ⎕WI 'XGetRows' [NumRows]
     'ADOR' ⎕wi 'Create' 'ADODB.Recordset'
ADOR
     'ADOR' ⎕wi '?GetRows' ⍝ This is ADO
XGetRows method:
  Result ← ⎕WI 'XGetRows' [Rows@Long [Start [Fields]]]
     'ADOR' ⎕wi 'Delete'
```

So, what do the records look like?

```
    a[1;]
1.5 UK 34603
```

The third value, OrderDate looks anomalous. In fact, it is an **Access** internal representation of the date: **Access** stores dates as a count of the number of data from a reference date. In order to investigate this, change the query as shown in **Figure 11-7 Query dates**.

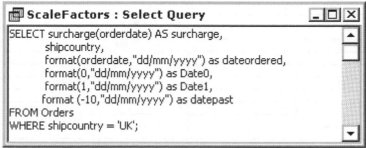

Figure 11-7 Query dates

After GETQuery has been run and the rows retrieved in the same variable, the first row looks as follows:

```
    a[1;]
1.5 UK 26/09/1994 30/12/1899 31/12/1899 20/12/1899
```

The reference date, 1, is 31/12/1899; earlier dates are kept as a scalar lower than 1 and later dates as a scalar higher than 1. This also illustrates the value of **Access** functions used in queries—the format function has been used to return dates in a manner that is meaningful to APL.

Some observations on the SQL statement are necessary:

● It is valid to use the name of a function as a columns name (surcharge) but a calculated column (orderdate) needs to be renamed.

● It is necessary to terminate the instance of **Access** started by the GetQuery function, by using:

```
    'ac' ⎕wi 'XQuit'
```

11.2.3 Deleting Access objects

The 'DoCmd' property or the **Access** object provides the 'DeleteObject' method that allows objects from a database to be deleted. The syntax is:

```
    Do.Cmd.DeleteObject objectType, objectName
```

The objectType name/value pairs are listed in **Table 11-1 Object types**.

Constant	Value
AcDefault	-1
Actable	0
acQuery	1
acForm	2
acReport	3
acMacro	4
acModule	5
acDataAccessPage	6
acServerview	7
acDiagram	8
acStoredProcedure	9

Table 11-1 Object types

In APL+Win, objectType must be referred to by value. An example is:

```
    ∇ DeleteObject
[1]   ⍝ System Building with APL+Win
[2]   ⎕wself←'ac' ⎕wi 'Create'
          'Access.Application'
[3]   ⎕wi 'XOpenCurrentDatabase' 'c:\nwind.mdb'
[4]   ⎕wi 'xDoCmd.DeleteObject' 1
          'ScaleFactors'
    ∇
```

The function DeleteObject removes the object ScaleFactors from C:\NWIND.MDB.The object to be deleted must exist in the target database: a runtime error will occur if it does not.

11.3 JET Engine types

Table 11-2 JET Engine type lists the possible values for 'Engine Type'

Data Source	Value	Description
Unknown	0	
JET10	1	
JET11	2	
JET20	3	Access 2.0
JET3X	4	Access 97
JET4X	5	Access 2000
DBASE3	10	
DBASE4	11	
DBASE5	12	
Excel30	20	
Excel40	21	
Excel50	22	
Excel80	23	
Excel90	24	
Exchange4	30	
LOTUSWK1	40	
LOTUSWK3	41	
LOTUSWK4	42	
PARADOX3X	50	
PARADOX4X	51	
PARADOX5X	52	
PARADOX7X	53	
TEXT1X	60	
HTML1X	70	

Table 11-2 JET Engine type

The JET provider may be used on a variety of data sources and its 'Engine Type' property may explain vagaries in its behaviour. In order to retrieve the information about 'Engine Type', query the 'xConnectionString' of the connection object:

```
    'ADO' ▢wi 'xConnectionString'
Provider=Microsoft.Jet.OLEDB.4.0;Password="";User ID=Admin;Data
Source=c:\nwind.mdb;Mode=Share Deny None;Extended Properties="";Jet
OLEDB:System database="";Jet OLEDB:Registry Path="";Jet OLEDB:Database
Password="";Jet OLEDB:Engine Type=5;Jet OLEDB:Database Locking Mode=0;Jet
OLEDB:Global Partial BulkOps=2;Jet OLEDB:Global Bulk Transactions=1;Jet
OLEDB:New Database Password="";Jet OLEDB:Create System Database=False;Jet
OLEDB:Encrypt Database=False;Jet OLEDB:Don't Copy Locale on
Compact=False;Jet OLEDB:Compact Without Replica Repair=False;Jet
OLEDB:SFP=False
```

11.4 Access—below the surface

Access can work with either file-based native databases, MDB files, or SQL Server based databases, ADP—**Access** Data Project—files. MDB files can be used with the JET 4.0 provider and ADP files require the SQLOLEDB provider; each provider supports their own SQL enhancements, which are exclusive. MDB files can contain native tables—managed within the **Access** application—or tables linked to a database—managed by the database; none or some of the tables may be linked and tables may be linked to different data sources. On the other hand, ADP files provide a means of working with SQL Server databases *as if* they were **Access** databases.

The Microsoft Data Engine (**MSDE**) provides an alternative route to working with SQL Server databases. MSDE is included on the Office Premium and Professional CD but is not installed automatically. As to be expected, it does have some limitations regarding size, concurrent usage, etc.

From the developer's point of view, this complex scenario can be simplified by using the ADO object with SQL. This not only provides a standard way of dealing with all tables but also ensures that experience of this technology acquired with other data sources is still relevant.

A tangible benefit of using **Access** is that it provides a friendly interface for managing and visualizing data; this is especially useful for server databases.

Figure 11-8 Filters

This facility is available either from the toolbar or by right mouse click on an open table or query: see **Figure 11-8 Filters**. The 'Filter For' box allows the value to be selected, from the current column, to be specified An example of a filtered view is shown in **Figure 11-9 Using 'Filter' in Access**.

The use of the 'filter' facility on an open table or query enables complex data scenarios—interactively—that can provide valuable insight into the data.

The underlying table or query is not changed to reflect the filter that is in place and, regrettably, the SQL behind the filter view cannot be saved. If accurate notes are made while configuring the filter view, an SQL statement to retrieve the same data can be written easily.

titles : Table					_ □ ×
title_id	title	type	pub_id	price	ac
BU1032	The Busy Exe	business	1389	£19.99	£
▶ BU7832	Straight Talk	business	1389	£19.99	£

Record: I◄ ◄ 2 ► ►I ►* ⊗ ►! of 2 (Filtered)

Figure 11-9 Using 'Filter' in Access

Using the first button, it is possible to extract all records from the table **Titles** where *type* is *business* and *price* is *19.99*. This is equivalent to the following SQL statement:

```
'SELECT * FROM TITLES WHERE TYPE='BUSINESS' AND PRICE = '19.99'
```

Access using ADP files provides a client/server solution. For APL+Win, the underlying MDF file—or SQL Server database or ADF file—can be used directly.

11.4.1 MDB files

A native **Access** (MDB) database can contain native tables or tables linked to another source; depending on the source, some properties of the linked table may be inaccessible from **Access**.

The database, MDBFILE.MDB, contains two tables; **Native** is a native **Access** tables whereas 'DBO_AUTHORS' is a table linked to the SQL Server **AUTHORS** table in the **PUPS** database. The table icon visually distinguishes one from the other: see **Figure 11-10 Table types**.

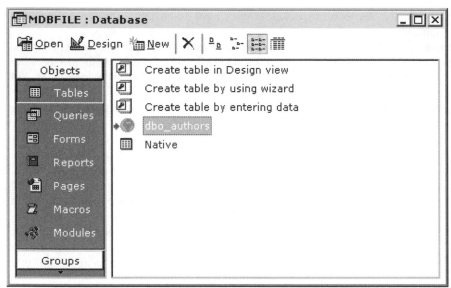

Figure 11-10 Table types

The collection of objects exposed by a native MDB database is the same whether or not it contains linked tables.

11.4.2 ADP files

An **Access** Data Project (**ADP**) file is one that **Access** uses to manage a SQL Server MDF file natively. Any changes made to the MDF file from the **Access** interface, where permitted, are permanent; that is, are written back to the MDF file.

This database is pointing to the SQL Server 'AUTHORS' database: see **Figure 11-11 An ADP file:**

● The collection of objects is different from that shown in an MDB database.

● An ADP database does not permit linked tables; however, new tables may be added either via the **Insert | Table** or **File | Get External Data | Import…** menu options.

● The **Queries** of MDB files and **Views** of ADP files are similar as far as they are a representation of data in underlying tables.

● A data project has an *.ADP file managed by **Access** and the database file *.MDF managed by SQL Server.

Note the differences in the Objects box in **Figure 11-11 An ADP file**. In particular, 'Stored Procedures' is an option.

Stored procedures are akin to SQL statements that are stored in the database itself; they are capable of accepting runtime parameters and execute on the database server.

Although stored procedures are a feature of SQL Server databases and unavailable with **Access** MDB files**, Access** allows them to be managed from within its own interface. The critical bonus for the user is that **Access** transparently manages its object hierarchies, including menus, depending on whether an MDB or ADP file is open.

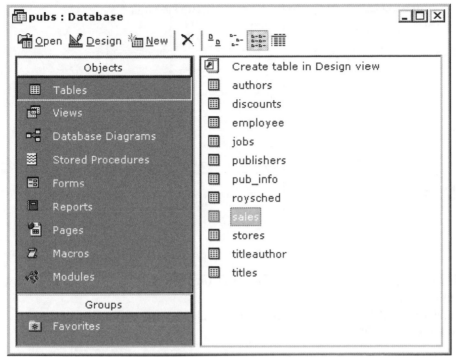

Figure 11-11 An ADP file

11.4.3 The MDB/ADP file menu

In addition to the different collection of objects exposed by the two types of databases, the **Access File** menu is different, shown in **Figure 11-12 MDB File menu** and **Figure 11-13 ADP File menu**.

The MDB menu does not have the **File | Connection...** option and the ADP menu does not have the **Link Tables...** option. The **File | Get External Data** menu exposes different options. Some other common MDB menu options may be restricted when working with ADP files.

The ADP **File | 'Connection...'** option provides a convenient method for examining or constructing the connection string to external tables; for new connections, this invokes the 'Microsoft SQL Server Data Wizard' that allows SQL Server databases to be created via user interaction.

Access user interface and common menu options, such as **File | Export** and **Tools | Office Links,** provide powerful means of analysing the data in both types of databases, MDB and ADP: this includes the facility for exporting tables to **Excel** and **Word**.

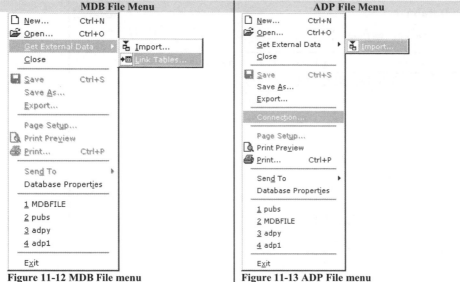

Figure 11-12 MDB File menu   **Figure 11-13 ADP File menu**

11.4.4 Linking tables

The **File | Get External Data | Link Tables…** option allows other external data sources to be linked to an **Access** MDB table.

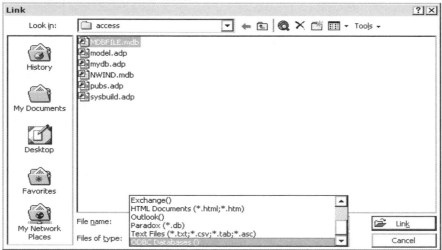

Figure 11-14 Linking tables

In order to create a table linked to a server database, select 'ODBC Databases()' from the **Files Of Type** box on the **Link** dialogue box, shown in **Figure 11-14 Linking tables**. Further dialogues will be displayed to collate information for linking a table to an external source; this depends on the data source selected.

The external data sources may be another **Access** MDB file, an **Excel** workbook, a text file, or another server database—all subject to access permissions:

● A linked table allows the source data to be viewed within the **Access** interface and any changes are permanent.

● A linked table cannot be linked again.

11.5 Working with many data sources

Access offers four compelling reasons for promoting its use:

● It offers a user-friendly interface for reviewing tables.

● It allows tables from disparate sources to be linked. For example, an MDB file may contain tables from one or more Oracle databases and one or more SQL Server databases.

● **Access** SQL, or Queries, works seamlessly with linked tables. The **Access** SQL parser has a powerful SQL formatting tool that formats SQLs neatly,

● Queries—or SQL statements—can be constructed using the 'Design' view or the wizard or simply typed.

Of course, the penalty is working with the **Access** object hierarchy or with the **Access** interface.

In order to enable seamless access to link tables, it is recommended that the 'Save password' option, see **Figure 11-15 Linking tables with password**, be enabled when creating a linked table.

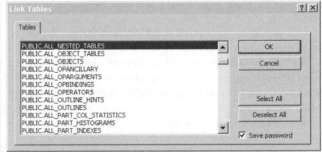

Figure 11-15 Linking tables with password

Access will prompt for the password when working within its user interface.

11.5.1 Troubleshooting databases with linked tables

If an external table is linked with or without the 'Save Password' option enabled, and that table is being accessed via an ADO connection string, an error occurs. This is because the connection to the database is established with one string and another, held in the MSysObjects table, is required to access that particular table. **Access** manages this logical leap internally—and will prompt for a missing user id and password if required—but ADO does not.

11.5.1.1 Linked tables: using the ODBC driver

The connection problem can be explored with a small function:

```
      ∇ Sql MDBLTODBC ConnectionString;⎕wself
[1]     ⍝ System Building with APL+Win
[2]     ⎕wself←'ADORS' ⎕wi 'Create' 'ADODB.RecordSet'
[3]     ⎕wi 'XOpen' Sql ConnectionString 3 3 1
[4]     'This table has ',(⍕⎕wi 'RecordCount'), ' records'
[5]     ⎕wi 'XClose'
      ∇
```

The left-hand argument specifies the SQL statement and the right-hand argument specified the connection string; in this instance, the ODBC driver will be used. What is the record count of the native table **NATIVE**?

```
Sql←'SELECT * FROM NATIVE;'
ConnectionString←'Provider=MSDASQL;Driver={Microsoft Access Driver
(*.mdb)};DBQ=c:\ajay\access\mdbfile.mdb;'
      Sql MDBLTODBC ConnectionString
This table has 1 records
```

This works fine! Next, what is the connection string maintained by **Access** for the linked table DBO_AUTHORS?

```
Sql←'SELECT Connect FROM MSysObjects WHERE NAME="DBO_AUTHORS"'
      Sql MDBLTODBC ConnectionString
⎕WI ERROR: Microsoft OLE DB Provider for ODBC Drivers exception 80040E09
[Microsoft][ODBC Microsoft Access Driver] Record(s) cannot be read; no
read permission on 'MSysObjects'.
MDBLTODBC[3] ⎕wi 'XOpen' Sql ConnectionString 3 3 1
          ∧
```

This fails because of lack of permission. The same SQL executes correctly within the access user interface. Finally, what is the record count of the linked table DBO_AUTHORS?

```
Sql←'SELECT * FROM DBO_AUTHORS'
      Sql MDBLTODBC ConnectionString
⎕WI ERROR: Microsoft OLE DB Provider for ODBC Drivers exception 80004005
[Microsoft][ODBC Microsoft Access Driver] ODBC--connection to 'SysBuild'
failed.
```

This too fails because of a connection failure; however, the SQL evaluates successfully as a query within the **Access** Query interface.

With an MDB file, it appears that an automation client may not query the system table **MSysObjects** and user tables linked to an external data source. This is anomalous since both these tables can be queried from within **Access**. It would seem that **Access** maintains another connection string internally. The following function returns this string:

```
    ∇ Z←BaseConnection Database;⎕wself
[1]   ⍝ System Building with APL+Win
[2]   ⎕wself←'ac' ⎕wi 'Create' 'Access.Application'
[3]   ⎕wi 'XOpenCurrentDatabase' Database
[4]   Z←(⎕wi 'xCurrentProject') ⎕wi 'xBaseConnectionString'
[5]   ⎕wi 'XCloseCurrentDatabase'
[6]   ⎕wi 'XQuit'
    ∇

      ConnectionString←BaseConnection 'C:\AJAY\ACCESS\MDBFILE.MDB'
      ConnectionString
PROVIDER=Microsoft.Jet.OLEDB.4.0;DATA SOURCE=C:\AJAY\ACCESS\MDBFILE.MDB;
PERSIST SECURITY INFO=FALSE;Jet OLEDB:System Database=
C:\PROGRA~1\COMMON~1\System\SYSTEM.MDW
```

Yes! **Access** does maintain another connection string. Note the presence of a reference to a system database. The default location of this database can be found in the value of:

```
      'ac' ⎕wi 'xDBEngine.SystemDB'
```

If **Access** needs to be started with an alternative path for the system database, it must be started as follows:

path\access.exe / *systempath*\system.mdw

Now, the same set of queries can be tried with the new connection string. The first query:

```
      Sql←'SELECT * FROM NATIVE'
      Sql MDBLTODBC ConnectionString
This table has 1 records
```

The second query:

```
      Sql←'SELECT Connect FROM MSysObjects WHERE NAME="DBO_AUTHORS"'
      ConnectionString←BaseConnection 'C:\AJAY\ACCESS\MDBFILE.MDB'
      Sql MDBLTODBC ConnectionString
This table has 1 records
```

The third query:

```
      Sql←'SELECT * FROM DBO_AUTHORS'
```

```
     Sql MDBLTODBC ConnectionString
This table has 23 records
```
All three queries work. With an additional line [4.5] ▢wi 'GetString' inserted into MDBLTODBC, it is possible to see the underlying connection string for the linked DBO_AUTHORS table:
```
     Sql MDBLTODBC ConnectionString
This table has 1 records
DSN=SysBuild;Description=AUTO SETUP;UID=sa;APP=Microsoft« Access;
     WSID=AJAYLAP;DATABASE=pubs;Network=DBMSSOCN
```
The mysterious reference to 'SysBuild' is now explained; it is a system DSN.

11.5.1.2 Linked tables: using the JET provider

The audit trail for troubleshooting the three SQL statements shown in the previous section applies equally well when using the JET provider. The JET connection string is:
```
     ConnectionString←'Provider=Microsoft.Jet.OLEDB.4.0;Data
Source=C:\AJAY\ACCESS\MDBFILE.MDB'
```

11.6 Troubleshooting data projects

Access Data Projects (**ADP**) can be created for SQL Server databases only. For other data sources, the only option is linked tables. The easiest method for creating **Access** Data Projects is via the **File | New** menu option. It presents the dialogue shown in **Figure 11-16 Creating ADP files**.

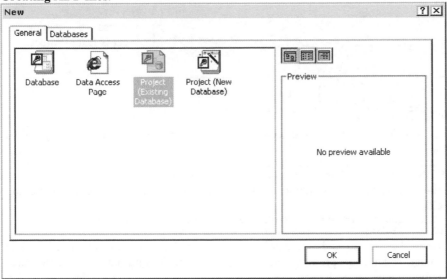

Figure 11-16 Creating ADP files

Select '**Project(Existing Database)**' for linking to an existing SQL Server database or '**Project(New Database)**' for creating a new ADP file AND a new MDF file, SQL Server database. **Access** prompts for the names of the ADP file, both for '**Project(Existing Database)**' and '**Project(New Database)**'; this is followed by the Universal Data Link (**UDL**) wizard, shown in **Figure 11-17 Connection for ADP files**, that collates the connection string to the underlying SQL server database.

The underlying connection string can be retrieved using the BaseConnection function. Alternatively, the connection string can be queried from within **Access**: press **ALT + F11, Ctrl + G** and, in the **Immediate Window**, type:
```
     ?Application.CurrentProject.BaseConnectionString
```

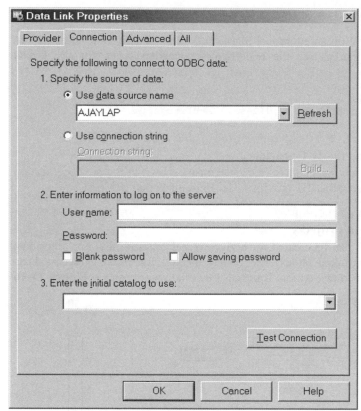

The string returned will depend on the options selected in the UDL dialogue. In order to make SQL queries seamless, the user id and passwords must be specified or constructed at runtime.

Note the facility for attaching a database. With a distributed application, the installation procedure will be expected to copy a database (MDF file) to an appropriate location and then that file can be attached to the SQL Server during the process of creating an **Access** Data Project.

Figure 11-17 Connection for ADP files

If the 'Use Windows NT Integrated security' option is used, the connection string will be something like:

```
'PROVIDER=SQLOLEDB.1;INTEGRATED SECURITY=SSPI;PERSIST SECURITY
INFO=FALSE;INITIAL CATALOG=MASTER;DATA SOURCE=AJAYLAP'
```

And, if the 'Use a specific user name and password' option is used, the connection string will be something like:

```
'PROVIDER=SQLOLEDB.1;Password=SPHYNX;PERSIST SECURITY INFO=TRUE;USER
ID=AA;INITIAL CATALOG=SALES03;DATA SOURCE=SQLSERVM'
```

In this case, the user id and password are hard coded in the connection string. The application must collate, or prompt for, a specific user id and password and make appropriate substitutions for each user.

The **'Project(Existing Database)'** and **'Project(New Database)'** menu options correspond to the 'XNewAccessProject' and 'XCreateAccessProject' methods of the **Access** application object.

11.6.1 Using an existing MDF file

A new ADP file using an existing SQL Server database may be created programmatically with this function:

```
      ∇ Z←MDFFile CreateExistADP ADPFile;⎕wself
[1]    ⍝ System Building with APL+Win
[2]    Z←'PROVIDER=SQLOLEDB.1;INTEGRATED SECURITY=SSPI;PERSIST SECURITY
          INFO=FALSE;INITIAL CATALOG=*;DATA SOURCE=AJAYLAP'
[3]    Z←∊((~Z='*')⊂Z),¨MDFFile ''
```

```
[4]     □wself←'ac' □wi 'Create' 'Access.Application'
[5]     Z←□wi 'XNewAccessProject' ADPFile Z
[6]     □wi 'Delete'
      ∇
```

The left-hand argument specifies the name of the existing SQL Server database and the right-hand argument specifies the new ADP file name. The code will fail if the ADP file exists in the default file path or if the SQL Server is not attached to the server. The following line creates the project 'LocalMaster.ADP', linking it to the 'Master' database:

```
      'MASTER' CreateExistADP 'LocalMaster.ADP'
```

Note the technique used in line [3] to substitute a runtime parameter in a connection string; this can be used for user id and password substitution at runtime.

11.6.2 Using a new MDF file

This uses the 'XCreateAccessProject' method of the **Access** object. Although the syntax of this method is the same as that for 'XNewAccessProject', the deduction of the connection string is complicated. This function will create a new data project, as follows:

```
      ∇ CreateNewADP;□wself
[1]     ⍝ System Building with APL+Win
[2]     □wself←'ADO' □wi 'Create' 'ADODB.Connection'
[3]     □wi 'XOpen' 'PROVIDER=SQLOLEDB.1;INTEGRATED SECURITY=SSPI;PERSIST
          SECURITY INFO=FALSE;DATA SOURCE=AJAYLAP'
[4]     □wself←'ac' □wi 'Create' 'Access.Application'
[5]     □wi 'XCreateAccessProject' 'adpy.adp' (((↑'ADO' □wi 'Execute'
          'CREATE DATABASE NEWMDF2') □wi
          'xActiveConnection.ConnectionString'),';Initial Catalog=newmdf2')
[6]     □wi 'XQuit'
[7]     'ADO' □wi 'Delete'
      ∇
```

A two-stage process is involved. First, the SQL Server database is created using ADO, line [2] to line [3], and then the **Access** object is used to create the **Access** Data Project, line [4]. In line [5], both the ADO and **Access** object are used. The code assumes that the ADP and MDF files do not exist already; if either does, the   code will fail.

11.6.3 Data project: using the ODBC driver

The audit trail of section **Linked tables: using the ODBC driver** applies.

11.6.4 Data project: using the JET provider

The audit trail of section **11.5.1.2 Linked tables: using the JET provider** applies. A further complication is that it is necessary to verify whether the ADP project is connected to its underlying data source, using the following function:

```
      ∇ Z←IsConnected DBFile;□wself
[1]     ⍝ System Building with APL+Win
[2]     □wself←'ac' □wi 'Create' 'Access.Application'
[3]     □wi 'XOpenCurrentDatabase' DBFile
[4]     Z←□wi 'xCurrentProject.IsConnected'
[5]     □wi 'xCloseCurrentDatabase'
[6]     □wi 'XQuit'
      ∇
```

This function will return 1 if the ADP file is connected to its underlying MDF file and 0 otherwise.

11.7 The Jet compromise

The JET provider works fine with native tables in an MDB file but not with linked tables unless the 'baseconnectionstring' stored in the database is used; an ADP files is a special case of linked tables. The question that needs to be addressed is whether it is sensible to work with linked tables using the JET driver. The answer must be 'no' on two counts: first, the code becomes unnecessarily complicated, requiring the use of the **Access** object, and second, there is a performance sacrifice in having **Access** sit between the application and the original data source. It would be more efficient to access the source data directly.

A further issue needs to be addressed with linked tables; the connection string to the underlying data is held in an un-encrypted format in the MSysObjects system table within the MDB file. This may present an unnecessary security risk. It is possible to have **Access** prompt for user ids and passwords at runtime; that is, to create connection strings without the 'Save password' option enabled—refer to **Figure 11-15 Linking tables with password**. When working within its user interface, **Access** can prompt for missing information.

The dialogue prompting for a user name and password varies from data source to data source. The dialogues for Oracle and SQL Server are shown in **Figure 11-18 Oracle prompt** and **Figure 11-19 SQL Server prompt,** respectively.

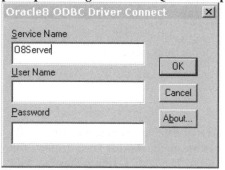

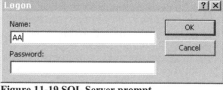

Figure 11-19 SQL Server prompt

Figure 11-18 Oracle prompt

However, this fails when working with an **Access** data source with ADO.

11.8 Unified approach with ADO and SQL

Access uses the JET provider for MDB files and the SQLOLEDB provider for ADP files. The provider in use can be queried programmatically:

```
    ∇ Z←WhichProvider DBFile;⎕wself
[1]   ⍝ System Building with APL+Win
[2]   ⎕wself←'ac' ⎕wi 'Create' 'Access.Application'
[3]   ⎕wi 'XOpenCurrentDatabase' DBFile
[4]   Z←⎕wi 'xCurrentProject.BaseConnectionString'
[5]   ⎕wi 'xCloseCurrentDatabase'
[6]   ⎕wi 'XQuit'
    ∇
```

This function creates an instance of the **Access** object, loads a file, and returns the provider that is being used:

```
    WhichProvider 'adpy.adp'
PROVIDER=SQLOLEDB.1;INTEGRATED SECURITY=SSPI;PERSIST SECURITY
INFO=FALSE;DATA SOURCE=AJAYLAP;INITIAL CATALOG=newmdf2
    WhichProvider 'MDBFILE.MDB'
PROVIDER=Microsoft.Jet.OLEDB.4.0;DATA SOURCE=c:\ajay\access\MDBFILE.MDB;
PERSIST SECURITY INFO=FALSE;Jet OLEDB:System
Database=C:\PROGRA~1\COMMON~1\System\SYSTEM.MDW
```

The implication of this is that developer must be aware of the provider being used and make allowances for the differences. This is particularly relevant for queries in that the two providers support different dialects of SQL.

Although this seems like a very convenient method for retrieving the base connection string, it may cause problems with databases that have auto-starting forms and macros: the connection string cannot be returned silently and without user interaction. For instance, the NWIND.MDB database displays a start-up banner that requires the **OK** button to be clicked.

11.8.1 ADOX:ADO Extension for data definition and security

One difference is that the JET provider supports ADOX 'Create' whereas the SQLOLEDB provider does not. First, verify that the database C:\NEW.MDB does not exist:

```
⎕wcall 'PathFileExists' 'c:\new.mdb'
```
0

A zero result implies that it does not. Next, run the function that can create the database:

```
ADOXCreateJET
```

Now, the file should exist, as proven by:

```
⎕wcall 'PathFileExists' 'c:\new.mdb'
```
1

The file C:\NEW.MDB is created without recourse to the **Access** object by the function ADOXCreateJET:

```
    ∇ ADOXCreateJET;⎕wself
[1]   ⍝ System Building with APL+Win
[2]   ⎕wself←'ADOX' ⎕wi 'Create' 'ADOX.Catalog'
[3]   0 0⍴⎕wi 'XCreate' 'Provider=Microsoft.Jet.OLEDB.4.0;Data
        Source=c:\new.mdb;Jet OLEDB:Engine Type=5;'
[4]   ⎕wi 'Delete'
    ∇
```

The method for creating a new database using the **Access** object is:

```
⎕wcall 'PathFileExists' 'c:\new2.mdb'
```
0
```
AccessCreate
⎕wcall 'PathFileExists' 'c:\new2.mdb'
```
1

The function AccessCreate is:

```
    ∇ AccessCreate;⎕wself
[1]   ⍝ System Building with APL+Win
[2]   ⎕wself←'ac' ⎕wi 'Create' 'Access.Application'
[3]   0 0⍴(⎕wi 'xDBEngine') ⎕wi 'XCreateDatabase' 'c:\new2.mdb'
        ';langid=0x0409;cp=1252;country=0'
[4]   ⎕wi 'XQuit'
    ∇
```

The SQLOLEDB provider does not support the ADOX 'Create' method; the method for creating a new SQL Server database is shown in the function CreateNewADP.

11.9 Access SQL

Access has its own SQL dialect: APL+Win can use either an ODBC driver or the JET provider for working with **Access** Tables and Queries. From within the **Access** user interface, queries can be specified using either the native user interface or SQL statements. Note the following:

● Like Tables, Queries can be used as a data source in an SQL statement when using an ODBC driver or the JET provider.

• Whilst Queries *within* **Access** may be specified using VBA reserved keywords and VBA user-defined functions, the **Access** ODBC driver or the JET provider do not support some of these keywords and any VBA user-defined functions. Thus, SQL statements tested within the **Access** user interface are not guaranteed to work when the database is accessed via ODBC or JET.

11.9.1 Access SQL with user-defined function

From an **Access** session using the NWIND.MDB database, specify a user-defined function as follows:

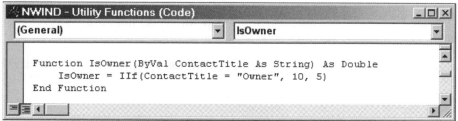

Figure 11-20 Query1 User-defined function

Next, specify a query using this function.

The SQL view is shown in **Figure 11-21 Query1 SQL statement**.

Figure 11-21 Query1 SQL statement

As can be seen in **Figure 11-22 Query1 results** this SQL statement works within the **Access** user interface. However, it is neither possible to specify a user-defined function when using ODBC or JET nor is it possible to access a Query that is based on such a function:

```
     ∇ UDFFail;Cnn;Sql
[1]    ⍝ System Building With APL+Win
[2]    Cnn←'Provider=Microsoft.Jet.OLEDB.4.0;Data Source=c:\nwind.mdb;'
[3]    Sql←'Select * from Query1'
[4]    ⎕wself←'ADORS' ⎕wi 'Create' 'ADODB.Recordset'
[5]    ⎕wi 'XOpen' Sql Cnn
     ∇
```

This function produces a runtime error:

```
     UDFFail
⎕WI ERROR: Microsoft JET Database Engine exception 80040E14 Undefined
function '
     IsOwner' in expression.
UDFFail[5] ⎕wi 'XOpen' Sql Cnn
          ∧
```

The JET provider uses the database file, NWIND.MDB, and its SQL engine but not the **Access** application itself: as seen in **Figure 11-22 Query1 results,** the query works fine within **Access** itself.

SQL is discussed in **Chapter 14 *Structured Query Language*.**

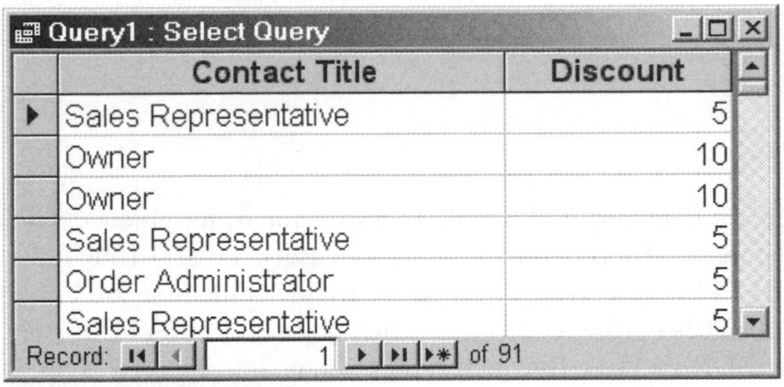

▦ Query1 : Select Query	_□×|
Contact Title	**Discount**
▶ Sales Representative	5
Owner	10
Owner	10
Sales Representative	5
Order Administrator	5
Sales Representative	5

Record: ◄◄ ◄ | 1 | ► ►◄ ►✳ | of 91

As can be seen, the Discount column has been calculated correctly.

Figure 11-22 Query1 results

11.10 Database filing

APL has neglected to exploit the advantages of storing its workspace objects—unlocked functions and variables—and its files—created by APL or custom files used by APL—in a database. Two of the key advantages are:

● Several versions of APL objects can coexist in the same database, each tagged with a version number and a timestamp; this facilitates version tracking and the reversion to an earlier state.

● Database filing protects objects, especially files, from accidental modification or deletion.

The size limits of the database used dictates the number of objects that can be stored: for server database, there is no practical limit other than those imposed by hardware resources.

11.10.1 Using an Access database

Any database is suitable for filing if it supports the IMAGE or Binary Large Objects (**BLOB**) data types: the examples that follow use an **Access** database. With server databases such as SQL Server, ORACLE, and DB2, database administrators will create the database. With **Access**, APL can create a database programmatically, using this function:

```
        ∇ Z←CreateMDB R;sql;cns;⎕wself
  [1]   ⍝ System Building with APL+Win
  [2]   ⍝ R is a fully qualified ACCESS database file name e.g C:\SBWA.MDB
{1[3]   :if 0=⎕wcall 'PathFileExists' R
  [4]       Z←'CREATE_DB-',R,' General' ⍝ sort order = 'General'
  [5]       Z←⎕wcall 'sqlConfigDataSource' θ 4 'Microsoft Access Driver
          (*.mdb)' Z ⍝ 4 = ODBC_ADD_DSN
1}[6]   :endif
  [7]   cns←'Provider=Microsoft.Jet.OLEDB.4.0;Data Source=',R,';'
  [8]   ⎕wself←'∆ADOC' ⎕wi 'Create' 'ADODB.Connection'
  [9]   ⎕wi 'XOpen' cns
  [10]  sql←''
  [11]  sql←sql,"CREATE TABLE APLOBJ("
  [12]  sql←sql,"        USERID CHAR VARYING (50) WITH COMPRESSION,"
  [13]  sql←sql,"        OBJID CHAR VARYING (255) WITH COMPRESSION,"
  [14]  sql←sql,"        OBJTYPE CHAR VARYING(10),"
  [15]  sql←sql,"        VERSION FLOAT,"
  [16]  sql←sql,"        CONTENT IMAGE,"
  [17]  sql←sql,"        UPDATED DATE DEFAULT NOW(),"
  [18]  sql←sql,"        CREATED DATE DEFAULT NOW(),"
  [19]  sql←sql,"        CONSTRAINT FILES_PK PRIMARY KEY (USERID,OBJID)"
  [20]  sql←sql,"                    );"
  [21]  Z←⎕wi 'XExecute' sql
```

```
[22]    ⎕wi 'XClose'
[23]    ⎕wi 'Delete'
[24]    Z←⎕wcall 'PathFileExists' R
    ∇
    CreateMDB 'c:\SBWA.MDB'
```
1

   This APL+Win function creates a database and a table within it for filing objects. The table contains the following columns: USERID (the name of the user), OBJID (the fully qualified name of the object), OBJTYPE (the type of object, Variable, Function, or File), VERSION (version number), CONTENT (the contents of the object), UPDATED (the timestamp of the last update), and CREATED (the timestamp when the object was first added).

   Two vital facilities in APL+Win enable database filing, namely, its ⎕dr system function, which is capable of left-hand arguments, 'wrapl' and 'unwrapl', and its '#' ⎕wi VT (*value*) (*conversion*) system method.

11.10.2 Storing variables and functions

Any type of workspace variable can be stored: numeric or character and scalars, vectors, or arrays, whether simple, nested, or mixed nested. The storage process uses complex ADO and APL features: refer to the comments in the function, *Chapter 12 – Working with ActiveX Data Objects*, and the APL+Win help files for guidance.

```
          ∇ Z←StoreNL23 R;sql;cns;⎕wself;VERSION;Values;Data
   [1]    ⍝ System Building with APL+Win
   [2]    ⍝ R = NAME of variable or unlocked function
{1[3]    :select θρ⎕nc ,R
= [4]         :case 2
   [5]            (Z Data)←2 ('wrapl' ⎕dr ⍕R)
= [6]         :case 3
{2[7]          :if 0∊ρData←⎕cr R
   [8]             → 0×Z←¯1 ⍝ Function is locked, ignore
+ [9]          :else
   [10]             (Z Data)←3 ('wrapl' ⎕dr Data)
2}[11]         :endif
+ [12]       :else
   [13]          → Z←0 ⍝ name is not a variable or function
1}[14]   :endselect
   [15]   ⍝ Get Connection object
   [16]   ⎕wself←'⍙ADOC' ⎕wi 'Create' 'ADODB.Connection'
   [17]   cns←'Provider=Microsoft.Jet.OLEDB.4.0;Data Source=C:\SBWA.MDB;'
   [18]   ⎕wi 'XOpen' cns
   [19]   ⍝ Get highest version number for R in database
   [20]   sql←"SELECT MAX(VERSION) AS VERSION FROM APLOBJ WHERE USERID =
          '",GetUser,"' AND OBJID = '",R,"';"
   [21]   ⍝ Get Recordset object
   [22]   0 0ρ'⍙ADORS' ⎕wi 'Create' 'ADODB.Recordset'
   [23]   '⍙ADORS' ⎕wi 'XOpen' sql cns
   [24]   ⍝ Increment VERSION number by 1, default 1
{1[25]   :if ('#' ⎕wi 'null')≡'⍙ADORS' ⎕wi 'xFields().Value' 'VERSION'
   [26]       VERSION←'1'
+ [27]   :else
   [28]       VERSION←⍕1+'⍙ADORS' ⎕wi 'xFields().Value' 'VERSION'
1}[29]   :endif
   [30]   '⍙ADORS' ⎕wi 'XClose'
```

```
[31]   A Create a record in database for next entry (Date fields have
         defaults: no need to supply values)
[32]   Values←"'",GetUser,"','",R,"','",(∈(2 3∈⎕nc R)/'Variable'
         'Function'),"','",VERSION
[33]   0 0ρ⎕wi 'XBeginTrans'
[34]   sql←'INSERT INTO APLOBJ (USERID,OBJID,OBJTYPE,VERSION)
         VALUES(',Values,');'
[35]   0 0ρ⎕wi 'XExecute' sql
[36]   ⎕wi 'XCommitTrans'
[37]   A Select and lock the record for updating with the contents of
         object R
[38]   sql←"SELECT * FROM APLOBJ WHERE  USERID = '",GetUser,"' AND OBJID
         = '",R,"' AND VERSION =",VERSION,";"
[39]   '∆ADORS' ⎕wi 'XOpen' sql cns (⎕wi '=adOpenDynamic') (⎕wi
         '=adLockOptimistic')
[40]   '∆ADORS' ⎕wi 'xFields().AppendChunk' 'CONTENT' ('#' ⎕wi 'VT' Data
         8209)
[41]   '∆ADORS' ⎕wi 'XUpdate' A Update database
[42]   (⊂'∆ADORS' '∆ADOC') ⎕wi ¨¨⊂¨'XClose' 'Delete'
     ∇
```

The function returns the name classification for all objects written successfully to the database or failure codes 0 (not a variable or function) or ¯1 (locked functions). The right-hand argument of the function is a literal, representing the name of a function or variable.

This functions stores variables and unlocked functions as a byte array; see line [40]: variables are reduced to vectors, see line [5], and the canonical representation of functions; see line [10]. Some examples of syntax are:

```
xx←(⎕nl 2 3) 'SystemBuilding' (10) (0/0) (2 3ρ6?100) (⊂¨2 2ρ'AB')
StoreNL23 'xx' A Variable
     StoreNL23 'RetrieveFile' A Function
```

11.10.2.1 Retrieving Variables

The function stores variables cumulatively: that is, every call to StoreNL23 adds a new entry to the database with a different version number. This illustrates one approach to filing. Create a variable and test its retrieval:

```
     xx≡RetrieveNL2 'xx' A Should match
1
```

The workspace variable matches its representation retrieved from the database. The function RetrieveNL2 is as follows:

```
     ∇ Z←RetrieveNL2 R;sql;⎕wself
  [1]    A System Building with APL+Win
  [2]    sql←"SELECT CONTENT FROM APLOBJ WHERE OBJID = '",R,"' AND OBJTYPE
         = 'Variable' AND VERSION=(SELECT MAX(VERSION) FROM APLOBJ WHERE
         OBJID='",R,"');"
  [3]    cns←'Provider=Microsoft.Jet.OLEDB.4.0;Data Source=C:\SBWA.MDB;'
  [4]    ⎕wself←'∆ADORS' ⎕wi 'Create' 'ADODB.Recordset'
  [5]    ⎕wi 'XOpen' sql cns
{1[6]    :if ⎕wi 'xEOF'
  [7]        Z←''
+ [8]    :else
  [9]        Z←'unwrapl' ⎕dr ⎕wi 'xFields(0).GetChunk' (⎕wi
         'xFields(0).ActualSize')
1}[10]   :endif
  [11]   (⊂'∆ADORS') ⎕wi ¨¨'XClose' 'Delete'
     ∇
```

This function retrieves the latest version of a given variable and returns the value.

11.10.2.2 Retrieving Functions

The following function retrieves that latest version of a given function and returns its canonical representation: it does not fix the function.

```
        ∇ Z←RetrieveNL3 R;sql;cns;⎕wself
   [1]   ⍝ System Building with APL+Win
   [2]   sql←"SELECT CONTENT FROM APLOBJ WHERE OBJID = '",R,"' AND OBJTYPE
         = 'Function' AND VERSION=(SELECT MAX(VERSION) FROM APLOBJ WHERE
         OBJID='",R,"');"
   [3]   cns←'Provider=Microsoft.Jet.OLEDB.4.0;Data Source=C:\SBWA.MDB;'
   [4]   ⎕wself←'∆ADORS' ⎕wi 'Create' 'ADODB.Recordset'
   [5]   ⎕wi 'XOpen' sql cns
{1[6]   :if ⎕wi 'xEOF'
   [7]       Z←''
+ [8]   :else
   [9]       Z←'unwrapl' ⎕dr ⎕wi 'xFields(0).GetChunk' (⎕wi
         'xFields(0).ActualSize')
1}[10]  :endif
   [11]  (⊂'∆ADORS') ⎕wi ¨'XClose' 'Delete'
        ∇
```

In order to fix the function, use the system function ⎕fx:

```
        ⎕fx RetrieveNL3 'StoreFile'
StoreFile
```

11.10.3 Storing Files

Any files can be stored: text files, component files, or other custom format files such as *.XLS, DOC, etc. The files may be created by APL or other applications under the control of APL or independently. The function is:

```
        ∇ Z←L StoreFile R;cns;sql;Size;Loop;Chunk
   [1]   ⍝ System Building with APL+Win
   [2]   ⍝ Ensure file (R) is available for exclusive access
   [3]   R←⎕wcall 'CharUpper' R
{1[4]   :if 1≤Z←⎕wcall '_lOpen' R 16 ⍝ Returns ‾1 if file does not exist
   [5]       Size←⎕io⊃⎕wcall 'GetFileSize' Z ''
   [6]       0 0ρ⎕wcall '_lClose' Z
+ [7]   :else
   [8]       → Z←0,0ρ'File ',R,' is in use.'
1}[9]   :endif
   [10]  Chunk←64536 ⍝ Read in 64K chunks to minimise WS Full
   [11]  Loop←(2/Chunk)⊤Size
   [12]  ⎕wself←'∆ADOS' ⎕wi 'Create' 'ADODB.Stream' ('Type' 1)
   [13]  ⎕wi 'XOpen'
   [14]  ⎕wi 'XLoadFromFile' R
   [15]  cns←'Provider=Microsoft.Jet.OLEDB.4.0;Data Source=C:\SBWA.MDB'
   [16]  ⎕wself←'∆ADORS' ⎕wi 'Create' 'ADODB.RecordSet'
   [17]  sql←"DELETE FROM APLOBJ WHERE USERID ='",GetUser,"' AND
         OBJID='",R,"';"
   [18]  ⎕wi 'XOpen' sql cns
   [19]  sql←"INSERT INTO APLOBJ (USERID,OBJID,VERSION,OBJTYPE)
         VALUES('",GetUser,"','",R,"',1,'File');"
   [20]  ⎕wi 'XOpen' sql cns
   [21]  sql←"SELECT * FROM APLOBJ WHERE USERID='",GetUser,"' AND
         OBJID='",R,"';"
```

```
   [22]   ⎕wi 'XOpen' sql cns (⎕wi '=adOpenDynamic') (⎕wi
          '=adLockOptimistic')
{1[23]   :while 0≠1↑Loop
   [24]      ⎕wi 'xFields().AppendChunk' 'CONTENT' ('#' ⎕wi 'VT' ('∆ADOS'
          ⎕wi 'XRead' Chunk) 8209)
   [25]      Loop←Loop-1 0
1}[26]   :endwhile
   [27]   Loop←1↓Loop
{1[28]   :if 0≠Loop
   [29]      ⎕wi 'xFields().AppendChunk' 'CONTENT' ('#' ⎕wi 'VT' ('∆ADOS'
          ⎕wi 'XRead' Loop) 8209)
1}[30]   :endif
   [31]   ⎕wi 'XUpdate'
   [32]   (⊂'∆ADORS' '∆ADOS') ⎕wi ¨¨⊂¨'XClose' 'Delete'
        ∇
```

Unlike the StoreNL23 function, which stores cumulatively, the function StoreFile stores files uniquely. This illustrates a different approach to filing; either method is viable depending on the needs of the application. An example of usage is:

```
'c:\myAPL.SF' ⎕xfcreate 1 ⍝ Create a Component file
(10?100) ⎕fappend 1 ⍝ Store some random numbers
⎕funtie 1
StoreFile 'c:\myapl.sf'
```

11.10.3.1 Retrieving Files
The following function retrieves files from the database:

```
       ∇ Z←L RetrieveFile R;cns;sql;Chunk;Loop;⎕wself
   [1]   ⍝ System Building with AP+Win
   [2]   R←⎕wcall 'CharUpper' R
{1[3]   :if 0=⎕nc 'L'
   [4]      L←R
1}[5]   :endif
   [6]   0 0ρ⎕wcall 'DeleteFile' L
   [7]   cns←'Provider=Microsoft.Jet.OLEDB.4.0;Data Source=C:\SBWA.MDB'
   [8]   sql←"SELECT CONTENT FROM APLOBJ WHERE OBJTYPE = 'File' AND
          USERID='",GetUser,"' AND OBJID='",R,"';"
   [9]   Chunk←64536
   [10]  ⎕wself←'∆ADORS' ⎕wi 'Create' 'ADODB.RecordSet'
   [11]  ⎕wi 'XOpen' sql cns
   [12]  Loop←(2/Chunk)⊤⎕wi 'xFields().ActualSize' 'CONTENT'
   [13]  0 0ρ'∆ADOS' ⎕wi 'Create' 'ADODB.Stream' ('Type' 1)
   [14]  '∆ADOS' ⎕wi 'XOpen'
   [15]  0 0ρ⎕wcall 'DeleteFile' L ⍝ Delete target file from filing system
{1[16]  :while 0≠1↑Loop
   [17]     '∆ADOS' ⎕wi 'XWrite' ('#' ⎕wi 'VT' (⎕wi 'xFields().GetChunk()'
          'CONTENT' Chunk) 8209)
   [18]     Loop←Loop-1 0
1}[19]  :endwhile
   [20]  Loop←1↓Loop
{1[21]  :if 0≠Loop
   [22]     '∆ADOS' ⎕wi 'XWrite' ('#' ⎕wi 'VT' (⎕wi 'xFields().GetChunk()'
          'CONTENT' Loop) 8209)
1}[23]  :endif
   [24]  '∆ADOS' ⎕wi 'XSaveToFile' L ('∆ADOS' ⎕wi '=adSaveCreateOverWrite')
```

```
[25]   Z←⎕wcall 'PathFileExists' L
[26]   (⊂'∆ADORS' '∆ADOS') ⎕wi ¨¨¨'⊂'¨'XClose' 'Delete'
    ∇
```

The optional left-hand argument specifies an alternative name for the file.

```
'c:\newcopy myapl.sf' RetrieveFile  'c:\myapl.sf'
'c:\newcopy myapl.sf' ⎕xftie 1
'c:\myapl.sf' ⎕xftie 2
(⎕fread 2,1)≡⎕fread 1,1
```
1

11.10.4 Deploying database filing

Database filing not only protects an application's objects but also ensures that they are in a single location, the database, and offers version control facilities. The functions listed above are prototypes of the functions that an application can deploy: in practice, an application will secure a connection to the database at start up and sever the connection on exit. The functions will not connect and disconnect to the database with every read or write; this ensures that there are minimum overheads to database filing.

• The database table APLOBJ is subject to the constraint that the combination of columns USERID, OBJID, and VERSION produces a unique value.

• The functions ensure that only the user who stores objects can retrieve them: this may be an unnecessary restriction by modifying the functions.

• Another modification might relate to the need for retrieving older versions of a variable or function. For example, in order to retrieve the penultimate version of a variable, modify the SQL statement in RetrieveNL2 as follows:

```
[2] sql←"SELECT CONTENT FROM APLOBJ WHERE OBJID = '",R,"' AND OBJTYPE =
        'Variable' AND VERSION=((SELECT MAX(VERSION) FROM APLOBJ WHERE
        OBJID='",R,"')-1);"
```

11.11 Automation issues

Unfortunately, **Access** does not have a macro recorder. As a consequence, it is much harder to use it as an automation server than either **Excel** or **Word**. Nonetheless, **Access** is potentially easier to use: there is no need to manage the sophisticated formatting syntax like those of **Excel** and **Word**.

Chapter 12

Working with ActiveX Data Object (ADO)

The Microsoft ActiveX Data Object (**ADO**) and the Microsoft Data **Access** Software Development Kit (**SDK**) are available for download from the Microsoft site. ADO is the ActiveX control that allows access to data sources. The SDK provides the documentation on ADO, Open Data Base Connectivity (**ODBC**) and Object Linking and Embedding for Data Bases (**OLEDB**). The SDK is an essential resource that enables effective use of ADO; it provides documentation of data access and contains many examples in various languages.

12.1 Translating code examples into APL+Win

APL is not one of the languages documented within the SDK, but it can use the ADO control and access data sources much as other languages—provided that the syntax can be translated to APL. Invariably, language features in the documentation present issues in APL; however, these can usually be resolved—usually, if it works with **VB**, it will work with APL.

A connection object is an ADO object that must be opened, thereby establishing a connection to a data source. Thus, two steps are involved; first, a connection object must be created and second, it must be opened. Typically, this is illustrated in **VB** as follows:

```
 1 Function GetConnectionVB(ByVal Connection As String) As Boolean
 2     On Error Resume Next
 3     Set ADO = CreateObject("ADODB.Connection")
 4     If 0 = Err.Number Then
 5         ADO.Open Connection
 6             If 0 = Err.Number Then GetConnection = True
 7     Else
 8         Err.Clear
 9     End If
10 End Function
```

This function returns **TRUE** when a connection is successfully established or **FALSE** if the connection fails. A connection may fail either because ADO is unavailable or because the connection string is invalid.

12.1.1 Simulating On Error Resume Next

Although APL does not have the 'On Error Resume Next' construction, it can be emulated by `'→ ⎕lc+1'`; this assumes that the APL code does *not* contain lines with multiple statements. This forces execution to continue with the statement following the statement that encounters a runtime error. In addition, APL does not have the error object present in **VB**.

12.1.1.1 The error collection

VB has an error object, Err; it has Number, Description, and Source as properties and Clear, Raise as methods. APL does not have a direct equivalent; however, most of the workings of the error object can be emulated within APL without compromise.

'Err.Number' returns an integer as an error code. An error code of 0 indicates no error. 'Err.Description'—corresponding to ⎕dm—provides a meaningful translation of the error number. 'Err.Source' holds the name of the module or class in which an error is encountered—*not* in the function or subroutine. With APL, the name of the function and the line at which an error is encountered is included in ⎕dm. The name of the function may be extracted with this code:

```
(+/∧\~⎕si[⎕io;]='[')↑⎕si[⎕io;]
```

'Err.Clear' clears the current error condition. There is no equivalent of 'Err.Clear' in APL; however, it is unnecessary as error trapping can be localised in APL and, out of scope, ⎕elx reverts to its original setting, which is ⎕dm by default. 'Err.Raise' arbitrarily causes a specific error condition; it is used during development to simulate a runtime error and to test the attendant error trapping code. A similar technique may be used with APL+Win although there is no way of creating a specific error.

One further observation is that a variable that is declared in **VB** as a type other than Variant, is implicitly assigned a value—thus, the function GetConnectionVB which returns a Boolean result is assigned **FALSE** by default—whereas in APL, return values must be explicitly assigned. The APL+Win equivalent of the GetConnectionVB function is:

```
     ∇ Z←GetConnectionAPL Connection;⎕elx
[1]    ⎕elx←'→ ⎕lc+1'
[2]    ⍝
[3]    ⎕wself←'ADO' ⎕wi 'Create' 'ADODB.Connection'
[4]    :if Z←0=ρ⎕dm
[5]        ⎕wi 'XOpen' Connection
[6]        Z←0=ρ⎕dm
[7]    :endif
     ∇
```

With both the **VB** and APL+Win functions, runtime errors are likely in lines [3] and [5] of the corresponding functions. An error at line [3] indicates that the ADO ActiveX control is missing; an error at line [5] indicates that the connection string is incorrect.

12.1.2 Simulating On Error Goto

Although the construction 'On Error Goto' is discouraged on grounds that it makes code harder to read, it is still widely used in **VB**. Where necessary, this too can be emulated in APL: see **Table 12-1 VB/APL Error handling**.

VB	APL
1 Sub ErrTrapGoto	∇ OnErrorGotoAPL;⎕elx;ResumeNext
2 On Error Goto L0	[1] ⎕elx←'ResumeNext←⎕lc+1 ◇ → L0' ⍝ On Error
3 a=2/0	Goto L0
4 MsgBox a	[2] a←2÷0
5 Exit Sub	[3] Msg a
6 L0:	[4] → 0
7 Err.Clear	[5] L0:
8 On Error Goto 0	[6] ⎕elx←'⎕dm' ⍝ Err.Clear
9 a=0	[7] a←0
10 Resume Next	[8] → ResumeNext
11 End Sub	∇

Table 12-1 VB/APL Error handling

Line [6] of the APL function achieves both lines [7] and [8] of the **VB** subroutine. In **VB**, 'Resume Next' causes processing to restart with the statement following the one that encountered a runtime error; this is implemented in APL by holding the absolute number of the next line in a local variable. In the illustration above, the error handler forces resumption of execution at the line following the line that encountered an error. In **VB**, it is possible to force re-evaluation of the statement of the line that caused the error—by using Resume instead of Resume Next—this is achieved by setting Resume to → ⎕lc in APL+Win. Unless used in a context where a statement fails because of timing, Resume or → ⎕lc may introduce an indefinite loop; therefore, it must be used with great care. The statements 'Exit Sub', 'Exit Function', 'Exit Do', and 'Exit Property' are equivalent to → 0 in APL. Code using branching to absolute line numbers is harder to maintain and prone to unexpected bugs and must be avoided.

12.2 The connection object

A connection object is an instance of the ADO connection object. An *active* connection object is an instance of an ADO connection object that has been opened; one that has successfully established a connection to a data source. If ADO is installed, a connection object is created as follows:

```
⎕wself←'ADO' ⎕wi 'Create' 'ADODB.Connection'
```

12.2.1 Creating an active connection object

The **VB** syntax for creating a connection is:

```
connection.Open ConnectionString, UserID, Password, Options
```

The Open method is applied to an existing connection object whose state must be closed. Thus, if a connection object is to be reused to create another connection, it must be closed first. Furthermore, a time-period can be specified for the connection to be made, in seconds—if unsuccessful within that duration, the connection fails causing a runtime error. A duration of 0 means no time out period. The following example establishes a connection to an **Access** database:

```
     ∇ Connect;⎕wself
[1]    ⍝ System Building with APL+Win
[2]    ⎕wself←'ADO' ⎕wi 'Create' 'ADODB.Connection'
[3]    ⍝ Connection object has been re-created; therefore it is closed.
[4]    ⍝ The following lines can verify the state of the object and close
          it if necessary.
[5]    ⍝ :if 0≠⎕wi 'xState'
[6]    ⍝     ⎕wi 'XClose'
[7]    ⍝ :endif
[8]    ⎕wi 'xConnectionTimeOut' 0 ⍝ indefinite, i.e. no timeout.
[9]    ⎕wi 'XOpen' 'Provider=MSDASQL;Driver={Microsoft Access Driver
          (*.mdb)};DBQ=c:\nwind.mdb;'
     ∇
```

In line [2], the connection object exists without an active connection; an active connection is established in line [9]. Each active connection object has a unique collection of properties. Each available property holds a value that indicates a characteristic of the connection. When a connection object is created, it does not have a link to a data source and the connection is not active. It may be activated in using two types of syntax.

12.2.2 Database connection using a connection string—syntax I

The Open method is invoked with a string argument that represents the information needed to connect to the data source, as follows:

```
⎕wi 'XOpen' 'Provider=MSDASQL;Driver={Microsoft Access Driver
(*.mdb)};DBQ=c:\nwind.mdb;'
```

The argument of the 'XOpen' method is a connection string; it specifies several parameters separated by semi-colon, in the format *argument = value*, that allows connection to a data source. The arguments specified depend on the 'provider' used to make the connection.

In this context, a service provider is "Software that encapsulates a service by producing and consuming data, augmenting features in your ADO applications. It is a provider that does not directly expose data; rather it provides a service, such as query processing. The service provider may process data provided by a data provider"; a data provider is "Software that exposes data to an ADO application either directly or via a service provider."

12.2.2.1 Typical connection strings

A connection string contains enough information for the connection object to establish an active connection to a data source. This is a read/write property of the connection object. The provider may alter or supplement the contents of the connection string property—the string as seen by the connection object can be queried thus:

```
       'ADO' □wi 'xConnectionString'
Provider=MSDASQL.1;Extended Properties="DBQ=c:\nwind.mdb;Driver={Microsoft
Access Driver (*.mdb)};DriverId=281;FIL=MS Access;
MaxBufferSize=2048;PageTimeout=5;"
```

Note that supplementary 'argument = value' pairs have been added—occasionally, it may be necessary to override the default values by explicitly specifying the relevant 'argument = value' item. The connection string property of an active connection may provide valuable clues in debugging its behaviour, particularly when this is different on different computers. In order to promote consistent behaviour, the connection string property of an active connection may be used to replace the connection string used to create the active connection in the first place. That is, line [9] of the Connect function can be modified such that the argument of the XOpen method uses the connection string returned by 'ADO' □wi 'xConnectionString'. The connection strings to typical data sources are listed in **Table 12-2 Sample connection strings**.

```
1    SQL Server 2000 ODBC Driver
DSN=SQL2K;UID=SYSSQL;PWD=sa
2    Oracle 8.0 ODBC Driver
DSN=SALES;UID=MAILSHOT;PWD=sa
3    Text Driver
Provider=MSDASQL;Driver={Microsoft Text Driver (*.txt; *.csv)};DBQ=c:\;
4    Excel Driver
Provider=MSDASQL;Driver={Microsoft Excel Driver (*.xls)}; DBQ=c:\Adoconn.xls;
5    Access Driver
Provider=MSDASQL;Driver={Microsoft Access Driver *.mdb)};DBQ=c:\nwind.mdb;
6    Jet 4.0 Provider
Provider=Microsoft.Jet.OLEDB.4.0;Data Source=c:\nwind.mdb;
7    MSDAORA Oracle Provider
Provider=MSDAORA;Data Source=SALES;User ID=MAILSHOT; Password=sa;
8    SQLOLEDB SQL Server Provider
Provider=SQLOLEDB;Data Source=sql2000;Initial Catalog=sysdat;User
        Id=sysdat;Password=sa;
9    Microsoft Data Engine MSDE
Provider=MSDASQL;Extended Properties="DRIVER=SQL Server;SERVER=AJAY
        ASKOOLUM;UID=sa;WSID=AJAY ASKOOLUM;DATABASE=pubs"
```

Table 12-2 Sample connection strings

For a Microsoft Data Engine (MSDE is a cut-down version but fully compatible of SQL Server) connection , both SERVER and WSID are the name of the personal computer; the UID and password may be omitted when the option 'Use Trusted Connection' is set to **TRUE**. MSDE 2.0 is a scaled down version of Microsoft SQL Server 2000 but the databases are interchangeable—subject to size.

12.2.2.2 UserID

This optional parameter comprises a character literal—that contains a user name—to use when establishing the connection. Server-based databases such as SQL Server and Oracle databases usually require this parameter; file-based databases such as **Access**, **Excel,** etc., generally do not. For MSDE, the default user id is *sa*—system administrator.

Applications that have their own custom password protected access procedures generally use the same default User ID for all users when connecting to the data source, internally. In such circumstances, the data source is configured to grant access to the default user—usually, the name of the database and that of the default user is identical. In other words, access to the application implies automatic access to the data source. The alternative is to control access to the data source and allow automatic access to the application; in this scenario, each user has a unique id.

12.2.2.3 Password

This optional parameter comprises a character literal—that contains a password—to use when establishing the connection. For MSDE, there is no password for the default user.

12.2.2.4 Options

This optional parameter comprises a numeric value—that contains one of two possible values from **Table 12-3 Connection modes**—to use when establishing the connection. The values are held as ADO constant—unless explicitly defined, these constants are unavailable in APL+Win and the numeric values must be explicitly specified, as follows:

Constant	Decimal Value	Effect
adConnectUnspecified	⁻1	Default. Opens the connection synchronously—waits until the connection is established.
adAsyncConnect	16	Opens the connection asynchronously—does not wait for the connection to be established.

Table 12-3 Connection modes

It is usual to omit this parameter thereby causing it to take the default value of ⁻1.

12.2.3 Database connection using properties—syntax II

The properties of the instance of the connection object may be individually populated before invoking the Open method. Each data source requires a different subset of these properties to be initialised. Refer to the documentation of the 'provider' or 'driver' intended to make a connection for documentation of the properties that need to be populated. In this context, a driver is software that creates an interface between an application and a data source; each specific data source has its own driver. The complete list of the properties of a connection object may be enumerated thus:

```
      ∇ Z←ConnectionProperties;⎕wself
[1]   ⍝ System Building with APL+Win
[2]   ⎕wself←'ADO' ⎕wi 'Create' 'ADODB.Connection'
[3]   Z←⊃((⎕wi 'xProperties') ⎕wi ¨(⊂⊂'xItem'),¨((⍳(⎕wi 'xProperties') ⎕wi
      'xCount')-⎕io)) ⎕wi ¨⊂'xName'
      ∇

      ConnectionProperties
Password
Persist Security Info
User ID
Data Source
Window Handle
Location
Mode
Prompt
```

```
Connect Timeout
Extended Properties
Locale Identifier
Initial Catalog
OLE DB Services
General Timeout
```
Only a closed instance of a collection object has the above read and write properties. The state of a connection object may be queried by:
```
'ADO' ⎕wi 'xState'
```
The possible values of the 'state' property are shown in **Table 12-4 Object state values**.

Constant	Decimal Value	Meaning
adStateClosed	0	Object is closed
adStateOpen	1	Object is open
adStateConnecting	2	Object is connecting
adStateExecuting	4	Object is executing a command
adStateFetching	8	Object is retrieving its rows

Table 12-4 Object state values

The parameters for connecting to the same data source may be specified as follows:
```
(⎕wi 'xProperties("Extended Properties")') ⎕wi 'xValue'
'Provider=MSDASQL;Driver={Microsoft Access Driver
(*.mdb)};DBQ=c:\nwind.mdb;'
```
Then, the connection object may be opened by:
```
'ADO' ⎕wi 'XOpen'
```
Upon being opened, the connection object becomes active and its dynamic properties are exposed. The list properties that are exposed depend on the provider used to establish the connection. The full list of properties, in the order exposed, can be retrieved as follows:
```
⊃((⎕wi 'xProperties') ⎕wi ¨(⊂⊂'xItem'),¨((⍳(⎕wi 'xProperties') ⎕wi
'xCount')-⎕io)) ⎕wi ¨⊂'xName'
92 34
```
The corresponding values of those properties may be retrieved, similarly by:
```
⊃((⎕wi 'xProperties') ⎕wi ¨(⊂⊂'xItem'),¨((⍳(⎕wi 'xProperties') ⎕wi
'xCount')-⎕io)) ⎕wi ¨⊂'xValue'
```
The 'Open' method of the connection object expects literal—or human readable—values as it arguments; if these arguments are stored centrally in an unprotected database, in an encrypted format, the application must decrypt them internally before passing them to the 'Open' method.

Altogether, there are some 154 dynamic properties. Each connection supports its own SQL variant. In an ideal scenario, ANSI SQL—supported by all connection objects—would imply a single SQL statement regardless of the data source. In practice, an APL+Win application that is independent of its data source needs to provide the functional logic to elicit the optimal SQL for each type of connection.

Note, again, that the DBMS Name for the MSDE and SQL Server connections is identical. If an active connection to a data source exists in an APL+Win object called ADO, a function such as *GetSQL* can return the appropriate SQL. Look at the following code:
```
    ∇  Z←GetSQL
  [1]   ⍝ System Building with APL+Win
{1[2]     :select case 'ADO'  ⎕wi 'xProperties(←DBMS Name←).Value'
* [3]        :case 'Microsoft SQL Server'
  [4]            Z←...
* [5]        :case 'Oracle'
```

```
      [6]                 Z←...
★   [7]                :case 'EXCEL'
      [8]                 Z←...
★   [9]                :case 'ACCESS'
      [10]                Z←...
★   [11]               :case 'TEXT'
      [12]                Z←...
★   [13]               :case 'MS Jet'
      [14]                Z←...
+   [15]               :else
      [16]                Z←0×Msg 'Unknown data source'
1}[17]      :endselect
        ∇
```

All the dynamic properties of an active connection can be enumerated, comprising of a nested vector of two elements—names and values—thus:

```
      dynprop←((⊂(⎕wi 'xProperties') ⎕wi ¨(⊂⊂'xItem'),¨((⍳(⎕wi
'xProperties') ⎕wi 'XCount')-⎕io))) ⎕wi ¨¨¨⊂¨'Name' 'Value'
```

The Item property of the Properties collections works in option base 0—that is, index origin 0. Each connection exposes its properties in a different order, as shown below with the sqlserver and oracle list of dynamic properties:

`(⎕io⊃sqlserver)⍳⊂'DBMS Name'`	`(⎕io⊃oracle)⍳⊂'DBMS Name'`
12	11

```
      Properties of an active connection can be queried in several ways:
      'ADO' ⎕wi 'xProperties("DBMS Name").Value'
ACCESS
      'ADO' ⎕wi 'xProperties.Item("DBMS Name").Value'
ACCESS
      'ADO' ⎕wi 'xProperties.Item(11).Value'
ACCESS
```

It is advisable to refer to a member of the properties collection by name—as shown in the function GetSQL—instead of its ordinal position as this avoids the overhead of converting the ordinal position to base 0 and the need for connection specific code. The complete list of properties of a particular active connection—in the order exposed by it—can be returned, as a character array, by:

```
      list←⍕⊃¨⍪¨¨¨((⊂(⎕wi 'xProperties') ⎕wi ¨(⊂⊂'xItem'),¨((⍳(⎕wi
'xProperties') ⎕wi 'XCount')-⎕io))) ⎕wi ¨¨¨⊂¨'xName' 'xValue'
```

12.3 The record object

A record object contains sets of zero or more rows—or records—extracted from a database using an SQL statement—it may contain one or more columns—or fields—from a single or multiple tables. The columns in each record object may include existing columns as well as columns calculated from existing columns. A record object that is open is a result set or the record set.

The characteristics of the record object depend on the provider or driver used in the connection string and on the method used to create it. In APL+Win, the record object can be created in two ways.

12.3.1 Record object using redirection

This method relies on an active connection that returns a record object when it executes an SQL statement:

```
      ∇ RecordObjectRedirection;⎕wself
[1]    ⍝ System Building with APL+Win
[2]    ⎕wself←'ADO' ⎕wi 'Create' 'ADODB.Connection'
[3]    ⎕wi 'XOpen' 'Provider=MSDASQL;Driver={Microsoft Access Driver
      (*.mdb)};DBQ=c:\nwind.mdb;'
```

```
[4]      ⎕wi 'Execute>ADO.Recordset' 'SELECT * FROM customers;'
[5]      ⎕wself←'ADO.Recordset'
[6]      ⎕wi 'XClose'
[7]      ⎕wi 'xCursorType' (⎕wi '=adOpenStatic')
[8]      ⎕wi 'XOpen'
[9]      ⍝ Instances of both objects continue to exist
     ∇
     RecordObjectRedirection
ADO
     'ADO.RecordSet' ⎕wi 'XRecordCount'
91
```

Line [3] creates an active connection; line [4] creates the record object using the APL+Win redirection syntax. The record object has 91 records.

12.3.2 Record object without connection object

This method does not rely on an active connection object. A record object is created directly and opened with the connection string as a parameter. Unlike the redirection method where two objects are required, this method requires a single object.

Like the active connection object, the record set object also exposes dynamic properties. Consider the following code:

```
     ∇ RecordObjectDirect;⎕wself
[1]      ⍝ System Building with APL+Win
[2]      ⎕wself←'ADORS' ⎕wi 'Create' 'ADODB.Recordset'
[3]      ⎕wi 'XOpen' 'Select * from customers'
             'Provider=MSDASQL;Driver={Microsoft Access Driver
             (*.mdb)};DBQ=c:\nwind.mdb;'
[4]      ⎕wi 'XClose'
[5]      ⎕wi 'xCursorType' (⎕wi '=adOpenStatic')
[6]      ⎕wi 'XOpen'
[7]      ⍝ Instance of object continues to exist
     ∇
     RecordObjectDirect
ADORS
     'ADORS' ⎕wi 'XRecordCount'
91
```

In general, a connection object is used in circumstances when a method applied to it does not return a record object; a record object may be used directly when a record collection is returned.

12.3.3 Creating a record object

The syntax of the Open method of the record object is:

```
recordset.Open Source, ActiveConnection, CursorType, LockType, Options
```

The SDK provides full documentation of the syntax for creating record objects; the syntax used directly affects the behaviour and efficiency of the record object. For example,

```
     ∇ Z←CursorTypeImpact;⎕wself
[1]      ⍝ System Building with APL+Win
[2]      ⎕wself←'ADORS' ⎕wi 'Create' 'ADODB.RecordSet'
[3]      ⎕wi 'XOpen' 'Select * from customers'
             'Provider=MSDASQL;Driver={Microsoft Access Driver
             (*.mdb)};DBQ=c:\nwind.mdb;'
[4]      Z←⎕wi 'xRecordCount'
     ∇
     CursorTypeImpact
‾1
```

This does not return the expected result, 91, because the default cursor type is 0—adOpenForwardOnly. The value of ⁻1 indicates that the record count is indeterminate, given the type of cursor in use:

```
'ADORS' ⎕wi 'xCursorType'
0
```

12.3.3.1 Source

The source parameter specifies what data is returned. It is a query against existing data; typically, this is an SQL statement. The SDK provides the following definition:

```
"Optional. A Variant that evaluates to a valid Command object, an SQL statement, a
table name, a stored procedure call, a URL, or the name of a file or Stream object
containing a persistently stored Recordset".
```

Note that this parameter is deemed optional. Line [6] of *RecordObjectDirect* calls the XOpen method without specifying any parameters.

12.3.3.2 Active connection

The active connection specifies the means of connecting to the data source. As already seen, this may be an active connection object or a connection string. The SDK provided the following definition is:

```
"Optional. Either a Variant that evaluates to a valid Connection object variable
name, or a String that contains ConnectionString parameters".
```

12.3.3.3 Cursor type

The value of this parameter affects the efficiency of data retrieval and the properties of the record object. The SDK definition is:

```
"Optional. A CursorTypeEnum value that determines the type of cursor that the
provider should use when opening the Recordset. The default value is
adOpenForwardOnly."
```

The CursorTypeEnum value specifies the type of cursor used in the record object; the optional values documented in the SDK are shown in **Table 12-5 Cursor options**.

Constant	Decimal Value	Description
adOpenUnspecified	⁻1	Does not specify the type of cursor
adOpenForwardOnly	0	Default. Uses a forward-only cursor. Identical to a static cursor, except that you can only scroll forward through records. This improves performance when you need to make only one pass through a Recordset.
adOpenKeyset	1	Uses a keyset cursor. Like a dynamic cursor, except that you can't see records that other users add, although records that other users delete are inaccessible from your Recordset. Data changes by other users are still visible.
adOpenDynamic	2	Uses a dynamic cursor. Additions, changes, and deletions by other users are visible, and all types of movement through the Recordset are allowed, except for bookmarks, if the provider doesn't support them.
adOpenStatic	3	Uses a static cursor. A static copy of a set of records that you can use to find data or generate reports. Additions, changes, or deletions by other users are not visible.

Table 12-5 Cursor options

12.3.3.4 Lock type

The Lock Type property arising from the CurTypeImpact function is:	`'ADORS' ⎕wi 'xLockType'` `1`

The LockTypeEnum value specifies the type of lock placed on records during editing; the optional values are shown in **Table 12-6 Lock type options**.

Constant	Decimal Value	Description
adLockUnspecified	1	Does not specify a type of lock.
AdLockReadOnly	1	Indicates read only records.
adLockPessimistic	2	Indicates pessimistic locking, record by record. The provider does what is necessary to ensure successful editing of the records, usually by locking records at the data source immediately after editing.
AdLockOptimistic	3	Indicates optimistic locking, record by record. The provider uses optimistic locking, locking records only when you call the Update method.
adLockBatchOptimistic	4	Indicates optimistic batch updates. Required for batch update mode.

Table 12-6 Lock type options

The SDK definition for Lock Type is:

"Optional. A LockTypeEnum value that determines what type of locking (concurrency) the provider should use when opening the **Recordset**. The default value is **adLockReadOnly**."

12.3.3.5 Options

The SDK definition of this parameter is:

"Optional. A **Long** value that indicates how the provider should evaluate the Source argument if it represents something other than a **Command** object, or that the **Recordset** should be restored from a file where it was previously saved. Can be one or more CommandTypeEnum or ExecuteOptionEnum values, which can be combined with a bitwise AND operator."

The CommandTypeEnum value specifies how the provider should interpret a command argument; the optional values documented in the SDK are shown in **Table 12-7 Command type options**.

Constant	Decimal Value	Description
adCmdUnspecified	⁻1	Does not specify the command type argument.
adCmdText	1	Evaluates CommandText as a textual definition of a command or stored procedure call.
adCmdTable	2	Evaluates CommandText as a table name whose columns are all returned by an internally generated SQL query.
adCmdStoredProc	4	Evaluates CommandText as a stored procedure name.
adCmdUnknown	8	Default. Indicates that the type of command in the CommandText property is not known.
adCmdFile	256	Evaluates CommandText as the file name of a persistently stored Recordset. Used with Recordset.Open or Requery only.
adCmdTableDirect	512	Evaluates CommandText as a table name whose columns are all returned. Used with Recordset.Open or Requery only. To use the Seek method, the Recordset must be opened with adCmdTableDirect.

Table 12-7 Command type options

The ExecuteOptionEnum value specifies how a provider should execute a command; the optional values documented in the SDK are shown in **Table 12-8 Command execution options**.

Constant	Decimal Value	Description
adOptionUnspecified	⁻1	Indicates that the command is unspecified.
adAsyncExecute	16	Indicates that the command should execute asynchronously.
adAsyncFetch	32	Indicates that the remaining rows after the initial quantity specified in the CacheSize property should be retrieved asynchronously.

Constant	Decimal Value	Description
adAsyncFetchNonBlocking	64	Indicates that the main thread never blocks while retrieving. If the requested row has not been retrieved, the current row automatically moves to the end of the file.
If you open a Recordset from a Stream containing a persistently stored Recordset, adAsyncFetchNonBlocking will not have an effect; the operation will be synchronous and blocking. The adAsynchFetchNonBlocking setting has no effect when the adCmdTableDirect option is used to open the Recordset.		
adExecuteNoRecords	128	Indicates that the command text is a command or stored procedure that does not return rows (for example, a command that only inserts data). If any rows are retrieved, they are discarded and not returned.
The setting adExecuteNoRecords can only be passed as an optional parameter to the Command or Connection Execute method.		
adExecuteStream	1024	Indicates that the results of a command execution should be returned as a stream.
The setting adExecuteStream can only be passed as an optional parameter to the Command Execute method.		
adExecuteRecord	2048	Indicates that the CommandText is a command or stored procedure that returns a single row, which should be returned as a Record object.

Table 12-8 Command execution options

In APL, the decimal values may be used, as the constants are not available unless explicitly resolved in the workspace: this is possible for APL+Win and APLX.

12.3.4 Cloning a record object

The *Clone* method of a record set object creates a duplicate of the record set object; the clone may be locked as read only. The original record set and its clone become separate objects for the following reasons:

- Any *Filter* applied to the original record set is not applied to the clone.

- Either may be closed without affecting the other.

- Each record set has its own current record.

Thus, different records can be accessed from each record set. Any changes made to the original record set can be undone readily by resorting back to the clone record set.

12.3.4.1 Cloning method error

After cloning, each record set object exists in its own right; deleting one does not affect the other. However, the clone record set is somehow tied to the original record set and inherits any deletions made to it. In order to illustrate this, a record is added to the categories table of the NWIND.MDB database:

```
      ∇ AddRecord;⎕wself;Cnn;Sql
[1]     ⍝ System Building with APL+Win
[2]     Cnn←'Provider=MSDASQL;Driver={Microsoft Access Driver
          (*.mdb)};DBQ=c:\nwind.mdb;'
[3]     Sql←"INSERT INTO CATEGORIES (CategoryName) VALUES('Dummy');"
[4]     ⎕wself←'ADOCon' ⎕wi 'Create' 'ADODB.Connection'
[5]     ⎕wi 'XOpen' Cnn
[6]     ⎕wi 'XExecute' Sql  1
[7]     ⎕wi 'Delete'
      ∇
```

The incremented number of records in *Categories* can be established using the following code:

```
    ∇ BeforeDelete;Cnn;Sql
[1]    ⍝ System Building with APL+Win
[2]    Cnn←'Provider=MSDASQL;Driver={Microsoft Access Driver
          (*.mdb)};DBQ=c:\nwind.mdb;'
[3]    Sql←'SELECT * FROM categories;'
[4]    ⎕wself←'∆ADORS' ⎕wi 'Create' 'ADODB.RecordSet'
[5]    ⎕wi 'XOpen' Sql Cnn (⎕wi '=adOpenStatic') (⎕wi
          '=adLockBatchOptimistic') (⎕wi '=adCmdText')
    ∇

       '∆ADORS' ⎕wi 'xRecordCount'
9
```

The function Clone creates a clone of the record set:

```
    ∇ Clone;⎕wself;Cnn;Sql
[1]    ⍝ System Building with APL+Win
[2]    Cnn←'Provider=MSDASQL;Driver={Microsoft Access Driver
          (*.mdb)};DBQ=c:\nwind.mdb;'
[3]    Sql←'SELECT * FROM categories;'
[4]    ⎕wself←'∆ADORS' ⎕wi 'Create' 'ADODB.RecordSet'
[5]    ⎕wi 'XOpen' Sql Cnn (⎕wi '=adOpenStatic') (⎕wi
          '=adLockBatchOptimistic') (⎕wi '=adCmdText')
[6]    '∆ADORSClone' ⎕wi 'Create' (⎕wi 'XClone' (⎕wi '=adLockReadOnly'))
[7]    '∆ADORS' ⎕wi 'xFilter' "categoryname= 'Dummy'"
[8]    '∆ADORS' ⎕wi 'XDelete' (⎕wi '=adAffectCurrent')
[9]    '∆ADORS' ⎕wi 'UpdateBatch'
[10]   '∆ADORS' ⎕wi 'xFilter' (⎕wi '=adFilterNone')
    ∇
```

The record set created in line [5] has 9 records; it is cloned in line [6] and the record where category name is *Dummy* is deleted in lines [7] to [9] from the original record set. At this point, the original record set must have 8 records and the clone must also have 8 records: this is established as follows:

```
'∆ADORS' ⎕wi 'xRecordCount'              '∆ADORS.Clone' ⎕wi 'xRecordCount'
8                                         8
```

In fact, both record sets have the same number of records; that is, both record sets have remained associated.

12.3.4.2 Cloning with Stream object

First, the dummy record must be added to the table using the *AddRecord* function. The following function creates a clone record set:

```
    ∇ StreamClone;Cnn;Sql
[1]    ⍝ System Building with APL+Win
[2]    Cnn←'Provider=MSDASQL;Driver={Microsoft Access Driver
          (*.mdb)};DBQ=c:\nwind.mdb;'
[3]    Sql←'SELECT * FROM categories;'
[4]    ⎕wself←'∆ADORS' ⎕wi 'Create' 'ADODB.RecordSet'
[5]    ⎕wi 'XOpen' Sql Cnn (⎕wi '=adOpenStatic') (⎕wi
          '=adLockBatchOptimistic') (⎕wi '=adCmdText')
```

At this point, the record set contains all 9 records from the **Categories** table in NWIND.MDB:

```
[6]    '∆ADOStream' ⎕wi 'Create' 'ADODB.Stream'
[7]    ⎕wi 'XSave' ('∆ADOStream' ⎕wi 'obj')
```

Line [6] creates a stream object to which line [7] saves the record set to the stream object:

```
[8]    'ΔADORSClone' Δwi 'Create' 'ADODB.RecordSet'
[9]    'ΔADORSClone' Δwi 'XOpen' ('ΔADOStream' Δwi 'obj')
[10]   'ΔADOStream' Δwi 'Delete'
```

Line [8] creates a new record set; line [9] opens the original record set—the new record set is now a clone of the original record set. Line [10] deletes the *Stream* object, as it is no longer needed, as shown by:

```
[11]   'ΔADORSClone' Δwi 'xFilter' "categoryname= 'Dummy'"
[12]   'ΔADORSClone' Δwi 'XDelete' (Δwi '=adAffectCurrent')
[13]   'ΔADORSClone' Δwi 'UpdateBatch'
[14]   'ΔADORSClone' Δwi 'xFilter' (Δwi '=adFilterNone')
     ∇
```

The *Dummy* record is removed from the clone record set in lines [11] to [14]. The record counts of the original and clone record sets confirm that the two are dissociated. This is established b querying the record counts of each, as follows:

```
'ΔADORS' Δwi 'xRecordCount'           'ΔADORSClone' Δwi 'xRecordCount'
9                                     8
```

12.3.5 Tables in a data source

In most circumstances, a record object that contains a selection of records from one or more tables may be created directly without recourse to an existing active connection object. However, an active connection has other information that cannot be retrieved via a record object. For example, if the names and types of tables in a data source are required, an active connection is necessary:

```
     ∇ Z←OpenSchema;Δwself
[1]    A System Building with APL+Win
[2]    Δwself←'ADO' Δwi 'Create' 'ADODB.Connection'
[3]    Δwi 'XOpen' 'Provider=MSDASQL;Driver={Microsoft Access Driver
          (*.mdb)};DBQ=c:\nwind.mdb;'
[4]    Δwself←'.Recordset' Δwi 'Create' (Δwi 'XOpenSchema()' (Δwi
          '=adSchemaTables'))
[5]    Z←((⊂Δwself) Δwi ¨(⊂⊂'xFields.Item().Name'),¨⁻1++\(Δwi
          'xFields.Count')/1)
[6]    Z←Z,[Δio]Δwi 'XGetRows'
[7]    A Instances of both objects continue to exist
     ∇

     'ADO.RecordSet' Δwi 'XRecordCount'
32
```

This returns 32 rows and 9 columns, namely, TABLE_CATALOG, TABLE_SCHEMA, TABLE_NAME, TABLE_TYPE, TABLE_GUID, DESCRIPTION, TABLE_PROPID, DATE_CREATED, and DATE_MODIFIED.

The current record object contains all types of tables; a particular type of table, say TABLE, can be extracted from it:

```
     'ADO.RecordSet' Δwi 'xFilter' "TABLE_TYPE='TABLE'"
     'ADO.RecordSet' Δwi 'XGetString'
```

The result is shown in **Table 12-9 Filtering schema table**.

TABLE_CATALOG	TABLE_NAME	TABLE_TYPE
C:\NWIND	Categories	TABLE
C:\NWIND	Customers	TABLE
C:\NWIND	Employees	TABLE

TABLE_CATALOG	TABLE_NAME	TABLE_TYPE
C:\NWIND	Order Details	TABLE
C:\NWIND	Orders	TABLE
C:\NWIND	Products	TABLE
C:\NWIND	Shippers	TABLE
C:\NWIND	Suppliers	TABLE

Table 12-9 Filtering schema table

With the filter in place, the record object reports a subset of its records:

```
'ADO.RecordSet' ⎕wi 'XRecordCount'
8
```

The filter may be switched off and the original record count restored by executing the following code:

```
'ADO.RecordSet' ⎕wi 'xFilter' ''
'ADO.RecordSet' ⎕wi 'XRecordCount'
32
```

12.3.5.1 Verifying the existence of a table

The *Filter* property of the record object provides a convenient mechanism for verifying the existence of a particular row within in. A table, if it exists, is a row in the catalogue of a data source. Does a table named **Order Details** of type **TABLE** exist in the C:\NWIND.MDB data source? Try:

```
      ∇ Z←TableExists;⎕wself
[1]   ⍝ System Building with APL+Win
[2]   ⎕wself←'ADO' ⎕wi 'Create' 'ADODB.Connection'
[3]   ⎕wi 'XOpen' 'Provider=MSDASQL;Driver={Microsoft Access Driver
         (*.mdb)};DBQ=c:\nwind.mdb;'
[4]   ⎕wi 'XOpenSchema>ADO.Recordset' (⎕wi '=adSchemaTables')
[5]   ⎕wi 'xFilter' "TABLE_NAME='Order Details' AND TABLE_TYPE='TABLE'"
[6]   Z←0≠'ADO.Recordset' ⎕wi 'xRecordCount'
[7]   'ADO.Recordset' ⎕wi 'Delete'
      ∇
```

This method for checking whether a table exists will work with any data source—with slight modification—connected via ADO. File data sources such as XLS, CSV, and TXT files do not support the concept of a TABLE_TYPE; therefore, only the names of the files may be used to query their existence. For SQL Server and Oracle data sources, there are methods that are more direct and faster. For example:

```
      ∇ Z←SQLTableExists;⎕wself
[1]   ⎕wself←'ADORS' ⎕wi 'Create' 'ADODB.Recordset'
[2]   ⎕wi 'xCursorType' (⎕wi '=adOpenStatic')
[3]   ⎕wi 'XOpen' "select * from sysobjects where name='sales'"
         'Driver={SQL Server};Server=Ajay
         Askoolum;Database=pubs;UID=sa;PWD=;'
[4]   Z←0≠⎕wi 'xRecordCount'
[5]   ⍝ Object continues to exist
      ∇
```

The function *SQLTableExists* will work on SQL Server data sources. For Oracle, line [4] is different:

```
      ⎕wi 'XOpen' "Select * from all_tables where table_name = 'sales03'"
"Provider=MSDAORA;Data Source=Oracle03;User ID=Sales;Password=sa;"
```

Modified SQL statements may retrieve all table names:

```
'SELECT NAME FROM SYSOBJECTS' ⍝ SQL Server
'SELECT TABLE_NAME FROM ALL_TABLES' ⍝ Oracle
```

These functions are hard coded and depend on the cursor type being set to 3. In practice:

● The schema—or sysobjects or all_tables—of a data source will be retrieved from an existing connection; therefore, line [2] may be omitted. The record object may be created using the redirection syntax.

● The name—and type, where applicable—of the table being sought will be passed as arguments.

● Table names are unique within data sources; therefore, the record count after the filter is applied is either 0 or 1.

12.3.6 Working with record objects

Record objects indirectly resolve a number of APL+Win specific problems, namely:

● Code to read records from files is no longer necessary; this confers a huge advantage in that there is no longer any need to manage the length of records, to define the width, and types—and for date, the format—of fields.

● Since the record object lies outside of the APL+Win workspace, workspace management, in so far reading files is concerned, is also unnecessary. Thus, Windows resources permitting, APL+Win can process vast numbers of records.

The function *xls* reads the records from the worksheet shown in **Figure 12-1 Complete record set**.

```
      ∇ xls;⎕wself
[1]     ⍝ System Building with APL+Win
[2]     ⎕wself←'ADORS' ⎕wi 'Create' 'ADODB.RecordSet'
[3]     ⎕wi 'xCursorType' (⎕wi '=adOpenStatic')
[4]     ⎕wi 'XOpen' 'Select * from [mydata$]'
          'Provider=MSDASQL;Driver={Microsoft Excel Driver
          (*.xls)};DBQ=c:\Adox.xls;'
      ∇
```

That the record object does not use available workspace is easily verifiable—albeit, it contains 12 records, as follows:

```
      ⎕wa
1455068
      xls
      ⎕wa
1455068
      'ADORS' ⎕wi 'xRecordCount'
12
```

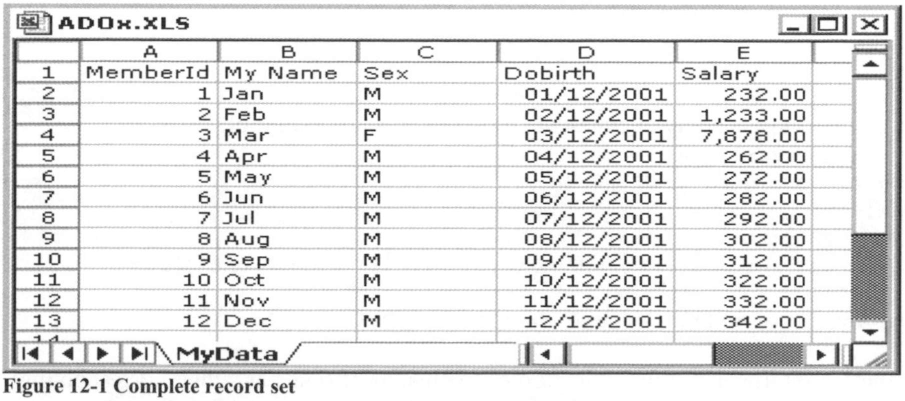

	A	B	C	D	E	
1	MemberId	My Name	Sex	Dobirth	Salary	
2	1	Jan	M	01/12/2001	232.00	
3	2	Feb	M	02/12/2001	1,233.00	
4	3	Mar	F	03/12/2001	7,878.00	
5	4	Apr	M	04/12/2001	262.00	
6	5	May	M	05/12/2001	272.00	
7	6	Jun	M	06/12/2001	282.00	
8	7	Jul	M	07/12/2001	292.00	
9	8	Aug	M	08/12/2001	302.00	
10	9	Sep	M	09/12/2001	312.00	
11	10	Oct	M	10/12/2001	322.00	
12	11	Nov	M	11/12/2001	332.00	
13	12	Dec	M	12/12/2001	342.00	

ADOx.XLS — MyData

Figure 12-1 Complete record set

However, there are some new problems, namely:

```
      ⎕←A←'ADORS' ⎕wi 'XGetRows'
```

```
 1  Jan M 37226   232
 2  Feb M 37227  1233
 3  Mar F 37228  7878
 4  Apr M 37229   262
 5  May M 37230   272
 6  Jun M 37231   282
 7  Jul M 37232   292
 8  Aug M 37233   302
 9  Sep M 37234   312
10  Oct M 37235   322
11  Nov M 37236   332
12  Dec M 37237   342
```

The internal representation of the date field, Dobirth, has been returned. Dates are a problem for APL+Win, as it does not have a date data type. However, the record object returns the numeric field, Salary, correctly—having stripped out the thousands separator. Moreover, this field is numeric and does not require conversion:

```
      A[;⎕io+4]+0
232 1233 7878 262 272 282 292 302 312 322 332 342
```

12.3.6.1 Field names in a record object

It is necessary to create the variables, record by record, contained in the record object. Ideally, field names should become variables in the workspace. The field names collection may be enumerated thus:

```
      ⎕←A←⊃⎕wi ¨(⊂⊂'xFields.Item().Name'),¨⁻1++\(⎕wi 'xFields.Count')/1
Membno
My Name
Sex
Dob
Salary
      ⎕nc A
0 4 0 0 0
```

This highlights another problem; the names of fields may not produce valid APL+Win variable names—in this example, 'My Name' is an invalid variable name. A quick solution is to remove embedded blanks from the field names:

```
      ⎕←A←⊃(⎕wi ¨(⊂⊂'xFields.Item().Name'),¨⁻1++\(⎕wi
'xFields.Count')/1)~¨' '
Membno
MyName
Sex
Dob
Salary
      ⎕nc A
0 0 0 0 0
```

Although this will generally alleviate the problem, it is not foolproof. Try:

```
      ∇ Z←CreateAPLVariables R;⎕wself
[1]    ⍝ System Building with APL+Win
[2]    ⎕wself←R
[3]    ⎕wi 'MoveFirst'
[4]    Z←⎕wi ¨(⊂⊂'xFields.Item().Name'),¨(⍳⎕wi 'xFields.Count')-⎕io
[5]    Z←⊂Z~¨' '
[6]    Z←Z,⊂⎕wi ¨(⊂⊂'xFields.Item().Value'),¨(⍳⎕wi 'xFields.Count')-⎕io
      ∇
```

This function returns a custom collection of the names of fields and their corresponding values. There is no guarantee that the ensuing strings become valid APL names; there may be other invalid characters, such as leading digit or other characters, and the resulting strings may represent names of variables already present in the workspace. Another solution is to assign the field values to arbitrary names. Try:

```
     ∇ Z←CreateAPLVariables2 R;⎕wself
[1]    ⍝ System Building with APL+Win
[2]    ⎕wself←R
[3]    ⎕wi 'MoveFirst'
[4]    Z←⎕wi ¨(⊂⊂'xFields.Item().Name'),¨(⍳⎕wi 'xFields.Count')-⎕io
[5]    R←~(⎕nc ⊃Z)∊0 2
[6]    (R/Z)←(⊂'Expr'),¨⊂[⎕io+1]'G<999>' ⎕fmt +\(+/R)/1
[7]    Z←(⊂Z),⊂⎕wi ¨(⊂⊂'xFields.Item().Value'),¨(⍳⎕wi 'xFields.Count')-⎕io
     ∇
```

Invalid names are replaced by Expr*nnn*, where *nnn* is a three digit sequential number starting with 1. For example:

```
     NewCol←CreateAPLVariables2 'ADORS'
     'GetNames' Collection NewCol
 Membno Expr001 Sex Dob Salary  2 2 2 2
```

Although this solves the problem of field names being invalid variable names, the collection must be examined in order to establish the new names. The problem with the date field, *Dobirth*, is still outstanding. The most convenient solution is to resolve both the name and date format problem at source, that is, in the SQL statement that collates the data:

```
     ∇ xls2;⎕wself;Sql;Cnn
[1]    ⍝ System Building with APL+Win
[2]    ⎕wself←'ADORS' ⎕wi 'Create' 'ADODB.RecordSet'
[3]    ⎕wi 'xCursorType' (⎕wi '=adOpenStatic')
[4]    Cnn←'Provider=MSDASql;Driver={Microsoft Excel Driver
         (*.xls)};DBQ=c:\Adox.xls;'
[5]    Sql←"SELECT [My Name] as Name,Sex,Format(Dob,'dd/mm/yyyy') as
         Dobirth,Salary, Membno as MembId from [mydata$];"
[6]    ⎕wi 'XOpen' Sql Cnn
     ∇
```

When the function is re-run, the new records are:

```
                'ADORS' ⎕wi 'xGetRows'
     Jan M 01/12/2001  232  1
     Feb M 02/12/2001 1233  2
     Mar F 03/12/2001 7878  3
     Apr M 04/12/2001  262  4
     May M 05/12/2001  272  5
     Jun M 06/12/2001  282  6
     Jul M 07/12/2001  292  7
     Aug M 08/12/2001  302  8
     Sep M 09/12/2001  312  9
     Oct M 10/12/2001  322 10
     Nov M 11/12/2001  332 11
     Dec M 12/12/2001  342 12
```

The collection may be safely assigned:

```
A←CreateAPLVariables 'ADORS'
'Assign' Collection A
```

The variable names are as specified in the SQL statement:

```
      ⎕io⊃A
Name Sex Dobirth Salary MembId
```

In the revised function, xls2, both the problem with variable names and with data values have been solved. Note that Dob has been renamed Dobirth—and the order of fields in the data source is not binding on the order in which the record object refers to them. Note that an explicit reference to a name that contains a space must enclose the name in square brackets.

12.3.6.2 Data types in a record object

The ADO data type returned may be queried thus:

```
      ∇ Z←GetADOVariablesTypes R;⎕wself
[1]    ⍝ System Building with APL+Win
[2]    ⎕wself←R
[3]    ⎕wi 'MoveFirst'
[4]    R←(⍳(⎕wi 'xFields') ⎕wi 'xCount')-⎕io
[5]    Z←((⎕wi 'xFields') ⎕wi ¨(⊂⊂'xItem'),¨R) ⎕wi ¨¨⊂'xName' 'xType'
      ∇
```

The field names and types, as shown on the right, are returned by the following expression:

```
   ⊃GetADOVariablesTypes 'ADORS'
```

Name	200
Sex	200
Dob	200
salary	5
MemberID	5

Thus, Dobirth is the same type as Name—known to be string. Unlike ADO, which has several data types (see the SDK documentation) APL has just two data types: character and numeric. Any ADO data type that does not transparently or seamlessly migrate into the APL data types will need to be coerced into a string and converted back as appropriate within the workspace.

12.3.6.3 Field name, value and type from a record set object

The name, value and type of each column in a record set can be referred to either by the name of the column or by its ordinal position. The ordinal position is specified in index origin 0 irrespective of its setting in APL+Win.

12.3.6.3.1 Reference by name

⎕wi "xFields.Item('Name').Name"	⎕wi "xFields.Item('Name').Value"
Name	Jan

For any given column, all three properties may be queries at once. For example:

```
      (⊂⎕wi "xFields.Item('Name')") ⎕wi ¨'Name' 'Value' 'Type'
Name Jan 200
```

As the column Name is the first column in the record set, its position is 0. The properties can be queried using this ordinal position.

12.3.6.3.2 Reference by ordinal position I

⎕wi 'xFields.Item(0).Name'	⎕wi 'xFields.Item(0).Value'
Name	Jan

For any given column, all three properties may be queries at once, using the ordinal position. See the following:

```
      (⊂⎕wi "xFields.Item(0)") ⎕wi ¨'Name' 'Value' 'Type'
Name Jan 200
```

12.3.6.3.3 Reference by ordinal position II

The syntax may be simplified further when using ordinal positions:

```
      (⊂⎕wi "(0)") ⎕wi ¨'Name' 'Value' 'Type'
Name Jan 200
```

As the ordinal position of any column is simply a number, all three properties of all columns can be returned at once:

```
      ⊃((⊂⎕wself)  ⎕wi ¨(⊂⊂'()'),¨ ̄1++\(⎕wi 'xFields.Count')/1)
⎕wi¨¨⊂'Name' 'Value' 'Type'
 Name           Jan 200
 Sex              M 200
 Dobirth 01/12/2001 200
 Salary         232   5
 MembId           1   5
```

12.3.6.4 Cursor type and location

At a superficial level, a cursor is simply a pointer to the current record within a record object. The data access SDK defines a cursor as "*a database element that controls record navigation, updatability of data, and the visibility of changes made to the database by other users.*" In other words, a cursor is complex software that manages the attributes of the record object. These attributes include:

● Concurrency management—how up to date the data presented in a record set is, especially in a multi-user access environment,

● The current position in the record object set.

● The number of rows returned.

● Whether or not the arbitrary movement—forward or backward—within the record object is enabled.

The more functionality a cursor delivers, the more complex the software and the greater the impact on performance; therefore, it is vital that the cursor is configured with performance in mind. The cursor requires the setting of:

● A cursor type, which is 0 by default.

● A cursor location, which is 2 by default.

The cursor location options, documented in the SDK, applicable to both connection and record objects are shown in **Table 12-10 Cursor location options**.

Constant	Decimal Value	Description
adUseNone	1	Does not use cursor services. (This constant is obsolete and appears solely for the sake of backward compatibility.)
adUseServer	2	Default. Uses data provider or driver supplied cursors. These cursors are sometimes very flexible and allow for additional sensitivity to changes make to the data source. However, some features of the Microsoft Cursor Service for OLE DB (such as disassociated Recordset objects) cannot be simulated with server-side cursors and these features will be unavailable with this setting
adUseClient	3	Uses client-side cursors supplied by a local cursor library. Local cursor services often will allow many features that driver-supplied cursors may not, so using this setting may provide an advantage with respect to features that will be enabled. For backward compatibility, the synonym adUseClientBatch is also supported.

Table 12-10 Cursor location options

The cursor location of a connection may be set only when the record set object is closed; the cursor location property becomes read only when the object is open. Either set the cursor location before opening the record object or, if it already opened, close it, set the cursor location and then re-open it. By default, a record object based on a connection object inherits the cursor location of the connection object. Consider this example:

```
     ∇ CursorLocation;⎕wself
[1]    ⍝ System Building with APL+Win
[2]    ⎕wself←'ADO' ⎕wi 'Create' 'ADODB.Connection'
[3]    ⎕wi 'xCursorLocation' (⎕wi '=adOpenStatic')
[4]    ⎕wi 'XOpen' 'Driver={SQL Server};Server=Ajay
       Askoolum;Database=pubs;UID=sa'
[5]    ⎕wself←'ADO.Recordset' ⎕wi 'Create' (↑⎕wi 'XExecute' 'SELECT * FROM
       SALES')
```

```
[6]    ⍝ Instances of both objects continue to exist
     ∇
```

When the function *CursorLocation* is run, a record object is created implicitly and it inherits the cursor location specified for the connection object:

`'ADO' ⎕wi 'xCursorLocation'`	`'ADO.Recordset' ⎕wi 'xCursorLocation'`
`3`	`3`
`'ADO' ⎕wi 'xState'`	`'ADO.Recordset' ⎕wi 'xState'`
`1`	`1`

If the cursor location of the connection object were changed, the record object retains its original cursor location. Look at:

```
      'ADO' ⎕wi 'xCursorLocation' 2 ⍝ adUseServer
      'ADO.Recordset' ⎕wi 'xCursorLocation'
3
```

Any attempt to alter these properties when the record object is open generates a runtime error:

```
      'ADO.Recordset' ⎕wi 'xCursorType' (⎕wi '=adOpenStatic')
⎕WI ERROR: ADODB.Recordset exception 800A0E79 Operation is not allowed
when the object is open.
      'ADO.Recordset' ⎕wi 'XClose'
      'ADO.Recordset' ⎕wi 'xCursorType' (⎕wi '=adOpenStatic')
      'ADO.Recordset' ⎕wi 'XOpen'
      'ADO.Recordset' ⎕wi 'RecordCount'
21
```

The cursor type options, documented in the SDK, applicable to record objects are shown in **Table 12-11 Cursor type options**

Constant	Decimal Value	Description
`adOpenUnspecified`	`¯1`	`Does not specify the type of cursor.`
`adOpenForwardOnly`	`0`	`Default. Uses a forward-only cursor. Identical to a static cursor, except that you can only scroll forward through records. This improves performance when you need to make only one pass through a Recordset.`
`adOpenKeyset`	`1`	`Uses a keyset cursor. Like a dynamic cursor, except that you can't see records that other users add, although records that other users delete are inaccessible from your Recordset. Data changes by other users are still visible.`
`adOpenDynamic`	`2`	`Uses a dynamic cursor. Additions, changes, and deletions by other users are visible, and all types of movement through the Recordset are allowed, except for bookmarks, if the provider doesn't support them.`
`AdOpenStatic`	`3`	`Uses a static cursor, a static copy of a set of records that you can use to find data or generate reports. Additions, changes, or deletions by other users are not visible.`

Table 12-11 Cursor type options

12.3.6.5 BOF

This property—Beginning of File (**BOF**)—is 1 or **TRUE** when the cursor position is before the first record in the record object; otherwise, it is 0 or **FALSE**. The APL+Win syntax is:

```
      'ADORS' ⎕wi 'xBOF'
0
```

12.3.6.6 EOF

This property—End of File (**EOF**)—is 1 or **TRUE** when the cursor position is after the last record in the record object; otherwise, it is 0 or **FALSE:**

```
      'ADORS' ⎕wi 'xBOF'
0
```

Both the BOF and EOF properties may <u>not</u> be simultaneously the same unless record object is empty.

12.3.6.7 Is record set empty?

If *both* the BOF and EOF properties are *simultaneously* **TRUE**, there are no records in the record object. A *RecordCount* of 0 also indicates that a record set is empty. This function returns **TRUE** if a record object has records or **FALSE** if it does not:

```
      ∇ Z←RSHasRecords R
[1]    ⍝ System Building with APL+Win
[2]    Z←×R ⎕wi 'xRecordCount'
[3]    :if ¯1=Z ⍝ Indeterminate
[4]        Z←~(R ⎕wi 'xEOF')∧R ⎕wi 'xBOF'
[5]    :endif
      ∇
```

12.3.6.8 MoveNext

This method moves the cursor location to the next record within the record object; this method does not return any result. An error occurs on attempting to move past the last record:

```
      'ADORS' ⎕wi 'XMoveNext'
```

Thus, it is necessary to ensure that the EOF property is not **FALSE** before invoking this method.

12.3.6.9 MovePrevious

This method moves the cursor location to the previous record within the record object; this method does not return any result. An error occurs on attempting to move before the first record:

```
      'ADORS' ⎕wi 'XMovePrevious'
```

Thus, it is necessary to ensure that the BOF property is **FALSE** before invoking this method.

12.3.6.10 MoveFirst

This method moves the cursor location to the first record within the record object; this method does not return any result. An error occurs on attempting to move to the first record of an empty record object:

```
      'ADORS' ⎕wi 'XMoveFirst'
```

Thus, it is necessary to ensure that the BOF property is **FALSE** before invoking this method.

12.3.6.11 MoveLast

This method moves the cursor location to the last record within the record object; this method does not return any result. An error occurs on attempting to move to the last record of an empty record object:

```
      'ADORS' ⎕wi 'XMoveLast'
```

Thus, it is necessary to ensure that the EOF property is **FALSE** before invoking this method.

12.3.6.12 Record loop

A control structure is required to loop through all available records in a record set; this does not require prior knowledge of the number of records it contains:

```
        ∇ RecordLoop R
   [1]     ⍝ System Building With APL+Win
   [2]     ⎕wself←R
{1[3]     :if (¯1≠⎕wi 'xRecordCount')∨~(⎕wi 'xEOF')∧⎕wi 'xBOF'
   [4]         ⎕wi 'XMoveFirst'
{2[5]         :while ~⎕wi 'xEOF'
   [6]             ⍝ ....................................
   [7]             ⍝ Application processing code goes here
   [8]             ⍝ ....................................
   [9]             ⍝ Query/Set number of records on each page:
   [10]            ⍝ ⎕wi 'xPageSize' ◇ ⎕wi 'xPageSize' n
   [11]            ⍝ where n is a non zero positive integer
   [12]            ⍝ Query page number on which current record is found:
   [13]            ⍝ ⎕wi  ¨'xAbsolutePage'
   [14]            ⍝ Query absolute record number:
   [15]            ⍝ ⎕wi 'xAbsolutePosition'
   [16]            ⎕wi 'XMoveNext'
2}[17]        :endwhile
1}[18]   :endif
        ∇
```

The function RecordLoop loops through the available records. The following function illustrates some of the provider or driver dependent properties of a record object:

```
        ∇ LoopFeedback;⎕wself
   [1]     ⍝ System Building With APL+Win
   [2]     ⎕wself←'ADO' ⎕wi 'Create' 'ADODB.Connection'
   [3]     ⎕wi 'xCursorLocation' (⎕wi '=adOpenStatic')
   [4]     ⎕wi 'XOpen' 'Driver={SQL Server};Server=Ajay
           Askoolum;Database=pubs;UID=
        sa'
   [5]     ⎕wself←'ADO.Recordset' ⎕wi 'Create' (↑⎕wi 'XExecute' "SELECT *
           FROM SALE
        S WHERE ORD_NUM LIKE 'P2%'")
   [6]     'RecordCount=' (⎕wi 'xREcordCount')
{1[7]     :if RSHasRecords ⎕wself
{2[8]         :while ~⎕wi 'xEOF'
   [9]             ('Page=' 'Record='),¨ ⎕wi¨ 'xAbsolutePage'
           'xAbsolutePosition'
   [10]            ⎕wi 'XMoveNext'
2}[11]        :endwhile
1}[12]   :endif
   [13]   '{Default} Pagesize=' (⎕wi 'xPageSize')
        ∇
        LoopFeedback
 RecordCount= 3
 Page= 1 Record= 1
 Page= 1 Record= 2
 Page= 1 Record= 3
 {Default} Pagesize= 10
```

The ADO record object always has the same set of properties; if a provider or driver does not support any of these properties, ADO returns ‾1. The SQL Server driver supports them but the JET provider does not. The record set has three records:

```
      'ADO.Recordset' ⎕wi 'xRecordCount'
3
```

12.3.6.13 RecordCount

The *RecordCount* property returns the number of records in a record object as a positive integer greater then or equal to 0. A value of ‾1 indicates either that the provider does not support this property or that the *CursorType* used makes the record count indeterminate.

A record count can be calculated in spite of provider or cursor type limitations: two SQL statements need to be executed, as shown here:

```
      ∇ Z←CursorLocation2;⎕wself
[1]    ⍝ System Building with APL+Win
[2]    ⎕wself←'ADO' ⎕wi 'Create' 'ADODB.Connection'
[3]    ⎕wi 'xCursorLocation' (⎕wi '=adOpenStatic')
[4]    ⎕wi 'XOpen' 'Driver={SQL Server};Server=Ajay
       Askoolum;Database=pubs;UID=sa'
[5]    0 0⍴'ADO.Recordset' ⎕wi 'Create' (↑⎕wi 'XExecute' 'SELECT Count(*)
       FROM SALES')
[6]    Z←'ADO.Recordset' ⎕wi '(0).Value'
[7]    ⎕wself←'ADO.Recordset' ⎕wi 'Create' (↑⎕wi 'XExecute' 'SELECT * FROM
       SALES')
      ∇
```

Line [5] creates a record object containing the count of the records. Line [6] extracts the value and in line [7] the record object is recreated to contain the actual records:

```
      CursorLocation2
21
```

12.3.6.14 AbsolutePosition

This is a read/write property. It returns the ordinal position of the record pointer in the record set object—usually a positive integer greater than or equal to 1. Other values are shown in **Table 12-12 Absolute position options**.

Constant	Value	Description
adPosUnknown	‾1	Either record set object is empty, or the current position is unknown, or the provider does not support the AbsolutePage or AbsolutePosition property.
adPosBOF	‾2	The BOF property of the record set object is True.
adPosEOF	‾3	The EOF property of the record set object is True.

Table 12-12 Absolute position options

12.3.6.15 AbsolutePage

This is a read/write property. It is a positive integer between 1 and PageCount if the provider supports this property or one of the following values. Setting this property to *n* where $1 \leq n \leq PageCount$ sets the pointer to the first record on page *n*. In order to loop through all pages, a loop is required in code:

```
      ∇ LoopByPage R
 [1]    ⍝ System Building with APL+Win
 [2]    ⎕wself←R
{1[3]   :for i :in +\(⎕wi 'xPageCount')/1
 [4]        ⎕wi 'xAbsolutePage' i
{2[5]       :for i :in ⍳⎕wi 'xPageSize'
 [6]           ⍝ ... code to process page i records
2}[7]       :endfor
```

```
1}[8]    :endfor
       ▽
```

The absolute page reference is based on index origin 1 irrespective of APL+Win index origin setting.

12.3.6.16 PageCount

This is a read only property of the record object. It is

```
          ⌈(⎕wi 'xRecordCount')÷⎕wi 'xPageSize'
```

This value is automatically refreshed. If the provider does not support the required properties, the value of page count is ¯1.

12.3.6.17 PageSize

This is a read/write property of the record set object with a default value of 10; it indicates the number of records on one logical page of the record set. The value of page size determines how many records are displayed at any one time when scrolling though a record set.

12.3.7 Navigating record set objects

When a record set object is opened, the record pointer is at the first record—assuming that the record object contains more than 0 records. In addition to the methods and properties discussed in the previous section, ADO provides other ways to reposition the record pointer.

12.3.7.1 Seek

Whether a provider has enabled supports for the Seek method can be queried programmatically. This method is available only when the provider supports both indexes on record set objects and the Seek method, like the JET provider. The syntax is:

```
recordset.Seek KeyValues, SeekOption
       ▽ Z←SeekACCESS;⎕wself
   [1]   ⍝ System Building with APL+Win
   [2]   ⎕wself←'ADORS' ⎕wi 'Create' 'ADODB.RecordSet'
   [3]   Cnn←'Provider=Microsoft.Jet.OLEDB.4.0;Data Source=c:\nwind.mdb;'
   [4]   ⎕wi 'xCursorLocation' (⎕wi '=adUseServer')
   [5]   ⎕wi 'XOpen' "Customers" Cnn (⎕wi '=adOpenKeySet') (⎕wi
         '=adLockReadOnly') (⎕wi '=adCMDTableDirect')
   [6]   Z←⊂'Absolute position at line [6] is ',⍕⎕wi 'xAbsolutePosition'
{1[7]   :if ∧/⎕wi¨ (⊂⊂'xSupports') ,¨(⎕wi ¨'=adSeek' '=adIndex')
   [8]       ⎕wi 'Index' 'PrimaryKey'
   [9]       ⎕wi 'Seek' 'PARIS' (⎕wi '=adSeekFirstEQ')
   [10]      Z←Z,⊂'Absolute position at line [10] is ',⍕⎕wi
         'xAbsolutePosition'
1}[11]  :endif
   [12]  Z←Z,⊂'Number of records ',⍕ ⎕wi 'xRecordCount'
       ▽
       ⊃SeekACCESS
Absolute position at line [6] is 1
Absolute position at line [10] is 57
Number of records 91
```

The documented seek options available are shown in **Table 12-13 Seek options**.

Constant	Value	Description
adSeekFirstEQ	1	Seeks the first key equal to KeyValues.
adSeekLastEQ	2	Seeks the last key equal to KeyValues.
adSeekAfterEQ	4	Seeks either a key equal to KeyValues or just after where that match would have occurred.
adSeekAfter	8	Seeks a key just after where a match with KeyValues would have occurred.
adSeekBeforeEQ	16	Seeks either a key equal to KeyValues or just before

Constant	Value	Description
		where that match would have occurred.
adSeekBefore	32	Seeks a key just before where a match with KeyValues would have occurred.

Table 12-13 Seek options

If Seek cannot find a matching record, absolute position will be a negative value. For example, if line [9] were

```
[9]        ⎕wi 'Seek' 'PARIIS' (⎕wi '=adSeekFirstEQ')
      ⊃SeekACCESS
Absolute position at line [6] is 1
Absolute position at line [10] is ¯3
Number of records 91
```

12.3.7.2 Find

Unlike Seek, Find does not require support for indexes; it locates the first record matching a given criteria. The syntax of this method is:

```
Find (Criteria, SkipRows, SearchDirection, Start)
```

Criteria is a conditional statement based on a single column. The optional argument *SkipRows* has a default value of 0; a non-zero integer specifies an offset from the current position where search will begin. The optional argument *SearchDirection* can be adSearchBackward (1) or adSearchForward (1). The optional argument *Start* is a bookmark that specified the start of the search. Try:

```
     ∇ Z←FindSQLServer;⎕wself
[1]   ⍝ System Building with APL+Win
[2]   ⎕wself←'ADORS' ⎕wi 'Create' 'ADODB.RecordSet'
[3]   ⎕wi 'xCursorLocation' (⎕wi '=adUseClient')
[4]   Cnn←'Driver={SQL Server};Server=Ajay
        Askoolum;Database=pubs;UID=SA;PWD=;'
[5]   ⎕wi 'XOpen' 'SELECT * FROM AUTHORS;' Cnn (⎕wi '=adOpenStatic') (⎕wi
        '=adLockReadOnly') (⎕wi '=adCMDText')
[6]   Z←⊂'Absolute position at line [6] is ',⍕⎕wi 'xAbsolutePosition'
[7]   ⎕wi 'XFind' "State = 'TN'"
[8]   Z←Z,⊂'Absolute position at line [8] is ',⍕⎕wi 'xAbsolutePosition'
[9]   Z←Z,⊂'Number of records ',⍕ ⎕wi 'xRecordCount'
     ∇
      ⊃FindSQLServer
Absolute position at line [6] is 1
Absolute position at line [8] is 11
Number of records 23
```

An absolute position that is a negative value indicates that a match was not found.

12.3.7.3 Filter

An ADO record set object has a Filter property that allows it to hide the records that do not satisfy a criteria based on one or more columns in it. For example:

```
     ∇ Z←FindSQLServer;⎕wself
[1]   ⍝ System Building with APL+Win
[2]   ⎕wself←'ADORS' ⎕wi 'Create' 'ADODB.RecordSet'
[3]   ⎕wi 'xCursorLocation' (⎕wi '=adUseClient')
[4]   Cnn←'Driver={SQL Server};Server=Ajay
        Askoolum;Database=pubs;UID=SA;PWD=;'
[5]   ⎕wi 'XOpen' 'SELECT * FROM AUTHORS;' Cnn (⎕wi '=adOpenStatic') (⎕wi
        '=adLockReadOnly') (⎕wi '=adCMDText')
[6]   Z←⊂'Absolute position at line [6] is ',⍕⎕wi 'xAbsolutePosition'
[7]   ⎕wi 'XFind' "State = 'TN'"
[8]   Z←Z,⊂'Absolute position at line [8] is ',⍕⎕wi 'xAbsolutePosition'
```

```
[9]    Z←Z,⊂'Number of records ',⍕ ⎕wi 'xRecordCount'
      ∇

      ⊃FilterSQLServer
Absolute position at line [6] is 1
Absolute position at line [8] is 1
Number of records 2
```

The record set object reports 2 records with the pointer set to the first record:

```
      (⊂⎕wself) ⎕wi ¨(⊂⊂'().Value'),¨⊂¨'job_id' 'lname'
 4 Chang
```

The next record can be queried after moving the record pointer:

```
      ⎕wi 'XMoveNext'
      (⊂⎕wself) ⎕wi ¨(⊂⊂'().Value'),¨⊂¨'job_id' 'lname'
 2 Cramer
```

12.3.7.4 Removing the filter

The filter is removed by setting it to adFilterNone (0), thus:

```
      ⎕wi 'xFilter' (⎕wi '=adFilterNone')
```

The record pointer moves to the first record:

```
      ⎕wi 'xAbsolutePosition'
1
```

This record has different values:

```
      (⊂⎕wself) ⎕wi ¨(⊂⊂'().Value'),¨⊂¨'job_id' 'lname'
 13 Accorti
```

And the record set object has a different number of records:

```
      ⎕wi 'xRecordCount'
43
```

A record set object that has its filter property set exposes only the records that satisfy the filter.

12.3.8 Working with complete record objects

In essence, the Seek and Find methods and the Filter properties of the record set object manipulate complete record sets, in memory; that is, there is no workspace memory overhead.

12.3.8.1 The Sort property

A further property, Sort, sorts the whole record set object by one or more columns in ascending or descending order. The cursor location must be adUseClient, as shown below:

```
      ∇ TextJETSort;⎕wself
[1]    ⍝ System Building with APL+Win
[2]    Cnn←'Provider=Microsoft.Jet.OLEDB.4.0;Data Source=c:\;Extended
         Properties="text;HDR=Yes;FMT=Delimited"'
[3]    Sql←'Select * from ABC.TXT'
[4]    ⎕wself←'ADORS' ⎕wi 'Create' 'ADODB.RecordSet'
[5]    ⎕wi 'xCursorLocation' (⎕wi '=adUseClient')
[6]    ⎕wi 'XOpen' Sql Cnn
      ∇
```

This function uses the JET provider to read a comma delimited text file before and after a Sort property is set. This results in:

```
⎕wi 'GetRows'                    | ⎕wi 'Sort' 'Type ASC,Class DESC'
                                 | ⎕wi 'GetRows'
 E377 04/12/2001 M               |  P108 01/09/1980 F
 F319 08/06/1988 M               |  T266 02/10/1987 M
 G309 12/02/1976 M               |  G309 12/02/1976 M
 T266 02/10/1987 M               |  F319 08/06/1988 M
 P108 01/09/1980 F               |  E377 04/12/2001 M
```

The Sort property is a string consisting of a column name followed by ASC or DESC—separated by a space. If multiple columns are specified, they must be separated by comma.

12.3.8.2 GetRows method

The syntax of the *GetRows* method is:

```
array = RecordObject.GetRows(Rows, Start, Fields )
```

All arguments to this method are optional. The Rows argument specifies the number of rows to retrieve. The Start argument can be 0 (from current record), or 1 (from first record) or 2 (from last record). The Fields argument is a single field name or ordinal position or a variant array of multiple field names and or ordinal positions. The Fields argument allows the retrieval of columns in an order different from the one in the record set object.

The result is *always* an APL+Win nested array of the values, shown in **Figure 12-2 ADO GETROWS**, column names are not returned. As the GETROWS method returns its result to the workspace, it may create a workspace full error when the record set is big.

In such circumstances, a looping solution is required in order to be able to read a subset of rows at a time or a subset of the columns or both.

```
.-+--------------------.
|.+--..-+--------..+.|
||P108||01/09/1980||F||
|'----''----------''-'|
|.+--..-+--------..+.|
||T266||02/10/1987||M||
|'----''----------''-'|
|.+--..-+--------..+.|
||G309||12/02/1976||M||
|'----''----------''-'|
|.+--..-+--------..+.|
||F319||08/06/1988||M||
|'----''----------''-'|
|.+--..-+--------..+.|
||E377||04/12/2001||M||
|'----''----------''-'|
'€--------------------'
```

Figure 12-2 ADO GETROWS

```
      ∇ GetRowsLoop R;⎕wself;Flds;Values
[1]     ⍝ System Building with APL+Win
[2]     ⎕wself←R
[3]     ⎕wi 'XMoveFirst'
[4]     Flds←((⊂⎕wself) ⎕wi ¨(⊂⊂'()'),¨¯1++\(⎕wi 'xFields.Count')/1)
        ⎕wi¨⊂'Name'
{1[5]   :while ~⎕wi 'XEOF'
[6]         Values←⎕wi 'XGetRows' 1 ('#' ⎕wi 'missing') ('#' ⎕wi
        'missing')
[7]         ⍝ Application code to process record(s)
1}[8]   :endwhile
      ∇
```

Line [4] collates all the field names. Line [6] reads one record from the current record pointer position and moves the record pointer to the next record. At line [7], the names of columns and their respective values are available to the application code. **Figure 12-3 One record in default order** shows an APL representation.

```
      ]display   Flds,[⎕io]Values
.+----------------------------------------------------------------------------------------.
|.+--.   .+------..+------..+--.    .+-----.     .+--.     .+--..+-.  .+-------.|
||au_id|  |au_lname||au_fname||phone|  |address|  |city|   |state||zip|  |contract||
|'-----'  '--------''--------''-----'  '-------'   '----'    '-----''---'  '--------'|
|.+--------..+--.   .+------.  .+---------..+---------.    .+---------..+-.  .+-----.  |
||172-32-1176||White|  |Johnson| |408 496-7223||10932 Bigge Rd.||Menlo Park||CA|  |94025| 1      |
|'-----------''-----'  '-------' '------------''---------------''----------''--'  '-----'         |
'€-----------------------------------------------------------------------------------------'
```

Figure 12-3 One record in default order

Flds is a nested vector of rank 1 whereas *Values* is a nested array of rank 2.	ρFlds 9 ≡Flds 2 ρρFlds 1	ρValues 1 9 ≡Values 2 ρρValues 2

First, values must be converted from a nested array to a nested vector with each column converted to a simple array, vector, or scalar and occupying one element. The result is:

```
      ρValues
1 9
      Values←⊃¨⊂[⎕io]Values
      ρValues
9
      ρ¨Values
 1 11   1 5   1 7   1 12   1 15   1 10   1 2   1 5   1
```

Second, each column or field name is available and its intrinsic type is automatically coerced into an APL type, character, or numeric.

12.3.8.3 Passing records to APL Grid

The result of the GetRows method can be passed directly to the APL Grid object, as it is compatible, by way of shape and rank, with the APL Grid object's data:

```
       ∇ GetRowsLoop2APLGrid R;⎕wself;Flds;Values
  [1]   ⍝ System Building with APL+Win
  [2]   ⎕wself←'MyForm' ⎕wi 'Create' 'Form' ('caption' 'GetRows to Grid')
        ('size' 12.6 50.8)
  [3]   '.Grid' ⎕wi 'Create' 'APL.Grid' ('where' (0,0,¯0.5+⎕wi 'extent'))
  [4]   ⎕wself←R
  [5]   ⎕wi 'XMoveFirst'
  [6]   'MyForm.Grid' ⎕wi 'Rows' (⎕wi 'xRecordCount')
  [7]   Flds←((⊂⎕wself) ⎕wi ¨(⊂⊂'()'),¨¯1++\(⎕wi 'xFields.Count')/1)
        ⎕wi¨⊂'Name'
  [8]   'MyForm.Grid' ⎕wi 'Cols' (ρFlds)
  [9]   'MyForm.Grid' ⎕wi 'xText' ¯1 (⍳ρFlds) Flds
{1[10]  :while ~⎕wi 'XEOF'
  [11]      Values←⎕wi 'XGetRows' 1 ('#' ⎕wi 'missing') ('#' ⎕wi
        'missing')
  [12]       'MyForm.Grid' ⎕wi 'xText' (¯1+⎕wi 'xAbsolutePosition')
        (⍳ρFlds) Values
1}[13]  :endwhile
  [14]  'MyForm.Grid' ⎕wi 'XFitCol' (⍳ρFlds)
  [15]  'MyForm' ⎕wi 'Wait'
       ∇
```

The APL Grid provides an ideal interface for editing values in a record set or for browsing the entire record set: this is shown in **Figure 12-4 GetRows to APL Grid**.

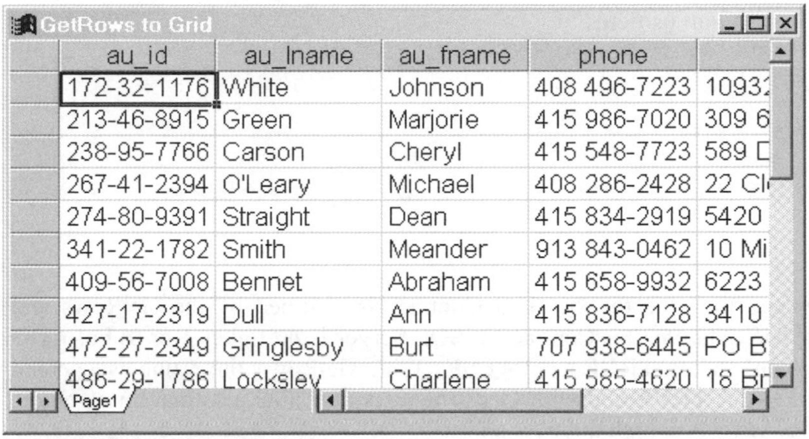

au_id	au_lname	au_fname	phone	
172-32-1176	White	Johnson	408 496-7223	10932
213-46-8915	Green	Marjorie	415 986-7020	309 6
238-95-7766	Carson	Cheryl	415 548-7723	589 D
267-41-2394	O'Leary	Michael	408 286-2428	22 Cl
274-80-9391	Straight	Dean	415 834-2919	5420
341-22-1782	Smith	Meander	913 843-0462	10 Mi
409-56-7008	Bennet	Abraham	415 658-9932	6223
427-17-2319	Dull	Ann	415 836-7128	3410
472-27-2349	Gringlesby	Burt	707 938-6445	PO B
486-29-1786	Locksley	Charlene	415 585-4620	18 Br

A loop structure is advisable in order to void workspace full errors—each record uses workspace memory before being written to the Grid.

Figure 12-4 GetRows to APL Grid

12.3.8.4 The GetRows syntax

The syntax used in line [4] is:

```
⎕wi 'XGetRows' 1 ('#' ⎕wi 'missing') ('#' ⎕wi 'missing')
```

The arguments of this method are positional. If an argument is omitted, all arguments following it must also be omitted—the second and third arguments may be omitted. While it is unlikely that the *Start* parameter will need to be specified, it is useful to be able to specify the third parameter either to retrieve a subset of columns or all the columns in a different order. Positional parameters cannot be omitted: therefore, in order to specify the third argument, the second argument must also be specified. There are 9 columns in this record set.

```
      ⎕wi 'XMoveFirst'
      ⎕rl←190254 ⋄ ColOrder←,,[0](9?9)-⎕io
      Values←⎕wi 'XGetRows' 1 ('#' ⎕wi 'missing') ('#' ⎕wi 'VT' ColOrder 8204)
      ]display   Flds[ColOrder+⎕io],[⎕io]Values
.→-----------------------------------------------------------------------------------------------.
↓.→------.        .→----.       .→----.       .→---.        .→-------..→----..→-------..→--.  .→-------.|
||address|        |phone|       |au_id|       |city|        |contract||state||au_lname||zip|  |au_fname||
|'-------'        '-----'       '-----'       '----'        '--------''-----''--------''---'  '--------'|
|.→--------------..→-----------..→-----------..→---------.               .→-.   .→----.    .→----..→------. |
||10932 Bigge Rd.||408 496-7223||172-32-1176||Menlo Park| 1              |CA|   |White|    |94025||Johnson| |
|'---------------''------------''------------''----------'               '--'   '-----'    '-----''-------' |
'∊-----------------------------------------------------------------------------------------------'
```

Figure 12-5 One record in desired order

Note the adjustment for index origin when the order of columns, *ColOrder*, is passed to the record set object and to APL+Win.

12.3.8.5 GetString method

The GetString method returns a simple string vector terminated by `⎕tclf`; each record is terminated by `⎕tcnl` and each field separated by `⎕tcht`. Its syntax is:

```
Variant = recordset.GetString(StringFormat, NumRows, ColumnDelimiter,
RowDelimiter, NullExpr)
```

The result can be transformed into a simple array with each row occupying one record, thus:

```
      ⎕wi 'MoveFirst'
      a←(⎕wi 'GetString')~⎕tclf
      a←⊃(a≠⎕tcnl)⊂a
      ⍴a
23 87
      ≡a
1
```

Each row can be split into its fields:
```
⊃(a[⎕io;]≠⎕tcht)⊂a[⎕io;]
172-32-1176
White
Johnson
408 496-7223
10932 Bigge Rd.
Menlo Park
CA
94025
-1
```

Note the difference in value for the field Contract as returned by the GetRows and GetString methods. The data type for Contract in SQL Server is *bit* (takes 1 or 0) but ADO converts it to Boolean—takes **TRUE** or **FALSE**. With GetRows the value is coerced correctly to the APL+Win **TRUE** and with GetString it is not: note also that an algebraic minus rather than the APL+Win high minus is shown.

As with the GetRows method, it may be necessary to retrieve a subset of records in order to avoid workspace full errors; unlike the GetRows methods, it is not possible to rearrange the ordinal position of the columns.

A typical use of this method might be to write the records in a CSV file, with or without column headers. The optional parameters of GetString enable records to be retrieved in the correct format. The *StringFormat* argument is asClipString. *NumRows* specifies the number of rows to retrieve from the current record pointer. The number of records retrieved is *NumRows* or a lesser number depending on the number of records available. The *ColumnDelimiter* specifies the character to be used to separate columns; by default it is ⎕tcht and for CSV files it need to be comma. The *RowDelimiter* specifies the character(s) used to separate records; it is ⎕tcnl by default and needs to be ⎕tcnl, ⎕tclf for CSV files. The final argument *NullExpr* specifies a string that is substituted for null column values.

The following function writes the record set to a native file in CSV format; column names are included:

```
      ∇ GetStringLoop R;⎕wself;Flds;Values;FileHwnd
 [1]    ⍝ System Building with APL+Win
 [2]    ⎕wself←R
 [3]    ⎕wi 'XMoveFirst'
 [4]    FileHwnd←¯1+⌊/0,⎕nnums,⎕xnnums
 [5]    'c:\temp\RS2CSV.CSV' ⎕xncreate FileHwnd
 [6]    Flds←((⊂⎕wself) ⎕wi ¨(⊂⊂'()'),¨¯1++\(⎕wi 'xFields.Count')/1)
        ⎕wi¨⊂'Name'
 [7]    (1⌽⎕tclf,(∈Flds,¨','),⎕tcnl) ⎕nappend FileHwnd
{1[8]   :while ~⎕wi 'XEOF'
 [9]        Values←⎕wi 'XGetString' (⎕wi '=adClipString') 5 ','
            (⎕tcnl,⎕tclf) ('#' ⎕wi 'null')
 [10]       Values ⎕nappend FileHwnd
1}[11]  :endwhile
 [12]   ⎕nuntie FileHwnd
      ∇
```

Lines [4] - [5] create a new native file—an error will occur if the file exists already. Line [6] converts the nested vector of column names to a string and writes it to the file—delimiters and terminating characters are added. Line [9] reads 5 records—as the record set has 23 records, only 3 records are returned in the final pass. The delimiter and terminating characters are automatically added by ADO in line [9]. The result is simply written to the file in line [10]. The resulting file can be opened in any text editor or directly in **Excel**.

Note the mechanism for passing missing and null values to ActiveX objects, illustrated in lines [6] and [9] of the function GetRowsLoop.

12.3.8.6 Update record set values

The Update method allows field values to be reassigned. This affects the current record and the record set object, in memory, and not the source. Use a lock type of adLockBatchOptimistic. The syntax is:

```
recordset.Update Fields, Values
```

The argument *Fields* is a nested vector of field names and *Values* is a nested vector of their corresponding new values. Use the function TextJETSort and add the following line:

```
[5.1]    ⎕wi 'xLockType' (⎕wi '=adLockBatchOptimistic')
```

Run the function and set the record pointer to record 3: ⎕wi 'xAbsolutePosition' 3

The values can be updated as follows:

```
      ⎕wi 'xUpdate' ('#' ⎕wi 'VT' ('Class' 'Type') 8204) ('#' ⎕wi 'VT'
('NewClass' 'NewType') 8204)
```

The record set is shown on the right. It contains the new values for record 3. The field names may be specified in any order and the values should correspond to the field names and types. Any fields that retain their original values are excluded from the fields list.

```
        ⎕wi 'XMoveFirst'
        ⎕wi 'XGetRows'
E377       04/12/2001 M
F319       08/06/1988 M
NewClass 12/02/1976 NewType
T266       02/10/1987 M
P108       01/09/1980 F
```

12.3.8.7 Adding a new record to a record set object

The AddNew method allows new records to be appended to a record set object if the underlying provider supports it. Try:

```
      'ADORS' ⎕wi 'XSupports()' (⎕wi '=adAddNew')
1
```

The AddNew method automatically sets the record pointer to the new record; new records are added at the end of the record set, thus:

```
      ⎕wi 'XAddNew'
      ⎕wi 'xAbsolutePosition'
6
```

One or all fields can be updated using the Update method. Use the following code:

```
      ⎕wi 'xUpdate' ('#' ⎕wi 'VT' ('Type' 'StartDate') 8204) ('#' ⎕wi 'VT'
('New' '01/01/2003') 8204)
```

In this example, the field Class was not updated and will have a default value of Null.

```
      ⎕wi 'xAbsolutePosition' 6
      ρ¨⎕wi 'XGetRows' 1
  0   10   3
```

```
        ⎕wi 'XMoveFirst'
        ⎕wi 'XGetRows'
E377 04/12/2001 M
F319 08/06/1988 M
G309 12/02/1976 M
T266 02/10/1987 M
P108 01/09/1980 F
     01/01/2003 New
```

Refer to the *Fabricating a record set object* section for further illustration of the ADDNew and Update methods. The ADO Update method and the SQL Update directive work in different contexts and are not the same, except in name only.

12.3.8.8 Deleting records from a record set

The Delete method deletes records from the record set objects. The syntax is:

```
recordset.Delete AffectRecords
```

The AffectRecords argument is one of the values shown in **Table 12-14 Affected records**.

Constant	Value	Description
adAffectAll	3	Affects all visible records.
adAffectAllChapters	4	Affects all records, including ones hidden by the Filter property.
adAffectCurrent	1	Affects only the current record.
adAffectGroup	2	Affects only records that satisfy the current Filter property setting.

Table 12-14 Affected records

```
      ⎕wi 'XAbsolutePosition' 6
      ⎕wi 'XDelete' (⎕wi '=adAffectCurrent')
```

In this example, the sixth record is deleted from the record set object. The underlying data source is not affected.

```
      ⎕wi 'XMoveFirst'
      ⎕wi 'GetRows'
E377 04/12/2001 M
F319 08/06/1988 M
G309 12/02/1976 M
T266 02/10/1987 M
 P108 01/09/1980 F
```

The effect of this method can be perverse: refer to the SDK documentation.

12.3.8.9 UpdateBatch records

The UpdateBatch method writes the modified record set object back to the underlying data source if the provider supports this facility.

```
      ⎕wi 'XUpdateBatch'
```

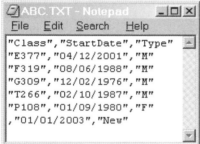

Figure 12-6 Updated data source

Several records may be added using the AddNew method—all the changes take place within the record set object in memory. All the changes are committed with a single UpdateBatch call.

The source data file C:\ABC.TXT is shown with NOTEPAD after the UpdateBatch method is applied, see **Figure 12-6 Updated data source**. The first value is empty as expected.

The JET provider using a text file as the data source does <u>not</u> support this method if any records have been deleted from the underlying data source. In order to test this, revert back to the state of the record set object at the end of section *12.3.8.6 Update record set values*.

12.3.8.10 Multiple record objects

A provider may enable the facility for retrieving multiple record objects from *the same data source* via a single Open method applied to a record object; SQLOLEDB is an example of such a provider. The relevant code is:

```
      ∇ MultipleRecordsets;⎕wself
[1]   ⍝ System Building with APL+Win
[2]    ⎕wself←'ADO' ⎕wi 'Create' 'ADODB.Connection'
[3]    ⎕wi 'xCursorLocation' (⎕wi '=adOpenStatic')
[4]    ⎕wi 'XOpen' 'Driver={SQL Server};Server=Ajay
         Askoolum;Database=pubs;UID=sa'
[5]    ⎕wi 'XExecute>ADO.Recordset' 'Select * from publishers;select * from
         sales;select * from authors;'
      ∇
```

In this example, the SQL statement specifies two queries, each terminated by semi-colon. When compound queries are used, the record object returns the records from the first query, in this case 'SELECT * FROM PUBLISHERS;'. The subsequent queries are evaluated on invoking the *XNextRecordSet* method of the record object:

```
      'ADO.RecordSet' ⎕wi 'xRecordCount'
8

      'ADO.RecordSet' ⎕wi 'XNextRecordSet>ADO.RecordSet'
      'ADO.RecordSet' ⎕wi 'xRecordCount'
21
```

In this example, the ADO.Recordset object is overwritten; the result set of the first query is lost. The new ADO.Recordset object contains the result sets from the remaining queries and it contains the result set from the second query. Consider an example with three queries, as shown by:

```
      ∇ MultipleRecordsets;⎕wself
[1]    ⍝ System Building with APL+Win
[2]    ⎕wself←'ADO' ⎕wi 'Create' 'ADODB.Connection'
[3]    ⎕wi 'xCursorLocation' (⎕wi '=adOpenStatic')
[4]    ⎕wi 'XOpen' 'Driver={SQL Server};Server=Ajay
         Askoolum;Database=pubs;UID=sa'
[5]    ⎕wi 'XExecute>ADO.Recordset' 'Select * from publishers;select * from
         sales;select * from authors;'
      ∇

      MultipleRecordsets ⍝ Run queries again
      'ADO.Recordset' ⎕wi 'XNextRecordSet>ADO.Recordset1'
      'ADO.Recordset' ⎕wi 'XNextRecordSet>ADO.Recordset2'
      'ADO.Recordset2' ⎕wi 'xRecordCount'
23
      'ADO.Recordset1' ⎕wi 'xRecordCount'
21
      'ADO.Recordset' ⎕wi 'xRecordCount'
8
```

The result set of the query in line [5] is returned in the record object ADO.Recordset; subsequent result sets may be created from ADO.Recordset by invoking the XNextRecordSet method. The results of all three queries are then available in different record objects: ADO.Recordset, ADO.RecordSet1, and ADO.RecordSet2. There is no way of moving forward or backward in a record object containing compound queries. The 'Multiple Results' property of an *active* connection indicates whether multiple record sets are supported. Look at:

```
      'ADO' ⎕wi 'xProperties("Multiple Results").Value'
1
```

SQL Server supports compound queries but none of the other drivers mentioned in this chapter do. The circumstances where compound queries may be useful are not obvious. One such occasion is when using the **Excel** driver; this driver restricts the number of fields or columns to 256. It would be useful to use compound queries to read, say, two sheets to work around this limitation—but the driver does not support compound queries.

```
      ⎕wself←'ADO' ⎕wi 'Create' 'ADODB.Connection'
      ⎕wi 'XOpen' 'Provider=MSDASQL;Driver={Microsoft Excel Driver
(*.xls)};DBQ=c:\Adox.xls;'
      ⎕wi 'xProperties("Multiple Results").Value'
0
```

12.3.8.11 Saving into a DOM tree

A record object can save its record set directly to an XML DOM tree. The SDK provides the following **VB** code:

```
[1] Dim xDOM As New MSXML.DOMDocument
[2] Dim rsXML As New ADODB.Recordset
[3] Dim sSQL As String, sConn As String
[4]
[5] sSQL = "SELECT customerid, companyname, contactname FROM customers"
[6] sConn="Provider=Microsoft.Jet.OLEDB.4.0;Data Source=D:\Program Files" & _
[7]          "\Common Files\System\msadc\samples\NWind.mdb"
[8] rsXML.Open sSQL, sConn
[9] rsXML.Save xDOM, adPersistADO    'Save Recordset directly into a DOM tree.
```

The **VB** code may be translated into APL+Win as follows:

```
    ∇ SavetoDOM;⎕wself
[1]    ⍝ System Building with APL+Win
[2]    ⎕wself←'ADORS' ⎕wi 'Create' 'ADODB.Recordset'
[3]    ⎕wi 'XOpen' 'SELECT customerid, companyname, contactname FROM
          customers' 'Provider=Microsoft.Jet.OLEDB.4.0;Data
          Source=C:\NWind.mdb'
[4]    0 0⍴'DOM' ⎕wi 'Create' 'MSXML.DOMDocument'
[5]    ⎕wi 'XSave' ('DOM' ⎕wi 'obj') (⎕wi '=adPersistXML')
[6]    ⎕wi 'Delete'
    ∇
```

The connection string has been modified to reflect a different location for the data source. In this example, a DOM tree is created dynamically without recourse to an XML file. Once the result set has been saved into a DOM tree, the DOM object can manipulate it in the usual way:

```
    SavetoDOM ⍝ run the function
    ⍴'DOM' ⎕wi 'xxml'
9478
```

The optional formats for saving results sets, as documented in the SDK, are shown in **Table 12-15 Persisted format options**.

Constant	Decimal Value	Description
adPersistADTG	0	Indicates Microsoft Advanced Data TableGram (ADTG) format.
adPersistADO	1	Indicates that ADO's own Extensible Markup Language (XML) format will be used. This value is the same as adPersistXML and is included for backwards compatibility.
adPersistXML	1	Indicates Extensible Markup Language (XML) format.
adPersistProviderSpecific	2	Indicates that the provider will persist the Recordset using its own format.

Table 12-15 Persisted format options

The XML representation of the complete result set is saved to C:\AA.XML; the target file must not exist already:

```
    'DOM' ⎕wi 'xsave' 'c:\aa.xml'
```

The DOM tree may be recreated from the saved XML file:

```
    ∇ LoadDOM;⎕wself
[1]    ⍝ System Building with APL+Win
[2]    ⎕wself←'DOM' ⎕wi 'Create' 'MSXML.DOMDocument'
```

```
[3]    0 0ρ⎕wi 'Xload' 'c:\aa.xml'
     ∇
      ρ'DOM' ⎕wi 'xxml'
9478
```

12.3.8.12 Saving in XML format

A record set can be saved directly to a file on disk in XML format:

```
     ∇ SavetoXML;⎕wself
[1]    ⍝ System Building with APL+Win
[2]    ⎕wself←'ADORS' ⎕wi 'Create' 'ADODB.Recordset'
[3]    ⎕wi 'XOpen' 'SELECT customerid, companyname, contactname FROM
          customers' 'Provider=Microsoft.Jet.OLEDB.4.0;Data
          Source=C:\NWind.mdb'
[4]    ⎕wi 'XSave' 'c:\aaxml.dat' (⎕wi '=adPersistXML')
[5]    ⎕wi 'Delete'
     ∇
```

In line [4], the name of the file is specified—it can have any extension; if the specified file name already exists, an error occurs.

12.3.8.13 Saving in ADTG format

The Microsoft Advanced Data Table Gram (**ADTG**) is another proprietary format (binary) for saving record sets to disk. This functionality is implemented as follows:

```
     ∇ SavetoADTG;⎕wself
[1]    ⍝ System Building with APL+Win
[2]    ⎕wself←'ADORS' ⎕wi 'Create' 'ADODB.Recordset'
[3]    ⎕wi 'XOpen' 'SELECT customerid, companyname, contactname FROM
          customers' 'Provider=Microsoft.Jet.OLEDB.4.0;Data
          Source=C:\NWind.mdb'
[4]    ⎕wi 'XSave' 'c:\aaADTG.dat' (⎕wi '=adPersistADTG')
[5]    ⎕wi 'Delete'
     ∇
```

The target file must not exist already. This format stores the information in binary format; the files are smaller and the process faster than saving to other formats.

12.3.8.14 Disconnected record object

A record set that is saved to disk represents a disconnected record set. A file saved in XML format may be browsed using any text editor; however, the ADTG format is not human readable.

Disconnected record objects may be quite valuable when data from a central data source is required for offsite working; this includes working on a laptop and making the data available to a third party. Note:

• A disconnected record set may be used much as a connected record set using the MSPERSIST provider, as follows:

```
     ∇ ConnectDRO;⎕wself
[1]    ⍝ System Building with APL+Win
[2]    ⎕wself←'ADORS' ⎕wi 'Create' 'ADODB.Recordset'
[3]    ⎕wi 'XOpen' 'c:\aaADTG.dat' 'Provider=MSPERSIST'
     ∇
```

• At the point of creation, the disconnected record set is identical to the source from which it was created.

```
     ConnectDRO
     'ADORS' ⎕wi 'xRecordCount'
91
```

- Any changes made to the record object may be re-submitted to the original data source using the *UpdateBatch* method of the record object, if the provider supports this method and the appropriate lock type, 4 (adLockBatchOptimistic), was used to create the record set.

- A persisted record set object is a clone of the original record set and, unlike a cloned record set, it does not reside in memory.

12.4 The data source catalogue

The ADO method *OpenSchema* queries the catalogue of a data source. The syntax of the *OpenSchema* method is:

```
Set recordset = connection.OpenSchema (QueryType, Criteria, SchemaID)
```

The 'QueryType' parameter is a constant whose possible values are shown in **Table 12-16 Schema options**.

OpenSchema Constant	Decimal Value	Constraint Columns	Description
adSchemaProcedures	16	PROCEDURE_CATALOG PROCEDURE_SCHEMA PROCEDURE_NAME PROCEDURE_TYPE	Returns the names of procedures defined in the catalogue.
adSchemaTables	20	TABLE_CATALOG TABLE_SCHEMA TABLE_NAME TABLE_TYPE	Returns the names of tables (including views) defined in the catalogue.
adSchemaViews	23	TABLE_CATALOG TABLE_SCHEMA TABLE_NAME	Returns the names of views defined in the catalogue.

Table 12-16 Schema options

The possible values for PROCEDURE_TYPE and TABLE_TYPE are shown in **Table 12-17 Procedure/Table types**.

PROCEDURE TYPE		TABLE TYPE
Value	Description	ALIAS
1	DB_PT_UNKNOWN—Unable to determine whether it returns value.	TABLE SYNONYM
2	DB_PT_PROCEDURE—Procedure: does not return value.	SYSTEM TABLE
3	DB_PT_FUNCTION—Function: returns value.	VIEW GLOBAL TEMPORARY LOCAL TEMPORARY SYSTEM VIEW

Table 12-17 Procedure/Table types

All queries are accessible by the current user. The optional 'Criteria' parameter is an array based on the 'Constraint Columns' of each query type. The 'SchemaID' parameter is superfluous unless the query type is adSchemaProviderSpecific, decimal value ‾1. The sample **VB** code for getting a list of all the system tables in a data source is:

```
Set objRS = objConn.OpenSchema (adSchemaTables, _
                    Array(Empty,Empty,Empty),_
                    "SYSTEM_TABLE")
```

VB Data Type	APL+Win Equivalent
Empty	'#' ⎕wi 'VT' 0 0

Array(Empty,_
    Empty, _
    Empty, _
    "SYSTEM_TABLE")
	Assign Empty to a variable: Empty←'#' ⎕wi 'VT' 0 0
	Then:
	'#' ⎕wi 'VT' ((3/Empty) "SYSTEM_TABLE") 8204

Table 12-18 Translating data types

The problem for APL is that it does not natively have the data types 'Empty' and 'Array'. However, the GUI system object (#) has a method 'VT' (variant type) that can be used for passing these data types; **Table 12-18 Translating data types** has some examples:

The optional 'Criteria' parameter simply allows one or all the constraint columns to be specified. Consider **Table 12-19 Scenario implications**.

Scenario	Third Element	Third Argument	Description
1	0	1	Return all names of type specified
2	1	0	Return name specified, irrespective of type
3	1	1	Return name and type specified

Table 12-19 Scenario implications

In practice, the first argument, the third element of the second argument, and, where available, the third argument is specified:

● The syntax in scenario 1 can be used to return all names of a specific type: for instance, when populating a combo box.

● The syntax in scenario 2 can be used to verify whether a name exists as any type. As query type 20 returns both tables and views and names have to be unique within the database, this is probably the most useful call.

● The syntax in scenario 3 can be used to verify if a particular name exists as a particular type: this may be used to determine if the name needs to be dropped before being recreated or created before being populated.

Of course the whole database catalogue can be returned by simply omitting the 'Criteria' parameter altogether and using query type 20.

12.5 Learning ADO

This chapter illustrates the basic application of ADO in handling data in the workspace. ADO itself can be complex to deploy in a client server environment and a grasp of this technology is fundamental.

Microsoft data access SDK provides several help files full of working examples. The menu options are shown in **Figure 12-7 MDAC documentation.**

Figure 12-7 MDAC documentation

Subject to end user licence agreement, the latest version of MDAC may be freely downloaded from www.microsoft.com.

Data Source Connection Strategies

During development, it is quite simple to connect to a data source using known items of information that comprise the connection string. It is expedient to create the complete connection string, including sensitive information such as USER ID and PASSWORD, on an ad hoc basis or to hard code it. However, the connection string presents complex design issues for deploying an application, especially to remote sites, namely:

• It is unnecessarily restrictive to coerce all clients to use the same data source; applications should allow clients autonomy in selecting their own data source. In practice, clients do not have complete freedom but have a choice from a list of data sources prescribed by the application.

• A connection method that relies on a system Data Source Name (**DSN**) will require that DSN to be created on each computer using the application—this may present an unacceptable burden on system administrators; the application may need to create a system DSN programmatically.

• A connection based on a user DSN: unlike a system DSN that is available to all users of a computer, a user DSN is only available to the user profile; only the user who created it can use it.

• A connection that requires a file DSN will require the distribution of that file together with the application—the problem of different data sources and changing connection parameters remains an issue.

• A connection that requires a Universal Data Link (**UDL**) presents the same issues as one that uses a file DSN.

• It is unrealistic to assume that USER ID and PASSWORD will never change—therefore these items cannot be hard coded. This raises the issue of how the application will request this information at runtime.

• A connection that does not rely on a DSN can be hard coded—this is a DSN-less connection; the problem of getting the USER ID and PASSWORD remains an issue.

• It may be unacceptable to store USER ID and PASSWORD in a UDL or DSN files in a human readable format as this compromises data security.

An application that does not connect to its data source simply does not run. Usually, an application either presents a data connection interface—perhaps to elicit a USER ID and PASSWORD—or silently establishes connection before the user interface is visible.

Of course, an application may allow complete flexibility in the method of connecting to data sources by permitting users to select either a DSN, or a DSN-less or an OLE DB connection, as shown in **Figure 13-1 Connection methods**.

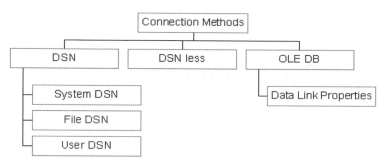

Figure 13-1 Connection methods

Each of these methods of connection will require further configuration for the specific data source—SQL Server, Oracle, **Access**, **Excel**, comma separated values (**CSV**) format text files, etc. Each combination of connection method and data source presents issues relating to design, management, and performance of the application—and, the testing overhead increases rapidly with the number of combinations.

DSN connections may be configured manually. Click Start | Settings | Control Panel and then find Administrative Tools | Data Sources (ODBC). The dialogue shown in **Figure 13-2 ODBC Data Source Administrator** appears. This permits user, system, and file DSNs to be added, removed, and configured.

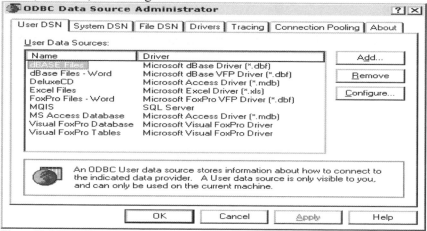

Figure 13-2 ODBC Data Source Administrator

This dialogue also provides facilities for testing connections and for enabling connection pooling. Each tab shows the DSNs already created.

The location at which each connection method is stored is shown in **Table 13-1 Location of ODBC data sources.** A user DSN is available exclusively to the user who set it up. A system DSN is not user specific; it is available to any user having access to the underlying data source. A file DSN is also available to any user. File DSNs are stored in the filing system whereas user and system DSNs are stored in the **Registry**.

DSN Type	Registry Location
User	HKEY_CURRENT_USER\SOFTWARE\ODBC\ODBC.INI\ODBC Data Sources
System	HKEY_LOCAL_MACHINE\SOFTWARE\ODBC\ODBC.INI\ODBC Data Sources
File	\Program Files\Common Files\ODBC\Data Sources

Table 13-1 Location of ODBC data sources

The default location for file DSNs is stored in the **Registry**.

```
     RegRead 'HKEY_LOCAL_MACHINE\SOFTWARE\ODBC\ODBC.INI\ODBC File
DSN\DefaultDSNDir'
C:\Program Files\Common Files\ODBC\Data Sources
```

Click the **Set Directory** button on the File DSN tab—shown in **Figure 13-3 Default DSN directory**—if the default directory must be changed.

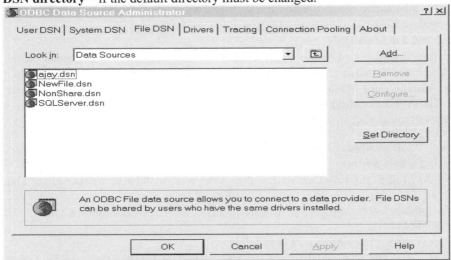

Figure 13-3 Default DSN directory

A warning shown in **Figure 13-4 File DSN warning** is given before any changes are made permanent. It is not recommended that this location be changed: any DSNs stored at the old location are not automatically moved to the new one.

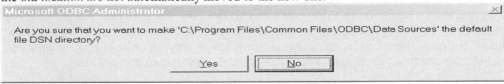

Figure 13-4 File DSN warning

13.1 The application handle

All the SQL APIs require the handle of the host application; if this argument is invalid or 0, the dialogues are not shown. In the development system, the handle is returned by `'#' ⎕wi 'hwndmain'`; in a runtime system, this returns 0 and the handle of a form within the application must be used. `appForm ⎕wi 'hwnd'`.

In this chapter `'#' ⎕wi 'hwndmain'` is used throughout; if the application will be deployed with the APL+Win runtime system, the handle of an applications form must be used.

13.2 The DSN overhead

An application that relies on a DSN relies on that DSN being present on the target machine—the **Registry** is interrogated to determine whether the DSN exists. A file DSN may be distributed with the application—and installed therewith. A user or system DSN may be a little more problematic.

One solution is to export the DSN key from the **Registry** and distribute the **Registry** file with the application and then to import it on the target machines. This may be problematic in that it may create conflicts with the existing DSN of the same name. An alternative is to provide documentation to system administrators on setting up the DSN. The overriding implication is that application set-up involves the overhead of having to configure every computer—not only to empower clients to select their own DSN name but also to enable the correct database server name and passwords to be included. It is impractical to assume that a DSN by a particular name is not already in use or that a client's database server name and password would be identical to those used by the application provider.

13.2.1 Acquiring a default DSN

If an application is hard coded to use a specific DSN, its installation routine should create that DSN. Alternatively, when an application runs for the first time, it can invoke the ODBC Data Source Administrator Manager *automatically* and allow the user to create a DSN of the same name. The API declaration is:

```
SQLManageDataSources=B(D hwndParent) ALIAS SQLManageDataSources LIB
    ODBCCP32.DLL
        ⎕wcall 'SQLManageDataSources' ('#' ⎕wi 'hwndmain')
```

13.2.2 Creating a data source

The Window API call SQLCreateDataSource allows the user to add a new data source. The API call is:

```
SQLCreateDataSource=B(D hwndParent,*C lpszDS) ALIAS SQLCreateDataSource LIB
    ODBCCP32.DLL
```

The APL+Win code to call this API is as follows:

```
        ⎕wcall 'SQLCreateDataSource' ('#' ⎕wi 'hwndmain') ⊖
```

The lpszDS parameter is not specified: the dialogue shown in **Figure 13-5 Creating a DSN** prompts for the name to which to save the new data source. Any of the three types of DSN may be selected.

Note that the dialogue indicates the scope of each type of DSN. Although file DSNs are indicated to be machine independent, not all such DSNs are necessarily so. For instance, if a file DSN simply refers to a user or system DSN, it is not machine independent, as it requires the DSN that is referenced, which exists on the machine only. In order to make it apparent, include the words 'not shareable' in the name of the file DSN when it uses a DSN.

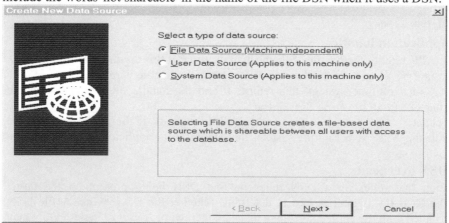

Figure 13-5 Creating a DSN

The **Next** button shows other dialogues that allow the connection to be fully configured and saved on the local machine.

13.3 Automating user/system DSN creation

If a DSN is to be used to connect to a data source, a proactive approach that allows the application to create its own DSN programmatically is preferable. An application that enables users to create their own DSN, on first use, gives the client complete flexibility in choosing the type of DSN that will be used.

13.3.1 With an API call

The following API call enables a DSN to be created programmatically:

```
SQLConfigDataSource=L(D hwndParent,D fRequest,*C lpszDriver,*C lpszAttributes) ALIAS
    SQLConfigDataSource LIB ODBCCP32.DLL
```

The hwndParent parameter determines whether the API call allows user interaction—seamlessly within the application environment—or whether the DSN is created silently. **Table 13-2 Parent window handles** shows the possible values for this parameter. In practice, the value should be the dynamic handle of the application form from which this API is called.

hwndParent	Description
0	No user interaction or suppress interaction.
'#' ⎕wi 'hwndmain'	Provide interaction using parent handle.
appForm ⎕wi 'hwnd'	Provide interaction using parent handle.

Table 13-2 Parent window handles

The fRequest parameter determines whether a user or system DSN will be created, edited or deleted. The potential values are shown in **Table 13-3 DSN management constants**.

Decimal Values	Constant	Description
0	ODBC_ADD_DSN	Add a user data source
1	ODBC_CONFIG_DSN	Edit/Configure a user data source
2	ODBC_REMOVE_DSN	Remove a user data source
4	ODBC_ADD_SYS_DSN	Add a system data source
5	ODBC_CONFIG_SYS_DSN	Edit/Configure a system data source
6	ODBC_REMOVE_SYS_DSN	Remove a system data source
7	ODBC_REMOVE_DEFAULT_DSN	Do not use; see text below.

Table 13-3 DSN management constants

The lpszDriver parameter specifies the name of the DSN. An application that elects to create a DSN silently, that is without user interaction, must ensure that it does *not* overwrite or re-configure a DSN already present on the user's computer.

The lpszAttributes parameter specifies the information that allows connection to the database. Under some circumstances, it may be possible to presume the contents of this parameter; for example, when the application provider also provides and manages the database on behalf of the client.

13.3.1.1 Enumerating existing user/system DSNs

If the list of existing DSNs can be established, it is quite straightforward to verify whether one by a particular name exists. User and system DSNs are stored in the **Registry**; they can be enumerated with the RegEnumKeyEX API function:

```
      ∇ Z←L RegEnum R
[1]    ⍝ System Building with APL+Win
[2]    R[R⍳'\']←⎕tcnul
[3]    R←(R≠⎕tcnul)⊂R
{1[4]  :if 0=↑Z←⎕wcall (⊂'RegOpenKeyEx'),R[⍳2],0 'KEY_ALL_ACCESS'
       (4⍴⎕tcnul)
[5]        L←⊖ρ323 ⎕dr (⎕io+1)⊃Z
```

```
  [6]        Z←''
{2[7]          :while 0=↑R←⎕wcall 'RegEnumKeyEx' L (↑ρZ) (260ρ' ') (82 ⎕dr
       260) 0 0 0
  [8]              Z←Z,⊂(323 ⎕dr (⎕io+2)⊃R)↑(⎕io+1)⊃R
2}[9]          :endwhile
  [10]     0 0ρ⎕wcall 'RegCloseKey' L
1}[11]   :endif
         ∇
```

This function returns the names of existing DSNs:

```
      ⊃RegEnum 'HKEY_LOCAL_MACHINE\SOFTWARE\ODBC\odbc.ini'
ODBC File DSN
ODBC Data Sources
sysdsn
      ⊃RegEnum 'HKEY_CURRENT_USER\SOFTWARE\ODBC\odbc.ini'
ODBC Data Sources
Visual FoxPro Database
Visual FoxPro Tables
dBase Files - Word
FoxPro Files - Word
MQIS
MS Access Database
dBASE Files
Excel Files
```

The list returned will almost certainly vary from computer to computer.

13.3.1.1.1 Does DSN exist?

It is a simple matter to check a particular DSN name against the list of existing DSNs—note that although DSN names are stored as entered, they are not case sensitive. Try:

```
      ∇ Z←L DSNExists R
  [1]   ⍝ System Building with APL+Win
  [2]   R←,⎕wcall 'CharUpper' R
{1[3]   :select L
★ [4]       :caselist 'System' 'SYSTEM'
  [5]           Z←⊃RegEnum 'HKEY_LOCAL_MACHINE\SOFTWARE\ODBC\ODBC.INI'
★ [6]       :caselist 'User' 'USER'
  [7]           Z←⊃RegEnum 'HKEY_CURRENT_USER\SOFTWARE\ODBC\ODBC.INI'
1}[8]   :endselect
  [9]   Z←⎕wcall 'CharUpper' Z
  [10]  L←(0↓ρZ)⌈0↓ρR
  [11]  Z←×+/0,(L↑R)∧.=⍉((ρZ)⌈0,L)↑Z
        ∇
```

This function enumerates the list of user or system and verifies whether a given name exists in that list—the result is 1 if it does and 0 otherwise. It also allows for the contingency that the existing list of DSNs could be empty and uses the API call CharUpper to convert its arguments to uppercase. Note:

```
      'System' DSNExists 'sysdsn' ⍝ or 'SYSTEM' DSNExists 'sysdsn'
1
      'User' DSNExists 'sysdsn' ⍝ or 'USER' DSNExists 'sysdsn'
0
```

13.3.1.1.2 Get a unique DSN name

A simple strategy may be used to generate unique DSN names; one such strategy is to use a string—perhaps an acronym for the application—followed by three digits comprising milliseconds from the system clock. The algorithm can loop until a unique name is found, using the dynamically changing system clock. Type the following code:

```
        ∇ Z←L GetUniqueDSN R
   [1]    ⍝ System Building with APL+Win
{1[2]    :repeat
   [3]        Z←⎕wcall 'CharUpper' (R,⍕0⍳⎕ts)
{2[4]        :select L
★  [5]            :caselist 'System' 'SYSTEM'
   [6]                R←'HKEY_LOCAL_MACHINE' RegEnum
        'SOFTWARE\ODBC\ODBC.INI'
   [7]                Z←'SYS',Z
★  [8]            :caselist 'User' 'USER'
   [9]                Z←'USR',Z
   [10]               R←'HKEY_CURRENT_USER' RegEnum 'SOFTWARE\ODBC\ODBC.INI'
2}[11]       :endselect
   [12]        R←⎕wcall ¨ (⊂⊂'CharUpper'),¨⊂¨R
1}[13]   :until 0=(⊂Z)⍳R
        ∇

   'System' GetUniqueDSN 'MKVA' ⍝ or 'SYSTEM' GetUniqueDSN 'MKVA'
SYSMKVA270
   'User' GetUniqueDSN 'MKVA' ⍝ or 'USER' GetUniqueDSN 'MKVA'
USRMKVA600
```

Note that this function prefixes the name of the DSN by SYS or USR depending on whether a user or system DSN name is requested. In addition, a different method—from the function DSNExists—for seeking a match against existing names is used. A strategy such as this ensures that a unique lpszDriver name is used where necessary.

In order that an application recognises an arbitrarily named DSN, it needs to record that name either in a key within the **Registry** or within an INI file for subsequent use.

13.3.1.2 Enumerating existing file DSNs

File DSNs are stored, by default, in \Program Files\Common Files\ODBC\Data Sources. The location of file DSNs is stored in the following key:

```
HKEY_LOCAL_MACHINE\SOFTWARE\ODBC\ODBC.INI\ODBC File DSN\DefaultDSNDir
```

A connection object can use a file DSN at any location, not just those found at the default location. Therefore, it may be necessary to scan all volumes in order to compile an exhaustive list of the file DSNs present. For practical purposes, it can be assumed that file DSNs will be in the default location. Try the following:

```
        ∇ Z←GetFileDSNs R;⎕wself
   [1]    ⍝ System Building with APL+Win
   [2]    ⎕wself←'FSO' ⎕wi 'Create' 'Scripting.FileSystemObject'
   [3]    ⎕wself←'FSO.F' ⎕wi 'Create' ((⎕wi 'XGetFolder' R) ⎕wi 'Files')
   [4]    Z←''
   [5]    0 0⍴⎕wi 'EnumStart'
{1[6]    :while 0≠⍴R←⎕wi 'EnumNext'
   [7]        Z←Z,⊂R ⎕wi 'xName'
1}[8]    :endwhile
   [9]    ⎕wgive 0
   [10]   0 0⍴⎕wi 'EnumEnd'
```

```
[11]   'FSO' ⎕wi 'Delete'
       ∇
       folder←'HKEY_LOCAL_MACHINE' RegQuery key
       ⊃GetFileDSNs folder
myfiledsn.dsn
```

The enumeration of DSNs returns a nested vector of DSN names; such a list can be used to populate a drop-down box on a user form that permits the selection of a DSN at runtime.

13.3.2 With the 'Prompt' property of a connection object

The prompt property of a closed connection object allows connection strings to be constructed on the fly. This property is used to specify whether a dialogue box is displayed as a prompt for missing parameters when opening a connection to a data source; this provides a convenient resolution of problems at run time. The default value is 4—no dialogue box should be displayed.

The choice of the prompt value may depend on sensitivity of the data source. **Table 13-4 Connection prompt** shows the possible values.

Constant	Decimal Value	Description
adPromptAlways	1	Prompts always.
adPromptComplete	2	Prompts if more information is required.
adPromptCompleteRequired	3	Prompts if more information is required but optional parameters are not allowed.
adPromptNever	4	Never prompts.

Table 13-4 Connection prompt

The function ADOPrompt1 invokes a series of dialogue boxes allowing the developer to construct a connection string dynamically—this is a flexible way to experiment with connection strings during the development phase.

```
      ∇ ADOPrompt1;⎕wself
[1]    ⍝ System Building with APL+Win
[2]    ⎕wself←'ADO' ⎕wi 'Create' 'ADODB.Connection'
[3]    ⎕wi 'xProperties("Prompt").Value' (⎕wi '=adPromptAlways')
[4]    ⎕wi 'XOpen'
      ∇
```

A new data source name may be specified or an existing one may be selected from the first dialogue, shown in **Figure 13-6 Select data source,** which is presented. The parameters of a file data source are stored in a file; those of a machine data source, that is a system or user DSN, are stored in the **Registry**. The 'Look in' box points to the default location; the file DSNs at this location are listed. If the file DSN is stored elsewhere, click the arrow of the combo box to select the appropriate location.

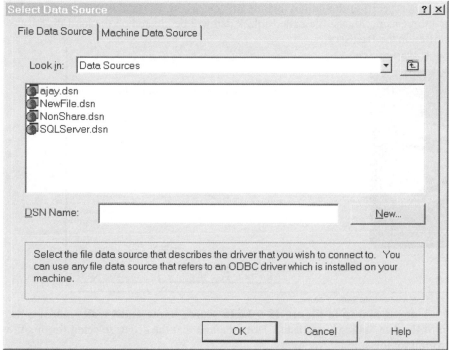

Figure 13-6 Select data source

Access to any data source depends upon the availability of a driver on the computer; the available drivers are displayed, as shown in **Figure 13-7 Selecting a driver**.

If the data source is known, its driver can be chosen from the list. All installed drivers appear in this list.

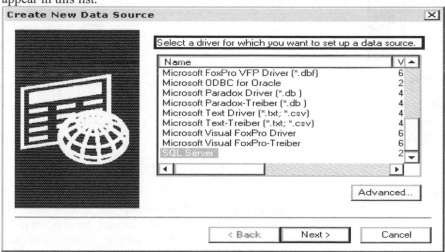

Figure 13-7 Selecting a driver

The SQL Server driver is selected; this driver was installed with MSDE. The next dialogue, **Figure 13-8 Creating a new data source name**, prompts for the name of the DSN. Note that if a file DSN is being created, a file extension will be added to the name specified: the name specified should not include this extension. 'SQL2k' is the name given to this file data source name; thus a file called sqk2k.dsn will be created on the hard disk at the location specified in the **Registry**.

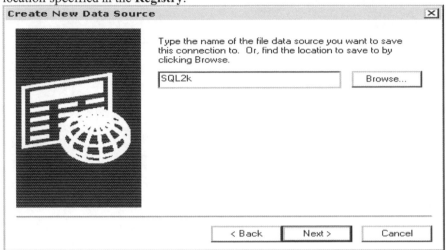

Figure 13-8 Creating a new data source name

The next dialogue, shown in **Figure 13-9 Data source name audit trail**, confirms the specification of the data source just created. Note the hint that the driver selected for this file DSN may prompt for missing information.

Figure 13-9 Data source name audit trail

In this case the SQL Server driver has been selected. It is obvious that several items of information are missing, for example, the names of the server, the database, the user, as well as the user's password. The subsequent dialogues prompt for information required by the driver—in this case the SQL Server driver. At each stage, the wizard verifies the information specified and warns if it encounters any problems.

The dialogue shown in **Figure 13-10 SQL Server as a new data source** is presented.

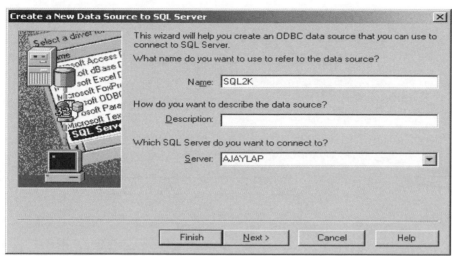

Figure 13-10 SQL Server as a new data source

A description may be given to the data source and the name of a server needs to be specified. The server name box is a combo box; it will contain any SQL servers that are recognised on the computer. If a server name does not appear in the combo box, a new name may be specified—the wizard will validate the server name before proceeding to the next step, invoked by clicking the **Next** button.

The next step prompts for the USER ID and PASSWORD using the dialogue shown in **Figure 13-11 Specifying access parameters**. Note the following:

● On a computer running the appropriate version of Windows, NT authentication may be used, if desired.

● It is advisable to use NT authentication only when <u>all</u> the data sources supported by the APL application support this feature.

● No information needs to be entered on this dialogue if the driver does not require a USER ID or PASSWORD; an example is the **Excel** ODBC driver.

In this case, NT authentication is used—no USER ID or PASSWORD will be required—indicated by these boxes being greyed out.

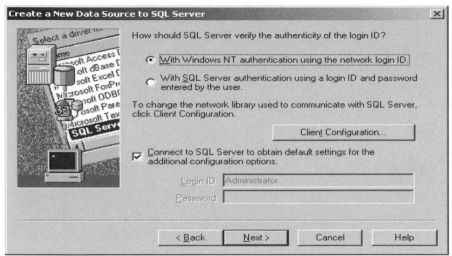

Figure 13-11 Specifying access parameters

The default database that this DSN will use can be selected from the drop-down box—refer to the dialogue shown in **Figure 13-12 Specifying default database**. Note that the wizard provides a list of the databases available—this confirms that the wizard validates what is specified at each stage.

If SQL Server authentication is used, the Login Id and Password must be specified if the connection is to be tested. Otherwise, these items may be omitted and specified at runtime.

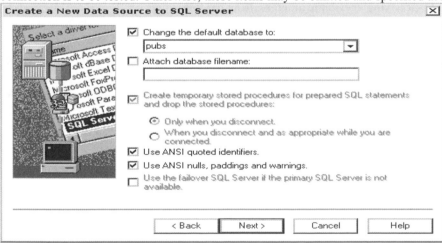

Figure 13-12 Specifying default database

The default database to use when a connection is established is PUBS; this default may be overridden at runtime. Note the facility for attaching any database that may exist

On clicking Next, the next dialogue, shown in **Figure 13-13 Specifying session parameters**, allows preferences to be set.

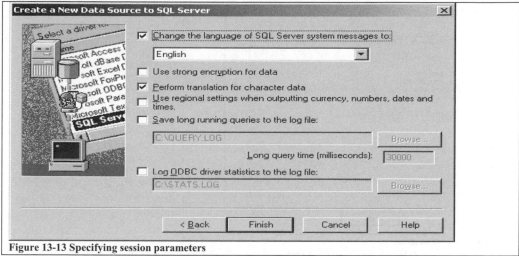

Figure 13-13 Specifying session parameters

The language is changed to **English**.

Finally, on clicking **Finish** the dialogue shown in **Figure 13-14 Testing data source** is shown. It confirms the information collated by the wizard. Usually the **Cancel** button reverts to an earlier dialogue thereby allowing corrections to be made.

If all necessary information is specified, the connection can be tested using the **Test Data Source** button. If some information has been omitted, the test cannot be carried out—the missing parameters can be specified at runtime.

Figure 13-14 Testing data source

At this stage, click either the **OK** or the **Test Data Source** button. In this instance the latter button is clicked and the information specified is tested successfully as confirmed in **Figure 13-15 Test outcome**.

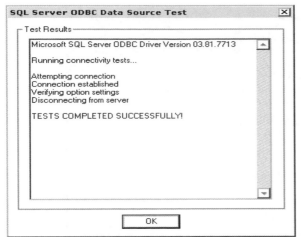

A file called SQK2K.DSN is
created. It contains:

```
[ODBC]
DRIVER=SQL Server
UID=Administrator
Trusted_Connection=Yes
LANGUAGE=British English
DATABASE=pubs
WSID=AJAYLAP
APP=APL+Win
SERVER=AJAYLAP
Description=SQL Server (MSDE)
```

The content will vary—it
depends on the driver
selected.

Figure 13-15 Test outcome

On clicking **OK**, the function ADOPrompt1 has run successfully and has created a
connection string:

```
      'ADO' ⎕wi 'xConnectionString'
Provider=MSDASQL.1;Extended Properties="Description=SQL Server
(MSDE);DRIVER=SQL Server;SERVER=AJAYLAP;APP=APL+Win;
WSID=AJAYLAP;DATABASE=pubs;LANGUAGE=British; Trusted_Connection=Yes"
```

The connection string may be archived—in a global variable or a file—and reused in the
future without the wizard. In other words, the developer can construct the connection string
using the wizard and distribute the constructed connection string when the application is
deployed, thus:

```
      sql2k←'ADO' ⎕wi 'xConnectionString'
```

Henceforth, a connection can be established as follows:

```
    ∇ ADOPrompt3;⎕wself
[1]    ⍝ System Building with APL+Win
[2]    ⎕wself←'ADO' ⎕wi 'Create' 'ADODB.Connection'
[3]    ⍝ ⎕wi 'xProperties("Prompt").Value' 1
[4]    ⎕wi 'XOpen' sql2k
    ∇
```

The file SQL2K.DSN may also be reused:

```
    ∇ ADOPrompt2;⎕wself
[1]    ⍝ System Building with APL+Win
[2]    ⎕wself←'ADO' ⎕wi 'Create' 'ADODB.Connection'
[3]    ⍝ ⎕wi 'xProperties("Prompt").Value' 1
[4]    ⎕wi 'XOpen' 'filedsn=c:\program files\common files\odbc\data
         sources\SQL2K.DSN'
    ∇
```

In line [3], a connection to the data source is established by direct reference to the file
DSN. Note that the location of file DSNs may vary depending on local configuration. Unless
a fully qualified path name is included with a file DSN name, Windows will look in the
default location only. The file must be distributed with the application and must exist on
target PCs on which the applications would run. This option is unavailable if a machine data
source is selected, as the information is held in the **Registry**—corresponding entries will
need to be created in the **Registry** of all target machines.

13.4 The ODBC Data Source Administrator

The general method for creating DSNs is Start | Settings | Control Panel | Administration Tools + Data Sources (ODBC). On some versions of Windows, click Start | Settings | Control Panel | ODBC Data Sources. This starts a wizard; the first dialogue is shown in **Figure 13-16 ODBC data source administrator**.

This allows the following:

● Full management—create, delete, and modify—of all user, system, and file DSNs.

● Visual verification of the list of drivers installed on the computer.

● Fine tuning of the runtime behaviour of the connection made with a DSN, including tracing and connection pooling.

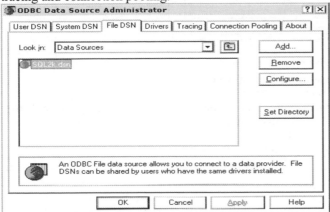

In this example, the file DSN tab is shown—note the additional **Set Directory** button.

This changes the value of DefaultDSNDir, the default location for file DSNs, in the **Registry**.

Figure 13-16 ODBC data source administrator

This dialogue can be invoked programmatically within the APL session, via an API function. This function will display the dialogue and return 1 if a valid handle is supplied; otherwise, no dialogue is displayed and 0 is returned. This requires the following entry in APLW.INI.

```
SQLManageDataSources=B(D hwndParent) ALIAS SQLManageDataSources LIB
     ODBCCP32.DLL
```

The APL+Win syntax is:
```
⎕wcall 'SQLManageDataSources' ('#' ⎕wi 'hwndmain')
```

Although the dialogue is displayed within the application session, the application is unaware of what activity takes place. If the application needs to know what DSNs, if any, have been removed or added, it will be necessary to compile a list of each type of DSN and to record the DefaultDSNDir setting before the dialogue is invoked and then again after it is closed. A comparison of the two lists will identify DSNs that have been added or removed. This does not establish whether a DSN has been modified. In practice, this does not matter whether a DSN has been modified. An application simply needs to use a DSN and determine if the connection succeeded or not; if the connection fails, the application needs to re-invoke the ODBC Data Source Administrator.

A key aspect of creating a DSN is the choice of the driver. The list of available drivers is shown in **Figure 13-17 Installed drivers;** all available drivers are listed. Note that the dialogue box scrolls to the right to reveal other columns of information. If a required driver does not appear in the list presented, it needs to be installed on the computer.

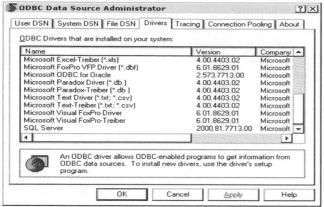

It is essential that the name of the driver specified exists in the 'Name' column of the 'Drivers' tab. The entry in the 'Name' column rather than the one under the 'File' column—to be found further to the right—is specified.

The file is SQLSVR32.DLL.

Figure 13-17 Installed drivers

13.4.1 Enumerating installed drivers

The list of installed drivers is found in the **Registry** at
HKEY_LOCAL_MACHINE\SOFTWARE\ODBC\ODBCINST.INI.

On this computer there are twenty-nine drivers; the list of drivers is presented in **Table 13-5 Enumerating installed drivers** has been enumerated with this line of code:

```
⊃RegEnum 'HKEY_LOCAL_MACHINE\SOFTWARE\ODBC\ODBCINST.INI'
```

	Driver name
1	MS Code Page Translator
2	ODBC Translators
3	SQL Server
4	ODBC Drivers
5	Microsoft Access Driver (*.mdb)
6	Microsoft Text Driver (*.txt; *.csv)
7	Microsoft Excel Driver (*.xls)
8	Microsoft dBase Driver (*.dbf)
9	Microsoft Paradox Driver (*.db)
10	Microsoft Visual FoxPro Driver
11	Microsoft ODBC for Oracle
12	Microsoft dBase VFP Driver (*.dbf)
13	Microsoft FoxPro VFP Driver (*.dbf)
14	MS Code Page-Übersetzer
15	Microsoft Access-Treiber (*.mdb)
16	Microsoft Text-Treiber (*.txt; *.csv)
17	Microsoft Excel-Treiber (*.xls)
18	Microsoft dBase-Treiber (*.dbf)
19	Microsoft Paradox-Treiber (*.db)
20	Microsoft Visual FoxPro-Treiber
21	Conversor de pßgina de c^digo MS
22	Driver do Microsoft Access (*.mdb)
23	Driver da Microsoft para arquivos texto (*.txt; *.csv)
24	Driver do Microsoft Excel(*.xls)
25	Driver do Microsoft dBase (*.dbf)
26	Driver do Microsoft Paradox (*.db)
27	Driver para o Microsoft Visual FoxPro
28	ODBC Core
29	Microsoft FoxPro Driver (*.dbf)

Table 13-5 Enumerating installed drivers

Predictably, the APL+Win fonts cause a problem—note, for instance, line 21—since the APL character set is unable to represent the characters correctly. In the **Registry**, this appears as 'Conversor de página de código MS'. This would not be a problem when creating a DSN using the ODBC Administrator; however, if the driver name were being specified in APL+Win code in the non English speaking world, it would be problematic.

13.5 System DSN connection

In the previous section a system DSN, SQL2K.DSN, was created, tested, and saved via the 'prompt' property of a connection object. This method, like the ODBC Data Source Administrator, leaves the application unaware of what happens to the list of DSNs on the computer.

An alternative means of creating a system DSN, independently of a connection object, is via the SQLConfigDataSource API call.

```
SQLConfigDataSource=L(D hwndParent,D fRequest,*C lpszDriver,*C lpszAttributes) ALIAS
    SQLConfigDataSource LIB ODBCCP32.DLL
```

13.5.1 Creating a SQL Server system DSN

The advantage in using this API is that no user interaction may be required and the application is aware of what DSN is created. The API call suppresses the 'Create new data source' dialogues when the hwndParent parameter is 0. The following function uses the APL+Win handle; this will show the dialogues and on completion of all the steps, create a new system DSN:

```
      ∇ Z←CreateSysDSN R
[1]    ⍝ System Building with APL+Win
[2]    (Z R)←(⎕io⊃R) (1↓R)
[3]    R←∊R,¨⎕tcnul
[4]    Z←⎕wcall 'SQLConfigDataSource' ('#' ⎕wi 'hwndmain') 4 Z R ⍝
          4=ODBC_ADD_SYS_DSN
      ∇

      CreateSysDSN 'SQL Server' 'DSN=SysBuild' 'SERVER=AJAY ASKOOLUM'
'DESCRIPTION=Local SQL Server' 'DATABASE=pubs'
```

That the DSN has been created can be verified programmatically.

```
      'System' DSNExists 'SysBuild'
```
1

Figure 13-18 Creating a system DSN also confirms that a DSN named SysBuild has been created.

lpszDriver: The first element of the right-hand argument is the drive description.

lpszAttributes: The other elements specify name/value pairs

Figure 13-18 Creating a system DSN

The name/value pairs within *lpszAttribute* may be specified in any order. Line [4] provides hard coded values for *hwndParent* ('#' ⎕wi 'hwndmain') and *fRequest* (4); 0 may be specified for *hwndParent*. On a Windows 2000 computer, where trusted connections may be used, the right-hand argument would be specified as:

```
     CreateSysDSN 'SQL Server' DSN=SysBuild' 'SERVER=AJAYLAP'
'DESCRIPTION=Local SQL Server' 'DATABASE=pubs' 'Trusted_Connection=Yes'
```

The DSN must be tested with an ADO connection object with:

```
     'ADO' ⎕wi 'Create' 'ADODB.Connection'
     'ADO' ⎕wi 'xProperties("Prompt").Value' (⎕wi '= adPromptComplete')
     'ADO' ⎕wi 'XOpen' 'DSN=SysBuild'
```

Setting the 'Prompt' property to 2—ask for information, if required—ensures that the connection object asks for any missing information, as shown in **Figure 13-19 SQL Server login I**.

Figure 13-19 SQL Server login I

The connection may fail if information is missing or incorrect. In this case, the logon id shown is the current user name. The default user id for the pubs database is *sa* and does not have a password—this was changed accordingly before clicking **OK**. The connection string of the object is now:

```
     'ADO' ⎕wi 'xConnectionSTring'
Provider=MSDASQL.1;Extended Properties="DSN=SysBuild; Description=Local
SQL Server;UID=sa;PWD=; APP=APL+Win;WSID=AJAYLAP;DATABASE=pubs"
```

Note that the default USER ID has been added. In addition, the DSN has been modified to include the UID parameter.

The **Options>>** button on the dialogue shown in **Figure 13-19 SQL Server login I** allows further configuration of the connection, notably, a different database can be used with the same DSN. The dialogue expands as shown in **Figure 13-20 SQL Server login II**.

Note that the API call is aware of *hwndParent*: it has identified the application name as APL+Win.

Additional information is sought within the 'Option' frame.

Figure 13-20 SQL Server login II

In fact, the DSN can be modified dynamically at connection time, as follows:

```
'ADO' ⎕wi 'XOpen' 'DSN=sysbuild;UID=sa;PWD=;DATABASE=master'
'ADO' ⎕wi 'xConnectionString'
Provider=MSDASQL.1;Extended Properties="DSN=sysbuild; Description=Local
SQL Server;UID=sa;PWD=; APP=APL+Win;WSID=AJAYLAP;DATABASE=master"
```

In other words, even when the 'prompt' property is set to request additional information, a connection string that specifies the missing information has the effect of overriding existing information in the DSN and of suppressing the dialogue that prompts for the missing information. The supplementary information specified is *not* written back to the DSN—in this instance, note that the database has been changed to 'master'.

13.5.2 Creating an Oracle system DSN

The process for setting up DSN's varies depending on the driver used. Oracle presents a single dialogue for creating system DSNs, see **Figure 13-21 Oracle DSN creation**.

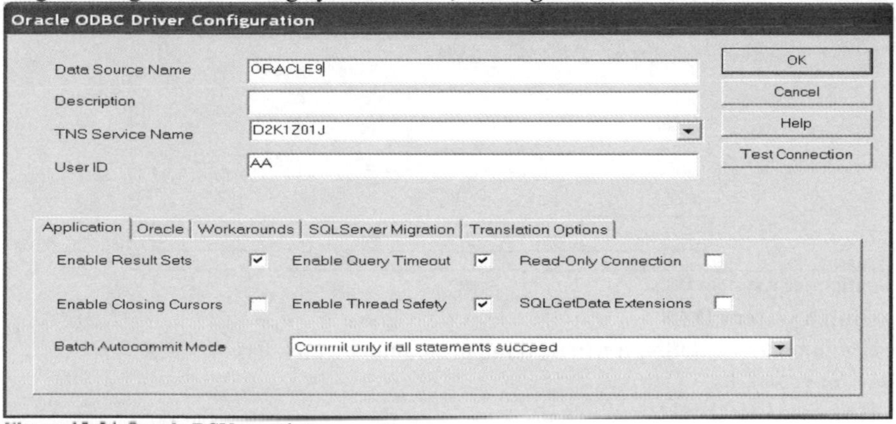

Figure 13-21 Oracle DSN creation

If the set-up procedure or wizard has an option to test a DSN, it is good practice to use it: this will save errors at runtime.

13.5.3 Configuring a system DSN

An existing DSN may fail to yield a successful connection for a variety of reasons. In the event of failure, the existing DSN may be deleted and a new one by the same name can be created. A better strategy is to allow the DSN to be modified and tested: this would enable the cause of the failure to be identified.

The following function invokes the configuration dialogue shown in **Figure 13-22 Configuring a system DSN**. It has the effect of clicking the **Configure** button on the ODBC Data Source Administrator dialogue:

```
     ∇ Z←ConfigSysDSN R
[1]   ⍝ System Building with APL+Win
[2]   (Z R)←(⎕io⊃R) (1↓R)
[3]   R←∊R,¨⎕tcnul
[4]   Z←⎕wcall 'SQLConfigDataSource' ('#' ⎕wi 'hwndmain') 5 Z R ⍝
5=ODBC_CONFIG_SYS_DSN
     ∇
```

The syntax is:

```
 ConfigSysDSN  'SQL Server' 'DSN=SYSBUILD'
```

The first element of the right-hand argument, *lpszDriver,* is the name of the driver. The second element, *lpszAttributes*, is a name/value pair but only the DSN name is required. The name of the system DSN is not case sensitive.

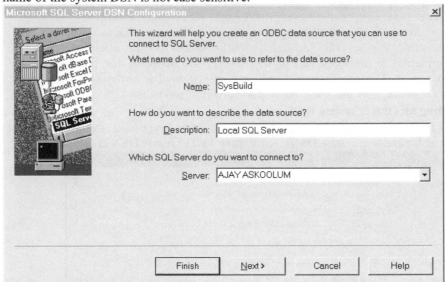

Figure 13-22 Configuring a system DSN

13.5.4 Removing a system DSN

If it is necessary to remove a DSN, it can be done programmatically. Try:

```
     ∇ Z←RemoveSysDSN R
[1]    ⍝ System Building with APL+Win
[2]    (Z R)←(⎕io⊃R) (1↓R)
[3]    R←∈R,¨⎕tcnul
[4]    Z←⎕wcall 'SQLConfigDataSource' ('#' ⎕wi 'hwndmain') 6 Z R ⍝
          5=ODBC_REMOVE_SYS_DSN
     ∇
```

The right-hand argument of *RemoveSysDSN* is identical to that of *ConfigSysDSN*:

```
     RemoveSysDSN  'SQL Server' 'DSN=SYSBUILD'
```

13.6 User DSN Connection

The SQLConfigDataSource can be used to create, configure, and remove user DSNs in the same way as system DSNs. Apart from user rather than system DSNs being managed, the main difference is that the fRequest parameter is different.

Rather than create comparable individual functions for user DSNs, it is preferable to create one function to manage all system and user DSNs with the programme flow being controlled by the left-hand argument. That is, a function with methods will keep the workspace tidy. Consider the following:

```
     ∇ Z←L DSNManager R;hwndParent
  [1]    ⍝ System Building with APL+Win
  [2]    (Z R)←(⎕io⊃R) (1↓R)
  [3]    R←∈R,¨⎕tcnul
{1[4]    :if 'AUTO'≡⎕wcall 'CharUpper' (4↑L)
  [5]        L←4↓L
```

```
   [6]         hwndParent←0
+  [7]    :else
   [8]         hwndParent←'#' ⎕wi 'hwndmain'
1}[9]    :endif
{1[10]   :select L
★ [11]       :case 'CreateUsrDSN' ∧ 1=ODBC_ADD_DSN
   [12]          Z←⎕wcall 'SQLConfigDataSource' hwndParent 1 Z R
★ [13]       :case 'ConfigUsrDSN'  ∧ 2=ODBC_CONFIG_DSN
   [14]          Z←⎕wcall 'SQLConfigDataSource' hwndParent 2 Z R
★ [15]       :case 'RemoveUsrDSN' ∧ 3=ODBC_REMOVE_DSN
   [16]          Z←⎕wcall 'SQLConfigDataSource' hwndParent 3 Z R
★ [17]       :case 'CreateSysDSN' ∧ 4=ODBC_ADD_SYS_DSN
   [18]          Z←⎕wcall 'SQLConfigDataSource' hwndParent 4 Z R
★ [19]       :case 'ConfigSysDSN'  ∧ 5=ODBC_CONFIG_SYS_DSN
   [20]          Z←⎕wcall 'SQLConfigDataSource' hwndParent 5 Z R
★ [21]       :case 'RemoveSysDSN' ∧ 6=ODBC_REMOVE_SYS_DSN
   [22]          Z←⎕wcall 'SQLConfigDataSource' hwndParent 6 Z R
★ [23]       :case 'ODBC_REMOVE_DEFAULT_DSN'
   [24]          ∧ Z←⎕wcall 'SQLConfigDataSource' hwndParent 7 θ θ
1}[25]   :endselect
       ∇
```

The left-hand argument is specified as one of the literals used on the case statements. If the left-hand argument includes the string 'auto', the API calls do not display any dialogues in the application session. WARNING: Lines [18] and [19] are untested in this mode; line [19] will permanently remove the following keys from the **Registry** and is deliberately commented. DefaultDSNDir specifies the location where file DSNs are stored, usually at these locations:

```
HKEY_LOCAL_MACHINE\SOFTWARE\ODBC\ODBC.INI\ODBC File DSN\DefaultDSNDir
HKEY_LOCAL_MACHINE\SOFTWARE\ODBC\ODBC.INI\ODBC\ODBCINST.INI\ODBC Drivers
```

13.7 DSNManager syntax summary

The full syntax for the DSNManager function is summarised in **Table 13-6 DSNManager syntax**. When the left-hand argument starts with the string 'Auto', dialogues are suppressed by forcing the parent handle to 0; otherwise, the system handle is used to force user interaction with the setting-up dialogues.

The right-hand argument is always a nested vector; the driver name must be the first element followed by the DSN name; the remaining arguments apply to the 'Create' method only and are optional. Note:

● The DSN will not be functional unless the other arguments are specified or the DSN has been tested.

● The driver name must be specified precisely; the available list of drivers can be enumerated—refer to 13.3.1.2 Enumerating existing file DSNs.

• The left- and right-hand arguments for the DSNManager function is shown below:

Left-hand argument	Right-hand argument
AutoCreateUsrDSN | CreateUsrDSN	Driver (DSN=DSNName) [(attname=att)…(attname=att)]
AutoConfigUsrDSN | ConfigUsrDSN	Driver (DSN=DSNName)
AutoRemoveUsrDSN | RemoveUsrDSN	Driver (DSN=DSNName)
AutoCreateSysDSN | CreateSysDSN	Driver (DSN=DSNName) [(attname=att)…(attname=att)]
AutoConfigSysDSN | ConfigSysDSN	Driver (DSN=DSNName)
AutoRemoveSysDSN | RemoveSysDSN	Driver (DSN=DSNName)

Table 13-6 DSNManager syntax

13.8 File DSN Connection

A file DSN is like a system DSN and is available to all users, but the connection parameters are stored in a text file with a .DSN extension rather than the **Registry**. This text file can be created using NOTEPAD or any text editor or using the ODBC Data Source Administrator.

The best approach is to use the SQLCreateDataSource API call: see **Figure 13-5 Creating a DSN**.

```
⎕wcall 'SQLCreateDataSource' ('#' ⎕wi 'hwndmain') ⍬
```

If all the parameters are known, the SQLWriteFileDSN API can create a file DSN silently. This API is writing INI name/value pairs to a text file whose extension it sets to DSN and whose location it finds by reference to the **Registry**:

```
SQLWriteFileDSN=B(*C lpszFileName,*C lpszAppName,*C lpszKeyName,*C lpszString)
    ALIAS SQLWriteFileDSN LIB ODBCCP32.DLL
```

The lpszFileName is the name of the DSN file; the extension DSN is added if not present and the file is saved in the location held in

```
HKEY_LOCAL_MACHINE\SOFTWARE\ODBC\ODBC.INI\ODBC File DSN\DefaultDSNDir
```

The lpszAppName parameter is the string ODBC. The other two parameters, lpszKeyName and lpszString, specify name/value pairs. This API writes one name/value pair at a time. A cover function can write all the entries in a loop.

```
      ∇ Z←L CreateFileDSN R
 [1]    ⍝ System Building with APL+Win
 [2]    (Z L R)←1 (↑R) (1↓R)
 [3]    L←L 'ODBC'
{1[4]   :if 0∊R
 [5]        Z←Z∧⎕wcall (⊂'SQLWriteFileDSN'),L,0 0
+ [6]   :else
{2[7]       :while 0≠⍴R
 [8]            Z←Z∧⎕wcall (⊂'SQLWriteFileDSN'),L,(2↑('='≠⎕io⊃R)⊂⎕io⊃R)
 [9]            R←1↓R
2}[10]      :endwhile
1}[11]  :endif
      ∇
```

The syntax requires a file name followed by name/value pairs:

```
      CreateFileDSN FileName name=value ... name=value
      CreateFileDSN 'NewFile' 'DRIVER=SQL Server' 'UID=' 'PWD='
'DATABASE=pubs' 'SERVER=AJAY ASKOOLUM' 'Description=Created by
CreateFileDSN'
```

The file NEWFILE.DSN is created and stored at the default location.

The content of this file is shown in **Figure 13-23 NEWFILE.DSN**.

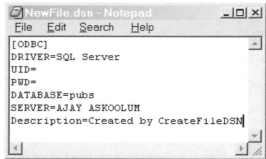

Figure 13-23 NEWFILE.DSN

The CreateFileDSN function can be used to add or modify any name/value pair:

• If a value is not specified, the existing value is deleted.

• If no name/value pairs are specified (denoted by 0), all entries from the file are deleted.

A text file can be transferred from one computer to another by simply copying it to the target, preferably in an identical location. If necessary, it can be modified easily using a text editor such as NOTEPAD and references to mapped drive letters, if any, can be replaced by their corresponding universal naming convention (**UNC**). A UNC name ensures that all users, irrespective of their respective drive mapping, can use the file DSN. In the example shown in **Figure 13-23 NEWFILE.DSN**, the last line will be modified to:

```
DefaultDir=\\<server name>\<share>
```

<server name> is the name of the network server and <share> is the shared folder on the network server. Mapped network drives are convenient but there are a finite number of drive letters and trouble free references usually requires that all users use the same letter for any given shared drive—in practice, this is rarely so.

13.8.1 Using a file DSN

The following function uses the file DSN NEWFILE.DSN:

```
      ∇ Z←UseNewFileDSN;⎕wself;Sql
[1]    ⍝ System Building with APL+Win
[2]    ⎕wself←'∆ADORS' ⎕wi 'Create' 'ADODB.Recordset'
[3]    Sql←"SELECT au_id,au_lname FROM authors WHERE au_lname like 'S%'"
[4]    ⎕wi 'XOpen' Sql 'Filedsn=newfile.dsn;UID=sa;PWD=;'
[5]    Z←⎕wi 'XGetString'
[6]    ⎕wi 'Delete'
      ∇
```

In line [4], the connection string is specified as NEWFILE.DSN and additional parameters that override the corresponding parameters in the DSN file, if present. For DSNs stored in the default location, the path does not need to be specified:

```
      UseNewFileDSN
341-22-1782    Smith
274-80-9391    Straight
724-08-9931    Stringer
```

Any parameter can be overridden, including the name of the database. Change line [3] and [4]. Line [3] specifies a new SQL statement. Line [4] supplies the name of a new database. For example,

```
[3]    Sql←"SELECT CustomerId,CompanyName FROM CUSTOMERS WHERE
          ContactTitle='Accounting Manager'"
[4]    ⎕wi 'XOpen' Sql 'Filedsn=newfile.dsn;UID=sa;PWD=;Database=northwind'
      UseNewFileDSN
```

```
BOTTM   Bottom-Dollar Markets
FISSA   FISSA Fabrica Inter. Salchichas S.A.
HANAR   Hanari Carnes
LILAS   LILA-Supermercado
QUEDE   Que Delⴰcia
QUICK   QUICK-Stop
ROMEY   Romero y tomillo
SUPRD   Supr⌊mes delices
VINET   Vins et alcools Chevalier
WARTH   Wartian Herkku
```

13.8.2 File DSN portability

A file DSN is portable if it literally contains all relevant information to establish a connection to a data source. A file DSN may contain reference to another DSN, which contains the actual parameters. Create a DSN that uses the system DSN SysBuild as follows:

```
      CreateFileDSN 'NonShare' 'DSN=SYSBUILD'
1

    ∇ Z←UseNonShareDSN;⎕wself;Sql
[1]    ⍝ System Building with APL+Win
[2]    ⎕wself←'∆ADORS' ⎕wi 'Create' 'ADODB.Recordset'
[3]    Sql←"SELECT au_id,au_lname FROM authors WHERE au_lname like 'S%'"
[4]    ⎕wi 'XOpen' Sql 'Filedsn=nonshare.dsn;UID=sa;PWD=;'
[5]    Z←⎕wi 'XGetString'
[6]    ⎕wi 'Delete'
    ∇

      UseNonShareDSN
341-22-1782    Smith
274-80-9391    Straight
724-08-9931    Stringer
```

This clearly works. However, the file NONSHARE.DSN is not portable—cannot be copied to another machine and expected to work—unless the system DSN SysBuild is also exported.

File DSNs are viable when they are self-contained and any reference to them is fully qualified in order to avoid **Registry** lookup for their location.

13.9 UDL connection

A Universal Data Link (**UDL**) connection is one based on a binary file—consisting of Unicode characters—that contains the data source connection properties and has extension UDL. Unlike file DSNs, there is no default location for UDL files; therefore, any reference to them must be fully qualified by a path name. Thus, UDL files are truly portable. Although UDL files can be browed within NOTEPAD, they cannot be edited or created using a text editor because they contain Unicode data.

On a Windows 2000 computer, such a file is created interactively: within **Windows Explorer**, right click, then click **New** and select **Microsoft Data Link**. A UDL file with a default name is created. Rename the file as appropriate but retain the UDL extension. Double click the file name to invoke the data link dialogue. A better approach is to create such a file programmatically; this works on any machine running Windows 95 or later.

A UDL file may be created from an existing ADO connection object. Try:

```
      ∇ Z←L UDL R;⎕wself
 [1]    ⍝ System Building with APL+Win
 [2]    ⎕wself←'∆UDL' ⎕wi 'Create' 'Microsoft OLE DB Service Component
        Data Links'
{1[3]   :if 0=⍴R~' '
 [4]        R←'U_D_L'
```

```
1}[5]     :endif
{1[6]     :if 0≠⍴R ⎕wi 'self'
+ [7]         :andif(R ⎕wi 'class')≡'ActiveObject ADODB.Connection'
{2[8]         :if Z←0≠↑⎕wi 'XPromptEdit' (R ⎕wi 'obj')
  [9]             Z←R ⎕wi 'xConnectionString'
2}[10]        :endif
```

Alternatively, it may be created without reference to any ADO connection object. See:

```
+ [11]    :else
  [12]        ⎕wi 'XPromptNew>ΔUDL'
{2[13]        :if Z←(⊂'xConnectionString')∊'ΔUDL' ⎕wi 'properties'
  [14]            Z←'ΔUDL' ⎕wi 'xConnectionString'
  [15]            'ΔUDL' ⎕wi 'Delete'
2}[16]        :endif
1}[17]    :endif
```

The UDL dialogue does *not* create any file; this has to be done with APL+Win code. If a file name is specified, the connection parameter is written to that file. If the **Cancel** button is clicked, the file is not updated if it exists or not created if it did not. Look at:

```
{1[18]    :if ~0≡Z
  [19]        Z←'[oledb]' '; Everything after this line is an OLE DB
        initstring' Z
  [20]        Z←(Φ⁻2↑⎕av),∊(∊Z,¨⊂⎕tcnl,⎕tclf),¨⎕tcnul
{2[21]        :if 0≠⎕nc 'L'
+ [22]            :andif 0≠⍴L~' '
  [23]            R←⁻1+⍳/0,⎕nnums,⎕xnnums
{3[24]            :if 0≠⎕wcall 'PathFileExists' L
  [25]                L ⎕xntie R
  [26]                L ⎕xnerase R
3}[27]            :endif
  [28]            L ⎕xncreate R
  [29]            Z ⎕nappend R
  [30]            ⎕nuntie R
  [31]            Z←1
2}[32]        :endif
1}[33]    :endif
        ∇
```

Otherwise, the connection parameter is returned to the workspace. If the **Cancel** button is clicked, the result is 0.

Although the UDL function will return the selected parameters to the workspace if a file name is not specified, there is little point in this: the returned value cannot be used as a connection string. Typical arguments of the UDL function are:

```
'C:\SqlServer.udl' UDL θ
'C:\SqlServer.udl' UDL 'adoc'
```

The data link dialogue has a multi-tab dialogue that allows all the properties of a connection string to be specified and tested. The **Provider** tab is shown **Figure 13-24 UDL Provider**. This dialogue has four tabs, namely:

- Provider
- Connection
- Advanced
- All.

Figure 13-24 UDL Provider tab shows the Provider tab. This allows the selection of an OLE DB Provider.

If a required Provider is not found in the list, it does not exist on the computer and must be installed: Provider information cannot be added manually.

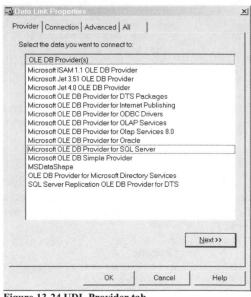

Figure 13-24 UDL Provider tab

Having selected the appropriate provider, click on the **Connection** tab.

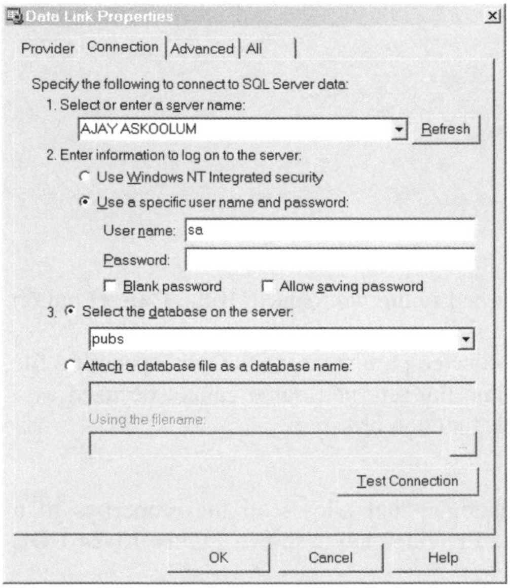

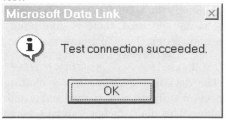

Figure 13-25 UDL Connection tab

Figure 13-25 UDL Connection tab shows the Connection tab.

The contents of the **Connection** tab vary depending on the selection made on the **Provider** tab.

Once the connection parameters have been specified, click the **Test Connection** button to verify that a connection may be established; this is confirmed in a message such as one shown in **Figure 13-26 UDL test**.

Figure 13-26 UDL test

The UDL function does not validate the file name—whether it has a UDL extension or whether it is valid.

Figure 13-27 UDL Advanced tab shows the Connection tab.

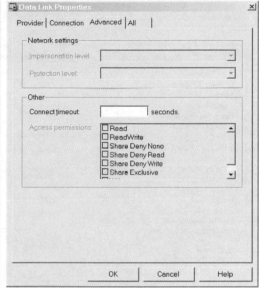

Figure 13-27 UDL Advanced tab

Connection timeout and access permission options may be specified.

● All the configuration information is then written to a file with extension UDL: All the configuration information is shown on the **All** tab, see **Figure 13-28 UDL All tab**.

● The **Edit Value** button allows the selected value to be changed.

Usually, the default value for any item of configuration is optimal: should any of the values be changed, the precise impact of such a change needs to be established first.

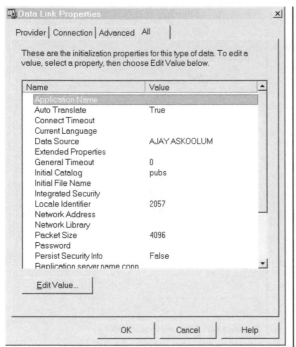

The example below uses an existing ADO connection object, adoc. The UDL file can be used either to open an ADO connection object or an ADO record set object:

Figure 13-28 UDL All tab

```
     □wself←'∆ADORS' □wi 'Create' 'ADODB.Recordset'
     □wi 'XOpen' 'Select * from authors' 'File
Name=c:\sqlserver.udl;UID=sa;PWD=;'

     □wself←'adoc' □wi 'Create' 'ADODB.Connection'
     □wi 'XOpen' 'File Name=c:\sqlserver.udl;UID=sa;PWD=;'
```

Additional parameters may be specified; these may override information already stored in the UDL file.

13.10 DSN-less connection

A DSN-less connection is one where all the connection parameters are specified directly without requiring these parameters to be stored in the **Registry** or a file. Quite literally, the phrase DSN= or FileDSN= or File Name= do not appear in the connection string.

A DSN-less connection would be expected to be faster than a DSN connection in that it eliminates the need to read the **Registry**. On the other hand, it means that connection details, including ones that may change periodically such as passwords, are hard coded within the application.

13.11 Server data sources

ODBC drivers for virtually any database server are available, as are OLE DB providers for most of them. The Windows platform installs ODBC drivers and providers for SQL Server, Oracle, etc. All tables and views as well as the schema are valid data sources.

13.12 Access data sources

Either the ODBC driver or OLE DB JET provider can be used with **Access** databases. All tables and views—or queries—as well as the schema are valid data sources.

13.13 Excel data sources

ADO can access **Excel** XLS files either via the OLE DB provider for ODBC or the OLE DB provider for Microsoft JET if the XLS file is not already open and a password is not required to open it. **Figure 13-29 Excel data sources** shows the driver and provider that can be used to access **Excel** workbooks: note the data sources options.

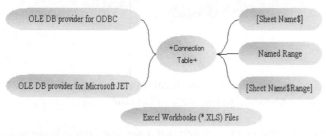

There are sets of names in an XLS file or workbook. Each set contains unique names: sheet names and the names of ranges.

Figure 13-29 Excel data sources

Worksheet names may not be valid APL+Win names—they may contain spaces or begin with an invalid character. On the other hand, the names of ranges within a worksheet are valid APL+Win names—the name of a range may also be used as the name of a workbook. In order to keep references to worksheet and range names distinct, four conventions are followed, namely:

• Worksheet names are followed by $; if the worksheet name is not a valid variable name, it is also enclosed within square brackets.

• Range names are enclosed within quotation marks—both providers tolerate single or double quotation names. Range names are not worksheet but workbook specific.

• If the source is a range on a specific sheet of a workbook, it must be fully qualified; that is the sheet name followed by $ and a range, specified using the **Excel** xlA1 reference style— that is A1:C4 rather than R1C1:R4C3—and enclosed within square brackets.

• If a range is used as the data source, it must be contiguous.

13.14 Text data sources

Traditionally, APL+Win applications have used text files for data acquisition and transfer and relied on the built-in native file handling functions. In other words, every aspect of the handling of text data sources has had to be handled by APL+Win code. Given that text files are inherently sequential and can only hold character data, a solution based on native file functions is slow and contain the overhead of bespoke code.

ADO provides a standard way for handling text data sources like any other data source. This solution works optimally the file is self-sufficient: that is, it contains records delimited by carriage return and line feed, columns within records are delimited by comma—if numeric columns use comma separators, character columns must be enclosed within double quotes—and column names are shown on the first row. **Figure 13-30 Text file data sources** shows the driver and provider that can be used to access **Excel** workbooks:

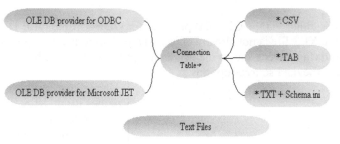

A SCHEMA.INI file supplements the data found in a text file. It must be found in the same location as the text file.

Figure 13-30 Text file data sources

If a file named SCHEMA.INI is present in the same location as the source file, both providers scan the file for an entry corresponding to the source file name.

13.14.1 The SCHEMA.INI file

The SCHEMA.INI file is itself a text file and can be created using any text editor. It contains the name of the text file in square brackets and details of each aspect of the content of the text file as a name/value pair.

This file provides information on up to five characteristics, in respect of each text file, namely:

- The name of the text file.
- The format of the text file.
- The names, widths, and types of each column.
- The character set used.
- The rules for handling the data found in the file.

While the SCHEMA.INI file provides more control over the manner in which the data in a text file can be handled, using such a file requires that the SCHEMA.INI file and the text file it relates to be handled such that they remain compatible. The SCHEMA.INI file must be located in the same folder as the text file.

The *name of the text file* starts a section SCHEMA.INI file:
```
[TEXTFILE.TXT]
```
This entry for the name of the text file is not case sensitive and may be a long file name. The name, excluding the path, is specified in full including the extension. The remaining characteristics are specified under the heading entry, without square brackets, *in any order*.

There are five configuration elements associated with the *format of the file*. First, the method of columns separation is indicated—when the *Delimited(character)* option is used, the character is specified without quotation marks. The possible mutually exclusive entries are:
```
Format=TabDelimited
Format=CSVDelimited
Format=Delimited(character)
Format-Delimited()
Format=FixedLength
```
Second, whether the text file contains column headers is indicated. The mutually exclusive entries are:
```
ColNameHeader=True
ColNameHeader=False
```
Third, the *names, widths, and types* of each column are specified on separate lines. Sample entries are:
```
Col1=MemberId Integer
```

```
Col2=Name Char width 30
Col3=FirstName Text Width 20
Col4="Date Of Birth" Date
...
Coln=Salary Float
```

These entries imply that **ColNameHeader** is **FALSE** and that the columns are delimited. Note that 'Char' and 'Text' may be used interchangeably. In addition, a column name such as Date Of Birth, which does not comprise a valid variable name, needs to be specified within double quotes. On the other hand, if **ColNameHeader** is **TRUE**, the name of the columns found in the header or first row will be overridden by the entries in the SCHEMA.INI file.

Fourth, the character set used in the file is specified. The mutually exclusive options are:

```
CharacterSet=ANSI
CharacterSet=OEM
CharacterSet=UNICODE
CharacterSet=A valid code page number (character set used in source file)
```

The default character set is ANSI. Usually, the entry is specified using the literal 'ANSI', 'OEM', or 'UNICODE'. A code page number may be used: the alternative form `CharacterSet=1252` is also valid.

Fifth, any special data type conversions are added, each on a separate line. Some examples are:

```
MaxScanRows=20
DateTimeFormat=dd/mm/yyyy
```

13.14.2 Creating SCHEMA.INI automatically

The SCHEMA.INI file can be created automatically to include all aspects of the specification except for some data type conversion rules, which must be created manually.

DSNManager can create a system or user DSN using the text driver; a system DSN is preferable as all users can access it. The DSNManager function requires the name of the DSN, the name of the driver and the location of the source data file—the DBQ parameter, as shown below:

```
'CreateSysDSN' DSNManager 'Microsoft Text Driver (*.txt; *.csv)'
'DSN=AutoSchema' 'Description=Auto generate SCHEMA.INI for all text files'
```

The dialogue shown in **Figure 13-31 Creating SCHEMA.INI automatically** appears:

• Uncheck the **Use Current Directory** flag and select the folder where the text source files are located. This will ensure that the SCHEMA.INI file is created in the correct folder.

• All the eligible files in the selected folder will be included in the SCHEMA.INI file.

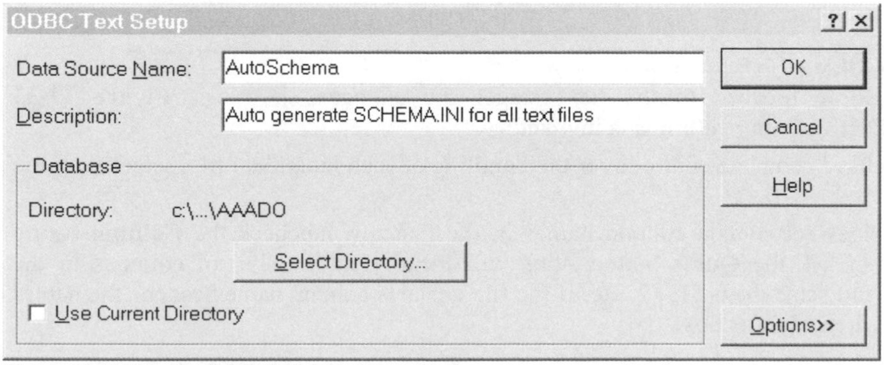

Figure 13-31 Creating SCHEMA.INI automatically

The **Options>>** button allows files with a particular extension to be explicitly included or removed. The dialogue shown in **Figure 13-32 SCHEMA.INI: Selecting files** is shown when **Options>>** is clicked.

Leave the extensions list as is. This facility creates basic SCHEMA.INI entries for all files with extension ASC, CSV, TAB, and TXT.

Note the **Help** button: it invokes further guidance on using the ensuing dialogues. Click the **Define Format** button in order to specify the handling of individual files found in the selected directory.

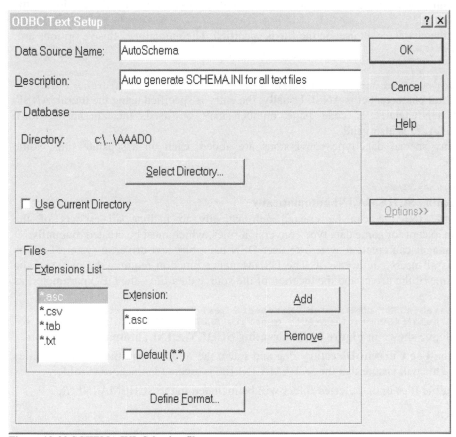

Figure 13-32 SCHEMA.INI: Selecting files

Click OK to proceed to the next stage; the dialogue shown as **Figure 13-33 SCHEMA.INI: Defining columns** is invoked:

● In order to have complete control over the handling of each individual file, select each file in turn.

● If the file does not contain column names in the first row, uncheck the **Column Name Header** flag. Click the **Guess** button—this will identify the number of columns in the selected file and name them F1, F2, etc. If the file contains column name headers, the names are shown in the **Columns** box.

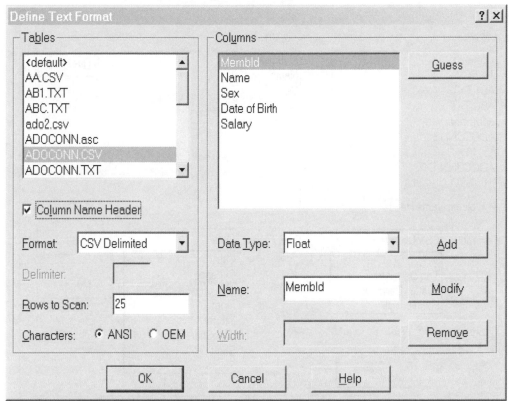

Figure 13-33 SCHEMA.INI: Defining columns

Any column names can be highlighted; the **Data Type** combo box shows its type and the **Name** box shows its name—when this is likely o be an invalid name, it may be changed by inserting a valid name in the **Name** box. The available options for **Format** and **Data Type** are shown on the right.

Although all files can be selected (*.*), the text driver can handle only the formats shown in the **Extension List** box, see **Figure 13-34 SCHEMA.INI: Formats and data types.** Namely:

● The **Guess** button retrieves the column headers from the selected file, when the Column Name Header checkbox is ticked.

● The **Row to Scan** value specifies the number of rows the driver will scan in order to determine the type of data in a column; a value of 0 forces the entire file to be scanned.

● The **Data Type** entry specifies the type that the driver determines; this can be overridden.

● The **Name** entry specifies the name of the column.

● Select the appropriate option from **Characters**—usually this will be ANSI.

● If a column that exists in the source file is not required, highlight it and click **Remove**.

● If a new column needs to be added, click **Add,** specify the name of the column in **Name**, and select its appropriate data type.

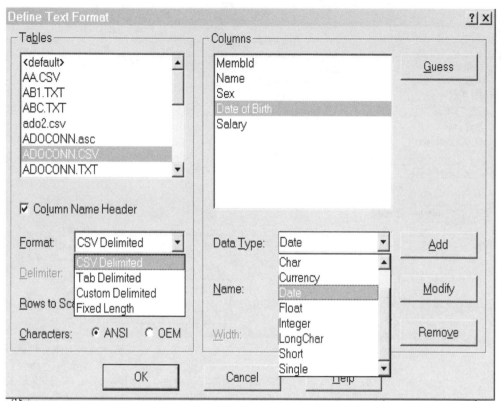

Figure 13-34 SCHEMA.INI: Formats and data types

Finally, click the **Modify** button: this updates the SCHEMA.INI file; the file is created if not already present.

A section of the file is shown in **Figure 13-35 SCHEMA.INI contents**. Default entries for all eligible files are created in the schema file using defaults.

Note the entries for the ADOCONN.CSV file; in particular, the name for column four is enclosed within double quotes. The name is an invalid APL+Win variable name but valid as a column name in an ADO record set object.

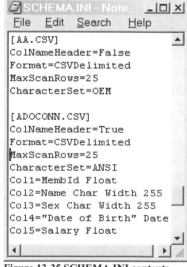

Figure 13-35 SCHEMA.INI contents

The DSN that started the dialogues for defining the SETUP.INI file is in fact not required to use the SCHEMA.INI file. It can be discarded by clicking the **Cancel** button. Alternatively, it can be deleted programmatically:

```
      'RemoveSysDSN' DSNManager 'Microsoft Text Driver (*.txt; *.csv)'
'DSN=AutoSchema'
```

13.14.3 Refining content of SCHEMA.INI file

Although the 'ODBC Text Setup' dialogues simplify the process of defining data sources, they do not offer any mechanism for specifying conversion rules for the underlying data. This has to be done manually.

Although the column Date of Birth was correctly identified as a date column, the driver needs to be made aware that this date is in yyyy/mm/dd format and needs to be converted into the regional settings date format.

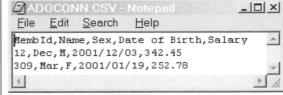

Figure 13-36 ADOCONN.CSV

The following function reads the ADOCONN.CSV file, shown in **Figure 13-36 ADOCONN.CSV**, using the text driver—not the DSN used for creating the SCHEMA.INI file. Note that there is no explicit reference to the SCHEMA.INI file but it is obvious that it is used. Try:

```
      ∇ Z←ReadADOCONN;⎕wself;Cnn;Sql
[1]     ⍝ System Building with APL+Win
[2]     ⎕wself←'∆ADORS' ⎕wi 'Create' 'ADODB.Recordset'
[3]     Cnn←'Provider=MSDASQL;Driver={Microsoft Text Driver (*.txt;
          *.csv)};DBQ=c:\system building with APL+win\aaado;'
[4]     Sql←'Select * from adoconn.csv'
[5]     ⎕wi 'XOpen' Sql Cnn
[6]     Z←⎕wi 'XGetString'
[7]     ⎕wi 'Delete'
      ∇
```

Without enforcing a conversion rule, the Date of Birth column has null values, as it is not valid in the UK regional date format:

```
      ReadADOCONN
12      Dec    M              342.45
309     Mar    F              252.78
```

In order to inform the driver of the date format used in the underlying data file, another entry is necessary in the SCHEMA.INI file. A rule for converting the date from the underlying format to the regional date format is necessary. The SCHEMA.INI file can be edited with NOTEPAD.

Note the DateTimeFormat entry in the file depicted in **Figure 13-37 Updated SCHEMA.INI file**.

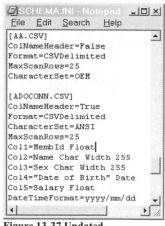

A conversion rule specified in the SCHEMA.INI file applies to <u>all</u> columns of the same type. Therefore, the text driver will not cope with a file that has two or more columns of the same type but in different formats.

Figure 13-37 Updated SCHEMA.INI file

13.15 Data source issues

All the requirements for data access exist on the Windows platform. APL+Win application design can separate data handling into a data tier quite effortlessly. The industry standard is to use database server data sources, such as SQL Server, Oracle etc, as these have intrinsic support not only for large volumes of data but also for concurrent usage by many users without degradation of performance. File databases such as **Access** provide similar facilities but performance is degraded as the number of users increases and there is a limit to the overall size of the data.

Both the ODBC driver and the JET OLEDB provider impose other restrictions that affect **Access**, **Excel,** and text file data sources, namely:

- A table may contain up to 255 fields only.
- The handling of null values varies, depending on the data source.
- Each column name is restricted to a maximum of 64 characters.
- Each column's content must not exceed 32,766 characters.
- Each record is restricted to a maximum of 65,000 characters.
- Concurrent usage by multiple users is not supported with **Excel** and text data sources.

Some of these restrictions do not apply to **Access** 2000 or later versions. With **Excel** data sources, the limit to the number of records is the physical limit to the number of rows on a worksheet, 65,535 allowing for a header row.

13.16 Inward APL+Win issues

ADO simplifies data acquisition and enables APL to handle large volumes of data. However, some APL specific issues reduce this simplicity.

13.16.1 Data types

ADO has many more data types than APL. ADO data types are transient in the sense that they only exist once a record set is extracted from a data source. RDBMS data sources, like SQL Server, supply the data types. For other sources, ADO scans a number of rows to guess the data types for each column or field. An application using ADO need only allow for the ADO data types as other data types implicit in the data source itself are discarded. ADO returns the data in one or another of its own data types and APL manages the coercion to an

APL type. For practical purposes, most of the ADO data types are transparently coerced into the APL data types.

13.16.1.1 Null data types

The null data type presents a particular problem for APL+Win. A null value is unique; it is not equal to any other value including another null value. The comparison operators cannot be used with a null value, as the result is also a null value.

Data types that APL does not recognise may be managed by coercion into another type that APL will recognise. Usually, null is coerced into a zero length vector of one element in APL+Win.

With text data sources, it may be necessary to remove or amend records with missing values, especially when such records occur within the range of records scanned by the driver to determine column types.

13.16.1.2 Date data types

APL+Win does not have a date data type either. Most data sources hold dates as a number whose integral part represents the number of days from a reference date and a fractional part that represents time as a portion of a day. A negative integral part will represent a date earlier than the reference date. The reference date varies from data source to data source:

- Oracle supports dates in the year range ⁻4713 to 9999.

- DB2 supports dates in the year range 1 to 9999.

- SQL Server has two date data types. The DATETIME type has the range 01/01/1753 to 31/12/9999 and the SMALLDATETIME type has the operational range 01/01/1900 to 06/06/2079.

- **Access** has a date range between year 100 and 9999.

- **Excel** uses 01/01/1900 as the default reference date. Optionally, this can be set to 01/01/1904. However, **Excel** does not recognise dates earlier than the reference date and will cause an error. **Excel** incorrectly treated the year 1900 as a leap year. The reference date is stored in the workbook. The latest date in **Excel** is 31/12/9999 irrespective of the reference date. However, depending on how an **Excel** worksheet is populated dates later than 2078 may not be recognised.

- The date range that is applicable to **Excel** files notionally applies to text files also.

APL receives a date value as an integer, a number of days from a reference date, or as a floating-point number representing both a number of days from a reference date and the time elapsed since midnight. Given that each data source may have its own reference date it is cumbersome to write APL code to format the date values. A sound solution is to format the date values as strings, in the short or long date regional format, within the SQL function that extracts the data in the first place. Wherever this is not possible, the VBScript FormatDateTime function can be used or the **VB** Format function, encapsulated as a method in an ActiveX DLL, may be called.

13.16.2 The atomic vector

The APL character set or atomic vector (□av) uses the extended character set in the range 128 to 255. European accented characters are also in this range as is the British currency symbol £. Some work around this is necessary to retrieve the characters in this range within the APL workspace.

The APL atomic vector is not an ANSI character set. However, there are some subtleties with the character set, as given below:

• Information copied and pasted from the APL+Win editor to another application is ANSI compliant—the target application will need to use an APL font to display correctly.

• Information copied and pasted from the active session is ANSI compliant except for numeric arrays.

• Information sent to the clipboard or to another application programmatically is not ANSI compliant unless converted to ANSI prior to dispatch.

In respect of the latter methods, the APL character set needs to be mapped to ANSI prior to transfer.

13.17 Outward APL issues

The issues highlighted in the previous section apply in reverse when APL writes data back to the underlying data source. More pertinently, APL needs to process arrays as rows, treating each as a record, when writing them to a database.

An APL development that partitions data into a separate tier requires a change in approach; arrays cannot simply be written to component files but need to be treated as records and handles sequentially.

13.17.1 CSV files issue

A comma separated values (**CSV**) file contains fields delimited by a comma and records delimited by carriage return and line feed characters. Strictly speaking, a character field must be enclosed within double quotation marks. And, because comma is a delimiter, it cannot also be used as a separator in numeric fields: that is, numeric fields cannot be formatted.

An **Excel** created CSV file does not enclose string values within double quotes. Ordinarily, this does not matter. It will do if a string column contains comma: for example, if a column Name contains values in the pattern Surname, Initials the CSV file will be seen to contain more column values than column headers.

13.18 The way forward with the data tier

Historically, text files have been a standard way for data acquisition and transfer in APL applications, using native file handling functions. ADO with the SCHEMA.INI file provides a more efficient way of using text files. However this approach leads to an unreliable application, as the possibilities for runtime errors are enormous given the variations in the format of text data files. There are too many variations in the decimal symbol, the currency symbol, negative numeric amounts, the precision of numeric fields, date formats, etc. A robust design must aim to isolate the data tier to a database server.

The Microsoft SQL Server Data Engine (**MSDE**) is available on any Microsoft Office compact disk (**CD**) that includes **Access**. It is not installed automatically via any installation configuration. It can be installed manually by double clicking on SETUPSQL.EXE, usually found in the \SQL\X86\SETUP folder. On installation, the target computer becomes a local database server.

MSDE does not have a user interface for database design or maintenance. It can be used from **Access** 2000 or from a programming environment such as APL. Databases created by MSDE are SQL Server compliant and thus can be manipulated with T-SQL. MSDE databases are upwardly compatible with SQL Server databases; unfortunately, MSDE imposes limits on the size of the database and the number of concurrent users. This does not stop application development using MSDE and deployment of the same databases with SQL Server.

At present, SQL SERVER Express 2005 is freely from Microsoft, subject to end user licence agreement.: it is a later release of the MSDE engine.

Chapter 14

Structured Query Language

The developer working on an application that supports a variety of data sources will rue the fact that the 'S' of SQL does not stand for 'Standard'. Variations in the SQL dialects supported by different databases add a costly overhead to development and testing. In spite of published standards for SQL, vendors maintain custom features in their particular SQL engine.

In 1986, the American National Standards Institute (**ANSI**) ratified IBM's SQL prototype, originally introduced as SEQUEL (Structured English Query Language) in 1974. SQL-89, published in 1989, is a revision of the original ANSI standard. A major revision, as published as SQL-92, followed in 1992. By this time, SQL was widely adopted as the data language of database vendors who had largely abandoned their own attempts to develop custom query languages. The latest standard, SQL:1999, was published in 1999, prompted partly by millennium dates issues; this is also known as SQL3. The latter naming convention, aligned to the International Organisation for Standards (**ISO**) convention saw the hyphen replaced by a colon and the use of a format comprising of all four digits in the year—this will simplify reference to follow-on standards published in this century.

Although SQL-92 is a widely recognised standard, several vendors continue to support proprietary data language features. In order to facilitate the adoption of the new standard, SQL-92 introduced three levels of conformance: entry, intermediate, and full. The SQL:1999 standard created the concept of 'core' or 'entry level' SQL and introduced 'packages' of features that vendors could introduce at their discretion to reach a higher conformance level..

The SQL-92 entry level standard (ANSI SQL) *probably* works across all databases, and coexists with proprietary enhancements, which provide competitive advantage. There is no independent certification protocol for compliance with any of the published SQL standards, including the latest standard, SQL:2003.

14.1 SQL statements

An SQL statement is a string; although its formatting does not have any bearing on the results it produces, it is customary to observe certain conventions. These are:

• Terminate SQL statements with a semi-colon, although this is optional with some SQL dialects. In the case of providers that support multiple results sets, each individual SQL must terminate with a semi-colon. Each SQL statement returns a different set of records.

Enclose strings within the SQL expression within single quotation marks, although some dialects also allow double quotes. Within a programming environment, all SQL statements are themselves enclosed within quotation marks as they are strings. As APL permits single and double quotation marks interchangeably, enclose the whole expression within double quotation marks and the embedded substrings within single quotation marks in order to avoid replication of embedded quotation marks.

System Building with APL + Win   A. Askoolum
© 2006 Research Studies Press Limited

● Show every SQL keywords in uppercase and on a new line: this makes the expressions more readable, especially in documentation.

● If an SQL statement contains a string that has an embedded quotation mark, replicate the embedded quotation mark. For example, in order to select the record for someone named M'Shea, the SQL expression is "SELECT * FROM TABLE WHERE NAME = 'M" Shea'".

● Some dialects of SQL are case sensitive when dealing with string values.

14.2 SQL prime culprits

Applications designed to work with a variety of data sources require special care with SQL. Some dialects of SQL insist on the terminating semi-colon while others are indifferent. It is good practice to include the semi-colon always; however, it is an error to add semi-colon to a sub query—a query embedded in another query. Vendor-specific SQL extensions that do not have a corresponding construction in other SQL dialects might imply that it may be impossible to derive the same results in all data sources.

14.2.1 Handling NULL values

The *null* data value is the prime culprit for runtime SQL errors; it affects virtually all SQL statements. A null value is a missing value and is not equal to any other value, not even another null value. Test for null values with the 'IS' or 'IS NOT' keywords: no comparison operators may be used. The operator 'IS' is followed by a space.

Null values are a constant source of unexpected runtime errors. In VB, 'Error 9 Invalid use of null', domain error in APL, simply aborts an application. Unlike VB, APL does not have a 'IsNull' function that detects a null value; APL does not have a null data type.

SQL Server also has a keyword 'ISNULL'; this permits the arbitrary replacement of null values at runtime. This is one word, not be confused with the SQL 'IS NULL' where the two words are separated. With advance knowledge of a column's data type, an SQL statement can replace any null value by another appropriate value. Consider this example:

```
SELECT   ISNULL(STATE,'**') AS STATE,
         CITY
    FROM  PUBLISHERS;
```

Figure 14-1 SQL Server ISNULL shows the result: null values are replaced by '**'.
Note:

● The column to which 'ISNULL' is applied must be named: usually, the original name is reused.

● The replacement value must be consistent with the type and width of the column: in this case, the column 'state' will not accept more than 2 characters.

	state	city
1	MA	Boston
2	DC	Washington
3	CA	Berkeley
4	IL	Chicago
5	TX	Dallas
6	**	München
7	NY	New York
8	**	Paris

Figure 14-1 SQL Server ISNULL

Oracle's equivalent function is NVL:

```
SELECT   to_char(DOB,'yyyymmdd'),
         NVL(SURNAME,'*'
    FROM  BONUS;
```

Unless a column is explicitly named, Oracle uses the actual expression used to calculate it as its name.

Refer to the columns names in **Figure 14-2 Oracle NVL**.

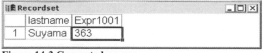

	TO_CHAR(DOB,'YYYYMMDD')	NVL(SURNAM,'*')
1	19360110	JORDAN
2	19500810	PEARCE
3	19500810	WILLIAMSON
4	19430723	TURNBULL
5	19470529	HUSSAIN

Figure 14-2 Oracle NVL function

14.2.2 SQL convention for column names

By default, the names of columns of data extracted with SQL are identical to the names at source, that is, on the database. The SQL dialect assigns an arbitrary name to any unnamed calculated column returned by an SQL statement. Column names in the database or those assigned by an SQL dialect may be problematic for APL because of a variety of reasons:

• The names may not be valid APL variable names.

• The name of a column may depend on its position in an SQL statement; this may easily change and cause errors in an application that uses the generated names.

• If SQL retrieves data for reports, the names may be cryptic or not descriptive enough for that purpose.

All columns returned by an SQL statement have a name: this may be the name at source, a name applied in the SQL statement, or one assigned by the SQL engine. The naming applied in an SQL statement does not change the names in the underlying data source. For SQL, a name that contains a space is valid if specified within square brackets. APL does not allow space in the name of any variable. The following SQL calculates the number of months between 'Birthdate' and 'Hiredate', without naming the result:

```
SELECT  Lastname,
        Datediff("m",Birthdate,HireDate)
  FROM  Employees
 WHERE  Lastname="Suyama";
```

Microsoft SQL Server uses the prefix 'Expr' followed by a three-digit sequential number starting with 1.

	lastname	Expr1001
1	Suyama	363

Figure 14-3 Generated names

The driver has assigned the name 'Expr1001' as the name of this column, see **Figure 14-3 Generated names**.

The naming convention increments the suffix '001' serially for unnamed columns in the same query. The **Excel** ODBC driver uses F followed by a number as the names of a column whose name is not explicitly specified.

14.2.3 SQL comments

Real world applications often include long and complicated SQL statements. From the point of view of maintenance, correctly formatted SQLs are essential for readability. Although SQL statements are always character vectors, it is also possible to add comments to the body of the SQL statement; these serve as in-line documentation. The conventions for in-line SQL comments depend on the SQL dialect. Most dialects support comments enclosed within /* and */ anywhere within the SQL statement but before the terminating semi-colon. The /* and */ pair may also include line feed characters thereby permitting multi-line comments. Besides the same conventions as for in-line comments being applicable, comments within

SQL statements must heed any limits that the SQL dialect imposes on the overall length of the whole SQL statement. In practice, the length of an SQL statement is not a constraint. Consider the following example:

```
SELECT   COUNTRY,
         SUM(A.TOTPUBS) AS TOTPUBS
   FROM  (/* SUB QUERY 2 */
SELECT   A.COUNTRY,
         SUM(A.NUMPUBS) AS TOTPUBS
   FROM   (/*SUB QUERY 1 */
SELECT   COUNTRY,
         1 AS NUMPUBS
   FROM  PUBLISHERS) AS A
GROUP BY  A.COUNTRY,
         A.NUMPUBS) AS A
GROUP BY  A.COUNTRY,
         A.TOTPUBS HAVING A.TOTPUBS >1;
```

SQL Server also allows -- (double dash) terminated by line feed (⎕tcnl) as comment lines.

```
SELECT   A.COUNTRY, SUM(A.NUMPUBS) AS TOTPUBS
   FROM  (--SUB QUERY, ADD A COUNT COLUMN
SELECT   COUNTRY, 1 AS NUMPUBS
   FROM  PUBLISHERS) AS A
GROUP BY  A.COUNTRY,A.NUMPUBS;
```

It is important to observe the convention for in-line comments; for an application supporting multiple data sources, this may imply several versions of a given SQL where the sole difference is the comments.

14.3 APL and SQL

Where would APL be today if it provided integral SQL support from the outset? Using ActiveX Data Object (ADO), APL has full access to SQL and all SQL compliant data sources; ADO also supports text files but not APL component files. Nonetheless, some issues require particular attention. Notoriously, APL does not have a date data type. The SQL comparison operators used with the WHERE clause are shown in **Table 14-1 SQL comparison operators**.

APL+Win	SQL	Applies to	Description
n = m	n = m	String, numbers and dates	Equals
n ≠ m	n <> m or n != m	String, numbers and dates	Is not equal
n < m	n < m	Numbers and dates	Is less than
~ n < m	n !< m	Numbers and dates	Is not less than
n > m	N > m	Numbers and dates	Is greater than
n ≤ m	n <= m	Numbers and dates	Is not greater than
n > m	N > m	Numbers and dates	Is greater than
n ≥ m	n >= m	Numbers and dates	Is not less than
~ n > m	n !> m	Numbers and dates	Is not greater than

Table 14-1 SQL comparison operators

SQL has close affinity with APL in the sense that it does not require loops for the calculation of results or for finding column totals. Predictably, SQL does not understand the APL comparison operators; use the **VB** operators instead. Moreover, SQL has the operators IS NULL and IS NOT NULL for testing whether a value is missing or not. An important difference is that the APL operators do not apply to dates whereas the SQL ones do, where indicated.

14.3.1 Coping with SQL variations

If a driver or provider is capable of performing the operation, the OpenSchema method of an ADO object returns details of the SQL supported by the provider. Consider this function:

```
     ∇ ADOSQLLanguages;⎕wself
[1]    ⍝ System Building with APL+Win
[2]    ⎕wself←'ADO' ⎕wi 'Create' 'ADODB.Connection'
[3]    ⎕wi 'XOpen' 'Provider=Microsoft.Jet.OLEDB.4.0;Data
          Source=c:\nwind.mdb;'
[4]    ⎕wi 'XOpenSchema>ADORS' (⎕wi '=adSchemaSQLLanguages')
     ∇
```

This fails with the JET provider: it does not support the OpenSchema method.

```
⎕WI ERROR: ADODB.Connection exception 800A0CB3 Object or provider is not
capable of performing requested operation.
```

Had it worked, the object ADORS would contain the information shown in **Table 14-2 SQL language**. The first column has a prefix SQLLANGUAGE.

Column name	Type indicator	Description
SOURCE	DBTYPE_WSTR	Should be "ISO 9075" for standard SQL.
YEAR	DBTYPE_WSTR	Should be "1992" for ANSI SQL-92-compliant SQL.
CONFORMANCE	DBTYPE_WSTR	One of the following: ENTRY, INTERMEDIATE, or FULL
_INTEGRITY	DBTYPE_WSTR	YES—Optional integrity feature is supported. NO—Optional integrity feature is not supported.
IMPLEMENTATION	DBTYPE_WSTR	NULL for "ISO 9075" implementation.
BINDING_STYLE	DBTYPE_WSTR	DIRECT for C/C++ callable direct execution of SQL.
PROGRAMMING LANGUAGE	DBTYPE_WSTR	NULL.

Table 14-2 SQL language

14.3.2 Using DMBS properties

All open connection objects have two extended properties, 'DBMS Name' and 'DBMS Version', which provide the means for coping with SQL variations. Query these properties for a given connection as shown in **Table 14-3 Runtime properties**.

APL+Win	'ADO' ⎕wi 'xProperties().Value' 'DBMS Name'
	'ADO' ⎕wi 'xProperties().Value' 'DBMS Version'
APL/W	(ADO.Properties.Item 'DBMS Name').Value
	(ADO.Properties.Item 'DBMS Version').Value

Table 14-3 Runtime properties

Usually, the version of the driver or provider is not an issue with SQL expressions. A cover APL function can return the correct SQL statement—if it is included in the function in the first place. Assume that an application executes two SQL statements, 'Member' and 'Sales', and that it supports text files, **Excel** ODBC driver, **Access** ODBC driver, JET provider, Oracle and Microsoft SQL Server as data sources. Given the connection object and the type of SQL required, this function returns the correct SQL within context:

```
      ∇ Z←Type GetSQL Connection;⎕wself
 [1]    ⍝ System Building with APL+Win
 [2]    ⎕wself←Connection
{1[3]   :select ⎕wi 'xProperties("DMBS Name").Value'
★ [4]      :case 'TEXT'
 [5]         ⍝ e.g. 'Provider=MSDASQL;Driver={Microsoft Text Driver
          (*.txt; *.csv)};DBQ=c:\;'
```

```
 {2[6]            :select Type
★ [7]                 :case 'Member'
  [8]                      Z←
★ [9]                 :case 'Sales'
  [10]                     Z←
+ [11]                :else
  [12]                     Z←0ρMsg 'Unable to recognise type for ',⎕wi
        'xProperties("DMBS Name").Value'
2}[13]            :endselect
★ [14]       :case 'EXCEL'
  [15]          ⍝ e.g. 'Provider=MSDASQL;Driver={Microsoft Excel Driver
        (*.xls)};DBQ=c:\Adoconn.xls;'
 {2[16]           :select Type
★ [17]                :case 'Member'
  [18]                     Z←
★ [19]                :case 'Sales'
  [20]                     Z←
+ [21]                :else
  [22]                     Z←0ρMsg 'Unable to recognise type for ',⎕wi
        'xProperties("DMBS Name").Value'
2}[23]            :endselect
★ [24]       :case 'ACCESS'
  [25]          ⍝ e.g. 'Provider=MSDASQL;Driver={Microsoft Access Driver
        (*.mdb)};DBQ=c:\nwind.mdb;'
 {2[26]           :select Type
★ [27]                :case 'Member'
  [28]                     Z←
★ [29]                :case 'Sales'
  [30]                     Z←
+ [31]                :else
  [32]                     Z←0ρMsg 'Unable to recognise type for ',⎕wi
        'xProperties("DMBS Name").Value'
2}[33]            :endselect
★ [34]       :case 'MS Jet'
  [35]          ⍝ e.g. 'Provider=Microsoft.Jet.OLEDB.4.0;Data
        Source=c:\nwind.mdb;'
 {2[36]           :select Type
★ [37]                :case 'Member'
  [38]                     Z←
★ [39]                :case 'Sales'
  [40]                     Z←
+ [41]                :else
  [42]                     Z←0ρMsg 'Unable to recognise type for ',⎕wi
        'xProperties("DMBS Name").Value'
2}[43]            :endselect
★ [44]       :case 'Oracle'
  [45]          ⍝ e.g. Provider=MSDAORA;Data Source=SALES;User
        ID=MAILSHOT; Password=sa;
 {2[46]           :select Type
★ [47]                :case 'Member'
  [48]                     Z←
★ [49]                :case 'Sales'
  [50]                     Z←
+ [51]                :else
```

```
  [52]                   Z←0ρMsg 'Unable to recognise type for ',⎕wi
         'xProperties("DMBS Name").Value'
2}[53]           :endselect
★ [54]        :case 'Microsoft SQL Server'
  [55]            ⍝ e.g. Provider=SQLOLEDB;Data Source=sql2000;Initial
         Catalog=sysdat;User Id=sysdat;Password=sa;
{2[56]           :select Type
★ [57]              :case 'Member'
  [58]                 Z←
★ [59]              :case 'Sales'
  [60]                 Z←
+ [61]              :else
  [62]                 Z←0ρMsg 'Unable to recognise type for ',⎕wi
         'xProperties("DMBS Name").Value'
2}[63]           :endselect
+ [64]        :else
  [65]            Z←0ρMsg 'Unable to recognise ',⎕wi 'xProperties("DMBS
         Name").Value'
1}[66]    :endselect
        ∇
```

The incomplete expression Z,, in the function is deliberate—it is not the intention to provide working SQL statements in this context. The developer may adopt several strategies for holding the SQL statements:

• Hard code them in a function and are inaccessible at runtime. Any revisions to the SQL statement will require the distribution of a workspace or a component file containing the function—depending on where the function was stored.

• Write them to a dedicated APL+Win component file, as data, and retrieve them at runtime. The statements are still inaccessible, but this approach has the advantage that revisions simply require the distribution of the component file.

• Write them to a dedicated table in the active connection. Although the table is not directly exposed, it is accessible from outside the application. Any deliberate changes to the SQL statements will require the database administrator at the client site to update the table.

Some examples of using the GetSQL function to retrieve particular SQL statements are:

```
'ADO' ⎕wi 'XExecute>ADO.Recordset' ('Market' GetSQL 'ADO')
'ADORS' ⎕wi 'XOpen' ('Market' GetSQL 'ADO') ⍝ ConnectionString
```

The object ADO.Recordset or ADORS will contain the records returned by the SQL statement returned by GetSQL. Should it be necessary to use provider or driver dependent versions of SQL statements, the function GetSQL may be adapted. The following code illustrates the principles for the SQL Server provider:

```
★ [53]        :case 'Microsoft SQL Server'
  [54]            ⍝ e.g. Provider=SQLOLEDB;Data Source=sql2000;Initial
         Catalog=sysdat;User Id=sysdat;Password=sa;
{2[55]           :select Type
★ [56]              :case 'Member'
{3[57]                 :select ⎕wi 'xProperties("DMBS Version").Value'
★ [58]                    :case '07.00.0842'
  [59]                       Z←
★ [60]                    :case '08.00.0194'
  [61]                       Z←
+ [62]                    :else
```

```
   [63]                        Z←0ρMsg 'Unable to recognise version ',⎕wi
          'xProperties("DMBS Version").Value'
3}[64]                    :endselect
★ [65]              :case 'Sales'
{3[66]                  :select ⎕wi 'xProperties("DMBS Version").Value'
★ [67]                     :case '07.00.0842'
   [68]                     Z←
★ [69]                     :case '08.00.0194'
   [70]                     Z←
+ [71]                     :else
   [72]                        Z←0ρMsg 'Unable to recognise version ',⎕wi
          'xProperties("DMBS Version").Value'
3}[73]                    :endselect
```

In practice, it will be impossible to enumerate all the version numbers.

14.3.3 ANSI SQL

It is tempting to conclude that an application especially one that can use several data sources should restrict SQL statements to the ANSI SQL standard. Consider a simple example: today's date is required on its own or as a column in a record set containing other information. However, vendor specific differences usually spoil this strategy, as seen below:

```
     ∇ Z←GetTimeStamp R;⎕wself;Cnn
   [1]   ⍝ System Building with APL+Win
   [2]   ⎕wself←'ADOR' ⎕wi 'Create' 'ADODB.RecordSet'
{1[3]  :select R
★ [4]      :case 'TEXT'
   [5]         Cnn←'Provider=MSDASQL;Driver={Microsoft Text Driver
          (*.txt; *.csv)};DBQ=;'
   [6]         ⎕wi 'XOpen' 'Select Now as Today' Cnn
★ [7]      :case 'EXCEL'
   [8]         Cnn←'Provider=MSDASQL;Driver={Microsoft Excel Driver
          (*.xls)};DBQ=C:\;'
   [9]         ⎕wi 'XOpen' 'Select Now as Today' Cnn
★ [10]     :case 'ACCESS'
   [11]        Cnn←'Provider=MSDASQL;Driver={Microsoft Access Driver
          (*.mdb)};DBQ=C:\NWIND.MDB;'
   [12]        ⎕wi 'XOpen' 'Select Now as Today' Cnn
★ [13]     :case 'MS Jet'
   [14]        Cnn←'Provider=Microsoft.Jet.OLEDB.4.0;Data
          Source=c:\nwind.mdb;'
   [15]        ⎕wi 'XOpen' 'Select Now as Today' Cnn
★ [16]     :case 'Oracle'
   [17]        Cnn←'Provider=MSDAORA;Data Source=SALES;User ID-MAILSHOT;
          Password=sa;'
   [18]        ⎕wi 'Xopen' 'Select SysDate from Dual as Today'
★ [19]     :case 'Microsoft SQL Server'
   [20]        Cnn←'Driver={SQL Server};Server=ajaylap;Database=;'
   [21]        ⎕wi 'XOpen' 'Select GetDate() as Today' Cnn
1}[22]  :endselect
   [23]  Z←⎕wi 'GetString'
   [24]  'ADOR' ⎕wi 'Delete'
     ∇
```

• The text, **Excel**, **Access**, and JET drivers/providers return the timestamp using the same SQL statement: the format of the timestamp is as specified in the Short Date and Time Format specified in regional settings.

• The ORACLE driver requires a completely different SQL statement that includes the FROM clause; Dual is a pseudo system table.

• The Microsoft SQL Server driver also requires a different SQL statement that does not require the FROM clause.

• There are other differences in the specification of the DBQ or Data Source parameter in the connection string.

The information returned, with the short date format set to yyyy-mm-dd on this computer, is:

```
2003-06-01 10:19:03
```

14.3.4 Date/Time handling in data sources

Databases store dates as a number whose integral part represents the number of days from a reference date and a fractional part that represents elapsed time as a portion of a day. A negative integral part will represent a date earlier than the reference date. The reference date varies from data source to data source.

The specification of literal dates in SQL statements may lead to unexpected results in a variety of circumstances. For example, 01/12/2000, can be either December1, 2000 or January 12, 2000. If the SQL dialect permits, reference to the month by the first three letters, thus, 01 Dec 2000 or 12 Jan 2000. By default, the format specified in the connection options of the server determines the format of dates; usually, this option refers to the regional timestamp (that is, date and time settings) settings.

14.3.4.1 Date formats

Were all dates to be in the format specified in the regional settings, APL will have a consistent way of dealing with dates, without recourse to an explicit data format. An alternative approach is to use the ISO date format yyyymmdd. If the dates arrive in the workspace as character, they are easily convertible to numbers and split any desired data format. Consider this example:

```
    □←a←⊃'19991203' '20030103' ⍝ Simulate a character array of ISO dates
19991203
20030103
```

Conversion of a character array of dates into the UK default format dd mm yyyy is straightforward:

```
    a←⍒⍴10000 100 100⊤⊃∊□fi ¨⊂[□io+1]a
 3 12 1999
 3  1 2003
```

14.3.4.2 ISO date format

An application destined for internationally use needs to recognise regional settings at runtime. The GetLocaleInfo API call can return the locale data format. Time is not usually relevant in date arithmetic. Establish the sort order of day, month, and year as follows:

```
    ⍋'ymd'⍳↑¨(b∊'ymd')⊂b
3 2 1
```

The following expression returns ISO dates (yyyy mm dd) in line with UK regional settings (dd mm yyyy) as follows:

```
    a[;⍋'dmy'⍳↑¨(b∊'dmy')⊂b]
 3 12 1999
 3  1 2003
```

The Microsoft SQL Server driver can return the date format applicable to the UK, irrespective of the regional settings with slight modification to the code. Try:

```
★ [19]      :case 'Microsoft SQL Server'
```

```
[20]              Cnn←'Driver={SQL Server};Server=ajaylap;Database=;'
[21]              ⎕wi 'XOpen' 'Select Convert(Char,GetDate(),103) as Today'
      Cnn
```

The third argument of the Convert function, 103, specifies the UK default date format. Specify 3, that is, 100 modulus the UK default date format value or 103 to show the year as two digits i.e. to drop the century from the year. If an argument of 112 is used, the ISO date format, yyyymmdd, is returned without the time.

In order to coerce the text, **Excel**, **Access**, and JET drivers/providers to use the timestamp format specified in regional settings, use the keyword *Format*—this rearranges the yyyy, mm, and dd values in the same order as the regional format:

```
★ [4]            :case 'TEXT'
  [5]            Cnn←'Provider=MSDASQL;Driver={Microsoft Text Driver
      (*.txt; *.csv)};DBQ=;'
  [6]            ⎕wi 'XOpen' "Select Format(Now) as Today" Cnn
```

The keyword *Format* also allows a second argument representing the data format: for instance, use yyyymmdd for ISO format:

```
  [6]            ⎕wi 'XOpen' "Select Format(Now,'yyyymmdd') as Today" Cnn
```

Note that an alternative construction swapping the quotation marks does not work with SQL statements—it fails at execution time:

```
  [6]    ⎕wi 'XOpen' 'Select format(Now,"yyyymmdd") as Today' Cnn
```

Predictably, the Oracle syntax is different:

```
    ⎕wi 'Xopen' "Select to_char(SysDate,'yyyymmdd') from Dual as Today"
```

Whether it is better to use SQL to retrieve the data that is required than to use ANSI SQL to extract the available data and work it within APL depends on the context. For example, it is simpler and more accurate to format a date in a regional format and return it to APL than to return the raw date value and format the date in APL+Win. The SQL view of the 'Current Product List' defined in NWIND.MDB is:

```
    SELECT   [Product List].ProductID, [Product List].ProductName
      FROM   Products AS [Product List]
     WHERE   ((([Product List].Discontinued)=No))
  ORDER BY   [Product List].ProductName;
```

Enclose names containing a space in square brackets; 'Product List', which is a temporary name, or alias, for the table **Products**, is enclosed in square brackets. In the second line of the query, 'AS [Products List]' establishes a temporary name. A reference to a column name in a table is specified as the table name followed by dot and then by the column name; square brackets may be used for both the dot notation and column name.

The Microsoft text driver names columns as F1, F2, etc., when using a SCHEMA.INI file. Since the text driver copes with a maximum of 256 columns, the driver assigned names will be in the range F1 to F256.

14.4 Learning SQL

An SQL statement is a character vector, albeit one delimited by linefeed. The data query language is easy to learn by example. For APL, SQL represents a 'graceful' way of making data available in the workspace beyond the constraints of the size of the workspace. Process each record retrieved in a record set object in a loop: the record set object resides outside the workspace. SQL complexity arises on two accounts, namely:

● Simple column selections are straightforward. The complexity of an SQL expression increases with the extent of data processing included in it.

● The differences in syntax of the SQL implementations of different vendors require an application to maintain several versions of SQL expressions to retrieve the same result.

14.4.1 The SQL Data Query Language

The DQL component of SQL is based solely on the keyword SELECT and as the name implies, it uses the data on a read only basis. This is also the component that is used most and with the most variation across SQL implementations. An application that uses a database relied on the SQL DQL for all access to the data. In order to explore SELECT,

• **Figure 14-4 AUTHORS table** shows the table AUTHORS from the SQL Server PUBS database.

• All SQL statements are assigned to an APL variable, sql, which is edited in the APL editor—linefeed characters are added by pressing the ENTER key.

• An ADO record set object, ADOR, is used to execute the SQL statements. The connection string and code to create the object are:

```
Cnn←'Driver={SQL Server};Server=ajaylap;Database=pubs'
'ADOR' ⎕wi 'Create' 'ADODB.Recordset'
```

• If the result returns a single row, the GetRows method is used to view the result or the xRecordCount property is used to query the number of record; where several rows are returned, the result is shown in the APL+Win grid object. The code is:

```
      ∇ APLGridDisplay R;⎕wself
 [1]    ⍝ System Building with APL+Win
 [2]    ⍝ R = Name of recordset object
 [3]    ⎕wself←'f1' ⎕wi 'Create' 'Form'  ('caption' 'Recordset')
 [4]    ⎕wself←'.gd' ⎕wi 'New' 'APL.Grid' ('where' (0 0,⎕wi 'size'))
 [5]    ⎕wi 'xFontSize' 0 0 15
 [6]    ⎕wself←R
 [7]    ⎕wi 'XMoveFirst'
 [8]    R←(⍳⎕wi 'xFields.Count')-⎕io
 [9]    R←((⎕wi 'xFields') ⎕wi ¨(⊂⊂'xItem'),¨R) ⎕wi ¨⊂'xName'
 [10]   'f1.gd' ⎕wi 'xRows' (⎕wi 'xRecordCount')
 [11]   'f1.gd' ⎕wi 'xCols' (⍴R)
 [12]   'f1.gd' ⎕wi 'xText' ¯1 (⍳⍴R) R
 [13]   R←1
{1[14]  :while ~⎕wi 'XEOF'
 [15]       'f1.gd' ⎕wi 'xText' R ¯1 (⍕R)
 [16]       'f1.gd' ⎕wi 'xText' R (⍳'f1.gd' ⎕wi 'xCols') (⍕¨⎕wi 'GetRows'
        1)
 [17]          R←R+1
1}[18]  :endwhile
 [19]   'f1.gd' ⎕wi 'XFitCol' (⍳9)
 [20]   ⎕wself←'f1'
 [21]   ⎕wi 'onShow' '''Set'' ReSize ⎕wself'
 [22]   ⎕wi 'Wait'
      ∇
```

The right-hand argument is the name of the record set object in quotation marks. This function acquires the basic data from a record set object (held outside of the workspace) row by row into the workspace and writes them to the APL Grid object (also held outside of the workspace). The row-by-row loop is necessary in order to minimize the incidence of 'Workspace Full' runtime errors. A number of examples in this chapter use the AUTHORS table from the SQL Server PUBS database; **Figure 14-4 AUTHORS table** shows its contents.

au_id	au_lname	au_fname	phone	address	city	state	zip	contract
172-32-1176	White	Johnson	408 496-7223	10932 Bigge Rd	Menlo Park	CA	94025	True
213-46-8915	Green	Marjorie	415 986-7020	309 63rd St. #4	Oakland	CA	94618	True
238-95-7766	Carson	Cheryl	415 548-7723	589 Darwin Ln.	Berkeley	CA	94705	True
267-41-2394	O'Leary	Michael	408 286-2428	22 Cleveland A·	San Jose	CA	95128	True
274-80-9391	Straight	Dean	415 834-2919	5420 College A·	Oakland	CA	94609	True
341-22-1782	Smith	Meander	913 843-0462	10 Mississippi D	Lawrence	KS	66044	False
409-56-7008	Bennet	Abraham	415 658-9932	6223 Bateman ;	Berkeley	CA	94705	True
427-17-2319	Dull	Ann	415 836-7128	3410 Blonde St	Palo Alto	CA	94301	True
472-27-2349	Gringlesby	Burt	707 938-6445	PO Box 792	Covelo	CA	95428	True
486-29-1786	Locksley	Charlene	415 585-4620	18 Broadway A·	San Francis	CA	94130	True
527-72-3246	Greene	Morningstar	615 297-2723	22 Graybar Hou	Nashville	TN	37215	False
648-92-1872	Blotchet-Halls	Reginald	503 745-6402	55 Hillsdale Bl.	Corvallis	OR	97330	True
672-71-3249	Yokomoto	Akiko	415 935-4228	3 Silver Ct.	Walnut Cre	CA	94595	True
712-45-1867	del Castillo	Innes	615 996-8275	2286 Cram Pl. ;	Ann Arbor	MI	48105	True
722-51-5454	DeFrance	Michel	219 547-9982	3 Balding Pl.	Gary	IN	46403	True
724-08-9931	Stringer	Dirk	415 843-2991	5420 Telegraph	Oakland	CA	94609	False
724-80-9391	MacFeather	Stearns	415 354-7128	44 Upland Hts.	Oakland	CA	94612	True
756-30-7391	Karsen	Livia	415 534-9219	5720 McAuley S	Oakland	CA	94609	True
807-91-6654	Panteley	Sylvia	301 946-8853	1956 Arlington I	Rockville	MD	20853	True
846-92-7186	Hunter	Sheryl	415 836-7128	3410 Blonde St	Palo Alto	CA	94301	True
893-72-1158	McBadden	Heather	707 448-4982	301 Putnam	Vacaville	CA	95688	False
899-46-2035	Ringer	Anne	801 826-0752	67 Seventh Av.	Salt Lake C	UT	84152	True
998-72-3567	Ringer	Albert	801 826-0752	67 Seventh Av.	Salt Lake C	UT	84152	True
								False

Record: [14] [4]   23 [▶] [▶I] [▶*] [◄][▶!.] of 23

Figure 14-4 AUTHORS table

14.4.1.1 Simple Select

The simplest of SQL statements comprise of just two keywords:

```
SELECT   *
   FROM   TABLENAME;
```

This statement selects all columns from the table *TableName* as they occur in the database. If a subset of columns is required or if the columns are required in a different order—perhaps for presentational purposes—enumerate all the column names in the desired order.

14.4.1.2 Counting the number of records

```
SELECT   COUNT(*)
   FROM   AUTHORS;
```

This statement returns the number of records in the table. This can be useful when the record set object reports that its xRecordCount property is indeterminate.

14.4.1.3 Sorting records using column names

```
SELECT   *
   FROM   AUTHORS
ORDER BY   STATE, AU_LNAME ASC;
```

This statement returns all records in ascending order of the State and au_lname column. If the reverse sorting order is requires, use the DESC qualifier instead.

14.4.1.4 Sorting records using column position

```
SELECT   *
   FROM   AUTHORS
ORDER BY   7 DESC,2 DESC;
```

The column position is in index origin 1 and is the position in the SELECT clause and NOT the position in the table specified in FROM.

14.4.1.5 DISTINCT rows

The distinct clause returns unique rows from one or more named columns. This example creates a table, from which it returns the unique ID column:

```
SELECT   DISTINCT(ID)
   FROM   (
           (SELECT 100 AS ID,'ER' AS TAG)
UNION ALL
           (SELECT 100 AS ID,'ER' AS TAG)
          )A
```

All columns is specified as * without round brackets, unlike the COUNT clause where all records is specified as (*). The Microsoft ODBC text driver and the JET provider has a DISTINCTROW * clause for the same purpose.

14.4.1.6 SORT order

The sort order is ASC, short for ASCENDING, by default. The reverse order is DESC, short for DESCENDING. The keyword applies only to the column name immediately preceding it. Thus, in the previous example, the table AUTHORS is returned in descending order of state and within it in ascending order of au_lname; no sort direction is specified for au_lname. Unless the default sort direction is required, specify the order for each column:

```
SELECT    *
   FROM   AUTHORS
ORDER BY  7 DESC,2 DESC;
```

14.4.1.7 Renaming/Adding columns

An SQL statement may rename a column in an existing table in its result; depending on the context, it may be appropriate to rename a column for clarity. This does not change the name of the column at source:

```
SELECT   AU_FNAME, 100 AS ARBITRARY, 50 AS [NEW COLUMN]
   FROM  AUTHORS
ORDER BY STATE, AU_LNAME ASC;
```

• This query returns one existing column, two arbitrary or calculated columns; the latter has a long name (invalid APL variable name) and is enclosed in square brackets.

• The keyword AS is used to specify the name or new name. If a calculated column is unnamed, it is named in accordance with the conventions used by the driver or provider.

• Other columns in the table may be used to sort the selected columns even though they are not included in the selection.

14.4.1.8 Conditional selection

Conditional selection of data is really the essence of working with relational data sources. It allows the selection of a subset of data from a source:

```
SELECT   AU_FNAME, 100 AS ARBITRARY, 50 AS [NEW COLUMN]
   FROM  AUTHORS
  WHERE  STATE IN('CA','TN')
         OR AU_LNAME LIKE 'SM%'
         OR (AU_LNAME LIKE 'GR_EN'
         AND CONTRACT = 1)
         AND AU_ID IS NOT NULL
         AND AU_FNAME BETWEEN 'ANN' AND 'DIRK'
         OR AU_FNAME LIKE '_[HMN]%'
ORDER BY STATE, AU_LNAME ASC;
```

Use the WHERE clause to specify simple or complex conditions for record selection:

• The IN keyword concisely specifies a numeric or literal list with each item being separated by comma; the effect is the same as (state = 'CA' OR state = 'TN'). The keyword NOT can precede the IN keyword—this will exclude items in the list.

• The OR, AND, and NOT keywords are like the APL+Win ∨, ∧ and ∼.

• The LIKE keyword allows wildcard selection; in this example, any au_lname beginning with SM is selected. If the condition were '%SM', any au_lname ending with SM is selected. If the condition were 'S%M', any au_lname starting with S and ending with M is selected. That is, % specifies any number of characters at the start or middle or end of a character column. Some SQL implementation use * like %.

• The symbol _ specifies a single character substitution; this may be repeated to specify multiple single characters. Some SQL implementations use ? for the same purpose.

• The square brackets specify a single character substitution from the list of characters within the square brackets. In this example, au_fname that has *h*, *m* or *n* in the second column is selected. The characters are matched irrespective of case. In order to exclude matches at the second column with any of the characters, h, m or n, specify a caret as the first character within the square brackets, [^hmn].

• The IS keyword is used when a comparison involves a NULL value. A NULL value is a missing value and not two NULL values are equal.

• The BETWEEN … AND construction allows the specification of a numeric or literal range. The NOT BETWEEN … AND construction excludes the range specified.

• Use round brackets to control the execution of the WHERE statement; the expression(s) within round brackets are evaluated first as intermediate results.

14.4.1.9 SOUNDEX selection

Both SQL Server and Oracle support the SOUNDEX keyword. SOUNDEX is a four character phonetic code of a given literal: it comprises of the initial letter of the literal followed by three digits. It enables a more general phonetic match of character columns:

```
    SELECT   AU_FNAME
      FROM   AUTHORS
     WHERE   SOUNDEX(AU_FNAME) LIKE 'A__0'
  ORDER BY   STATE, AU_LNAME ASC;
```

An APL function to generate SOUNDEX code for character literals is:

```
     ∇ Z←Soundex R
[1]    ⍝ System Building with APL+Win
[2]    Z←16 2 8 10 16 2 4 14 2 16 4 14 2 16/14⍴1 0
[3]    Z←Z\'AaEeHhIiOoUuWwYyBbFfPpVvCcGgJjKkQqSsXxZzDdTtLlMmNnRr'
[4]    Z←⌊|(Z⍳R)÷⁻18
[5]    R←1↑[(⍴⍴R)-~⎕io]R
[6]    Z←Z×Z≤6
[7]    Z←1↓[(⍴⍴Z)-~⎕io]Z×Z≠⁻1⌽Z
[8]    Z←R,'0123456'[⎕io+⊃3↑¨(⊂[(⍴⍴Z)-~⎕io]Z)~¨'0]
     ∇
```

This function produces the same codes as SQL Server SQL with these exceptions:

• The SQL Server dialect of SQL, also called Transact-SQL or T-SQL, does not discard two adjacent numeric codes that are identical whereas the APL+Win function does; as a result the code for Pfeister are P123 and P236 respectively.

• Transact-SQL discards all characters after the first invalid character in column two onwards but the APL+Win function does not.

• The APL function keeps the first letter as given—it is not turned to uppercase and is not discarded if it is not a letter of the alphabet.

An example of using the SOUNDEX function in an SQL statement is:

```
    SELECT   *
      FROM   AUTHORS
     WHERE   SOUNDEX(au_fname) = SOUNDEX('Smith')
```

In practice, the anomalies in different computations of the code are of no consequence in the same given environment. If the first letter must be uppercase, use the API call to transform the right-hand argument to uppercase before calling the function:

```
    Soundex ⎕wcall 'CharUpper' 'pfeister'
```

14.4.1.10 Other functions

SQL dialects support a number of functions for manipulating columns within SQL statements; for most part, the names of these functions vary but common equivalent functions are listed in **Table 14-4 Supported SQL string functions.**

	ODBC Text Driver	Jet Provider	SQL Server	Oracle	API Call
String to lower case	Lcase	LCase	LOWER	LOWER	CharLower
String to upper case	Ucase	UCase	UPPER	UPPER	CharUpper
Drop leading blanks	LTRIM	LTRIM	LTRIM	LTRIM	-
Drop trailing blanks	RTRIM	RTRIM	RTRIM	RTRIM	-
Length of string	LEN	LEN	LEN	LENGTH	-
Sub string	MID	MID	SUBSTRING	SUBSTR	-
Truncate at right	LEFT	LEFT	LEFT	-	-
Truncate at left	RIGHT	RIGHT	RIGHT	-	-

Table 14-4 Supported SQL string functions

In the endeavour to produce SQLs that will work with little or no modifications, it is advisable to use the common subset of functions only and to accomplish other functionality with APL code. The keywords SUBSTR, SUBSTRING, LEFT, RIGHT, and MID have a consistent syntax: *keyword(column,start,length)* and use index origin 1. Although Oracle does not support LEFT and RIGHT, the same effect can be achieved with SUBSTR.

The emerging pattern is that each driver and provider has its own set of functions that can be included in SQL statements. The text driver and JET provider allow most VBA functions within SQL statements; most of these also work with SQL Server. Oracle has the to_char, to_date, and to_number functions. For APL applications, a simple guideline for using vendor specific functions is:

• Use vendor specific functions whenever APL simply extracts the data and presents it in tabular form as a report.

• Extract the data using ANSI SQL—one that works across all databases—when it is going to be processed by APL.

There is one exception to these simple rules of thumb: when the data contains dates, use the vendor specific keyword to convert them into the appropriate regional format. **Table 14-5 Vendor SQL date formats** shows some typical formats.

Driver	SQL Statement
ODBC Text Driver	SELECT Format(Date(),'dd/mm/yyyy');
JET Provider	SELECT Format(Date(),'dd/mm/yyyy') ;
SQL Server	SELECT CONVERT(Char(10),GetDate(),103);
Oracle	SELECT(TO_CHAR,SysDate,'DD/MM/YYYY');

Table 14-5 Vendor SQL date formats

These examples return today's date; the names of a date column may be used instead of DATE(), GETDATE(), or SysDate.

14.4.1.11 SQL: Standard aggregate functions

The SELECT statement can use SQL Aggregate functions to return results from columns of a table; these functions are available in most SQL dialects. Most varieties of SQL support the following functions:

```
AVG       Returns the average value of a numeric column.
              SELECT  AVG(PRICE) AS AvgPrice
              FROM   TITLES;
```

This function ignores null values so it is not strictly $(+/,\mathrm{PRICE})\div\rho,\mathrm{PRICE}$; it is equivalent to:

```
SELECT   SUM(PRICE)/COUNT(*)
FROM     TITLES
WHERE    PRICE IS NOT NULL;
```

COUNT   Returns the number of records in a table, irrespective of the value of each cell. The result is NULL if there are no records.

MAX   Returns the maximum numeric value of a column or the last value of a character column when it is sorted in ascending alphabetical order.

```
SELECT   MAX(TITLE_ID) AS ALPHA,
         MAX(PRICE) AS Value
  FROM   TITLES;
```

MIN   Returns the minimum numeric value of a column or the first value of a character column when it is sorted in ascending alphabetical order.

If a column contains only null values, the MAX and MIN functions return NULL for both character and numeric columns.

SUM   Returns the sum of a numeric column.

All these function can be used in conjunction with a WHERE clause and apply only to rows where the condition, if specified, is **TRUE**. Most SQL implementations contain the provision of custom functions that may be included in SQL statements applicable to the vendor's database. In particular:

● Refer to the **Access** help file, search for *Microsoft Jet Database Engine SQL Reserved Words*, for other functions that can be used with the JET provider.

● **Excel** provides environment specific SQL enhancements; refer to *SQL Functions* in the **Excel** Help File.

● SQL Server has comprehensive help on its SQL dialect in its Transact-SQL Help file— also known as T_SQL

● SQL*PLUS, Oracle's equivalent to Query Analyser does not provide help on Oracle's SQL implementation. Oracle also offers PL/SQL, procedural SQL.

14.4.1.12 Grouping columns

List the state with two or more authors with contracts in descending order of state name, using the following:

```
SELECT   STATE,COUNT(*) AS NUMAUTHORS
  FROM   AUTHORS
 WHERE   CONTRACT = 1
GROUP BY STATE
HAVING   COUNT(*)>1
ORDER BY STATE DESC;
```

14.4.1.13 Collating columns by name

It might be desirable to assemble all or a subset of columns from a query in a particular order, perhaps for presentational purposes. Try:

```
SELECT   STATE,AU_FNAME
  FROM   AUTHORS
 WHERE   CONTRACT = 1
```

This is achieved by naming each column in the select clause. Note that the cardinal position of the columns cannot be used, as shown below:

```
SELECT   7,3
  FROM   AUTHORS
 WHERE   CONTRACT = 1
```

In this example, state is the 7th column and au_fname is the 3rd column. However, this creates two arbitrary columns with values 7 and 3 respectively, rather than return the 7th and 3rd columns.

14.4.1.14 Table ALIAS

In order to avoid confusion, it may be necessary specify column names with a prefix that is the name of the table, especially when multiple tables are involved.

Both the tables AUTHORS and TITLEAUTHOR contain a common column, AU_ID. Common columns may be used to link the tables. The following result is sought: the AU_ID and ROYALTYPER columns are required where AU_ORD in TITLEAUTHOR is greater than one.

A verbose composition of the SQL statement is:
```
SELECT   AUTHORS.AU_ID,
         TITLEAUTHOR.ROYALTYPER
  FROM   AUTHORS,
         TITLEAUTHOR
 WHERE   AUTHORS.AU_ID = TITLEAUTHOR.AU_ID
         AND TITLEAUTHOR.AU_ORD>1
```

This can be simplified, or shortened, by giving each table an alias. An alias is a temporary name assigned to a table. The same query using table aliases is:
```
SELECT   a.AU_ID,
         b.ROYALTYPER
  FROM   AUTHORS a,
         TITLEAUTHOR b
 WHERE   a.AU_ID = b.AU_ID
         AND b.AU_ORD>1
```

In this example, the tables AUTHORS and TITLEAUTHOR have aliases *a* and *b*, respectively. The SQL statement is shorter. Usually, a unique single letter is used as an alias if the intention is simply to make the SQL statement compact. However, if a table name contains spaces, an alias is necessary to keep the SQL simple.

For example, if the table TITLEAUTHOR were called TITLE AUTHOR, the verbose SQL statement would have been:
```
SELECT   AUTHORS.AU_ID,
         [TITLE AUTHOR].ROYALTYPER
  FROM   AUTHORS,
         [TITLE AUTHOR]
 WHERE   AUTHORS.AU_ID = [TITLE AUTHOR].AU_ID
         AND [TITLE AUTHOR].AU_ORD>1
```

With aliases for the table name, a simpler SQL statement is possible:
```
SELECT   a.AU_ID,
         b.ROYALTYPER
  FROM   AUTHORS AS a,
         [TITLE AUTHOR] b
 WHERE   a.AU_ID = b.AU_ID
         AND b.AU_ORD>1
```

14.4.1.15 The AS keyword

The optional AS keyword can be used to rename columns and tables; it makes the SQL more readable. However, it can be omitted if brevity of the SQL statement is imperative.

However, a table alias is essential when the same table is referred to more than once in the same SELECT statement. See *14.4.1.16.2 SELF JOIN* below.

14.4.1.16 Joining tables

A FROM statement that has more than one table involves the joining of those tables even when the word JOIN is not present in the SQL statement. In its simplest—perhaps least useful—form, an SQL JOIN produces a Cartesian product: the result has as many columns as the sum of columns in both tables and as many rows as the product of the number of rows in each table, irrespective of whether the tables have common column names. The SQL for a Cartesian product or simple join takes this form:

```
SELECT   *
    FROM   AUTHORS,
             TITLEAUTHOR
```

This type of JOIN is also described as a *cross join*. This APL+Win function creates an audit trail of the SQL:

```
      ∇ Z←SimpleJoin;⎕wself;Cnn;Sql
[1]     ⍝ System Building with APL+Win
[2]     ⎕wself←'∆ADORS' ⎕wi 'Create' 'ADODB.Recordset'
[3]     Cnn←'Driver={SQL Server};Server=AJAY
          ASKOOLUM;Database=pubs;UID=sa;PWD='
```

First, the table AUTHORS is queried:

```
[4]     Sql←'Select * from AUTHORS;'
[5]     Z←⊂Sql
[6]     ⎕wi 'XOpen' Sql Cnn (⎕wi '=adOpenStatic') (⎕wi '=adLockReadOnly')
          (⎕wi '=adCmdText')
[7]     Z←Z,⊂'Table AUTHORS has ',(⍕⎕wi 'xRecordCount'),' rows and ',(⍕⎕wi
          'xFields.Count'),' fields.'
[8]     Z←Z,⊂¯1↓∊(⎕wi ¨(⊂⊂'().Name'),¨⎕←¯1++\(⎕wi 'xFields.Count')/1),¨',¨','
[9]     ⎕wi 'XClose'
```

Second, the table TITLEAUTHOR is queried:

```
[10]    Sql←'Select * from TITLEAUTHOR;'
[11]    Z←Z,⊂Sql
[12]    ⎕wi 'XOpen' Sql Cnn (⎕wi '=adOpenStatic') (⎕wi '=adLockReadOnly')
          (⎕wi '=adCmdText')
[13]    Z←Z,⊂'Table TITLEAUTHOR has ',(⍕⎕wi 'xRecordCount'),' rows and
          ',(⍕⎕wi 'xFields.Count'),' fields.'
[14]    Z←Z,⊂¯1↓∊(⎕wi ¨(⊂⊂'().Name'),¨⎕←¯1++\(⎕wi 'xFields.Count')/1),¨',¨','
[15]    ⎕wi 'XClose'
```

Third, the two tables are joined:

```
[16]    Sql←'Select * from AUTHORS,TITLEAUTHOR;'
[17]    Z←Z,⊂Sql
[18]    ⎕wi 'XOpen' Sql Cnn (⎕wi '=adOpenStatic') (⎕wi '=adLockReadOnly')
          (⎕wi '=adCmdText')
[19]    Z←Z,⊂'Result Table AUTHORS,TITLEAUTHOR has ',(⍕⎕wi 'xRecordCount'),'
          rows and ',(⍕⎕wi 'xFields.Count'),' fields.'
[20]    Z←⊃Z,⊂¯1↓∊(⎕wi ¨(⊂⊂'().Name'),¨⎕←¯1++\(⎕wi 'xFields.Count')/1),¨',¨','
[21]    ⎕wi 'XClose'
      ∇

      SimpleJoin
Select * from AUTHORS;
Table AUTHORS has 23 rows and 9 fields.
au_id,au_lname,au_fname,phone,address,city,state,zip,contract
Select * from TITLEAUTHOR;
Table TITLEAUTHOR has 25 rows and 4 fields.
au_id,title_id,au_ord,royaltyper
```

```
Select * from AUTHORS,TITLEAUTHOR;
Result Table AUTHORS,TITLEAUTHOR has 575 rows and 13 fields.
au_id,au_lname,au_fname,phone,address,city,state,zip,contract,au_id,title_
id,au_ord,royaltyper
```

A simple join is usually of very limited practical value and presents two problems, namely:

● It introduces possible ambiguity about column names and values when the tables have identical column names.

● It takes a lot of resources; database vendors usually impose restrictions on the number of tables that can be joined in a single SQL statement.

Note that the clause JOIN was not used in performing the cross join.

14.4.1.16.1 INNER JOIN

The SQL JOIN can produce useful information when tables are joined conditionally. For example, AUTHORS and TITLEAUTHOR can be joined on condition that the common column AU_ID is equal. Some versions of SQL allow the words INNER JOIN to be used: requires the ON clause. See:

```
SELECT   *
   FROM   AUTHORS INNER JOIN TITLEAUTHOR
          ON AUTHORS.AU_ID = TITLEAUTHOR.AU_ID
```

A more general form of this SQL statement is:

```
SELECT   *
   FROM   AUTHORS,TITLEAUTHOR
   WHERE   AUTHORS.AU_ID = TITLEAUTHOR.AU_ID
```

The latter SQL statement is more intuitive: the WHERE condition is specified exactly as with the ON clause. This type of JOIN is called an *inner join* or an *equijoin*. The following explanation is given:

● Usually, a subset of the available columns from the inner join is retained: the SELECT clause specified the fully qualified name of the columns—that is, TableName.ColumnName

● Although the example uses a single condition, multiple columns may be used to specify the WHERE condition.

14.4.1.16.2 SELF JOIN

In an inner join, two separate tables are joined using one or more columns, of the same type and usually of the same name. A self-join is when a table is joined to itself to extract the relevant information.

Consider the text table shown **Figure 14-5 Self join table**; it shows details of a tutorial group where the student themselves act as session leaders. The LeaderID column holds a code to identify the student who is the leader. That is, the LeaderID and StudentID columns are synonymous.

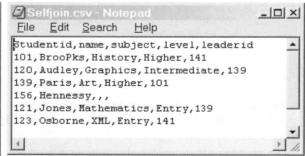

Figure 14-5 Self join table

In a database, the StudentID column may be a primary key—that is, it must be unique—and the LeaderID column must be contained within StudentID. Note:

● The requirement is to produce a table that shows all students who have elected a subject—thus, Hennessy would be excluded—alongside their subject choice and Leader name.

● A quick analysis suggests that LeaderId needs to be looked up StudentId to pick up the corresponding name.

● The column names that need to be cross-referenced are in a single table. Try:

```
SELECT    a.NAME AS STUDENT,
          a.STUDENTID AS ID,
          a.SUBJECT AS SUBJECT,
          a.LEVEL AS STAGE,
          b.NAME AS LEADER,
          '__:__AM/PM' AS START,
          '__:__AM/PM' AS FINISH,
          'R____' AS ROOM
  FROM    [SELFJOIN.CSV] AS a,
          [SELFJOIN.CSV] AS a
 WHERE    b.STUDENTID=a.LEADERID
ORDER BY  a.NAME"
```

In this SQL statement,

● The AS keyword is used to create aliases for both columns and tables.

● Three arbitrary columns are added, namely, START, FINISH, and ROOM.

The APL+Win function that executes this self-join is:

```
      ∇ SelfJoin;⎕wself
[1]   ⍝ System Building with APL+Win
[2]   ⎕wself←'∆ADORS' ⎕wi 'Create' 'ADODB.Recordset'
[3]   ⍝Cnn←'Provider=MSDASQL;Driver={Microsoft Text Driver (*.txt;
          *.csv)};DBQ=a:\;'
[4]   Cnn←'Provider=Microsoft.Jet.OLEDB.4.0;Data Source=c:\;Extended
          Properties="text;HDR=Yes;FMT=Delimited"'
[5]   Sql←"SELECT a.NAME AS STUDENT, "
[6]   Sql←Sql,"a.STUDENTID AS ID, "
[7]   Sql←Sql,"a.SUBJECT AS SUBJECT, "
[8]   Sql←Sql,"a.LEVEL AS STAGE, "
[9]   Sql←Sql,"b.NAME AS LEADER, "
[10]  Sql←Sql,"'__:__AM/PM' AS START, "
[11]  Sql←Sql,"'__:__AM/PM' AS FINISH, "
[12]  Sql←Sql,"'R____' AS ROOM "
[13]  Sql←Sql,"FROM [SELFJOIN.CSV] AS a, "
[14]  Sql←Sql,"[SELFJOIN.CSV] AS b "
[15]  Sql←Sql,"WHERE b.STUDENTID=a.LEADERID "
[16]  Sql←Sql,"ORDER BY a.NAME"
[17]  ⎕wi 'XOpen' Sql Cnn (⎕wi '=adOpenStatic') (⎕wi '=adLockReadOnly')
          (⎕wi '=adCmdText')
[18]  APLGridDisplay '∆ADORS'
[19]  ⎕wi 'XClose'
[20]  ⎕wi 'Delete'
      ∇
```

In line [3], the ODBC text driver is specified but is commented out; it can be used instead of the provider specified in line [4]. Lines [5] to [16] create the SQL statement. Furthermore:

● Embedded literals are enclosed within double quotes; this allows single quotes to be used within the string—SQL literals must be enclosed in single quotes.

● When constructing an SQL statement on multiple lines of APL+Win code, always add a trailing space to all but the final line.

This function will enhance SQL statements for readability:

```
      ∇ Z←L FormatSQL R
[1]    ⍝ System Building with APL+Win
[2]    Z←' ',⎕wcall 'CharUpper' (,∊R)
[3]    ((Z=⎕tcnl)/Z)←' '
```

The right-hand argument specifies the SQL string; this is transformed into a vector consisting of upper case characters, line feed characters; duplicate blanks are removed and linefeed characters are replaced by a space. Let's continue:

```
{1[4]    :if 0≠⎕nc 'L'
 [5]         ((Z=',')/Z)←⊂',',⎕tcnl
 [6]         Z←∊Z
1}[7]    :endif
```

If any left-hand argument is specified, a comma in the SQL statement is changed into a line feed character. This has the effect of showing column names on separate lines as an aid to readability especially when column aliases are used or created. This facility cannot be used if the SQL statement contains SQL functions whose arguments are separated by comma. The function cannot differentiate the context in which a comma occurs. Spaces are added where appropriate (for instance, comparison operators are separated by a space on either side), as shown:

```
[8]    ((Z∊'=<>+-⋆/()&|')/Z)←' ',¨((Z∊'=<>+-⋆/()&|')/Z),¨' '
[9]    Z←∊Z
```

In order to provide correct indentation, tag any UNION clauses present:

```
[10]   L←'UNION ALL' 'UNION DISTINCT' 'GROUP ' 'ORDER '⊆¨⊂Z
[11]   ((⁻5⌽∨/L)/Z)←':'
[12]   ((⁻5⌽'UNION '⊆Z)/Z)←'¦'
```

Add a space prefix to all SQL clauses for readability; use a linefeed prefix instead with those clauses that need to begin on a new line:

```
[13]   R←'UNION¦,UNION:DISTINCT,UNION:ALL,SELECT ,FROM
       ,GROUP:BY,ORDER:BY,WHERE ,INSERT , ON , USING'
[14]   R←R,', ALTER, CREATE, UPDATE, INSERT, DROP, HAVING,AND ,OR '
[15]   R←(R≠',')⊂R
{1[16]  :while 0≠⍴R
{2[17]      :if 1∊L←(⎕io⊃R)⊆Z
 [18]           Z[(L/⍳⍴L)∘.+(-⎕io)+⍳⍴⎕io⊃R]←' '
{3[19]           :select ⎕io⊃R
⋆[20]                :caselist 'AND ' 'OR '
 [21]                    (L/Z)←⊂⎕tcnl,⎕io⊃R
⋆[22]                :caselist 'UNION¦' 'UNION:ALL' 'UNION:DISTINCT'
 [23]                    ((':'=⎕io⊃R)/⎕io⊃R)←'¦'
 [24]                    (L/Z)←⊂⎕tcnl,(⎕io⊃R),⎕tcnl
+[25]                :else
 [26]                    (L/Z)←⊂⎕tcnl,((⎕io⊃R)~' '),'¦'
3}[27]               :endselect
 [28]               Z←∊Z
2}[29]      :endif
 [30]      R←1↓R
1}[31]  :endwhile
 [32]   Z←(~∧\Z∊' ',⎕tcnl)/Z
```

Finally, remove all superfluous spaces and other characters added to facilitate indentation:

```
{1[33]     :for R :in '¦ ' ' )' ' ,' ', ' ' ;' '; ' '( ' ', ' ' (' '- -' '/ /
            ⋆' '⋆ / /' '/ /' '< =' '> =' '< >'
{2[34]        :if 1∊L←R∊Z
  [35]            Z[(L/⍳⍴L)∘.+(-⎕io)+⍳⍴R]←'}'
  [36]            (L/Z)←⊂R~' '
2}[37]        :endif
  [38]        Z←(∊Z)~'}'
  [39]        Z←(~'   '≤Z)/Z
1}[40]     :endfor
  [41]     Z←(~¯1⌽'¦ '≤Z)/Z
  [42]     Z←(Z≠⎕tcnl)⊂Z
  [43]     Z←(+/¨∧\¨Z=' ')⌽¨Z
  [44]     L←((~⎕io)+Z⍳¨'¦')×'¦'∊¨Z
  [45]     Z←(((⌈/L)-L)⍴¨' '),¨Z
  [46]     Z←¯1↓∊((~' '∧.=¨Z)/Z),¨⎕tcnl
  [47]     ((Z∊':¦')/Z)←' '
  [48]     Z←Z,('; '=¯1↑Z)↓';'
        ▽
```

Each line of the result starts with an SQL keyword except when the left-hand argument is used. The function FormatSQL does not validate the SQL statement, and recognises only a common subset of SQL clauses; it simply introduces conventional formatting to most SQL statements. The SQL keywords SELECT, FROM, WHERE and ORDER BY are right aligned within a virtual column on the left, see **Figure 14-6 Formatted SQL I**.

```
  θ FormatSQL Sql
    SELECT A.NAME AS STUDENT,
           A.STUDENTID AS ID,
           A.SUBJECT AS SUBJECT,
           A.LEVEL AS STAGE,
           B.NAME AS LEADER,
           '__:__AM/PM' AS START,
           '__:__AM/PM' AS FINISH,
           'R___' AS ROOM
      FROM [SELFJOIN.CSV] AS A,
           [SELFJOIN.CSV] AS B
     WHERE B.STUDENTID=A.LEADERID
  ORDER BY A.NAME;
```
Figure 14-6 Formatted SQL I

When a left-hand argument is not specified, the formatting forces SQL keywords unto new lines, see **Figure 14-7 Formatted SQL II**; the presence of a left-hand argument also forces line breaks after commas, see **Figure 14-6 Formatted SQL I**.

```
      FormatSQL Sql
    SELECT A.NAME AS STUDENT, A.STUDENTID AS ID, A.SUBJECT AS SUBJECT,
       FROM [SELFJOIN.CSV] AS A, [SELFJOIN.CSV] AS B
      WHERE B.STUDENTID=A.LEADERID
    ORDER BY A.NAME;
```
Figure 14-7 Formatted SQL II

An SQL parser that validates and formats SQL is complex and expensive, not least because of the vendor specific deviations from ANSI SQL. A simple approach is to execute the SQL statement and to debug it interactively. A tool such as Query Analyser provides a good interface but it is exclusive to SQL Server.

Further formatting is necessary when an APL function incorporates SQL statements:

```
      ∇ Z←APLFormatSQL R
[1]    ⍝ System Building with APL+Win
[2]    Z←∊(⊂'Sql←"""',⎕tcnl),((⊂'Sql←Sql,"'),¨((⎕tcnl≠R)⊂R),¨'"'),¨⎕tcnl
      ∇
```

For example, consider this statement:

```
Sql←'Select a.* from invoices a, accounts b where a.customer=b.customer
and b.settled=0'
```

FormatSQL Sql returns:

```
      SELECT   A. *
        FROM   INVOICES A,ACCOUNTS B
       WHERE   A.CUSTOMER = B.CUSTOMER
               AND B.SETTLED = 0;
```

The APLFormatSQL function can construct APL code that retains the indentation and format:

```
        APLFormatSQL FormatSQL Sql
Sql←""
Sql←Sql,"SELECT A. * "
Sql←Sql,"   FROM INVOICES A,ACCOUNTS B "
Sql←Sql," WHERE A.CUSTOMER = B.CUSTOMER "
Sql←Sql,"         AND B.SETTLED = 0;"
```

SQL statements should always use single quotes; the APL code wraps individual lines of the SQL statement in double quotes. The linefeed character no longer exists because the physical APL lines provide the same effect. Moreover, all but the last line of the APL code has a trailing space; this ensures that lines do not run into each other upon concatenation.

14.4.1.17 Other Joins

SQL supports two other types of join, involving two tables. The first table is the left table; the second table is the right table. The two table names are separated by [LEFT JOIN | RIGHT JOIN] or [LEFT OUTER JOIN | RIGHT OUTER JOIN]; the word OUTER is optional, depending on the SQL dialect. In order to investigate the other types of join, consider two tables in an **Excel** workbook.

• There is one column common between the two tables; treat the AUTHORS table, shown in **Figure 14-8 AUTHORS table extract**, as the left table and the Books table, shown in **Figure 14-9 Books by author**, as the right table.

• The Name column in AUTHORS contains unique values; the author Puskin does not appear in the right table.

• The Name column in BOOKS contains some duplicate values by author—Shaw; the author Melville does not appear in the left table.

joins.xls:1

	A	B	C
1	**Name**	**FirstName**	**Lived**
2	Blake	William	1757-1827
3	Shaw	George Bernard	1856-1950
4	Moliere	Jean Baptiste P	1622-1673
5	Gogol	Nikolai V	1809-1852
6	Puskin	Alexander	1799-1837
7			

Sheet1 / Sheet2

Figure 14-8 AUTHORS table extract

joins.xls:2

	A	B	C	D
1	**Name**	**Book**		
2	Blake	Prophetic Books		
3	Shaw	Pygmalion		
4	Shaw	Saint Joan		
5	Moliere	Tartuffe		
6	Gogol	The Government Inspector		
7	Melville	Moby Dick		
8				

Sheet1 / Sheet2

Figure 14-9 Books by author

14.4.1.17.1 LEFT JOIN

The result set of a left outer join includes all the rows from the left table specified in the LEFT JOIN clause including the ones which do not have a match in the Name column of the right table. This is because:

- When a row in the left table has no matching rows in the right table, the associated result set row contains null values for all selected columns arising from the right table.
- In order to avoid replicated columns, select all columns from the left set and discard the common ones from the right set:

```
SELECT  A.*,B.BOOK
  FROM  [SHEET1$] A LEFT JOIN [SHEET2$] B
    ON  A.NAME=B.NAME;
```

The FROM line can be equally specified as:

```
  FROM  [SHEET1$] A LEFT OUTER JOIN [SHEET2$] B
```

This does not affect the result. Take care not to confuse the row numbers in the **Excel** bitmap and the APL Grid bitmap; the **Excel** bitmap shows the data from row two whereas the APL Grid starts on row 1. **Figure 14-10 LEFT OUTER JOIN** shows the result. Note the following:

- All the rows from the left table are returned, as expected.
- There were two matches for Shaw in the right table; this is also shown.
- Although there was no match for Puskin in the right table, it is still returned because all rows from the left table are returned.
- Missing values are returned as NULL.

For Puskin, the value for Book expected from the right table is, strictly speaking, NULL: the value is missing in the right table.

Recordset

	Name	FirstName	Lived	Book
1	Blake	William	1757-1827	Prophetic Books
2	Shaw	George Bernard	1856-1950	Saint Joan
3	Shaw	George Bernard	1856-1950	Pygmalion
4	Moliere	Jean Baptiste P	1622-1673	Tartuffe
5	Gogol	Nikolai V	1809-1852	The Government Inspector
6	Puskin	Alexander	1799-1837	

Page1

Figure 14-10 LEFT OUTER JOIN

14.4.1.17.2 RIGHT JOIN

A right outer join is the reverse of a left outer join. All rows from the right table are returned. Null values are returned for all rows in the right table that have no matching rows in the left table as shown:

```
SELECT   B.*,A.FIRSTNAME,A.LIVED
   FROM   [SHEET1$] A RIGHT JOIN [SHEET2$] B
     ON   A.NAME=B.NAME;
```

The result is shown in **Figure 14-11 RIGHT OUTER JOIN**. Note that all the rows from the right table and both rows for Shaw are returned.

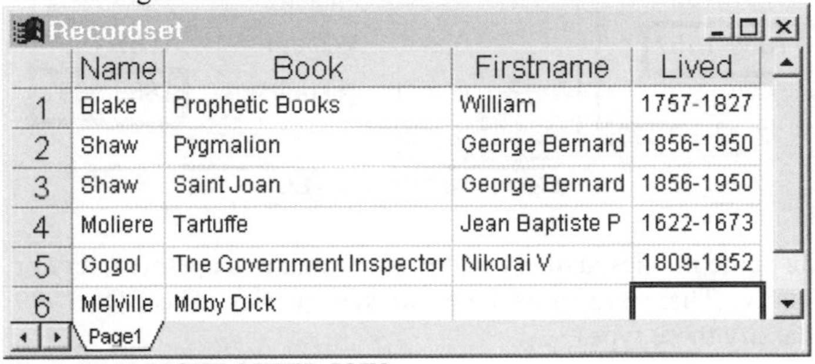

	Name	Book	Firstname	Lived
1	Blake	Prophetic Books	William	1757-1827
2	Shaw	Pygmalion	George Bernard	1856-1950
3	Shaw	Saint Joan	George Bernard	1856-1950
4	Moliere	Tartuffe	Jean Baptiste P	1622-1673
5	Gogol	The Government Inspector	Nikolai V	1809-1852
6	Melville	Moby Dick		

Figure 14-11 RIGHT OUTER JOIN

There was no match for Melville in the left table; it is still returned but the value for *Firstname* and *Lived* expected from the left table is, NULL.

14.4.1.18 UNION

The UNION clause requires at least two tables, the top result and the bottom result. Any number of UNION queries may be specified in a single SQL statement. Note the following:

• Each result must have the same number of columns and the type of the results must be the same at each ordinal column position in both sets.

• UNION does not match column names; it simply appends the first column from the bottom table to the first column from the top table and so on. Therefore it is necessary for the SELECT statements of both the top and bottom tables to re-order the columns in a consistent order; it is usually coincidental that this order is inherent.

• By default, UNION returns distinct rows only; in this case it also sorts the result in ascending order. Some SQL dialects require the DISTINCT or ALL qualifiers explicitly. The **Excel** driver only requires the ALL qualifier, if appropriate, and then the bottom table is simply appended to the top table.

Assuming that the tables TOPTABLE and BOTTOMTABLE exist, the following syntax is invalid:

```
TOPTABLE
      UNION
BOTTOMTABLE
```

It is not sufficient to specify just the names of the tables; the required columns of each table must be selected and the columns from the top table must be the same or compatible with those from the bottom table, in the order specified within the select clause. This is a correct construction:

```
SELECT   *
   FROM   [SHEET1$]
  WHERE   NAME = 'GOGOL'
UNION ALL
   SELECT   *
      FROM   [SHEET1$]
     WHERE   NAME IN ('GOGOL','BLAKE');
```

This is a trivial SQL statement; however, it does illustrate the difference between UNION, UNION DISTINCT, and UNION ALL. Note that the selection based on UNION DISTINT, shown in **Figure 14-12 UNION DISTINCT** is sorted whereas that based on UNION ALL is not, shown in **Figure 14-13 UNION ALL**.

UNION

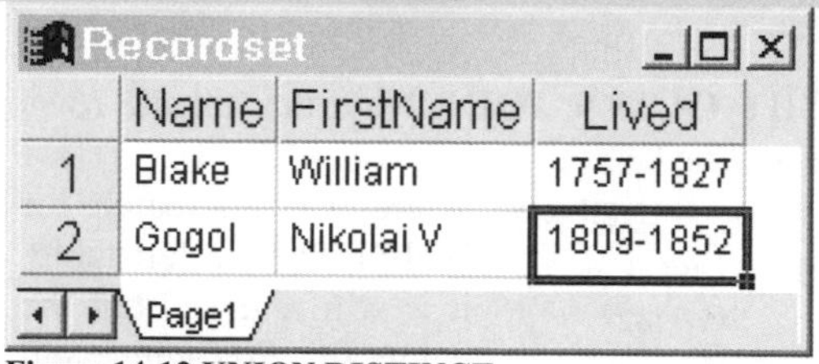

	Name	FirstName	Lived
1	Blake	William	1757-1827
2	Gogol	Nikolai V	1809-1852

Figure 14-12 UNION DISTINCT

UNION ALL

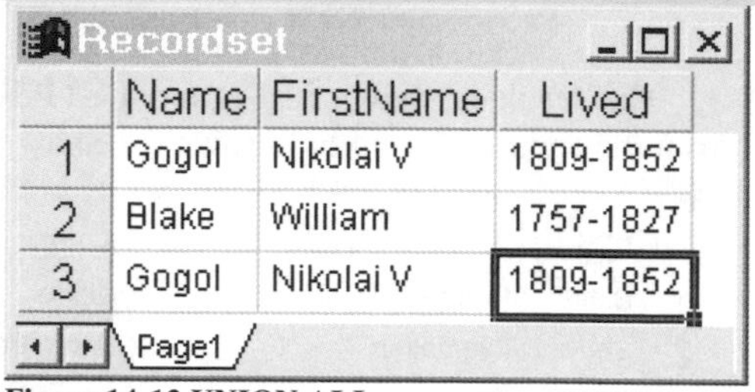

	Name	FirstName	Lived
1	Gogol	Nikolai V	1809-1852
2	Blake	William	1757-1827
3	Gogol	Nikolai V	1809-1852

Figure 14-13 UNION ALL

14.4.1.19 Sub Queries

A sub query is exactly what the name implies: a query that is subordinate to another query or one that is part of another query. There are three types of sub queries. The SQL:1999 standard does not attempt to classify these types.

14.4.1.19.1 Single row sub query

A single row sub query returns a single or atomic value that is incorporated into all rows of a query. A more descriptive name for queries of this type is scalar sub query. For example:

```
 SELECT  AU_FNAME,
(SELECT  MAX(PUBDATE)
   FROM  TITLES)
   FROM  AUTHORS;
```

The single row sub query is shown in italics. This is like selecting an explicitly specified scalar or constant that is included as a column on all rows of the query; the difference with a single row sub query is that the value of the constant is dynamically retrieved from any table in the same database. The flexibility of a SELECT statement is available to a single row sub query: the only constraint is that it must return a scalar value or a runtime error occurs.

14.4.1.19.2 Single column query

A single column query returns a single column, usually of the same name—or at least of the same type—as a column in the main query. It is used to restrict the selection in the main query: it is used with the WHERE clause. An example:

```
 SELECT  AU_FNAME,
         AU_LNAME
   FROM  AUTHORS
  WHERE  AU_ID IN
(SELECT  AU_ID
   FROM  TITLEAUTHOR
  WHERE  AU_ORD>1);
```

```
AU_IN NOT IN
```

The single column query is shown in italics; the sub query must return as single column or a runtime error occurs.

14.4.1.19.3 Table sub query

Table sub queries return single or multiple columns of zero or more rows and are used to combine columns from multiple tables or for rejecting the selection in the main query.

14.4.1.19.3.1 Conditional column collation

This involves the collation of columns from two or more sub queries based on multiple conditions involving one or more columns from each sub query or queries. Consider the following:

```
SELECT    A.AU_FNAME,
          A.AU_LNAME,
          B.ROYALTYPER
   FROM   AUTHORS A,
(SELECT   AU_ID,
          ROYALTYPER
   FROM   TITLEAUTHOR
  WHERE   AU_ORD>1) B
  WHERE   A.AU_ID=B.AU_ID;
```

In this example, three columns are collated from two tables based on the equality of one common column value. The sub query is shown in italics. Most JOIN and UNION queries are based on sub queries that are collated via one or more conditions that may be based on one or more columns common in each sub query.

14.4.1.19.3.2 Rejecting query selection

Any query that has one or more records with one or more columns can return either its selection or no records depending on whether a sub query returns one or more records or no records, respectively, when the main query is conditionally linked to the sub query with the EXIST clause. An example is:

```
SELECT    *
   FROM   AUTHORS
  WHERE   EXISTS                                      NOT EXISTS
(SELECT   *
  WHERE   0<>0);
```

The sub query is shown in italics; it will not return any records, as its WHERE condition can never be **TRUE**. Therefore, no records are returned: the highlighted WHERE clause will always be **FALSE**.

The EXISTS or NOT EXISTS clauses always return **TRUE** or **FALSE**; in this context, the sub query can be specified such that it selects records from one or more tables based on one or more conditions.

14.4.2 The SQL Data Manipulation Language

The DML component allows the underlying data to be changed from within the application.

14.4.2.1 INSERT and INTO

The INSERT clause updates a table either one row at a time or multiple rows as at time.

14.4.2.1.1 Insert single rows

For inserting single rows into an existing table, the syntax is:

```
INSERT   INTO TABLE (COL1,COLN,COL2)
VALUES   ('NEWSTRINGVALUE',100,'01/01/2000');
```

The column list (COL1,COLN,COL2) is optional. If omitted, values for all columns must be specified in the order in which the columns occur in the table. NULL may be specified for columns for which data is unavailable. This is dangerous as it is not obvious which values are being inserted. It is better to specify the column names—the column names are listed, separated by comma and without quotation marks—as this allows a subset of columns to be enumerated and this can be done in any order. However, the values must be in the order that the column names are specified—string values and dates are enclosed within single quotation marks.

14.4.2.1.2 Insert multiple rows

The INSERT clause also adds multiple records to a table. The general syntax is:

```
INSERT  INTO TABLE
SELECT  *
   FROM  TABLEX;
```

The selection of records may be as complex as required. INSERT requires that the table into which records are to be inserted exist already; if the target table does not exist, the SQL fails. The types of data selected must either match or be coercible into the type of their respective target column. If the target table does not exist and needs to be created on the fly, an alternative syntax is required. Try:

```
SELECT  * INTO AA
   FROM  AUTHORS;
```

The latter technique can be used to create an archive of a table instantly. It is also very useful in dealing with text data sources: the SQL statement may sort, select a subset of rows or a subset of columns or both and write the resulting record object to another table directly. If the table exists already, the SQL fails.

This technique fails with text files; depending on the format of the text file, it either fails completely or fails to write the file in the same format as the original file. It fails with files containing fixed length records also containing fixed length fields—padded with space.

14.4.2.1.3 Insert multiple rows conditionally

If two tables share a number of columns, that is, their names and types are the same, the values of those columns may be moved from one table to another. See the following statement:

```
INSERT  INTO TABLE (COLx, Coln)
        (
        SELECT    Colx
                  Coln,
        FROM      TABLEX
        WHERE     Col IN(2003,2004)
        )
```

• In this example, both TABLE and TABLEX must exist and the columns Colx, Coln must have the same or coercible data type in both of them.

• Either table may have other columns, which may or may not be common.

• As many rows as are selected from TABLEX are inserted into TABLE.

• The values for Coln and Colx from the selected rows in TABLEX are transferred to TABLE; the values of other columns that may exist in TABLE default to NULL.

14.4.2.1.4 INSERT INTO with Access

From the point of view of creating back up copies of tables, **Access** has a special feature that allows copies of tables to be made in a separate database. Try:

```
    ∇ AccessINTO;Cnn;SQl;⎕wself
[1]   ⍝ System Building with APL+Win
[2]   Cnn←'Provider=MSDASQL;Driver={Microsoft Access Driver
        (*.mdb)};DBQ=c:\nwind.mdb;'
[3]   Sql←"SELECT * INTO abc IN 'c:\copy.mdb' FROM customers;"
[4]   ⎕wself←'∆ADO' ⎕wi 'Create' 'ADODB.Connection'
[5]   ⎕wi 'XOpen' Cnn
[6]   ⎕wi 'XExecute' Sql
[7]   ⎕wi 'Delete'
    ∇
```

This function creates a copy of the customers table found in C:\NWIND.MDB as a table called ABC in a separate database C:\COPY.MDB.

14.4.2.2 Sorting a text file

If a text file is required in ascending order of the first and second columns, it can be accomplished using SQL. The function below explains it:

```
      ∇ L SortCSV R;⎕wself
[1]    ⍝ System Building with APL+Win
[2]    Cnn←'Provider=MSDASQL;Driver={Microsoft Text Driver (*.txt;
       *.csv)};DBQ=c:\'
[3]    Sql←'SELECT * INTO [', L, '] From [',R, '] order by 1 ASC, 2 ASC'
[4]    ⎕wself←'∆ADO' ⎕wi 'Create' 'ADODB.Connection'
[5]    ⎕wi 'XOpen' Cnn
[6]    0 0⍴⎕wi 'XExecute' Sql
[7]    ⎕wi 'Delete'
      ∇
```

The left- and right-hand arguments specify the target and source files, respectively. The target file must not exist already. The source file must be a valid data source: if necessary, it must be defined in the schema.ini file. This is a powerful technique for separating or reordering text data sources.

14.4.2.3 UPDATE

The UPDATE clause allows the value of named column(s) to be altered on existing rows. The general syntax is:

```
UPDATE    TABLE
   SET    COLNAME=NEWVALUE,NEXTCOLNAME='NEWVALUE'
 WHERE    ACOL <> CONDVALUE;
```

• The columns whose value is to be changed must exist in the table already; however, the name/value pair of existing columns can be specified in any order.

• The new values must be consistent with the column type.

• If the WHERE clause is omitted, all rows inherit the new values for the column specified.
The text driver does not support the UPDATE clause.

14.4.2.4 DELETE

The DELETE clause removes rows from the table. Try:

```
DELETE
   FROM    TABLE
  WHERE    ACOL <> CONDVALUE;
```

Note that the DELETE clause does not have an argument. If the WHERE clause is omitted, all the rows in the table are deleted; otherwise, only the rows for which the WHERE clause is **TRUE** are deleted. The Microsoft text driver does not support the DELETE clause.

The lack of support for UPDATE and DELETE clauses makes text data sources less efficient than other data sources A rather messy workaround is to UPDATE and DELETE values and records in the ADO record object and then to write the modified content of the record object to new files.

14.4.3 The SQL Data Definition Language

The DDL component of SQL allows permanent structural changes in a database from within an application. In practice, database administrators, using tools provided by the database vendor, usually carry out such changes; this ensures that there is some control over the database integrity.

It is inappropriate for applications to make destructive changes to a database.

14.4.3.1 CREATE

The CREATE clause allows database objects to be created; these include new databases, and new tables, procedures and triggers in a given database.

The CREATE DATABASE, PROCEDURE, and TRIGGER functionality is specific to the database driver or provider and it is unlikely that an application will have the requisite information to create them; examine the script for the pubs database:

```
CREATE DATABASE [pubs]  ON (NAME = N'pubs', FILENAME = N'D:\Program
Files\Microsoft SQL Server\MSSQL\data\pubs.mdf' , SIZE = 2, FILEGROWTH = 10%) LOG
ON (NAME = N'pubs_log', FILENAME = N'D:\Program Files\Microsoft SQL
Server\MSSQL\data\pubs_log.ldf' , FILEGROWTH = 10%)
```

As regards the CREATE TABLE functionality, the SQL statement is as follows:

```
CREATE    TABLE TABLENAME
          (COLUMN1 DataTypeForCOLUMN1,
          COLUMN2 DataTypeForCOLUMN2);
```

The data type literal is vendor specific. For example for a date type, Oracle will need DATE whereas SQL Server will need DATETIME, SMALLDATETIME, or TIMESTAMP.

The CREATE clause is of little or no value within application code because of these complications. A good practice is to maintain the structural aspects of a database with the vendor's administration tools and to manage the contents of tables with SQL.

14.4.3.2 ALTER TABLE

The ALTER TABLE clause allows columns to be added or dropped from an existing table. However, some implementations expressly disallow the ALTER TABLE clause. This clause is useful if a column is to be dropped. The general syntax is:

```
ALTER    TABLE MYTABLE
DROP    COLUMN MYCOLUMN;
```

In order to add a new column, the data type for the column must be specified. The following example adds a new column MYCOLUMN2 to an existing table MYTABLE in SQL Server:

```
ALTER    TABLE MYTABLE
   ADD    COLUMN MYCOLUMN2 VARCHAR(20) NULL;
```

The issue regarding the vendor specific data type names arises again as the type and constraints for the new column must be specified.

14.4.3.3 DROP TABLE

The DROP clause is useful for removing a table in a database. The general syntax is:

```
DROP    TABLE MYTABLE;
```

Note that this removes the table permanently. Likewise, subject to security implementations, it is possible to remove a database from SQL Server with a similar command:

```
DROP    DATABASE LAST YEAR;
```

14.4.4 The SQL Data Control Language

The DCL component of SQL deals with data access permissions and includes the ALTER PASSWORD, GRANT, REVOKE and CREATE SYNONYM clauses. These also make permanent changes to the database. It is inappropriate to control access permissions from within an application— database administrators are better placed to handle this aspect of administration.

14.4.5 The way forward with SQL

SQL is relatively easy to learn, as it is English-like.

The Microsoft Query Analyser, shown in **Figure 14-14 Microsoft Query Analyser**, provides a good interface for SQL drill; however, it works for SQL Server databases only and is proprietary software.

Should an APL+Win application support multiple databases, a custom interface that allows an SQL statement to be entered and executed must be developed. This will speed up the construction and debugging of SQL statements.

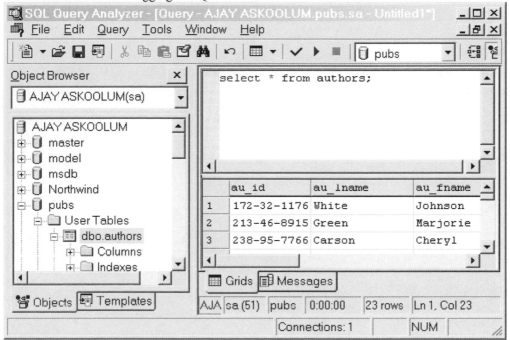

Figure 14-14 Microsoft Query Analyser

It requires a lot of practice to write SQL statements that produce the correct or expected result.

14.4.6 Debugging SQL

Unlike APL+Win code, an SQL statement that works is not necessarily correct and what might appear correct against one database may not be correct against a different database because of differences in the nature and volume of data.

14.4.7 Optimising SQL

Within limits, SQL dialects optimise the SQL statements for runtime efficiency. For a given task, SQL statements can be constructed in different ways; some would be more efficient than others. SQL statements involving sub queries are less efficient that those without.

Vendor tools such as Query Analyser have the ability to analyse the efficiency of SQL statements.

14.4.8 SQL dialect specialisation

Each SQL dialect supports exclusive features. If an application supports a single database, using the specialist SQL features is a viable option. However, if multiple databases are supported, SQL using vendor specific features is viable only when the corresponding syntax exists and is known for all databases. Otherwise, use ANSI SQL for data access and APL+Win to implement specialist processing.

Chapter 15

Application Evolution

Successful/profitable software continues to evolve _after_ release: failure to evolve leads to the software becoming less and less useful and finally disappearing. Usefulness is primarily a measure of user acceptance: users judge software firstly in respect of usability and secondly in respect of overall peripheral functionality, that is the ease with which the software delivers it core purpose. Traditionally, APL systems have tended to stagnate after delivery because they are not designed for change, as discussed below:

• The structure of evolving applications becomes more complex as incremental changes are added post delivery. Poor documentation of APL systems makes its structure appear more complex than it is to new programmers who inherit their maintenance.

• APL systems have tended to concentrate on functionality rather than usability: newer users expect usability to be comparable to other contemporary software.

These two factors combine to force the replacement of APL systems because of the inability of these systems to adapt readily to changes—partly because these systems pre-date the universal Windows GUI interface. Decision makers choose a development tool other than APL to avoid the sour experience of the existing APL system. If an APL application stores its data in component files, no amount of explanation will adequately answer user questions like, 'The cash flow data exists, why can't I export it to **Excel**?'

Incremental changes are necessary to cope with ever changing business needs, changes in Information Technology (**IT**) environment, legislation, and structural businesses changes: acquisitions and mergers happen, new departments are created, new product lines are launched, etc. IT innovations and improvements such as networks, client/server architectures and the Internet necessitate change. Legislation changes such as those relating to taxes or to pensions provision or changes imposed by sector regulatory bodies also necessitate change. A much more fundamental reason for change is client expectation which includes improvements is performance, the user interface, and the fixing of faults and anomalies, like year 2000 compliance, that come to light.

APL+Win applications are subject to the same maintenance pressures. Additionally, legacy applications may require migration from one platform to another, perhaps from mainframe to PC, or from DOS to Windows or from one version of APL+Win to a later version or from one evolution level to the next.

An APL application will has at least one main workspace with a latent expression to start the application. Typically, applications have several workspaces and several components files, each holding both data and functions. The master workspace dynamically copies in and removes groups of functions and data from ancillary workspaces and component files at runtime.

With legacy applications, the organisation of the application across several workspaces and component files is very likely to be unstructured. Legacy code will not incorporate control structures and likely to use several branching conventions. Code that follows no discernible convention for structure is difficult to read and maintain.

15.1 Application deployment

During the course of development, the workspace accumulates a lot of unseen clutter; the symbol table becomes large and functions inadvertently have trailing spaces on some lines. Tidy the workspace that is ready for release using the following steps:

- Ensure that the workspace is saved.
-)clear the workspace.
-)copy the workspace.
- Redefine each function: `⎕def ¨⎕cr ¨c[⎕io+1]⎕nl 3 ⍴` Resized.
- Reset system parameters as appropriate (`⎕io`, `⎕lx`, `⎕elx`, etc.)
-)wsid – rename the workspace as it was.
-)rsave the workspace ⍝ Symbol table has been tidied up.

Optionally, use the CSBlocks function to add wave pattern indentation to functions in order to improve readability.

15.2 The next release

The economics of commercial software is driven by the needs of its user base. An application that has been released represents an investment. Software is never perfect and maintenance releases inevitably follow. Newer releases are required to bring the application up to date, to incorporate new features, and to correct logic flaws reported by the client base.

This might seem to be cynical but new releases not only provide software vendors with marketing and pricing opportunities but also make their competitiveness visible in the market place. Every release provides an opportunity for profit and market share—for in-house applications, read increased budget and credibility, respectively.

15.2.1 The schedule of work

Each release is a project with associated costs. A project is a schedule of work that starts at a fixed time and is expected to finish at a later fixed time. Overshooting the completion date adds directly to the cost of the project not just in terms of the cost of the resources used in the extra duration but also in terms of the impact of the delay on existing and potential clients, internal or external.

15.2.2 Fault management

All applications have faults or bugs—a bug is a design flaw that escaped detection prior to release. It is necessary to investigate and log reported faults together with documentation of the circumstances that give rise to them. Faults will fall into two categories—critical or cosmetic. The critical faults are of a nature that prevent the application from functioning or cause it to function erroneously. Cosmetic faults are those that represent an irritation to users or faults for which a workaround can be found but do not stop the application from working correctly.

A review of the fault log will contribute to the schedule of work for the next release. It is usual to defer some faults to a subsequent future release in order to promote the timeliness of the release to hand.

15.2.3 Wish management

Likewise, an application should have a log of features requested by users. Some user wishes may reflect on their own inexperience or the quality of the training and documentation provided for the application. A review of this log will also contribute to the schedule of work.

Faults will affect all clients in more or less the same way. However, user- desired features may vary from client to client and indeed may conflict. Therefore it may not be possible, even if cost effective, to incorporate all wishes.

If an application has a user group, a meeting of the group provides a good forum for determining the key wishes of the user base. There is no better way to establish the credentials of an application than recommendation by its client base.

15.2.4 If it works, improve it!

The belief is that anything that works should be left alone, even if it works inefficiently. The rational for this approach is valid only when every thing else is constant—this is justified or tolerated on three grounds, namely:

• Firstly, changes may introduce new faults and it may not be possible to gauge the impact of changes on the application as a whole.

• Secondly, changes add to the project duration and costs.

• Thirdly, technological advancements or hardware upgrades or both usually eliminate efficiency bottlenecks.

This implies two things:

• Application changes are reactive rather than proactive.

• Application changes are deferred as long as possible, leading to the possibility of major expenses in the future.

Of course, every thing else is *not* constant on the Windows platform. Changes will happen and reactive changes are invariably associated with deadline pressures. Proactive changes give the application provider more scope for controlling the timeliness of changes. Failure to update an application in terms of its look and feel reduces the appeal or marketability of the product over time. Also, a reluctance to upgrade to a later version of the development tool or to revise code to make it more efficient makes the continued maintenance of the product much more difficult. The application may end up being based on technology that is no longer in tune with contemporary industry or technology standards. The industry trend is to innovate in order to retain market position and to extract the benefits of technology advancements.

15.2.5 Efficiency management

In addition to defining the next release just in terms of fault fixes and new functionality, particular modules of an application should be made demonstrably more efficient. This creates better user or client confidence and makes the application less vulnerable to competition.

Efficiency gains arise from the adoption of current technology and the replacement of legacy code by a more up to date version and from the removal or rewriting of system features that are perceived as problematic.

15.2.6 Small vs large scale improvements

The aggregate dividend from many small improvements is likely to be much bigger than that from any improvement on a large scale. Small improvements are also less risky in terms of their overall impact on the application user base as a whole.

A batch of small improvements will be more visible than one large radical improvement in an application. Small improvements, especially ones that are visible in the user interface, are less likely to upset the user base; any radical change to the application interface will imply that the user base must learn the application anew. This consideration will be a determining factor in the changes planned for a particular release.

Radical improvements are sometimes necessary. For example, most APL applications that rely on native and component files for their data tier can be overhauled via the deployment of databases for the same purpose.

15.3 Application workspace

During the course of development, workspaces tend to accumulate a lot of clutter in the symbol table. This may have been saved in a state where one or more functions are suspended, and may not have the environmental variables such as ☐lx, ☐elx, ☐dm, etc., set as required. Periodically, it is necessary to tidy the workspaces. Workspaces need to be tidied up prior to testing, documentation and release, by undertaking the following steps:

● Load the target workspace and remove all unnecessary variables, created by testing code in interactive mode, and functions, copied into the workspace to aid the development process.

● Clear the execution stack using)sic followed by <enter>.

● Save the workspace.

● Clear the active workspace and)copy the target workspace saved in the previous step.

● Set all environment variables as required.

● Rename the workspace to its original name and save it.

The workspace size after step 5 will usually be smaller than at either step 1 or 3 and the number of symbols, queried by)symb, which returns a two element vector showing the number of symbols allocated and the number in use, will in all likelihood also be less. APL+Win's symbol table is dynamic—it expands as required.

15.4 APL libraries vs UNC names

APL+Win still supports libraries—a facility for referring to a DRIVE:\FOLDER by an integer that can be used as a shortcut for managing workspaces and component files only—it does not apply to host files. Older versions of APL use library numbers as a symbolic reference to drives and folders and did not allow direct reference to drives and folders when referring to workspaces and component files.

The library facility inherited from APL*PLUS is anachronistic. It is useful in that it allows different DRIVE:\FOLDERS to be mapped to the same library number, one at a time, and thereby eliminates the need for changing hard coded DRIVE:\FOLDER references. In practice, applications normally use implicit references to DRIVE:\FOLDER—usually via an INI file:

```
[Folders]
Input=C:\Client X Input
```

The entry for Input is read at the start of the application and stored in a variable for use during the session. For example, if this is stored in a file called C:\SESSION.INI,

```
    Z←256ρ☐tcnul
    InputFolder←↑↑/☐wcall 'GetPrivateProfileString' 'Folders' 'Input'
'' Z 256  'c:\session.ini'
```

Then files will be referred to as (InputFolder,'\MYFILE.TXT'). It is preferable to use DRIVE:\FOLDER references rather than library numbers because library numbers cannot be used for native file operations or with Windows API calls. The maintenance of legacy code will be simpler if library numbers were replaced by a reference to DRIVE:\FOLDER.

One particular circumstance where library numbers might still be useful would be where there are no free drive letters for mapping network volumes. With APL+Win, a Universal Naming Convention (**UNC**) name cannot be mapped to a library number.

```
⎕libd '50 \\DEV\ADMIN\DATA'
```

Windows API calls work with UNC paths, as do some APL+Win functions such as ⎕n, ⎕xn, ⎕f and ⎕xf. However, some functions such as ⎕chdir and ⎕lib do not. From the point of view of future proofing application, library numbers need to be replaced by reference to UNC name or DRIVE:\FOLDER, respectively, for network as well as local drives.

15.5 Readability

A common complaint against APL is that its code is unreadable. This is usually justified on the premise that the code contains hieroglyphic characters (APL symbols) instead of English-like keywords. This argument is flawed in several respects: unreadable really means unintuitive rather than devoid of meaning.

The extent to which code is readable depends directly on the literacy level of the reader. The presence of English words in code gives nothing beyond the semblance of readability. Very few non-programmers will be able to tell the difference between the constructions in **Table 15-1 Loop construction**.

```
Do Until condition           │  Do
    statements               │       statements
Loop                         │  Loop Until condition
```
Table 15-1 Loop construction

A programmer with no knowledge of English will be able to understand a small fraction of any given **VB** code but will understand a much higher proportion of any corresponding APL code because the symbols will remain unchanged irrespective of the language the programmer speaks—readability is a direct function of literacy.

15.5.1 Style

The APL language has rich pathways—several ways of achieving the same result. The coding style of APL programmers varies not only from programmer to programmer but also for a given programmer over time. Several methods can be adopted to mitigate the variations in coding that ensue, namely:

• Unlike raw APL looping constructions, control structures provide a common understanding across programmers and avoid multiple exit points within a function.

• In practice, single statement lines are easier to debug than multiple statement lines.

• The use of standard APL idioms, whether adopted from standard libraries or developed internally, creates a degree of intuitiveness among programmers.

• The discipline of localising transient variables keeps the scope of functions clutter free.

• Assumptions about the APL environment need to be explicit. Wherever there is an assumption about index origin within a function, ⎕io must be localised therein. If the left-hand argument of a function is optional, the function must verify its presence and, if found missing, initialise it to an appropriate default.

The insertion of blank lines to separate blocks of code in a function does not add any clarity to the code. However, comments do and should be added where appropriate—failure to clarify code with comments will lead either to the programmer having to maintain the code or being constantly consulted during the application life cycle or to the application being scrapped prematurely. Neither consequence is in the interest of the APL or its programmer.

It is impossible to enforce any rules for in-line comments in code but the programmer should strive to keep them accurate and consistent with a house style. The purpose of a function may be clear during development, but it will not be so a few weeks on and comments provide the initial clues about the code when it needs to be investigated.

15.6 Global variables

An application may require a number of global variables. Such variables may be constant values or initial data required for the application to start and which will be modified at runtime.

With APL + Win, global variables may be assigned in two ways:

• Variables may be assigned arbitrarily and saved with the workspace.

• Variables may be assigned within a start-up function.

The latter method is preferable as it ensures that the application starts with known and consistent values.

15.6.1 Initial values

Typically, initial data will contain default values but will be modified at runtime either by the user or by the running application. Such variables can be assigned default values in the workspace at design time, thus:

```
InterestRate←5
LoanAmount←10000
Term←10
```

Variables created in this manner can be inadvertently re-assigned at runtime. A more robust approach is to create those variables within a function that will be executed at the start of the application. This will ensure that the required variables will always exist in the workspace at runtime and have the same initial expected values, as shown below.

```
    ∇ Initialise
[1]    InterestRate←5
[2]    LoanAmount←10000
[3]    Term←10
```

Such a function may also be used to set application specific parameters for the user interface, such as the application title and the name of its help file. These parameters might be set in user-defined properties of the system object, as:

```
[4]    '#' ⎕wi '∆ApplicationTitle' 'System Building with APL+Win'
[5]    '#' ⎕wi '∆ApplicationHelpFile' (⎕wsid,'.HLP') ⍝ Help file has same
          name as workspace, in the same location and uses WINHELP
    ∇
```

Should it be a requirement that such variables contain their last respective values on a subsequent use of the application, a convenient method is to write the variables to an INI file upon being reassigned and read from the same INI file at start up. Restarting a new session where a previous one terminated, provides a friendly touch; however, whilst this is easily implemented in single user applications, it may not be worthwhile to implement in multi-user applications. The INI file might look as follows:

```
[Initialise]
InterestRate=5
LoanAmount=10000
Term=10
    ⎕wcall 'WritePrivateProfileString' 'Initialise' 'InterestRate' (⍕5)
'c:\initialise.ini'
1
```

Similarly, values can be read using another standard Windows API call.

```
      ↑↑/⎕wcall 'GetPrivateProfileString' 'Initialise' 'InterestRate'  0
(256ρ⎕tcnl) 256  'c:\initialise.ini'
5
```

Both numeric and character values are read using the same function—numeric conversion requires ⍕, thus:

```
        ⍕↑↑/⎕wcall 'GetPrivateProfileString' 'Initialise' 'InterestRate' 0
(256ρ⎕tcnl) 256 'c:\initialise.ini'
5
```

The API calls do not require the file name to be short or to have the INI extension.

15.6.2 Constants

APL+Win does not support the constant data type. A typical approach is to create variables in the workspace at development time in interactive mode. However, this offers no safeguard against the variable being overwritten. A good workaround is to write a function that will return the constant. Rather than having a global variable, Version←'1.01.003', a function can be used, as shown:

```
   ∇ Z←Version
[1]    Z←'1.01.003'
   ∇
```

Any attempt to overwrite this constant by direct assignment outside the function will cause an error:

```
     Version←'1.01.004'
SYNTAX ERROR
     Version←'1.01.004'
     ∧
```

The function Version can be used as if it were a read only variable.

15.7 Using API calls

Unlike a development environment such as **VB** where every API call has to be specifically declared, most API calls are directly accessible from APL+Win without requiring declaration. It is likely that some API calls would create an error like:

```
     ⎕wcall 'PathFileExists' 'C:\'
⎕WCALL COMMAND ERROR: Undefined function: PathFileExists
     ⎕wcall 'PathFileExists' 'C:\'
     ∧
```

APL+Win may need to be restarted if the API definition is added by editing the INI file.

15.7.1 Adding missing API calls

In this instance, the PathFileExists has not been recognised by APL+Win because it has not been precompiled in the APLW.ADF file. This call can be added manually or using the W_Ini method. The API call is:

PathFileExists=L(*C pszPath) ALIAS PathFileExistsA LIB shlwapi.dll
```
     ⎕wcall 'PathFileExists' 'C:\'
1
```

15.8 Version control

VB has the facility for recording and tracking version numbers within EXE files; the version can be verified from **Windows Explorer** by querying the properties of the files. This is useful in tracking the version numbers of each file within an application. However, it does mean that an application version n.nnn.nnn comprises of several files each with their own version numbers. APL+Win files, workspaces or component file, do not have this automated facility.

The function ($\square$at) has the potential of providing rudimentary version management facilities at the level of each function. However, it will not work when functions are stored in files and dynamically fixed at runtime. In essence, using $\square$at can at best report that the timestamp of a function is the same or different from a reference timestamp—if the same, it will be safe to assume that the function has not been changed.

Any manual mechanism for version management will not be fail-safe and will not provide any means of reverting to an earlier version. Unlike VB, which can integrate version control software such as Microsoft Visual Source Safe (**VSS**) in the development environment, APL+Win does not have any means of version control.

15.9 Change management

Version control is a vital function, especially when several programmers are involved in a project and the ability to revert to earlier versions of functions is crucial in the investigations of bugs. The APL programmer has to rely on back ups of each generation of the application to do this. The Windows Briefcase offers very high level version control.

15.10 Legacy management

Aside of the APL mindset, APL+Win has inherited a legacy that includes the one or more of the following handicaps:

● Applications that have been migrated from earlier versions of APL; some of these might still rely on library numbers, evolution level 1, etc., and not benefit from the internal enhancements in the interpreter.

● Applications that do not have a standard GUI user interface; such applications compromise usability and credibility.

● Exclusive reliance on text or component files for data access.

● Inefficient workspace organisation.

Such applications are expensive to maintain: the ultimate cost is the disuse of APL altogether. The APL mindset is resistant to change and tends to isolate APL applications in a niche. The contemporary approach is not only to share the resources of the platform but also to behave like other applications.

Irrespective of their degree of compliance with contemporary standards, APL applications may appear to be or indeed may be indispensable. However, they are not irreplaceable. It is certain that their replacement will not use APL. That is a compelling incentive to bring application into line with contemporary standards.

15.10.1 Workspace organisation

As far as APL coding is concerned, the received wisdom is that functions should be no longer than the number of lines on the screen. This leans towards the break up of an application into many small functions. With APL+Win, there are some problems with this approach:

● APL interpreters for DOS were strictly procedural languages. APL+Win is both procedural and event driven. It requires a different approach—capturing each event for every object will create an unmanageable clutter of small functions.

● Under DOS, the screen was always a static size, 24 rows by 80 columns. With Windows, the screen size depends on the particular video card and monitor in use: there is no uniform external reference for screen size especially since the font size can be changed to accommodate more or less text on a screen of a given resolution.

• While a library of utility functions, which tend to be small, can promote the rapid assembly of code for a new application or new functionality in an existing one, there is a temptation to copy such functions and modify a small part of them to cope with something slightly exceptional. This is very expedient in the short term but, in the long term, it leads to a clutter of small functions and can be quite confusing especially if the names are reused.

• With previous generations of APL, the programmer had to achieve all functionality using native APL. No other options were available and APL had competitive advantage because it was a self-contained development environment. This gave rise to the need for many small utility functions. APL+Win supports the legacy features on which such utility functions were constructed. However, APL+Win now has richer native facilities that refute the need for utility functions. This is especially so since a large number of such functions can be replaced by calls to the standard Windows API.

Of course, some native functionality in APL+Win emulates platform functionality: 'emulate' is not synonymous with 'replace'. It may be convenient to use the APL functions in preference but at a cost: for example, the API GetSystemTime is not the same as ⎕ts nor is ⎕chdir the same as the GetCurrentDirectory API.

GetSystemTime=(>SYSTEMTIME) LIB Kernel32

```
      ⎕ts ◇ ⎕wcall 'GetSystemTime' ◊
2004 5 29 11 49 13 850
2004 5 6 29 10 49 13 850
```

GetSystemTime returns the system time is expressed in Coordinated Universal Time (**UTC**) and returns the day of the week as the third element. The ⎕chdir function returns the current directory but it fails if the current directory is not a mapped drive but refers to a UNC path.

The following examples illustrate how utility functions can be replaced by in-line APL code thereby eliminating the overheads introduced by such functions. Utility functions introduce their own scope and require their own error handling routines whereas in line code does not—it will share the error handling within context. On balance, utility functions require more resources and are liable to be less efficient.

15.10.2 Modernisation

Superficially, software modernisation or re-engineering is a slippery concept that is liable to mean different things to different people. The task to hand is clear-cut if it is simply taken to mean 'Make the APL application like any other application on the Windows platform.' The key to survival in the long term is to eliminate APL specific considerations in the choice of development tool. In particular:

• Add a standard GUI interface using the APL+Win GUI elements. The application is now a two-tier application comprising of the user tier and the business tier.

• Use ADO to deploy databases: the application has now acquired three-tier architecture; it has a user interface, a data tier, and the business tier.

• Use **Excel** and **Word** automation. The application is a Windows team player, like any other, and capitalises on users' existing skills.

• Use the Windows Registry, APIs, UNC names, implement XML and some HTM support. The application begins its metamorphosis into a strong team player without 'special' needs.

• Add scripting for bespoke application extension and the final vestige of APL, its keyboard, is relegated into insignificance.

• At this stage, the business tier appears incongruous: revamp this to include up-to-date user requirements using the experience accumulated thus far.

• Thereafter, document the application architecture. This will contain the industry acronyms that appeal to decision makers and, if issues arising from user consultation have been addressed, the user base is contented. Marketing will do the rest.

All of these changes require fundamental adjustments to the traditional APL+Win approach to system development. APL is a valuable but not the only, tool for application development: it needs to become aggressively collaborative.

15.11 Indentation

A wave pattern is commonly adopted for code structure; this is possible with APL+Win too provided that control structures are used and multi-statement lines are avoided. The pattern emerges with indentation at the left-hand margin. Indentation not only makes code more readable but also simplifies the tracking of the scope of control structures and emphasizes the code logic.

The APL+Win editor does not provide an automatic code indentation facility. However, it does provide crude support for indentation. A line added after a line that is indented starts at the same position as the preceding line. A block of code can be indented in one operation provided that the target block is selected prior to the tab or shift tab key being pressed within the editor, the tab, and shift tab keys simply insert or remove four spaces. This ensures that the indentation is identical on-screen and in print. APL+Win code does not contain the tab character and its presence, for instance when defining a function dynamically, will cause an error either during definition or at runtime.

In order to retain a discernible convention for code structure, changes to existing code need to follow the existing style or, where necessary, the new style must be applied to the entire code.

15.11.1 Limitations of indentation

A consistent indented layout of code is desirable but it may be of limited use with small and deeply nested functions. Indentation of short functions, one's that can be presented in their entirety in the editor and do not require scrolling, does not improve readability markedly.

The indentation of functions with deeply nested control structures may make them more difficult to read and maintain. Such functions may require continuous horizontal scrolling on-screen and in print or the session window, the lines may be very long and wrap on several lines with one or more blank lines at the start. A function nested five levels deep will have twenty spaces prefixing the fifth level and twenty four spaces prefixing each statement within it. Readability may be improved by replacing such functions by smaller functions designed to reduce the depth of control structures.

15.12 Documentation

Code, especially APL code, is not self-documenting. In-line comments that provide clues tend to become unsynchronised and misleading over time, as do functional specifications. Out-of-date documentation is a perilous hazard, for the following reasons:

• The mobility of the developer team is seriously compromised because the application cannot be passed on to different personnel: members of the existing team carry critical information in their head.

• It increases the likelihood of the developer introducing new bugs during routine enhancements or maintenance changes.

- It represents a constant source of frustration to users who respond by downgrading their esteem of the application.

Intentions to keep the documentation up to date are laudable but, in practice, rarely realised because coding and documentation are separate and inherently fragmented processes.

15.12.1 Context sensitive help and user manuals

The provision of context sensitive help is an un-negotiable hallmark of current Windows applications. The software tools that compile the context sensitive help files usually have facilities for producing user manuals too. Such tools automate the process of keeping user manuals up to date, more or less.

15.12.2 Auditing changes

APL+Win does not provide any tools for tracking changes with APL code. Whilst it is quite easy to determine whether two functions are different using the shape of the canonical representation, ⎕cr, of the functions and matching them, this is not a full proof method for concluding that they are identical. However, at the heart of any debugging exercise is the need to know what has changed. This is notoriously difficult to establish:

- Should differences in indentation be highlighted?

- Should the case of the code be considered? Arguably, it should be as APL+Win is case sensitive when it comes to string literals, and the names of variables, and functions. However, system quad functions are not case sensitive.

- Should blank lines be incorporated into the comparison?

- Should the comparison be made line by line or should the comparison identify missing and new blocks of code, deleted or inserted into a function?

In the absence of a more sophisticated tool, functions can be compared one at a time using **Word,** as follows:

- Save the canonical representation of the first copy of a function in a document, say, this is the target.

- Save the canonical representation of the second copy of a function in a different document, say this is the source.

Either document may be compared with the other. Open either the target or the source document. Click **Tools | Track Changes | Compare Documents** and select the other document. An example is shown in **Figure 15-1 Function comparison**.

- All changes are marked with a vertical bar in the left margin.

- Text that has been removed is shown with strike through formatting.

- New text is shown underlined and colour coded in order to distinguish it from underlined text in the document.

Unfortunately, the **Word** comparison facility cannot be customised for APL:

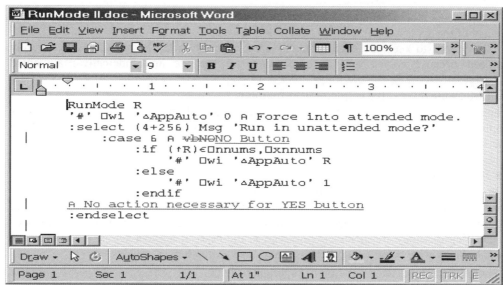

Figure 15-1 Function comparison

The comparison is case sensitive and not positional; that is, the content rather than the position (or style) of the content is compared. In this example there are three changes:

- The word 'vbNo' has been replaced by 'NO Button'
- A comment has been inserted before the ':endselect' line.
- A blank line has been inserted at the end.

Word is able to detect that no changes have been made to the ':endselect' line although its position has changed—it has simply been shifted down by one line. Had line numbers been included in the listing, all lines would be marked as different because the line numbers would be different. Several functions may be compared as long as they have been listed in the same order in both documents.

15.13 Testing

Application testing is not optional. This activity occurs at the end of the development cycle; there is always a temptation to bring a project back on schedule by limiting the duration of testing. If the duration is condensed, it follows that some testing is simply not undertaken. This can be very costly.

A maintenance release that turns out to be flawed will be very costly to the client and especially to the application provider. The client will have to revert to the previous version and become reluctant to upgrade in the future and the provider will need to re-release the flawed version and support many more legacy versions.

A common approach to testing involves two sets of tests. Firstly, every change is tested in isolation and secondly, the whole application is regression tested against the previous version. The wider the scope of the latter set of tests, the higher the confidence that the application will work as intended.

The credibility of an application depends on subsequent releases being backward compatible both in terms of the user interface and the results produced. Any radical change in either is liable to de-skill and antagonise the existing user base.

15.13.1 Functionality changes

In circumstances where changes are inevitable—such as those arising because of legislation—one or more systematic worked examples of the new functionality with clear reference to the source of change will restore user confidence.

15.13.2 Automatic migration

If it is necessary to change the structure of an application—for example, when replacing customised data handling based on APL component files by a database—the application needs to automate the transition. The changes should, as far as possible, be transparent to existing users. Users are liable to have valuable information already stored in the custom format and would be averse to recreating this in the new format and may be reluctant to upgrade. This may lead to the application provider having to maintain several versions of code; this makes the planning of future releases more complex.

In the event of a structural change, the application needs to incorporate a seamless migration routine for existing data accumulated by previous versions. Ideally, the migration routine must have the functionality to roll back changes. In practice, this may be impossible. A good strategy is to leave the source data unchanged and to write the migrated version to a different location. In case of an unexpected problem, the user can revert to the previous version.

Features such as migration of data will become increasingly more complex with the number of releases and do not form part of the data-handling module of the application itself. After all, a new client will not require them. From a design point of view, such features should be maintained independently of the application –they could be made available from the application via the application menu and appear to be integrated into the application.

15.14 Release

In practice, some parts of the schedule of work comprising a particular release are liable to be ready before others. For example, the default version of an application is likely to be ready before the bespoke versions. It is quite common to make an interim release to clients who use the default version and stage the release of bespoke versions. This may create an unmanageable version control problem; particularly where the completions of bespoke versions necessitate changes to the default version, and should be avoided. It is unlikely that a new release will have necessitated changes to every file comprising an application.

15.14.1 What's new?

A synopsis of the changes is an invaluable guide to clients as it assists them in the determination of the timing of the upgrade and in the assessment of the implication of the changes. A release document needs to contain sections describing faults—manifestations and resolutions—enhanced or new features and efficiency improvements.

Such a document may also be an effective means of publicising the application. A collection of these documents, if maintained on a web site, will provide a timeline history of the development of the product and may include a document that pre-empts the next release by providing advance information on what is intended. Good documentation enhances the credentials of the application provider.

15.14.2 Incremental upgrade

For existing clients, a release comprising of just the files that have changed may be a simple approach to upgrading their version of the application. Incremental upgrades can be transmitted to clients as attachments to electronic mail.

In practice, this approach may be problematic in that clients may end up with a hybrid system, one that has been partially or incorrectly upgraded, that behaves unpredictably. In such an event, it may be very time consuming (i.e. expensive) to resolve the issues.

15.14.3 Replacement upgrade

A full release of the application will need to be prepared for new or prospective clients. It is simpler to release the product in full to existing clients, as it will eliminate version incompatibility problems.

15.15 Application listings

A vital part of the maintenance cycle of an application that is ready for release is the production of a listing of all the functions it contains. A printed listing of an application not only provides an archive but also a ready to hand reference for the programmer, especially for APL+Win where code may be stored out of sight in several files. A paper copy of an application also provides a rudimentary and manual version control facility.

On some occasions, and for some people, a paper copy of a function is easier to follow than on-screen. A paper copy also acts as a version control document; comparison of functions is usually a good debugging tool when investigating changes in the behaviour of a function. Moreover, a paper copy can show the differences between two versions, say, the current version and the previous one. Version control software, such as **Visual Source Safe** (**VSS**), can do this on-screen—such software usually runs into difficulties with APL workspaces.

A workable document containing listings of an APL+Win application should have an alphabetical contents section, the functions listing (in any order) and an index of the function names and other key variables used within the application. A simple count of the number of pages on which a function name appears roughly indicates the number of times that function is used. **Word** is ideal for producing such a document.

15.15.1 Producing the listing

The function, WordList, uses **Word** to list the functions in a workspace. However, because of the nature of the APL character set, it is necessary to create a template—in the listing, APL.DOT is used—that contains four custom styles, namely APLHead, APLList, TOC1, and Index 1: the modification allow each style to show an APL font. The functions is:

```
      ∇ Z←L WordList R;⎕wself
 [1]     ⍝ System Building with APL+Win
{1[2]    :if 0 =ρ'WordList'⎕wi 'self'
 [3]        ⎕wself←'WordList' ⎕wi 'Create' 'Word.Application'
 [4]        0 0ρ⎕wi 'xDocuments.Add(Template)' 'APL.DOT'
 [5]        0 0ρ⎕wi 'xSelection.InsertBreak(Type)' (⎕wi
            '=wdSectionBreakNextPage')
 [6]        0 0ρ⎕wi 'xSelection.InsertBreak(Type)' (⎕wi
            '=wdSectionBreakNextPage')
 [7]        0 0ρ⎕wi 'xSelection.GoTo(What,Which,Count)' (⎕wi
            '=wdGoToSection') (⎕wi '=wdGoToFirst') 2
 [8]        0 0ρ⎕wi 'xActiveWindow.ActivePane.View.Type' (⎕wi
            '=wdPrintView')
 [9]        0 0ρ⎕wi 'xActiveWindow.ActivePane.View.SeekView' (⎕wi
            '=wdSeekCurrentPageFooter')
 [10]       Z←(⎕wi 'xSelection.Range') ⎕wi 'obj'
 [11]       0 0ρ⎕wi 'xApplication.NormalTemplate.AutoTextEntries("Page X
            of Y").Insert(Where)' Z
 [12]       0 0ρ⎕wi 'xActiveWindow.ActivePane.View.SeekView' (⎕wi
            '=wdSeekMainDocument')
```

```
   [13]      ⎕wi 'Δnf' 0
1}[14]    :endif
   [15]    ⎕wself←'WordList'
   [16]    ⎕wi 'xVisible' 0
   [17]    ⎕wi 'xScreenUpdating' 0
{1[18]    :select L
```

The left-hand argument of the function is either 'Insert' or 'Show'; the 'AV2ANSI' argument is used internally. The right-hand argument is a vector comprising the name of a single function or a two-dimensional array containing the names of several functions.

On being called for the first time, the function does the following:

- Creates an instance of **Word**.

- Adds a new document using the APL.DOT template.

- Creates three sections in the document; the first is for a table of contents, the second contains listings of functions, and the third contains an index of the names of functions added to the second section.

- Adds page numbering to the footer.

The **Word** session is invisible and an internal flag is set so that the **Word** application window is not updated dynamically but whenever **Word** determines suitable. This is a time consuming process and switching screen updating off minimizes the time taken. Try:

```
= [19]      :case 'Insert'
{2[20]       :if 1=ρρR
  [21]         R←,[0.5]R
2}[22]       :endif
  [23]       R←⊂[⎕io+1](3=⎕nc R)/R
  [24]       Z←ρR
  [25]       L←'AV2ANSI' WordList θ
{2[26]       :repeat
  [27]          0 0ρ⎕wi 'xSelection.Range.Style' ((↑⎕wi
           'xActiveDocument.Styles("APLHead")')⎕wi 'obj')
  [28]          ⎕wi 'xSelection.TypeText' ((⎕io⊃R)~' ')
  [29]          0 0ρ⎕wi 'xSelection.HomeKey(Unit,Extend)' (⎕wi
           '=wdLine') (⎕wi '=wdExtend')
  [30]          ⎕wi 'xActiveWindow.ActivePane.View.ShowAll' 1
  [31]          0 0ρ⎕wi
           'xActiveDocument.Indexes.MarkEntry(Range,Entry)'((⎕wi
           'xSelection.Range') ⎕wi 'obj') (⎕io⊃R)
  [32]          0 0ρ⎕wi 'xSelection.EndKey(Unit)' (⎕wi '=wdLine')
  [33]          ⎕wi 'xSelection.TypeParagraph'
  [34]          ⎕wi 'xActiveWindow.ActivePane.View.ShowAll' 0
  [35]          0 0ρ⎕wi 'xSelection.Range.Style' ((↑⎕wi
           'xActiveDocument.Styles("APLList")')⎕wi 'obj')
  [36]          ⎕wi 'xSelection.TypeText' (⎕av[(⎕io+L)[⎕avι⎕vr
           ⎕io⊃R]])
2}[37]          :until 0=ρR←1↓R
  [38]          ⎕wi 'Δnf' (Z+⎕wi 'Δnf')
  [39]          Z←(⍕⎕wi 'Δnf'),' Functions added to document.'
```

With a left-hand argument of 'Insert' the vector representation of the function(s) specified as the right-hand argument are added to section 2. The result is as follows:

```
'Insert' WordList 'IncludePicture' | 1 Functions added to document.
```

This call returns the number of functions added to the document so far. In case all the functions are not available in the workspace, they may be retrieved and listed by reiterative calls, as shown above. The code below completes the automation document.

```
= [40]        :case 'Show'
  [41]            0 0ρ⎕wi 'xSelection.GoTo(What,Which,Count)' (⎕wi
              '=wdGoToSection') (⎕wi '=wdGoToFirst') 1
  [42]            0 0ρ⎕wi 'xSelection.Range.Style' ((↑⎕wi
              'xActiveDocument.Styles("Normal")')⎕wi 'obj')
  [43]            ⎕wi 'xSelection.TypeText' 'Contents'
  [44]            ⎕wi 'xSelection.TypeParagraph'
  [45]            Z←((⎕wi 'xSelection.Range') ⎕wi 'obj') 1 1 1 3 1
              "APLHead,1"
  [46]            0 0ρ⎕wi
              (⊂'xActiveDocument.TablesOfContents.Add(Range,RightAlignPageNumbe
              rs,UseHeadingStyles,UpperHeadingLevel,LowerHeadi
            ngLevel,IncludePageNumbers,AddedStyles)'),Z
  [47]            0 0ρ⎕wi 'xSelection.GoTo(What,Which,Count)' (⎕wi
              '=wdGoToSection') (⎕wi '=wdGoToFirst') 3
  [48]            0 0ρ⎕wi 'xSelection.Range.Style' ((↑⎕wi
              'xActiveDocument.Styles("Normal")')⎕wi 'obj')
  [49]            ⎕wi 'xSelection.TypeText' 'Index'
  [50]            ⎕wi 'xSelection.TypeParagraph'
  [51]            Z←((⎕wi 'xSelection.Range') ⎕wi 'obj') (⎕wi
              '=wdHeadingSeparatorLetter') (⎕wi '=wdIndexIndent') 1 2
  [52]            0 0ρ⎕wi
              (⊂'xActiveDocument.Indexes.Add(Range,HeadingSeparator,Type,RightA
              lignPageNumbers,NumberOfColumns)'),Z
  [53]            ⎕wi 'xScreenUpdating' 1
  [54]            0 0ρ⎕wi 'xSelection.HomeKey(Unit)' (⎕wi '=wdStory')
  [55]            0 0ρ⎕wi 'xActiveDocument.EmbedTrueTypeFonts' 1
  [56]            ⎕wi 'xVisible' 1
  [57]            0 0ρ⎕wi 'xApplication.dialogs().Display' (⎕wi
              '=wdDialogFileSaveAs')
  [58]            ⎕wi 'Delete'
  [59]            Z←θ
```

With a left-hand argument of 'Show', the table of contents is inserted in section one and the table of indices in section three. Screen updating is switched on, and the application is made visible—the **Word** session should become visible in the **Taskbar**. The **Word** SaveAS dialogue is invoked and the APL object that is the instance of **Word** is terminated; **Word** continues to run.

If the **Word** dialogue does not appear as the top window, click on the **Word** session in the **Taskbar** to make to make it visible. At this point, it is advisable to save the document. The header and footer may be modified as deemed appropriate—for instance, the application title and a time stamp may be added to the header or footer.

A left-hand argument of 'AV2ANSI' returns the mapping of the ordinal positions of the atomic vector to the ANSI character set; see below.

```
= [60]        :case 'AV2ANSI'
  [61]            Z←θ
  [62]            Z←Z,   1 168 129 187 171 169 133    8    9   10   11
  [63]            Z←Z,156  13   14 157 182 230 231 237 145 250 243
  [64]            Z←Z,209 191 135 136 134   28 130 165 148 149   33
  [65]            Z←Z,  34  35   36   37   38   39   40   41   42   43   44
  [66]            Z←Z,  45   46   47   48   49   50   51   52   53   54   55
```

```
[67]       Z←Z, 56  57  58  59  60  61  62  63  64  65  66
[68]       Z←Z, 67  68  69  70  71  72  73  74  75  76  77
[69]       Z←Z, 78  79  80  81  82  83  84  85  86  87  88
[70]       Z←Z, 89  90  91  92  93  94  95  96  97  98  99
[71]       Z←Z,100 101 102 103 104 105 106 107 108 109 110
[72]       Z←Z,111 112 113 114 115 116 117 118 119 120 121
[73]       Z←Z,122 123 124 167 126 127 128 200 253 234 227
[74]       Z←Z,229 225 173 232 235 236 233 240 239 152 197
[75]       Z←Z,153 202 146 216 245 247 141 252 142 143 215
[76]       Z←Z,221 163 164 166 175 255 226 238 244 251 242
[77]       Z←Z,210 170 154 192 223 246 249 254 162 172 188
[78]       Z←Z,141 141 141 125 125 125 125  44  44 125 125
[79]       Z←Z, 44  44  44  44  44 193 194 195 196 197 198
[80]       Z←Z,199 200 201 202 203 204 205 206 207 208  46
[81]       Z←Z,210 211 212 213 214 215  44 217 218 219 220
[82]       Z←Z,221 222 125 168 185 224 189 241 168 140 132
[83]       Z←Z,131 179 181 139 155 147 180 186 158 174 144
[84]       Z←Z,138 137 150 151 248  35 177 178 160 190 159
[85]       Z←Z,176 125 161
[86]       Z←Z-⎕io
```

This is used to translate the vector representations of functions so that they appear correctly in **Word**. This call is for internal use only:

```
+ [87]       :else
  [88]           Z←⎕wself
1}[89]     :endselect
      ∇
```

APL expects each line of code to be executable: there is no facility for continuation lines as in **VB.**; however, the continuation lines are indented.

15.16 Epilogue

This book shows how current technology may be deployed with APL. In particular:

● Treat it as a dictionary of the building block prototypes that may be assembled selectively during system building.

● Use it as a means of enhancing your APL vocabulary and as a starting point for exploring system building elements, especially the Microsoft Office suite of applications and ADO, in greater detail.

● The Windows platform is an ever-changing landscape; developers need to update their awareness of new developments continually.

● The Microsoft .NET framework revolution is underway.

Not all solutions are appropriate in all situations. In system building, the objective is to solve the problems to hand, on a timely and efficient basis, and not to find problems that fit the available solutions. A sound strategy is one that implements solutions that can evolve. An advanced level of APL literacy together with an available collection of system building blocks promotes rapid application development.

BIBLIOGRAPHY

- *Microsoft MSDN Library* – http://msdn.microsoft.com
- *Microsoft Knowledge Base* – http://support.microsoft.com
- *Microsoft Newsgroup* – http://msdn.microsoft.com/newsgroups.
- APL+Win Version 4.0 Product Documentation
- APL+Win Grid documentation – www.apl2000.com
- APLX Product Documentation
- APL/W Product Documentation
- IBM System Journal Vol. 30 No 4, 1991, Twenty-Fifth Anniversary Issue
- *Vector*, Journal of the British APL Association – www.vector.org.uk
- *TryAPL2* – http://www14.software.ibm.com – IDIOMS workspacc.
- *FinnAPL Idiom Library* – http://www.pyr.fi/apl/texts/Idiot.htm
- *APL Advanced Techniques & Utilities* – Gary A Bergquist, 1987, Zark Incorporated
- *HTML Programmer's Reference* – Thomas A Powell & Dan Whitworth, Osborne/McGraw Hill, 1998, ISBN 0-07-882559-8
- *Advanced JavaScript Techniques* – Dan Livingston, Prentice Hall PTR, 2003, ISBN 0-13-047891-1
- *Essential JavaScript for Web Professionals* – Dan Barrett, Prentice Hall PTR, 2003, ISBN 0-13-100147-7
- *JavaScript Programmer's Reference* – Christian MacAuley & Paul Jobson, Osborne/McGraw Hill, 2003, ISBN 0-07-219296-8
- *VB 6 IIow to program* – II M Dietel, P J Dietel & T R Nicto, Prcnticc Hall , 1999, ISBN 0-13-456955-5
- *APL2 in Depth* – Norman D Thompson & Raymond P Povlika, Springler-Verlag, 1992, ISBN 0-387-94213-0
- *APL An Interactive Approach* – Leonard Gilman & Alan J Rose, John Wiley & Sons Inc, 1984, ISBN 0-471-09304-1
- *Learning and Applying APL* – B Legrand, John Wiley & Sons Ltd, 1984, ISBN 0-471-90243-8
- *Les APL étendus* – B Legrand, Masson, 1994, ISBN 2-225-84579-4
- *Internet Programming with VBScript and JavaScript* - Kathleen Kalata, Course Technology Thomson Learning, 2001, ISBN 0-619-01523-3
- *Mastering Microsoft Access 2000 Development* – Alison Baxter, Sams, 1999, ISBN 0-672-31484-3
- *Pure VB* – Dan Fox, Sams, 1999, ISBN 0-672-31598-X
- *Fun Web Pages with JavaScript* – John Shelley, Bernard Babani (publishing) LTD, 2002, ISBN 0-85934-520-3

• *SQL:1999 Understanding Relational Language components* – Jim Melton & Alan R Simon, Morgan Kaufman Publishers, 2002, ISBN 1-55860-456-1

• *SQL in 10 Minutes* – Ben Forta, Sams Teach Yourself, 2001, ISBN 0-672-32128-9

• *ADO:ActiveX Data Objects* – Jason T Roff, O'Reilly, 2001, ISBN 1-56592-415-0

• *Developing VB Addins* – Steven Roman, O'Reilly, 1999, ISBN 1-56592-527-0

• *VBScript in a Nutshell* – Matt Childs, Paul Lomax, & Ron Petrisha, O'Reilly, 2000, ISBN 1-56592-720-6

• *Excel 2000 VBA Programmer's Reference* – John Green, Stephen Bullen & Felippe Martin, Wrox Press Ltd ISBN 1-861002-5-48

• *Word 2000 VBA Programmer's Reference* – Duncan MacKenzie & Felipe Martins, Wrox Press Ltd, 1990, ISBN 1-861002-5-56

• *ADO 2.6 Programmer's Reference* – David Sussman, Wrox Press Ltd, 2000, ISBN 1-861-002-68-8

• *VBScript Programmer's Reference* – Susanne Clark, Antonio De Donatis, Adrian Kingsley-Hughes, Kathie Kingsley-Hughes, Brian Matsik, Erick Nelson, Piotr Prussack, Daniel Read, Carsten Thomsen, Stuart Updegrave & Paul Wilton, Wrox Press Ltd , 2000, ISBN 1-861002-71-8

• *APL Reference Manual*, Iverson Software, 1993, ISBN 1-895721-07

• *XML in 24 Hours* – Charles Ashbacher, SAMS Teach Yourself, 2000, ISBN 0-672-31950-0

• *XSLT Programmer's Reference* – Michael Kay, Wrox Press Ltd, 2003, ISBN 1-861005-06-7

Index

Erratum

For an updated version of the index please access the following URL:

wiley.com/go/askoolum

A
Access Control
File, 91
Folder, 90
Acronyms
ADO, 1
ADP, 333
ADTG, 381
ANSI, 16, 421
API, 3, 7
APL, 1
BLOB, 344
CBD, 143
COM, 13, 137
CSV, 420
DBA, 17
DCL, 18
DDL, 18
DLL, 8
DML, 18
DOM, 225
DQL, 18
GUI, 3, 7
HTML, 221
ISO, 421
OCX, 13
OLE, 13
RDBMS, 16
RPC, 142
SQL, 8
UNC, 7
UTC, 459
VB, 2
VSS, 22, 457
WSC, 227
XML, 225
XSL, 126
XSLT, 126
ActiveX
Argument Type, 175, 203Event handling, 137
GetObject, 201

Orphan, 163
APIINI
GetComputerName, 125
GetKey, 124
GetSection, 125
GetSectionKeysName, 125
GetSectionKeysValue, 125
GetSectionsName, 124
GetUserName, 125
WriteKey, 124
WriteSection, 124
APL Functions
ABC_Make, 81
AccessCreate, 342
AccessINTO, 447
AddBookmarks, 304
AddCode, 260
AddLink, 264
AddPicture, 309
AddQuery, 327
AddRecord, 359
AddTableBookmarks, 300
ADOPrompt1, 392
ADOPrompt2, 397
ADOPrompt3, 397
ADOSQLLanguages, 425
ADOXCreateJET, 342
APICallBack, 24
APIExists, 24
APIINI, 123
APIPointer, 24
APLEnum, 157
APLFormatSQL, 441
APLGrid, 144
APLGridDisplay, 430
APLinListView, 87
APLInput, 198
APLServer, 50
APLtoHTM, 222
App, 91
AutoBSTable, 31
Automate, 74

AutomationModelI, 269
BaseConnection, 337
BeforeDelete, 359
Benchmark, 40
BitAND, 35
BitEQV, 34
BitIMP, 34
BitNAND, 37
BitNOR, 37
BitOR, 36
BitXOR, 36
BlankArray, 299
BuildTreeList, 85
CallingFunction2, 31
CharArray1, 297
CharArray2, 297
CharArray3, 299
CloseExcel, 267
CloseWord, 320
COMEvent, 141
ConcurrentShare, 220
ConfigSysDSN, 402
ConformNumeric, 49
Connect, 351
ConnectDRO, 381
ConnectionProperties, 353
CopyFile, 29
CopyFileBS, 29
CreateAPLVariables, 364
CreateAPLVariables2, 364
CreateExistADP, 339
CreateFileDSN, 405
CreateMDB, 344
CreateNewADP, 340
CreateQueryJET, 324
CreateQueryODBC, 325
CreateShortcut, 113
CreateSysDSN, 400
CSBlocks, 55
CursorLocation, 367
CursorLocation2, 370
CursorTypeImpact, 356
DeleteBookmarks, 306
DeleteEnvironmentVariable, 113
DeleteFile, 116
DeleteObject, 331
DemoTable, 300
DemoXLEvent, 141
DisableExcelEvents, 252
DSNExists, 390
DSNManager, 403
DynProperties, 51
DynSyntax, 154

Email, 202
EnumWindowsCallback, 24
ExcelUDF, 260
ExecuteQueryJET, 324
ExecuteQueryODBC, 325
ExportCharts, 262
FabricateRS, 38
FabricateRS2, 39
FabricateRS3, 39
FileVersion, 188
FindSQLServer, 372
FixAPI, 23
FlowBS, 35
FlowBS2, 35
FlowIMP, 35
FormatSQL, 439
FormIE, 223
FormText, 318
FormText1, 319
FromISODate, 100
GenerateData, 46
GetADOVariablesTypes, 365
GetAuthors, 288
GetBookmark, 305
GetDateFormat, 103
GetEnvironmentVariables, 111
GetFileDSNs, 391
GetHandle, 268
GetLocaleFormat, 298
GetLocaleInformation, 20
GetObject, 201
GetQuery, 329
GetRangeAddress, 256
GetRecordset, 84
GetResult, 319
GetRowsLoop, 374
GetRowsLoop2APLGrid, 375
GetSpecialFolders, 111
GetSQL, 354, 425
GetStringLoop, 377
GetTimeFormat, 104
GetTimeStamp, 427
GetUniqueDSN, 390
GetVolumeInformation, 135
GetXLName, 250
Global, 137
Grid, 165
GRIDDate, 208
HandlingExcelMacros, 158
HelloWorld, 5
Help, 67
I2C, 51
IncludePicture, 310

IncludeText, 312
INItoXML, 134
InsertFormula, 308
Invoice, 33
IsConnected, 340
IsRPCAvailable, 142
JSArray, 246
LoadDOM, 381
LocaleDate, 106
LoopByPage, 370
LoopFeedback, 369
MDBTOODBC, 336
Msg, 73
MultipleRecordsets, 379
MyForm, 36
NativeHierarchy, 152
NestXML, 45
NotePad, 106
Now, 98
OnErrorGotoAPL, 350
OpenSchema, 361
PowersOf2, 36
ProcInput, 198
QueryDef, 327
QueryDef2, 328
ReadADOCONN, 417
RecordLoop, 41, 369
RecordLoop2, 41
RecordObjectDirect, 356
RecordObjectRedirection, 355
RedirEnum, 155
RedirEnum2, 156
RegDelete, 110
RegEnum, 389
RegionalFmt, 105
RegRead, 109
RegWrite, 109
RemoveSysDSN, 403
ReOrient, 303
ReSize, 89
Resolve, 161
RetrieveFile, 347
RetrieveNL2, 346
RetrieveNL3, 346
RSHasRecords, 368
RSinListView, 86
RSTable, 299
SampleFunction, 53
SampleRS, 49
SavetoADTG, 381
SavetoDOM, 380
SavetoXML, 381
Script, 48, 199

ScriptShare, 196
SecureAccess, 90
SecureFile, 91
SeekACCESS, 371
SeekFileAccess, 117
SelfJoin, 439
SendEmail, 201
SendMail, 201
SendToExcel, 43, 287
SetMode, 74
SetTranspose, 283
Show_APLWin, 85
Show_fmDemo, 84
ShowMessage, 70
ShowTranspose, 255
ShowTranspose2, 256
Signal, 74
SignalAPIError, 117
SimpleJoin, 437
SortCSV, 447
Soundex, 433
SpellCheck, 96
SQLTableExists, 362
StoreFile, 347
StoreNL23, 345
StreamClone, 360
Strip, 44
TableExists, 361
TestInXL, 217
TextJETSort, 373
Tree, 83
TreeView, 83
UDFFail, 343
UDL, 407
UPCASE, 98
UseNewFileDSN, 406
UseNonShareDSN, 406
UsingFlag, 251
VBArray, 247
VBDLL, 172
VBInput, 197
VTree, 79
WhatsThisHelp, 67
WhichAPL, 51
WhichProvider, 341
With, 4
Without, 4
WordList, 464
WordTable, 302
XLEvent, 140
xls, 362
xls2, 364
XML, 43

XML2, 45
XMLINI, 127
Application handle
Development, 387
Generic, 266
Run time, 387
C
Components
APL+Win Prefixes, 14
Automation client, 13
Automation server, 13
Dynamic binding, 14
Early binding, 14
In-process, 13
Late binding, 14
Out-of-process, 13
Static binding, 14
CSBlocks
GetIndent, 56
GetIndentX, 56
IDCS, 57
Indent, 56
List, 55
Strip, 55
StripX, 55
D
Date and Time
Days of Week, 178
FromISODate, 100
GetDateFormat, 102
GetTimeFormat, 102
Months of Year, 20, 203
Now, 98
ToISODate, 102
via APL Grid, 27, 208
via GetString, 43
Debugging
VBA code, 167, 215
Visual Basic code, 167
Drives
Connect, 119
Disconnect, 119
Network, 118
E
Email
mailto:, 202
SendEmail, 201
Enumerate
Charts, 262
Data types, 365
Drives, 35
DSNs, 389
Field names, 363

ODBC drivers, 399
Tables, 360
with APL+Win, 157
F
FSO, 136
G
GenerateData
BinaryString, 48
Codes, 47
Dates, 46
Floating, 47
Integers, 47
NumericVector, 48
String, 48
H
Help, 15
ActiveX, 15
Context Ids, 68
Context sensitive, 65
Events, 139
HTML, 69
Prompt, 66
Tooltip, 66
What's this, 67
WINHELP, 68
HTML, 221
from APL+Win, 222
Help, 69
I
Internet Explorer, 223, 239
L
Locale
LOCAL_SYSTEM_DEFAULT, 20
LOCAL_USER_DEFAULT, 20, 172
M
Major
Column, 218, 254
Row, 218, 254, 297
MessageBox
Msg, 71
MsgBox, 200
O
OOP, 11
Encapsulation, 12
Inheritance, 12
Object Oriented Programming, 11
Polymorphism, 12
Q
Quad primitives, 88
R
Registry, 18
REL40, 83

S

Script, 199
AddCode Method, 202
Error Trapping, 198
ShareAPLGrid, 219
ShareWSEngine, 210
Session Setting
Application title, 75
Auto mode, 74
Secure access, 90
Unattended mode, 73
ShellExecute, 202
ShowForm
Component Exe, 224
Strip
DeleteDB, 44
DeleteEB, 44
DeleteIB, 44
DeleteLB, 44
DeleteTB, 44
DeleteUB, 44

W

Windows
Desktop, 92
Regional settings, 19, 182
Registry, 18
Task Manager, 13
Windows API, 3, 97
DeleteFile, 114
EnumWindows, 24
FindExecutable, 114, 238
FindWindow, 266, 320
FormatMessage, 29
GetClassName, 268
GetCurrencyFormat, 104
GetCurrentDirectory, 121
GetDateFormat, 102
GetEnvironmentVariable, 113
GetLastError, 29
GetLocaleInfo, 23, 172
GetLogicalDrives, 35
GetModuleFileName, 51
GetNumberFormat, 104
GetPrivateProfileSection, 125
GetPrivateProfileSectionNames, 124
GetPrivateProfileString, 124
GetSystemDirectory, 110

GetSystemTime, 459
GetTempPath, 110
GetTimeFormat, 102, 104
GetUserName, 126
GetVersionEX, 190
GetVolumeInformation, 135
GetWindowsText, 268
HtmlHelp, 69
IsNetworkAlive, 118
IsWindow, 254
lClose, 30, 116
lOpen, 29, 115
MakeSureDirectoryPathExists, 98
MessageBox, 71
PathFileExists, 115, 457
PostMessage, 267
RegEnumKeyEX, 389
SendMessage, 267, 320
SetCurrentDirectory, 121
SetEnvironmentVariable, 112
SetLocaleInfo, 105
SQLConfigDataSource, 400
WinExec, 114
WinHelp, 65, 68
WritePrivateProfileSection, 124
WritePrivateProfileString, 124

X

XML, 225
DOMDocument, 380
FreeThreadedDOMDocument, 126
XSL, XSLT, 126
XMLINI, 123
DeleteKey, 130
DeleteSection, 132
GetKey, 130
GetSection, 131
GetSectionKey, 129
GetSectionKeysName, 132
GetSectionKeysValue, 133
GetSectionsName, 132
GetSignature, 128
Load, 128
Save, 133
SectionComment, 131
WriteKey, 129
WriteSection, 130